KT-442-115

A Companion to Economic Geography

Edited by

Eric Sheppard

University of Minnesota

and

Trevor J. Barnes

University of British Columbia

Blackwell
Publishing

© 2000, 2003 by Blackwell Publishing Ltd
except for editorial material and organization © 2000, 2003 by Eric Sheppard and Trevor J. Barnes

350 Main Street, Malden, MA 02148-5020, USA
108 Cowley Road, Oxford OX4 1JF, UK
550 Swanston Street, Carlton, Victoria 3053, Australia

All rights reserved. No part of this publication may be reproduced, stored in a retrieval system, or transmitted, in any form or by any means, electronic, mechanical, photocopying, recording or otherwise, except as permitted by the UK Copyright, Designs, and Patents Act 1988, without the prior permission of the publisher.

First published 2000 by Blackwell Publishing Ltd
First published in paperback 2003
Reprinted 2003

Library of Congress Cataloging-in-Publication Data

A companion to economic geography / edited by Eric Sheppard and Trevor J. Barnes.
    p. cm. — (Blackwell companions to geography ; 2)
    Includes bibliographical references and index.
    ISBN 0-631-21223-X (hbk : alk. paper) — ISBN 0-631-23579-5 (pbk : alk. paper)
    1. Economic geography. I. Sheppard, Eric S. II. Barnes, Trevor J. III. Series.

HF1025 .C66 2000
330.9—dc21   00-059898

A catalogue record for this title is available from the British Library.

Set in 10 on 12 pt Sabon
by Kolam Information Services Private Ltd, Pondicherry, India
Printed and bound in the United Kingdom
by TJ International Ltd, Padstow, Cornwall

For further information on
Blackwell Publishing, visit our website:
http://www.blackwellpublishing.com

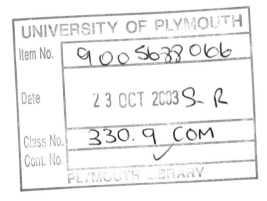

UNIVERSITY OF PLYMOUTH

Item No.	9 0 0 5638 066
Date	2 3 OCT 2003 S- R
Class No.	330. 9 COM
Cont. No.	

PLYMOUTH LIBRARY

Eric Sheppard would like to dedicate this book to the many undergraduates over the past twenty years in his Geography of the World Economy course, and the graduate students who took the pro-seminar The State, The Economy and Spatial Development, as well as the various research seminars in economic geography offered over the years. Trevor Barnes would like to dedicate this book to the hundreds of students at the University of British Columbia who have taken some version of his course, Introduction to Economic Geography. We have learnt much more from the students than we ever taught them.

Contents

Contributors

Ash Amin is Professor of Geography at Durham University. He has edited *Post-Fordism: A Reader; Globalisation, Institutions and Regional Development* (with Nigel Thrift); *Behind the Myth of European Union* (with John Tomaney); and *Beyond Market and Hierarchy: Social Complexity and Interactive Governance* (with Jerzy Hausner). He writes on regional development, cities, and democracy.

Trevor Barnes is Professor of Geography at the University of British Columbia. He is the author of *Logics of Dislocation*, and co-editor with Meric Gertler of *The New Industrial Geography*. His most recent research is on the history of geography's quantitative revolution.

Noel Castree is Reader in Geography at Manchester University. Co-editor of *Remaking Reality: Nature at the Millennium* and *Social Nature: Theory, Practice and Politics*, his main interests are in social theory approaches to environment and, more specifically, the greening of political economy.

J. K. Gibson-Graham is the pen name of Katherine Gibson, Professor of Human Geography in the Research School of Pacific and Asian Studies at the Australian National University, and Julie Graham, Professor of Geography at the University of Massachussetts, Amherst. JKGG is the author of *The End of Capitalism (as we knew it): a feminist critique of political economy* (Blackwell, 1996), and the co-editor of *Class and its Others* (Minnesota, 2000) and *Re/Presenting Class* (Duke, 2001).

Richard Grant is an Associate Professor at the University of Miami, Florida. He is editor of *The Global Crisis in Foreign Aid* (with Jan Nijman), as well as numerous articles on trade and economic globalization in the periphery of the world economy.

Nicky Gregson is Reader in Geography at the University of Sheffield. She is the author of *Servicing the Middle Classes* (with Michelle Lowe) and one of the WGSG collective responsible for *Feminist Geographies*, and has published numerous articles on both gender and consumption.

Dean M. Hanink is Professor and Head of Geography at the University of Connecticut. He is the author of *The International Economy: A Geographical Perspective* and *Principles and Applications of Economic Geography*. He has also written several articles on topics relating to environment and economic geography.

Susan Hanson is Professor of Geography at Clark University. She is the author of *Gender, Work, and Space* (with Geraldine Pratt) and editor of *The Geography of Urban Transportation*; in addition, she has written many articles on gender and urban labor markets and on urban transportation.

Roger Hayter is Professor of Geography at Simon Fraser University. He is the author of *Dynamics of Industrial Location: The factory, the firm and the production system*, as well as numerous articles on industrial geography and British Columbia's forest economy.

Andy Herod is Associate Professor of Geography at the University of Georgia. He is the editor of *Organizing the Landscape: Geographical Perspectives on Labor Unionism* (1998) and, with Gearóid Ó Tuathail and Susan Roberts, of *An Unruly World? Globalization, Governance and Geography* (1998). His research seeks to understand how workers and their organizations shape, and are shaped by, the geography of capitalism.

Helga Leitner is Professor of Geography at the University of Minnesota. She has written numerous articles and book chapters on the politics of immigration and citizenship, immigrant incorporation, and local economic development. Her current research interests are in geographies of governance and citizenship.

Andrew Leyshon is Professor of Economic and Social Geography at the University of Nottingham. He is the co-author of *Money/Space* (with Nigel Thrift) and co-editor of *The Place of Music* (with David Matless and George Revill). He is currently undertaking research on competitiveness in financial services, alternative systems of exchange, and the geographies of music technology.

Ron Martin is Professor of Economic Geography at the University of Cambridge. He has published numerous books, including *The Geography of Deindustrialization*, *Money, Power, and Space* (with S. Corbridge and N. Thrift), *Unemployment and Social Exclusion* (with P. Lawless and S. Hardy), *Money and the Space Economy*, and *The Economic Geography Reader* (with J. Bryson, N. Henry, and D. Keeble). He has also published more than 130 articles on the geography of labor markets, the geography of money, and regional political economy.

Katharyne Mitchell is an Associate Professor of Geography at the University of Washington. She has written *Transnationalism and the Politics of Space* (forthcoming) and numerous articles on migration, diaspora, and ethnicity.

Ann M. Oberhauser is an Associate Professor of Geography at West Virginia University. She has written numerous publications on economic restructuring and regional development in France, South Africa, and the United States. Her current research focuses on comparative analyses of gender and rural economic networks in Appalachia and southern Africa.

Brian Page is Associate Professor of Geography at the University of Colorado at Denver. He is the author of numerous articles on US agriculture, the meat packing industry, and historical regional development in the Midwest.

Joe Painter is Lecturer in Geography at the University of Durham. He is the author of *Politics, Geography and "Political Geography": A Critical Perspective*, and of numerous articles and book chapters on the geographies of regulation, governance, citizenship, and democracy.

Jamie Peck is Professor of Geography at the University of Wisconsin-Madison. An editor of *Antipode* and *Environment and Planning A*, he has research interests in labor geography, urban political economy, and economic regulation. He is the author of *Workfare States* and *Work-Place: The Social Regulation of Labor Markets*.

Paul Plummer is senior lecturer at the School of Geographical Sciences, University of Bristol. He has published research on the foundations of analytical political economy, competing theories of spatial competition, and the dynamics of regional growth and economic restructuring.

David Rigby is a Professor of Geography at the University of California, Los Angeles. His research interests focus on evolutionary economic models of competition, technological change, and regional growth. He is co-author, with Michael Webber, of *The Golden Age Illusion: Rethinking Postwar Capitalism*.

David Sadler is Reader in Geography at the University of Durham. He is author of *The Global Region: Production, state policies and uneven development*, co-author of *Approaching Human Geography: An Introduction to Contemporary Theoretical Debates* and *Europe at the Margins: New Mosaics of Inequality*, and editor of the journal *European Urban and Regional Studies*.

Erica Schoenberger is Professor of Geography at the Johns Hopkins University, with a joint appointment in anthropology. She is the author of *The Cultural Crisis of the Firm* and focuses her research on problems of industrial, technological, and social change, corporate culture and strategy, and the behavior of multinational corporations. She has an Australian Shepherd named Sasha.

Eric Sheppard is Professor of Geography at the University of Minnesota. He is the author of *The Capitalist Space Economy* (with Trevor Barnes) and *A World of Difference* (with Philip Porter), as well as numerous articles on regional political economy.

Peter Sunley is Senior Lecturer in Geography at the University of Edinburgh. He is author of *Union Retreat and the Regions* (with Ron Martin and Jane Wills). He has also published articles on geographies of labor and labor policy, and contributed to theoretical debates in economic geography.

Dr. Erik Swyngedouw is University Reader in Economic Geography at Oxford University and Fellow of St. Peter's College. He is also Associate Fellow of the Environmental Change Unit at Oxford University. He is co-editor of, among others, *The Urbanisation of Injustice* (with A. Merrifield) and *Towards Global Localisation* (with P. Cooke and others). He published numerous articles on the political economy of space, urban and regional development, and on political ecology.

Richard A. Walker is Professor of Geography, University of California, Berkeley. He is co-author of *The Capitalist Imperative* (with Michael Storper), *The New Social Economy* (with Andrew Sayer), and many articles on economic geography, urbanization, and California.

Barney Warf (PhD 1985, University of Washington) is Professor and Chair of Geography at Florida State University, Tallahassee. His research and teaching interests lie within the broad domain of regional development, straddling contemporary political economy and social theory on the one hand and traditional quantitative, empirical approaches on the other, particularly concerning producer services and telecommunications.

Michael Watts is the Chancellor's Professor of Geography and the Director of the Institute of International Studies at the University of California, Berkeley, where he has taught for more than twenty years. He has written extensively on the political economy of development, on agrofood systems, and on cultural theory, and has worked in West Africa, California, and Vietnam.

Michael Webber is Professor of Geography at the University of Melbourne. He is the author of *The Golden Age Illusion* (with David Rigby) and of *Global Restructuring: The Australian Experience* (with Bob Fagan), as well as numerous other books and articles on economic restructuring and the evolution of the global economy.

Figures and Tables

Acknowledgments

Eric Sheppard is deeply grateful to John Davey, who originally suggested the idea to him of a Companion to Economic Geography. Many students and colleagues have helped him understand economic geography better. It is hard to single out individuals, but he would particularly like to acknowledge (in no particular order) the influence of David, Dick, Les, Ros, Allen, Trevor, Phil, Gordon, Julie, Meric, Paul, Jim, Yu, Claire, Pádraig, Byron, Bill, Dmitri, Mary, Andrea, Abdi, Lucky, Michael, Franz, Herwig, Helga, Chris, Francis, Bongman, and Yeong-ki. Trevor Barnes would like to thank the Canada–US Fulbright Program for awarding him a fellowship for 1997–1998 during which period this book was begun. He would also like to thank the Department of Geography, University of Minnesota, for providing him with such a conducive physical and intellectual environment during that same year.

We both would like to acknowledge the support and dedication of staff at Blackwell Publishers, particularly Jill Landeryou, Sarah Falkus, and Joanna Pyke, without whom this book would not have been finished. We would also like to thank Andrew Murphy for assiduously compiling the index. We thank *Private Eye* for permission to reproduce figure 9.1.

Chapter 1

Introduction: The Art of Economic Geography

Trevor J. Barnes and Eric Sheppard

How do you turn economic geography into art? The Mexican painter, writer, and long-time communist Diego Rivera did it between 1932 and 1933 in painting the 25 panels of the mural "Detroit: Man and Machine" at the Detroit Institute of Arts (see cover). The panels tell a rich and complex economic geographical story: extracting resources from the ground, bringing together "hands" of workers from across the world, using various kinds of machinery – furnaces, stamping machines, drills, hoists, conveyor belts – in conjunction with the brawn of labor to manufacture and assemble, and finally produce, the finished product, in this case a car.

The mural was paid for by Edsel Ford, son of Henry, and modeled on perhaps the most famous factory of the twentieth century – the Rouge complex, the Ford automobile manufacturing plant located at Dearborn, Michigan, just outside Detroit. The complex was first used to produce automobiles from 1913. Twenty years later, by the time Rivera came to paint it, it was already a symbol of the modern machine age, and more generally of the energy and might of industrial capitalism. Operations at the Rouge complex were vertically integrated, that is, all the different component processes of manufacture were contained within a single site, and as a result the workforce requirements were immense. At one point 75,000 employees worked there. The "B" building alone was a quarter of a mile long, containing the entire assembly-line operation on a single level (and much of the focus of Rivera's artistic attention).

The assembly-line technique used in the "B" building was one of Henry Ford's most significant innovations, and was first introduced to the plant in 1927. Workers stood in place, and the work came to them on a conveyor belt. Once the work task was completed, the belt moved on, making the worker, as one Ford assembly-line employee put it, "nothing more or less than a robot" (quoted in Rochfort, 1987, p. 68). There is a clear sense of that in Rivera's mural. The workers do not simply tend the machines, but are fully integrated with them, their bodies synchronized with the swirling, ceaseless movement of industrial manufacture at the plant.

In many ways the Ford family were an unlikely patron of Rivera. Ford senior was vehemently anti-left, and anti-union. He once said, "people are never so likely to be wrong as when they are organized." What Rivera and Ford shared, however, was a fascination with machines, and the idea that the combination of "human and mechanical action" could produce a "creative power unparalleled in history" (Rochfort, p. 67). To capture that power in paint, Rivera spent his first three months in Detroit meticulously preparing before even picking up a paintbrush. He toured the plant, spoke to workers and engineers, and had hundreds of photographs taken of the different parts of the production process (see Detroit Institute of Arts, 1978). What emerges, in spite of Rivera's communist sympathies, is a relatively uncritical portrait of industrial capitalism at the Rouge. True, one of the workers is carrying a partially obscured placard with the words "We want." But the full placard would have read "We want beer," a popular anti-prohibition slogan at the time. In contrast, Rivera's next mural commission, which was for another industrial mogul, John D. Rockefeller, was more overtly political, and included Lenin as one of the onlookers. In this case Rockefeller balked, and ordered the finished mural to be chipped off the wall of the Rockefeller Center, New York City, where it had been painted.

For Rivera, the Rouge complex and its production of automobiles was an illustration of a broader phenomenon that he labeled a "wave-like movement" and found in "water currents, electric waves, stratifications of different layers under the surface of the earth, and in a general way throughout the continuous development of life" (Rivera, 1934, p. 50). Economic geography as it emerges from the contributions to this volume is a bit like this, too. The kinds of phenomena that economic geographers study – natural resources, manufacturing, information, money – are always on the move, continually undergoing transformation, morphing into new forms and identities. Of course, capturing that flux and movement is difficult. Diego Rivera did it by painting larger-than-life stories on walls. His 25 panels are a synchronic depiction of the economic geographical processes that produced and maintained the industrial behemoth of 1930s Detroit. But there are clearly other means to make that same depiction, some of which, we would argue, are found in the essays collected here. Of course, they are not couched in a visual idiom, relying on vivid colors or daring brush strokes, but in their own way, we contend, they are just as striking.

Although economic geographers are not in the business of daubing paint, at least in their professional lives, there is an art to what they do. For economic geographers of whatever stripe face the same difficulty as Diego Rivera – to represent on a flat surface economic geographical events that exist "out there." Rivera's medium was paint, and his flat surface was a wall, whereas the medium of economic geographers is writing, and their flat surface is a sheet of A4 or eight-and-a-half by eleven inch paper. In both cases, there is a need for appropriate techniques, sensitive interpretation, enthusiasm, dedication, adequate preparation, and prior training. One of the primary aims of this volume is to introduce you to the art of economic geography.

Our Approach

Economic geography has always been a mainstay within geography. But its history as a formal university-based discipline in Anglo-America is quite short, at best a hundred years. As a discipline it grew less out of concerns by economists to generalize and theorize, than the concerns of geographers to describe and explain the individual economics of different places, and their connections one to another. You can see this in the following quote from a very early German economic geographer, Karl Andree:

I sit at a mahogany table from Honduras. The carpet on which it stands has been manufactured at Kidderminster in England from wool brought by a sailor from the River Plate or New South Wales. The tea in a Berlin porcelain cup came from China or Assam, the coffee from Java, the sugar from Lower Saxony, Brazil or Cuba. I smoke Puerto Rican tobacco in my pipe whose stem grew in Hungary, the material for its Meerschaum bowl, carved in Thuringia, was dug in Asia Minor, the amber mouth piece came from the Baltic Sea, and the silver for the rim from the silver mines of the Erzgebirge, Harz or perhaps from Potosi [Peru]...

This passage was written in 1867, but the questions it raises still resonate. Where are things produced? Under what conditions? How is it that goods produced in one place end up at another? By what means? And who buys them? We're not claiming that economic geography is unaltered since 1867 – as the first section of our volume amply demonstrates, there have been profound changes in approach even over the last twenty years – but the broad kinds of questions that the discipline poses, and the interest in the ensuing answers, persists.

These kinds of questions have also attracted various forms of interest from outside the discipline. The state has always been interested. The beginnings of economic geography as a discipline in the late nineteenth century were directly connected with the belief of various European governments that knowledge of economic geography both buttressed their colonial projects and provided their domestic business class with a potential competitive advantage over rivals. More recently, attention has come from other social scientists. This is new. In the past, economic geography was the great borrower, taking ideas from others. Now, it has begun to lend. One reason is the interest in globalization. By its very nature, globalization is an economic geographical phenomenon, and traditional economic geographical ideas around spaces of flows, and places of control and production, are central to its understanding. While much of economic geography's recent appeal has been to sociologists and political scientists, some economists have also discovered geography. Perhaps the best-known example is the MIT economist Paul Krugman, who, in his work on trade and growth, moved away from the "wonderland of no dimensions" to the greater realism of at least three-dimensional Euclidean space. As he put it in 1991, "I realized that I spent my whole professional life...thinking and writing about economic geography, without being aware of it" (Krugman, 1991, p. 1). Admittedly, some economic geographers still wonder if he is doing economic geography (Martin, 1999), but at least it is a start.

We think, then, that there are some very good reasons to study economic geography, and to study it now. Undoubtedly this is the hubris of every generation of economic geographers. We believe, however, that two developments over the last twenty years – one external to the discipline, the other internal – make economic geography a central body of literature and knowledge for understanding the world at the turn of the millennium.

The first is that since the 1970s there have been some enormous changes affecting the economy, and society more generally: deindustrialization, industrial restructuring, the rise of information technology (IT) and computerization, the feminization of the labor market, and globalization. Such changes beg understanding, and indeed are already subject to much scrutiny. What emerges from that work is that in many cases these changes are inextricably bound up with issues of space and place. We don't mean merely that, like many processes, they take on particular geographical forms (although there is nothing mere about that), but that space and place are integral to the evolution of the processes themselves. In this sense, economic geographers are in exactly the right time and place to practice their art.

Consider again globalization. Arguably globalization has existed since the very first economic geographers were writing in the late nineteenth century, and likely well before then (certainly it is recognizable in Andree's ruminations). One aspect of globalization that has changed, though, is the means by which it is effected. Whereas in Andree's time it was bound up with Western European colonial policy and bureaucracy, now its prime bearers are multinational and transnational corporations (MNCs and TNCs). MNCs and TNCs are fundamentally economically geographical institutions. Their very definition and rationale is based on the idea of geographical differentiation: that different parts of their operations are located in different places. As such, they are ripe for economic geographical study. Others can and should study MNCs and TNCs, but the sensibility of economic geography is especially suited to the task and, given the role that they play in the globalization process, is ever more germane.

To take another example, the Fordist-style manufacture that Rivera depicted in his murals was increasingly abandoned from the 1970s onwards, and replaced by what was called post-Fordism, or flexible specialization. That move was not only about transforming a particular method of production – from assembly line and Taylorism to batch production and work teams – but also about transforming geography. Flexible specialization implied a set of new spatial arrangements that were part of its very definition. Greater physical interaction and closer proximity between firms became necessary by comparison to the more arm's-length relationship that had characterized Fordism in order to realize the benefits of flexible specialization. Subcontractors needed to be close to the production plant in order to respond quickly to changed demands; R&D staff were brought in from their research facilities, sometimes thousands of kilometers away, to work more centrally in the factory in order to consult with production staff; and workers were no longer spatially segmented along the assembly line, but enjoyed greater interaction and collaboration in work teams. Whatever else it was, the move from Fordism to flexible specialization entailed an economically geographical shift.

These are only a couple of examples, but we hope that they illustrate the broader argument. The recent changes in the economy, which have affected so many people in so many different ways, are not only economic but also fundamentally geographical. Economic geography is relevant not just as background atmospherics, to add ambience and color, but for understanding why economic change occurs at all. It enters into its very frame, its skeletal structure. And what is bred in the bone, comes out in the marrow.

The second reason for the renewed attention to economic geography is that it continues to undergo potent intellectual shifts, making it an arena of discursive ferment and vibrancy. Economic geography, in fact, has long been an important forum for trying out new ideas. Much of human geography's quantitative revolution of the 1960s, for example, was discussed within the context of economic geography. Since the late 1970s and early 1980s, the discipline has propeled discussions around political economy. As an approach, political economy is pervasive: it is how economic geography is now done. That said, political economy is no single, staid, monolithic tradition, but is multiple, dynamic, and differentiated. Debate is not always friendly, but through it new approaches, new theories, and new possibilities emerge, and constantly invigorate the field.

Political economy as a tradition began in the eighteenth century, but it is perhaps most closely associated with Marx's nineteenth-century writings about capitalism. In both cases, there was the insistence that the political and the economic are irrevocably bound; that the economy cannot be treated as sovereign and isolated, but must be understood as part of a set of wider social processes. While sharing these views, political economists and economic geographers have continued to debate two central questions: the definition of the social, and the nature of the connection between the social and the economic.

For Marx, the answers to those contentious questions were clear-cut. The social meant class relations, which under capitalism primarily consists of an antagonistic relationship between the working class and the capitalist class. Social classes, in turn, are connected to the economy by a functional relationship. This means that the form of the relationship among social classes is set by what contributes best to the development of the economy. Such a functional relationship is certainly not always smooth and unproblematic, but ultimately the economy prevails and produces those social relationships that are most appropriate for its development. If capitalism can best develop by a set of class relations consisting of workers and capitalists, that is what happens. The best, and the most well-known, economic geographical translation of Marx's views about social processes and their connection to the economy was carried out by David Harvey (1982), particularly in his writings during the 1970s and 1980s on the geography of accumulation.

Subsequently, other economic geographers, drawing upon other traditions of political economy, have developed alternative approaches, ones providing quite different answers to the two questions about the definition of the social, and its relation to the economy. This is not the place to provide a full-blown review, but those alternatives include:

- Doreen Massey's (1984) work on spatial divisions of labor. This is much more catholic than Harvey's in its definition of the social, and in the relationship it posits between the social and the economic. (Importantly, place itself partly determines the development of the economy.)
- Regulationist theory. This partly defines the social in terms of institutional and regulatory frameworks, and also attempts to move away from crude functionalism (Tickell and Peck, 1992).
- The analytical approach. This uses both mathematical reasoning and rigorous, formal statistical testing to determine logically how space and place make a difference both to the definition of social processes and to their relation to the economy (Sheppard and Barnes, 1990; Webber and Rigby, 1997).
- Most recently, the umbrella approach, sometimes called the "cultural turn." This brings together a variety of perspectives – post-Marxism, institutionalism,

economic sociology, feminist theory – trying both to widen radically what is included within the social, and also to move away from Marx's functionalism, and instead make use of notions like embeddedness (Martin, this volume), overdetermination (Gibson-Graham, this volume), or cultural performance (McDowell, 1997).

This review is necessarily very brief. Sustained examples of such political economic approaches can be found in the ensuing chapters. Our point, like the one we made about economic change, is that contemporary discussions around political economy make economic geography stimulating and provocative. There is a Chinese saying: "May you live in interesting times." Our argument is that they are here now in economic geography.

Purpose and Organization

Our broad purpose in commissioning the essays that make up this volume is to provide advanced undergraduate and graduate students, and faculty colleagues both inside and outside economic geography, with a sense of the state of the art of economic geography. We chose to focus on the ideas, concepts, and theories current within the discipline, rather than on empirical findings, because a discipline is characterized primarily by the questions it asks. We identified individuals who we thought characterized both the unity and diversity of economic geography, and we asked them to provide their own accounts. We asked only geographers to contribute because, as we argue above, economic geography is quite different from geographical economics (see also Sheppard, 2000). Certainly, exchanges of views between geographers and economists about economic geography are an important part of the evolution of the identity of the discipline, but we leave that task to others (Clark et al., 2000).

   The book is organized into five sections. Economic geography is a philosophically diverse discipline, employing and debating a variety of approaches. The first section, *Worlds of Economic Geography*, provides a sense of this diversity. Following the first essay, which provides a historical context for understanding the discipline, the ensuing chapters examine in turn the mathematical modeling tradition, Marxism, and three more recent variations on the political economy theme: feminism, institutional approaches, and poststructuralism. The essays not only convey a sense of the intellectual vitality of the field, but also show that accounts of economic geography, the questions taken to be important, and the methods used to find answers to them, depend on the broad philosophical and theoretical approach taken. In short, accounts of economic geography, as for any discipline, are *situated*; they depend on the point of view of the author, which in turn reflects how he/she is positioned intellectually, politically, and socially. As the volume turns to more substantive aspects of economic geography, we hope you will see how authors locate themselves on the intellectual terrain mapped in this section. As you read the essays in the other four sections, think about where the author is situated, how that affects the questions he or she asks about the topic of the chapter, and how those questions might differ if it had been authored by someone from a different perspective. We cannot include all perspectives on all topics, or the book would be five times as long, but the possibilities, and actual existence, of those different perspectives is part of what makes the discipline rich and exciting.

Because economic geography necessarily asks questions about things economic, the second section, *Realms of Production*, highlights classic themes in economics: production, work, agglomeration, competition and markets, economic growth, and technical change. In their own ways, each of these essays takes on the task of narrating how economic geographers have approached these topics, and how a geographical approach gives each of them a distinctive twist compared to their treatment in economics, even when economists ask questions about the geography of economic activity.

One of the things that makes economic geography a part of geography is a concern for human–environment, or nature–society, relations, and this is the focus of *Resource Worlds*, the third section. These essays include discussions about nature as an input to economic production; about agriculture, where that relationship is most stark; about political ecology, which has become the central theoretical approach for analyzing agricultural practices particularly in the Third World; about questions of the production of nature within economic geographical processes; and finally about resource towns, the non-agricultural places where production is most closely tied to nature.

Geography has always been a synthetic discipline, taking seriously the relationships between things studied separately in other disciplines (such as nature, and society). In *Social Worlds*, the fourth section, the essays synthesize across the human sciences, treating economic processes as part of other social processes, rather than separable from, superior to, or more fundamental than the rest of society. The themes addressed include: consumption and the family, social class, labor organizations (unions), questions of political governance, organization of producers (corporations), and social and ethnic networks. Each chapter seeks to show the importance of a geographical perspective in understanding the intersection of the economic with the social.

Last, but certainly not least, geography is not just about place but also about space: about how the distinctive economic characteristics of places, and the conditions of economic possibility that they face, depend on their economically geographical interdependencies with other places. In *Spaces of Circulation*, the final section, we turn attention to the geographical flows of commodities, money, and people, and to the transportation and communications systems produced to facilitate them. These essays suggest that even though the world is becoming smaller, one of the core metaphors for globalization, it is also an increasingly differentiated world, in part precisely because the interdependencies between places are increasing. The final essay links geographical thinking about globalization to similar work in international political economy.

In sum, we hope that these essays provide useful resources for further reading in economic geography, and also that some of you will join in this endeavor. We also hope that you are provoked to be critical: that you will reflexively re-examine your current thinking, and ask new questions about the adequacy of economic geography's representations and approaches highlighted here. If so, our goals in compiling this volume will have been achieved.

Bibliography

Andree, Karl 1867–77. *Geographie des Welthandels* (3 vols.) Stuttgart. [The quote is from Volume 1, p. 13 (first published in 1867)].

Clark, G., Gertler, M. and Feldman, M. 2000. *Handbook of Economic Geography.* Oxford: Oxford University Press.

Detroit Institute of Arts 1978. *The Rouge: The Image of Industry in the Art of Charles Sheeler and Diego Rivera.* Detroit: The Detroit Institute of Arts.

Harvey, D. 1982. *Limits to Capital.* Chicago: University of Chicago Press.

Krugman, P. 1991. *Geography and Trade.* Cambridge, MA: MIT Press.

Martin, R. L. 1999. The new "geographical turn" in economics: Some critical reflections. *Cambridge Journal of Economics,* 23, 65–91.

Massey, D. 1984. *Spatial Divisions of Labour: Social Structure and the Geography of Production.* London: Methuen.

McDowell, L. 1997. *Capital Culture.* Oxford: Blackwell.

Rivera, D. 1934. *Portrait of America.* London: George Allen & Unwin.

Rochfort, D. 1987. *The Murals of Diego Rivera.* London: Journeyman Press.

Sheppard, E. and Barnes, T. 1990. *The Capitalist Space Economy.* London: Unwin Hyman.

Sheppard, E. 2000. Geography or economics? In G. Clark, M. Gertler and M. Feldman (eds), *Handbook of Economic Geography.* Oxford: Oxford University Press (forthcoming).

Tickell, A. and Peck, J. 1992. Accumulation, regulation and the geographies of post-Fordism: Missing links in regulationist research. *Progress in Human Geography,* 16, 190–218.

Webber, M. and Rigby, D. 1997. *The Golden Age Illusion.* New York: Guilford Press.

Part I Worlds of Economic Geography

Chapter 2

Inventing Anglo-American Economic Geography, 1889–1960

Trevor J. Barnes

The American inventor Thomas Edison (1847–1931) said that his genius consisted of "1 per cent inspiration, and 99 per cent perspiration." In this chapter I make a similar argument about the invention of economic geography. While there were inspirational moments during its history, mostly the story of economic geography is about hard work, about the large number of specific concrete practical activities required to define and maintain an economic geographical order. Such organization was not instant, but occurred hesitantly, a consequence of particular human practices and accomplishments within specific geographical and historical settings. That is why "inventing" is in the chapter title. It signals that economic geography is contingent upon conscious acts of human creativity and effort rather than existing since time immemorial.

The idea that academic disciplines are human inventions, and in some cases very recent human inventions – economic geography, for example, was invented somewhere between the time of the first telegraph and the first radio – might seem disconcerting. This is because one connotation of the word invent is to make something up, to fictionalize, or, at worse, to lie. Under this interpretation to invent is to tell stories about a world that doesn't exist. In contrast, from its inception economic geography was (and is) about something real and substantial.

However, there is another connotation to inventing that better fits the aim of the chapter, and takes us back to Thomas Edison. Edison's inventions, such as the ticker tape machine (1870), the phonograph (1878), and the incandescent light bulb (1879), were "made up" in that they did not exist before he had them as ideas. But once they existed they produced definitive material effects, altering human practices and beliefs. Humans could do things that they could never do before: react to Wall Street's surges and slumps as they occurred, hear music originally played sometimes half a world away, and light up whole cities. Inventions in Edison's sense change the world, and give humans a different relationship to it.

My argument is that this second interpretation of inventing is useful in understanding the history of economic geography. Once economic geography was invented, and later reinvented, the world was altered. After economic geography existed, peoples' beliefs and actions were different: not only those of geographers, but of a variety of people.

There is one more parallel to draw. Inventing economic geography and the light bulb are context-dependent; that is, there is something about the peculiar combination of circumstances in which those inventions occurred that made them possible. For example, for Edison to invent the incandescent light bulb required knowledge about electricity and its properties, machines and specialized equipment that could generate electricity and transmit it across long distances, and a research laboratory containing instruments, materials, and trained technicians and scientists capable of undertaking such a project. These contextual factors made the incandescent light bulb possible, entering into its very construction, and shaping its form and capacities.

The same holds for economic geography. The discipline arose because of a set of contextual factors, and as those factors changed so did economic geography. Further, those factors, like the relationship between electricity and the light bulb, entered into economic geography's very constitution, shaping the questions it asked, the methods and tools it used, and the range of answers it deemed acceptable. For sure the discipline had its own geniuses, people who were brilliant in thinking up new ideas, but their creativity was always, like Edison's, informed and tempered by a wider context.

The chapter is divided into three main sections. The first describes the beginnings of the institutionalization of Anglo-American economic geography as a discipline in the late nineteenth and early twentieth centuries. Critical here was the presence of colonialism and the growth of global commercial trade. The second discusses the interval between the first and second world wars (1918–1939/41) when a regional approach to the discipline emerged. The final section focuses on the immediate post-war period which was the setting for the so-called "quantitative revolution" and "spatial science."

In a chapter as short as this it is not possible to cover everything. Instead, the chapter is organized around a set of vignettes marking distinct periods in economic geography's invention and reinvention. It is *a* history, and not *the* history. I should especially stress that the essay is limited to English-speaking, Anglo-North American economic geography, and to the voices of men. This is partly because of limitations of space, partly because of my own circumscribed abilities and biases, partly because of the kind of written sources that are available, and partly because of the very nature of economic geography itself. For good or bad, economic geography was historically a masculinist discipline, and linked to the European intellectual sensibility of the Enlightenment tradition (on both characteristics see Gibson-Graham's chapter in this volume). But to define the history of economic geography thus is not to argue for continuation in its old form. Indeed, the chapters that follow envisage something quite different.

Invention, Institutionalization, and Imperialism

It is very difficult to provide an exact date for the invention of economic geography. Possible contenders include: 1925, when the journal *Economic Geography* was first published; 1893, when economic geography courses were first taught at Cornell

University and the University of Pennsylvania; 1882, when the German geographer Götz distinguished between commercial and economic geography; and 1826, when *The isolated state*, for some the first classic treatise in economic geography, was written and privately published by the German landowner and farmer Johann von Thünen (1783–1850). These dates at least narrow the origins of economic geography to the nineteenth and early twentieth centuries. This period was important for the emergence of economic geography for two reasons.

The first is that this was an era when a number of academic disciplines, especially in the social sciences, were institutionalized within Western European and North American universities. Economic geography was one of those new disciplines (for other examples see Ross, 1991). These disciplines, such as sociology, or psychology, or economic geography, were defined by novel forms of representation which I will term "discourses." Discourse is a difficult word to define, but I use it to mean a network of concepts, statements, and practices that produce a distinct body of knowledge. A disciplinary discourse, for example, would include specialized vocabularies, conceptual and theoretical frameworks, diagrams, variables, and even tables of figures. The important point is that such discourses are crucial to creating – inventing – the world that is represented; that is, rather than simply mirroring a world that is out there, disciplinary discourses help to construct it. For example, up until at least the seventeenth century, if not later, the idea of the economy did not exist as such. Of course, humans have performed what we now call economic acts since Adam and Eve labored in the Garden of Eden. But it was only about 300 years ago that the economic was recognized as conceptually distinct. Partly allowing the economy to become visible – to be invented – were a novel set of practices derived from the emerging academic disciplinary discourse of economics (which I will call "discursive practices"), and composed of, for example, striking metaphors (Adam Smith's "invisible hand"), new concepts (Ricardo's comparative advantage), persuasive diagrams (Quesnay's *Tableau économique*), compelling equations (Say's Law), and meticulously calculated figures (William Petty's political arithmetic). Note that using the term discourse is not to deny the seriousness or materiality of the economic. The new economic world that was invented quickly affected people's most heartfelt beliefs and their most material actions.

In a similar fashion, albeit slightly later, economic geography was invented. Its discourse was composed of: new terms like "regions of production" (Chisholm, 1889) or "north–south axis" (Smith, 1913); new typologies and schemes of categorization that ordered the world, for example, linking global climate regimes and agricultural crops (Chisholm, 1889, pp. 57–153); and new maps and diagrams that made visible economic geographical connections, for example, between trade routes and modes of transportation (Smith, 1913). The result was the creation of new bits of reality that previously did not exist, like "commercial space" or "transactional centers," which, in turn, affected how people saw their position within the world, producing new beliefs and behaviors.

The second reason for the particular timing of economic geography is bound up with its relationship to colonialism, and especially commercial expansion. There is now much written about the relation between the emergence of academic geography and the rise of especially nineteenth-century Western European imperialism (Hudson, 1977; Peet, 1985; Livingstone, 1992). In particular, a link is often made between the environmental determinism that a number of early geographers espoused at the time, and the justification of imperialism. Environmental determinism is the idea that the natural environment determines everything about a given people, for example, that those living in tropical climates will be less energetic than those living in temperate ones (Livingstone, 1992, ch. 7). Such a thesis, in turn, was used as a justification for colonialism. If a people could not fully develop economically because of, say, the prevailing climatic regime in which they lived, it was then legitimate for Europeans to intervene and show them how: indeed, this was "the White man's burden."

While an important line of argument, these links between geography and environmental determinism are couched at the level of geography as a whole, and seem a bit too neat and mechanical (Livingstone's 1992 work, with its emphasis on "situated messiness," is an exception): if imperialism, then environmental determinism. In contrast, I will concentrate on only economic geography, and move away from such a tidy relationship. As we know even from Edison and his light bulb, inventing is messy and difficult. The same goes, I will argue, for economic geography.

In focusing only on economic geography, and its unkempt beginning, I start with two figures: George Chisholm (1850–1930) in Britain, and J. Russell Smith (1874–1966) in the United States. Before their writing there were odd facts about commerce and geography, dribs and drabs that didn't quite fit; afterwards there was a new discipline.

George G. Chisholm

The intellectual life of George G. Chisholm, author of the first English-language economic geography textbook, *Handbook of commercial geography* (1889), personifies those unkempt beginnings. Born in Scotland and attending Edinburgh University, he later moved to London, making his living primarily from writing and editing geographical textbooks, gazetteers, and atlases (Wise, 1975; MacLean, 1988).

Chisholm's *Handbook* contains everything anyone might want to know about world commodity production and the geographical conditions for trade. While Chisholm (1889, pp. iii–iv) claims in the Preface "to import an intellectual interest [to his inquiry] . . . and not to encumber the book with a multitude of minute facts," that interest is difficult to find, whereas minute facts litter every page. The *Handbook* contains neither an explicit theoretical statement nor organizational justification. His focus is the facts (especially evident in his very long statistical appendix). As he once said, "If . . . there is some drudgery in the learning of geography, I see no harm in it" (quoted in MacLean, 1988, p. 25).

In 1908 Chisholm left London to become a lecturer in geography at Edinburgh (the first such position in Scotland), and it was there that he first wrote programmatic statements for the new discipline of economic geography:

I would say then that it is the function of geography with respect to any class of phenomena that have a local distribution to explain that distribution in so far as it can be explained by variations connected with place in the operation of causes whose operation varies according to locality or according to the relation of one locality to another (Chisholm, 1908, pp. 568–9).

In so far as this statement is understandable it points to the importance of local, place-based factors and their interrelationship in determining the geography of economic activities. There is certainly no indication here that Chisholm is some lackey to imperialism mouthing environmental determinism. That said, the context of imperialism in which Chisholm did his inventing clearly influenced his work, and his conception of economic geography. Those connections were complex, however.

First, and most directly, the *Handbook* was a celebration of imperial commerce, the technology that made it possible, and a global trading system with Britain at its center. The late nineteenth century when Chisholm wrote the *Handbook* was a period in which Britain was the workshop of the world, "heavily dependent upon international trade and at the hub of a world-wide empire" (MacLean, 1988, p. 21). In part this was made possible by "improvements in the means of transport and communication" (Chisholm, 1889, p. 47), but also in part because of free trade and Britain's military might. Chisholm was not only extolling the economic benefits of trade in his volume, though, but also economic geography. For him, trade is primarily geographical. Chapter 1 begins: "The great geographical fact on which commerce depends is that different parts of the world yield different products, or furnish products under unequally favorable conditions" (Chisholm, 1889, p. 1). In this sense, to acclaim trade is simultaneously to acclaim "the great geographical fact" of commercial or economic geography.

A second connection is that Chisholm's book, through its use of maps, figures, and tables of numbers, made visible, and therefore knowable, the complex filaments of the imperial project. In this sense, Chisholm's book constructed imperialism as well as representing it. This might seem a strange claim to make. I am not suggesting that Chisholm made up imperialism, but I am saying that without his work and other similar works people could not have known imperialism as such. Imperialism is so large and complex that it only becomes knowable when it is reduced to the printed page, which is precisely what Chisholm's maps, figures, and tables accomplish. This also speaks to the hard work and mundane quality of the effort required to invent (and celebrated by Edison's aphorism). One can hardly imagine a more tedious and time-consuming task than Chisholm assembling massive amounts of data in tabular form, searching down every last fact, and constructing detailed maps. But these mundane and meticulous acts contributed to the reproduction of imperialism, and helped to invent the new discipline of economic geography.

A final connection between imperialism and Chisholm's work is found in education. Chisholm begins his Preface by saying that he wants to teach "those entering on commercial life" (p. iii); that is, to make them better prepared and competitive in their business activities by furnishing them with the right knowledge. In carrying out that task, Chisholm was prompted and guided by two contextual factors. The first was that during the late nineteenth century Britain was facing stiff commercial competition from Germany and its empire. Chisholm believed Germany's competitiveness stemmed directly from the superior educational prowess of its business class who among other things were steeped in economic geography. The second was a report published in 1885 by a British Royal Commission headed by Scott Keltie documenting the poor state of geographical education in Britain (a response in part to the 1870 Education Acts). As an academic field, however, geography lacked qualified teachers, suitable textbooks, and even a curriculum (MacLean, 1988, p. 23).

Chisholm's book met both needs. It provided the British business class with commercial geographical information, and British schools with a curricular template, a source of classroom information, and something to be displayed both to students and to school governors and inspectors. Chisholm, however, was also speaking to an emerging community of academic geographers. His work became part of the broader process of institutionalizing geography as a discipline, especially after his appointment at Edinburgh University. Through Chisholm's role as an educator, economic geography as a newly invented discipline diffused, taking on efficacy: economic geography began changing the world.

Chisholm's book was intimately connected to his life and times. The colonial project was certainly helped on its way by Chisholm's writings, but it wasn't there working behind his back ensuring that he wrote only the right words or drew only the right maps. In fact, later in his life Chisholm (1921, p. 186) made a plea against the exploitation of colonial labor, which indicates again the untidy character of invention. Imperialism could not simply call forth actors such as Chisholm to provide instant legitimation. It was always more complicated.

By the time Chisholm died in 1930, the legitimacy of economic geographical discourse, and the object it constituted, were established. His work in part set the economic geographical die, stamped, as we've seen, by such characteristics as a wariness of theory, a concern with empirical detail, a celebration of numbers, a predilection for geographical categorization made visible by the map, and a tracing of relations among places through the media of various kinds of economic flows, especially that of the commodity. If none of this seems exceptional it is because we are the heirs of Chisholm's more than century old legacy. At the time, though, he was making things up – inventing – as he went along.

J. Russell Smith

On the other side of the Atlantic, the process of invention was slightly different. The main impetus came from economics, and in particular the dissatisfaction of American economists with the discipline's abstractness (Fellmann, 1986). As a reaction, some turned to the German Historical School which emphasized context and concrete detail. As the economist Edward Van Dyke Robinson explained at the time (1909, p. 249, and quoted in Fellmann, 1986, p. 316): "After the [German] Historical School of economists had introduced the idea of relativity as to time and place, the necessity was apparent not only for a historical but also for a regional treatment for economic phenomena – in other words for economic history and economic geography."

Consequently, in the United States economic geography was initially taught by economists or those in business schools. An early site was the Wharton School at the University of Pennsylvania. In 1903 it hired one of its doctoral students, a transportation economist, J. Russell Smith, who three years later founded within the School the Department of Geography and Industry. In 1913 Smith cemented both his own reputation and that of US economic geography by publishing *Industrial and commercial geography*. It was both a new and improved as well as an American version of Chisholm's *Handbook* (Starkey, 1967, p. 200)

Divided into two sections which correspond to the terms of its title, the first (and longest) part of *Industrial and commercial geography* discusses the production of particular resources and manufactured goods, and the second part discusses world trade. Throughout there is a dynamism and movement that is absent in Chisholm's work, and which is a product of Smith's focus on technological changes around transportation and communications. In particular, such changes for Smith facilitated a "world [commodity] market" (Smith, 1913, p. 16) defined by a distinct geography of control and production. Control is located in one "mere corner of the world" (Smith, p. 874), i.e. NW Europe and the NE seaboard of the US, because it "has capital to spare" (Smith, p. 874), and because of its "transactional" role as a "place where bargains are made" (Smith, p. 867). In contrast, the rest of the world is defined by its role as a producer.

The discourse that emerges from Smith's book is similar to that of Chisholm's. Again there is a focus on empirical detail, the shunning of theory (although environmental determinism appears in the introductory chapter to Smith's book, it was dropped in later editions), and a penchant for geographical categorization. Smith is perhaps even more emphatic than Chisholm in emphasizing exchange and trade. For J. Russell Smith exchange and trade are irresistible impulses. Whenever one place has a surplus of one good, and another place a surplus of a different good, trade naturally unfolds between them. In addition, Smith joins an emphasis on trade with a stress on technology. With such innovations as the steamboat, railway, telegraph, and telephone, trade need no longer be locally constrained but becomes global. In this sense Smith's book reads as a primer on globalization, albeit written at the beginning of the twentieth century and not at its end.

As with Chisholm's *Handbook*, the sources of success of Smith's *Industrial and commercial geography* are multiple. There is the role of the Historical School which gives rise to the very idea of economic geography, and, in turn, to economic and commercial geography courses that had never existed before. Moreover, by a happy coincidence, this development occurred just as geography itself was becoming institutionalized. When the influence of the Historical School began to wane by the 1910s as economics became more analytical and less empirical, and as a result economic geography courses were dropped by departments of economics, they were picked up by newly-formed geography departments such as Smith's (Fellmann, 1986, p. 319).

There is also the peculiar conjuncture of Smith's own interests. By focusing on transportation and communications technology Smith could simultaneously celebrate human creativity, the power of the machine, trade and exchange, and the importance of geography, while at the same time not giving the discipline away to physical geography (which was the inclination of perhaps the most famous US geographer at the time, William Morris Davies at Harvard). Smith could sustain such a position in part because of the compelling narrative he used to structure his book which joined the benefits of exchange with the benefits of technological innovation. It was a story that resonated perfectly with the free-enterprise and machine-age sentiments of an early twentieth-century America beginning to engage in global trade.

Finally, as with Chisholm, there is Smith's role as an educator. His influence was not confined to the students he taught while at Wharton (and at Columbia

University in New York from 1919 onwards). He was, like Chisholm, a dry lecturer, his poor delivery exacerbated by "a slight lisp and a low, monotonous way of speaking" (Rowley, 1964, p. 87). But he shone as a writer. His book was enormously successful, garnering compliments by geographers, and, perhaps even more importantly, by university administrators who decided the fate of geography departments (Rowley, pp. 50–1). Like Chisholm's *Handbook*, Smith's volume was more than just a student text to be read to pass final exams. It became an active component in the very invention of economic geography as a discipline.

By focusing on Chisholm and Smith I've delineated some of the elements of the newly invented discipline of economic geography and its world. The discipline was concerned with empirical detail, global geographical categorization based upon commodity specialization, and the spatial patterns and conditions of commercial trade. Economic geography did not have to turn out in the way it did, for example, it could have taken on an abstract theoretical bent (which was the trajectory of economics). That it did not do so goes to the importance of historical and geographical context. That context was not mere background atmospherics, but entered in the very body of the discipline's knowledge. The context of commercial expansionism and imperial control made a difference. But as I've also stressed, early economic geographers were not the mere dupes of that context: there was individual inspiration, along with perspiration.

Regions, Differentiation, and Richard Hartshorne

Within fifteen years or so of Smith's book being published, economic geography began to change. There are again parallels with what happens to inventions. Often they undergo further alteration, sometimes taking forms that are barely recognizable compared to their original incarnation. Contrast, for example, early computers that took up the space of many rooms with the present lap-top. These kinds of developments or reinventions, whether it be computers or economic geography, require much labor and reorganization.

In the case of economic geography it became noticeably different by the 1930s as focus shifted from the general commercial relations of a global system to the geography of narrowly bounded, unique regions, especially those close to home. In part, this move to the region stemmed from a long-standing debate in the economic geographical literature. Smith's book was structured for the most part thematically, around commodities, trade, and transportation, and not regionally. For some American economic geographers, such as Ray Whitbeck, this was a mistake. As he argued in his critical review of Smith's book, there is "a distinction between a text book of commerce and industry and one on commercial and industrial geography" (Whitbeck, 1914, p. 540). For Whitbeck (1915–16, p. 197) the emphasis should be geographical: "the unit should be the country and not the commodity."

From the end of World War I in 1918, Whitbeck's position became increasingly popular, particularly in US economic geography (which is the focus of this section). The result was a regional perspective that made the region, and its unique features, the focus of inquiry. That perspective is readily disclosed in the various textbooks that were published from the mid-1920s onwards. Ray Whitbeck's own *Economic geography*, co-written with his University of Wisconsin colleague Vernor Finch, is typical (and first published in 1924). Justifying economic geography on the grounds that it provides "a kind of knowledge that educated people need and use," the key organizing idea is "areal differences," the interpretation of which "requires the reasoned association of facts of many kinds" (Whitbeck and Finch, 1935, p. v). Those facts are then neatly ordered under an identical fourfold typology for each of the regions investigated: agriculture, minerals, manufacture, and commercial trade, transportation and communications.

Another example is Clarence Fielden Jones's (1935) textbook *Economic Geography* which begins: "Everyone likes to travel. Most of us wish to visit distant lands" (quoted in Berry, Conkling and Ray, 1987, p. 27). The eightfold typology used by Fielden Jones to organize the facts taken from his travels is more finely variegated than Whitbeck's and Finch's, but it performs the same role: a typological grid for sorting observations that can then be photographed, mapped, tabulated, or, most likely, merely listed under the appropriate classificatory heading. By comparing the facts of the different regions by using the same typological grid, geographical differences are immediately seen, and areal differentiation shines by its own light.

More broadly, this regionalist perspective represented a shift in the economic geographical discourse, changing the object of enquiry. Let me explore that discourse by briefly discussing the work of Richard Hartshorne (1899–1992), an American geographer who perhaps more than any other geographer is associated with the regionalist perspective. Not that Hartshorne was the first to conceive of regionalism, but he systematically codified it, and provided an intellectually rigorous justification. In so doing he gave economic geography a critical role.

In the 1920s Hartshorne carried out work in economic geography, but he is best known for his writings on the history and philosophy of the discipline, and, in particular, the book, *The Nature of Geography* (1939), which drew upon the German geographer Alfred Hettner (1859–1941). One purpose of that book was to define a geographical unit that could be used as a basis to organize and integrate the often scattered and multifarious pieces of information collected by geographers. Hartshorne's answer: the region. It allowed geographers to integrate otherwise disparate geographical facts, rendering the complexity of the world comprehensible.

But how should the region be defined? Regions based on watersheds, for example, would be quite different from regions based upon economic specialty. For Hartshorne the economic ruled supreme. He thought the activities of keeping "body and soul together" (Hartshorne, 1939, p. 334) are the most central to human life, and should be used to delineate the region. More specifically, within the wider set of economic activities that occurs he thought those around agriculture, and the family farm in particular, are especially important for regional delimitation.

Hartshorne, therefore, justifies his regionalist perspective on the basis of economic geography: "there is no boundary between economic and regional geography" (Hartshorne, 1939, p. 408). For him the world is made up of a patchwork quilt of economically defined regions, where each region is a closely defined complex of interconnected elements including, for example, land and plants, buildings and livestock, tools and production methods, as well as various invisible components like prices and markets, and practical and abstract knowledge. Examples might include: the American corn belt, the Po plain in Italy, or the Vale of Evesham in

Southwest England. In each case the complex interaction among the different elements within a region is critical, producing a unique regional entity.

Let me note some of the principal differences between this regionalist economic geographical discourse and the older one of Chisholm and Smith. It will help clarify the nature of the economic geographical reinvention that occurred.

First, regionalism emphasizes the geographically unique. While the same type of elements are found in different economic regions, how they are combined together and relate one to another "occurs but once on the earth" (Hartshorne, 1939, p. 393). For Hartshorne two related implications follow. That no law-like statements can be made about regions, implying that economic geography cannot be a predictive science (this is especially important given that from the mid-1950s onwards economic geography makes strenuous bids to become exactly a predictive science). And that a regional approach, or what amounts to the same thing, an economic geographical approach, must be descriptive. If economic geographers cannot invoke explanatory laws, they must be "concerned with the description and interpretation of unique cases..." (Hartshorne, 1939, p. 449). In contrast, Chisholm and Smith, while recognizing the importance of geographical difference, and relying heavily on description, are at pains to emphasize the wider geographical and economic system that revolves around world production and trade, and that connects places and regions together. This is missing in much of the regionalist economic geography. The region is interesting in and of itself.

Second, the regional uniqueness emphasized by Hartshorne provides justification for the rapid spread of regional typologies that typified economic geography. If a region is defined by its combination of different economic elements, then to show its uniqueness we must classify those elements, and then, using the typologies constructed, demonstrate that those elements appear in different mixtures in different places. In contrast, in Chisholm's and Smith's works, there is less concern with typologies as such. Places are categorized, but only by the commodities they produce and trade, with the purpose of emphasizing geographical connectivity: the global system comes first, its constituent regions defined by commodity production came second.

Lastly, for Hartshorne the complexity of the region requires sustained, on-the-ground field research. Geographers must muddy their boots if they are to understand the world. Again, in contrast, neither Chisholm's nor Smith's work required them to be physically present: in fact, given the global ambitions of their project, it would have been impossible for them to do so. Collecting and organizing numbers, and drawing maps based upon them, was sufficient. Of course, they still went into the field – Chisholm was an especially ardent field tripper – but it was not an integral component of their enterprise.

The exact reasons for economic geography's regionalist reinvention are difficult to locate. A number of causes seem plausible. There is a changing historical and geographical context. Both the slowing of colonialism, and the effects of the 1930s Depression on world trade and commerce, rendered Chisholm's and Smith's projects less pressing. There is Hartshorne's intellectual engagement with German geography and in particular with Hettner's. (In 1938–9 Hartshorne spent a sabbatical year in Austria writing *The Nature of Geography*; Hartshorne, 1979.) There are also internal sociological reasons, of which the most cited is the embarrassment caused by the discipline's earlier affair with environmental determinism. A focus on regional study represented if not "greener pastures" at least "smaller safer pastures" (Mikesell, 1974, p. 1). But also the discipline of geography itself was becoming much more professionalized and institutionalized within a university setting. As it did so the stakes became higher, with individual academic reputations hinging on the successful promulgation of new ideas and discourses such as the regional one.

Whatever the reasons, external or internal, the upshot was quite a different economic geography from what went before, with such emphases as fieldwork and regional boundary making, the construction of new typologies, and a distinctive view of the purpose of geography. Furthermore, regionalist economic geography remained intact for at least twenty years after *The Nature of Geography* was published. Partly this was because of Hartshorne's status and power within American geography, but it was also a legacy of his wartime activities in US military intelligence in the Office of Strategic Services (forerunner of the Central Intelligence Agency) which involved him supervising a large number of geographers. According to Butzer (1989, p. 50), the close-knit relationships forged during that period "crystallized into a coherent paradigm, reinforced by close proximity and strong personal ties, and that endured for many years as an elite club." From the mid-1950s onwards, however, that club began to be criticized in large part from a younger generation who were never members.

The Quantitative Revolution and Spatial Science

That younger generation, at least in the United States, initially gathered at two principal sites: the University of Iowa, Iowa City, and the University of Washington, Seattle. It was from Iowa that the first voice openly opposed to Hartshorne was heard. In 1953 Fred K. Schaefer (1953), a political refugee from Nazi Germany and an inaugural member of the Department of Geography at Iowa, published in the flagship periodical, *Annals, Association of American Geographers*, "Exceptionalism in geography." It was both a repudiation of Hartshorne's position, and a call for a scientific approach to geography based upon the search for geographical laws (the ultimate form of a scientific generalization). Unfortunately, Schaefer died in June 1953, before his article even appeared in print, and so he was never able to elaborate his argument, nor defend himself from Hartshorne's (1955) subsequent attack. But the article became a rallying point for the younger generation of economic geographers who were intent on reinventing the discipline as a science, or "spatial science" as it was later dubbed.

To do so required the techniques, logic, and vocabulary of scientific practice: that is, a new disciplinary discourse was needed, and consequently a new object of inquiry. Here a colleague of Schaefer's at Iowa, his boss, Harold McCarty, played a critical role. While suspicious of Schaefer's grand philosophizing, and rankled by his combativeness, McCarty concurred with Schaefer that economic geography should move away from regionalism and become more scientific. In 1940 McCarty had published *The geographic basis of American economic life*, on the surface a conventional regional account of US economic geography. Its intellectual point of departure was fundamentally different from Hartshorne's, however. Regions were McCarty's unit of analysis neither because of the authority of Hettner, nor because they were convenient integrative units, but because market forces made economic geographical reality that way (McCarty, 1940, p. 19). There is a hard economic conceptual logic driving McCarty's analysis based upon market forces, which was new to the discipline, and represented by economic theory. As he put it: "Economic geography derives its concepts largely from the field of economics and its method largely from the field of geography" (McCarty, 1940, p. xiii).

McCarty's use of that word method is very broad because the specific technical methods he later employed came not from geography per se, but from statistics. In particular, he pioneered within economic geography the use of the statistical techniques of regression and correlation analysis (McCarty et al., 1956), conceiving the very discipline in their image (Barnes, 1998b). Such practices marked Iowa as one of the first quantitative economic geography departments in the US.

At the other "center of calculation" in US economic geography, the University of Washington, the two pivotal faculty members were Edward Ullman, and, perhaps more importantly, William Garrison. Garrison had joined the department in 1950, following a PhD at Northwestern University, and before that wartime service in the US Airforce where he took courses in statistics and mathematics as part of his training as a meteorologist (an experience common to other pioneering quantifiers).

Statistics and mathematics were to define the program at Washington. Garrison gave the first advanced course in statistics in a US geography department in 1954. It was not only numbers but also machines that were important. There were the large, cumbersome electrical mechanized Friden calculators, but more significant was the even larger, more cumbersome, computer. In an early advertisement for the department, Donald Hudson (1955) boasted about the departmental use of an IBM 604 digital computer, also a national first. The programming technique of so-called "plug wiring" involving plugging wires into a circuit board was crude and inefficient, but it helped define and consolidate the new vision of the discipline as based on science and the latest technology.

Even at this point we begin to see the shape of a different kind of economic geography emerging. It was radically different from Hartshorne's field-based, typological, descriptive one centered on the region, and sometimes only on the farm. Instead, the new economic geography was primarily undertaken at a desk and involved calculators, computers, graphs, numbers, and spread sheets; it employed an increasingly abstract, theoretical vocabulary taken primarily from economics; it concerned itself with finding causes and explanations, and not simply with classification; and it was not content with written descriptions of the unique, but focused also on the conceptual and numerical analysis of the general. Economic geography was becoming a fully fledged social science stressing the social over the natural, and scientific analysis over "mere" description. More generally, as economic geographical discourse was reinvented an entirely different economic geographical world was emerging defined by abstract space, geometrical axioms, Greek symbols, and regression lines.

But it was hard work, and often contested. This can be seen in Britain where a similar process unfolded, albeit slightly later. There "the revolution" was associated especially with Peter Haggett, who was at Cambridge and later Bristol University, and Richard Chorley, a physical geographer at Cambridge. They were given the moniker the "terrible twins" because of their role in introducing a scientific approach to the discipline. In Haggett's (1965) case that approach was partly worked out within the context of economic geography. He argued that through a process of abstraction economic geographers should construct an ideal simplification of reality – a model – and then using statistical methods test it against the real world. Examples of modeling are found in Paul Plummer's chapter, which follows this one, and it is now a well-known strategy. But at the time it was seriously opposed by those holding on to regionalism. No wonder. It presaged an entirely different kind of economic geographical practices (see Chorley's hilarious 1995 account of Haggett being told off by his Head for showing a regression coefficient).

Another movement that reinforced economic geography's trajectory towards spatial science during the 1950s was regional science. Child of the energetic and ambitious American economist Walter Isard, regional science was born in Detroit in December 1954, when the first meeting of the Regional Science Association was convened. Isard's wider purpose (1956, p. 25) was to add spatial relationships to the hitherto "wonderland of no dimensions" that constituted the economist's world. The strength of his theoretical treatise, *Location and space economy* (1956) (immediately used by Garrison at Washington for his economic geography seminar), was such that by 1958 Isard had established the first Department of Regional Science, perhaps appropriately at the University of Pennsylvania, Russell Smith's *alma mater*. Certainly, during these early years there was theoretical, personnel, and institutional cross-pollenization between regional science and the new economic geography which benefited both. But it was always a relationship pregnant with potential conflict because the two projects claimed the same academic turf. Isard's attitude that economic geographers were simply the hewers and bearers of data for regional scientist theoreticians didn't help, despite the fact that it was close to McCarty's original view.

By 1960 these different elements – the new economic geography in the USA and the UK, and regional science – had come together to form a coherent network. It was characterized by a specific set of geographical nodes such as Iowa City, Seattle, Philadelphia, and Cambridge (UK), and connected by: the circulation of paper – early on, graduate students at Washington initiated their own Discussion Paper series and sent it to kindred souls around the world; flows of money – especially important in the USA were grants from the Office of Naval Research that favored the large-scale, collaborative, and practical projects pursued by the new economic geographers (Pruitt, 1979); and the movement of people – visitors were brought in sometimes from far afield, and as graduate students obtained their PhDs they were often hired by like-minded departments. There was no single organizing force to the network, but it quickly enrolled new people and things into its organization, defining both a new discipline and new relations around the world.

But why did this reinvention of economic geography happen when it did? The post-war years were a period of massive expansion of higher education in both North America and Britain. In part, this occurred because of wider structural changes in economy and society turning on the rise of managerial Fordism and a Keynesian welfare state. The result was a changed mind-set that stressed higher education, and a material affluence that allowed for appropriate funding (Scott, 2000). In addition, the 1950s was a period of optimism about science and technology, and an accompanying instrumental "can do" attitude. All problems could be solved by applying scientific principles, using appropriate technology. This was certainly the case with the new economic geographers. Much of the early work was about using mathematics and computer-generated results to devise practical solutions to problems in transportation, or urban planning, or devising optimal solutions for the location of firms, shops, and residents.

Given its scientific aspirations it may seem odd to claim that spatial science is an invention. After all, scientific knowledge is supposedly impeccable. My argument, however, is that the invented nature of spatial science is evident in its social and institutional origins which were hesitant, complex, and messy. Spatial science did not arise in one blinding moment, with people suddenly seeing the light of scientific truth. Rather, it was a result of some economic geographers making use of hitherto unused vocabularies: in the idiom of statistics and mathematics, in the axioms of Euclidean geometry, in the abstruse language of economic theory, and in the "if, then" statements of computer programming. One shouldn't underestimate the difficulty of this process. It required perseverance, command of difficult corpuses of knowledge, and many late nights. It required much perspiration. The results were new pieces of reality, and a new object of inquiry. That object of inquiry, however, was not (at last) the "real" economic geography because, as should already be clear, there is no such thing. What is taken to be real is contingent upon the prevailing disciplinary discourse, and as the latter changes so does the former.

Conclusion

The purposes of this chapter were twofold. First, to present briefly some of the leading approaches and personalities within the history of Anglo-American economic geography, 1889–1960. Trying to fit seventy-one years of history into a short chapter necessarily results in both substantive and interpretive omissions. Substantively much is missing – including, for example, the influence of non-Anglo-American intellectual traditions such as the German location school (Barnes, 1998a), or dissenting voices (Hepple, 1999), or legions of unnamed economic geographers who carried out bread-and-butter empirical research. Interpretively I made nothing of the fact that my history was only of "great men" living in metropolitan powers. From feminist and postcolonial perspectives a very different historical narrative could have been written about those who were excluded. That said, I hope my history conveys a sense of flux and contingency as the discipline is invented and reinvented.

This leads to my second purpose, which is to make a methodological argument about economic geography by using the metaphor of invention. As a metaphor "invention" is useful precisely because it is unsettling. Whether you accept it or not, the metaphor makes you pause and think about the kind of entity that is economic geography. My argument is that the inventiveness of economic geography is in the detailed practices of its disciplinary discourses: drawing maps, making tables, interviewing informants, or staring at a computer monitor. Those discursive practices may not seem like much when we do them, but they are central to defining a particular object of inquiry, and hence delineating the nature of the discipline. For example, when I was an undergraduate student in the mid-to-late 1970s, learning statistics, punching computer cards, and formally deriving abstract models formed the basis of my study; it was what you had to do to be an economic geographer. Now, few if any of these discursive practices are required of students, but other practices are, such as learning qualitative research methods, knowing the lexicon of political economy, and reading exotic social theory written by Parisian intellectuals. Economic geography, as a result, is quite different. But that difference, and indeed the enormous changes occurring in economic geography since spatial science and reflected in the following chapters, are not to be bemoaned, but celebrated. Thomas Edison never thought he had made the perfect invention, and neither should we.

Acknowledgment

I would like to thank Eric Sheppard for his astute comments, and also those of my Introduction to Economic Geography students Alex Lefebvre and Chris Neidl. The future of the discipline is in safe hands if they, and others like them, continue in the subject.

Bibliography

Barnes, T. J. 1998a. Envisioning economic geography: Three men and their figures. *Geographische Zeitschrift*, 86, 94–105.

Barnes, T. J. 1998b. A history of regression: Actors, networks, machines and numbers. *Environment and Planning*, A 30, 203–23.

Berry, B. J. L., Conkling, E. C., and Ray, D. M. 1987. *Economic Geography: Resource Use, Locational Choices, and Regional Specialization in the Global Economy*. Englewood Cliffs, NJ: Prentice-Hall.

Butzer, K. W. 1989. Hartshorne, Hettner and *The Nature of Geography*. In J. N. Entrikin and S. D. Brunn (eds), *Reflections on Richard Hartshorne's The Nature of Geography*. Occasional publications of the Association of American Geographers, Washington, D.C., 35–52.

Chisholm, G. G. 1889. *Handbook of Commercial Geography*. London and New York: Longman, Green, and Co.

Chisholm, G. G. 1908. The meaning and scope of geography. *The Scottish Geographical Magazine*, 24, 561–75.

Chisholm, G. G. 1921. The drift of economic geography. *The Scottish Geographical Magazine*, 37, 184–6.

Chorley, R. J. 1995. Haggett's Cambridge: 1957–66. In A. D. Cliff, P. R. Gould, A. G. Hoare, and N. J. Thrift (eds). *Diffusing Geography: Essays for Peter Haggett*. Oxford: Blackwell, 355–74.

Fellmann, J. D. 1986. Myth and reality in the origin of American economic geography. *Annals, Association of American Geographers*, 76, 313–30.

Haggett, P. 1965. *Locational Analysis in Human Geography*. London: Arnold.

Hartshorne, R. 1939. *The Nature of Geography. A Critical Survey of Current Thought in Light of the Past*. Lancaster, PA: AAG.

Hartshorne, R. 1955. "Exceptionalism in geography" re-examined. *Annals, Association of American Geographers*, 45, 205–44.

Hartshorne, R. 1979. Notes towards a bibliography of *The Nature of Geography*. *Annals, Association of American Geographers*, 69, 63–76.

Hepple, L. W. 1999. Socialist geography in England: J. F. Horrabin and a workers' economic and political geography. *Antipode*, 31, 80–109.

Hudson, B. 1977. The new geography and the new imperialism: 1870–1918. *Antipode*, 9, 2, 12–19.

Hudson, D. 1955. University of Washington. *The Professional Geographer*, 7, 4, 28–9.

Isard, W. 1956. *Location and Space Economy*. London: Wiley.

Israel, P. 1998. *Edison: A Life of Invention*. London: Wiley.

Jones, C. F. 1935. *Economic Geography*. New York: Henry Holt & Co.

Livingstone, D. N. 1992. *The Geographical Tradition: Episodes in the History of a Contested Enterprise*. Oxford: Blackwell.

MacLean, K. 1988. George Goudie Chisholm 1850–1930. In T. W. Freeman (ed), *Geographers Bibliographical Studies*, volume 12. London: Infopress, 21–33.

McCarty, H. H. 1940. *The Geographic Basis of American Economic Life*. New York: Harpers & Brothers.

McCarty, H. H., Hook, J. C., and Knox, D. S. 1956. *The Measurement of Association in Industrial Geography*. Iowa City: Department of Geography, University of Iowa.

Mikesell, M. W. 1974. Geography as the study of the environment: an assessment of some new and old commitments. In I. R. Manners and M. W. Mikesell (eds), *Perspectives on Environment*. Washington DC: Association of American Geographers, 1–23.

Peet, R. 1985. The social origins of environmental determinism. *Annals, Association of American Geographers*, 75, 309–33.

Pruitt, E. L. 1979. The office of naval research and geography. *Annals, Association of American Geographers*, 69, 103–8.

Ross, D. 1991. *The Origins of American Social Science*. Cambridge: CUP.

Rowley, V. M. 1964. *J. Russell Smith: Geographer, Educator, and Conservationist*. Philadelphia: University of Philadelphia Press.

Schaefer, F. K. 1953. Exceptionalism in geography: a methodological introduction. *Annals, Association of American Geographers*, 43, 226–49.

Scott, A. J. 2000. Economic geography: the great half century. In G. Clark, M. S. Gertler and M. Feldman, *A Handbook of Economic Geography*. Oxford: Blackwell.

Smith, J. Russell. 1913. *Industrial and Commercial Geography*. New York: Henry Holt and Co.

Starkey, O. P. 1967. Joseph Russell Smith, 1874–1966. *Annals, Association of American Geographers*, 57, 198–202.

Van Dyke Robinson, E. 1909. Economic geography: An attempt to state what it is and what it is not. *Publications of the American Economic Association*, 3rd series 10, 247–57.

Von Thünen, J. H. 1966. *The Isolated State*, an English translation of *Der Isolierte Staat* by C. M. Wartenberg (ed. P. Hall). Oxford: Pergamon Press (originally published in 1826).

Whitbeck, R. H. 1914. Review of J. Russell Smith's industrial and commercial geography. *Bulletin of the American Geographical Society*, 46, 540–1.

Whitbeck, R. H. 1915–16. Economic geography: its growth and possibilities. *Journal of Geography*, 14, 284–96.

Whitbeck, R. H. and Finch, V. C. 1924. *Economic Geography*. New York: McGraw-Hill.

Whitbeck, R. H. and Finch, V. C. 1935. *Economic Geography*. New York: McGraw-Hill, third edition.

Wise, M. J. 1975. A university teacher of geography. *Transactions, Institute of British Geographers*, 66, 1–16.

Chapter 3

The Modeling Tradition

Paul S. Plummer

As part of a widespread engagement with contemporary cultural and social theory, many economic geographers are abandoning the search for analytical or "general" explanations of the processes determining the evolution of the economic landscape in favor of contextual or "local" explanations that are designed to take account of the complex construction of societal processes. In the wake of this broadly defined social and cultural turn in economic geography, it is now conventional to conceive of the modeling tradition as part of an historical legacy resulting from a particular set of cultural and institutional practices that constituted the quantitative and theoretical revolution of the 1960s (Cloke et al., 1991; Barnes, this volume). This historical moment consisted of a distinctive set of presuppositions about the way in which the social world is constructed (ontology) and how we know that social world (epistemology). It is commonly supposed that, within the modeling tradition, the objective is to account for observed patterns of spatial economic activity using mathematical models and quantification. In addition, these mathematical models should be grounded in the competitive economic processes that operate between individual producers and consumers.

The supposition that we should employ mathematical models and quantification to reveal the spatial configuration of the economic landscape finds its rationale in a *positivist* conception of science (Johnston, 1986). According to this conception of inquiry, our knowledge of the world is grounded in experience or observation, and justified through a process of empirical validation. Empirical validation, or testing, can entail either corroborating the truth (confirmation) or establishing the falsity (falsification) of our mathematical models on the basis of empirical evidence (Giere, 1984). In practice, model validation has often consisted of analyzing map patterns in order to establish general principles (morphological laws) governing the structure of the economic landscape (Harvey, 1969). Worse still, for many economic geographers, the modeling tradition has come to be associated with endless statistical analyses that are supposedly conducted in the "search" for empirical relationships between some phenomenon and a set of potentially "explanatory" variables (Massey and Meegan, 1985).

Whilst a variety of mathematical models have been proposed for economic geography, the history of modeling has been dominated by assumptions derived from *neoclassical economics*. According to neoclassical orthodoxy, economic outcomes are the result of the rational action of individuals, operating through a market mechanism to produce optimal outcomes that achieve an equilibrium (i.e. a position such that no individual has an incentive to change his/her current decision). In economic geography, neoclassical economics has become translated into the supposition that observed patterns in the economic landscape are the outcome of competitive processes operating between individual consumers and producers.

For many, the quantitative and theoretical revolution has now passed. In the last thirty years, both positivism and neoclassical economics have been subject to a widespread set of criticisms that question the degree to which mathematical models are capable of providing either understanding or explanation of the evolving economic landscape (Sayer, 1984; Gregory, 1978). The resulting consensus appears to be that the modeling tradition should be consigned to the dustbin of intellectual history, to be replaced by an array of approaches encompassing Marxian alternatives, feminist theory, social and institutional approaches, and postmodern interventions (cf. subsequent chapters in this section).

In this chapter, I attempt to provide a more sympathetic overview of the place of modeling within contemporary economic geography. My objective is to provide a constructive rather than destructive critique of contemporary modeling approaches. My hope is that a new generation of economic geographers will be encouraged to take up the challenge of developing model-based explanations in economic geography. To justify a claim for the modeling tradition in economic geography, I critically examine both the philosophical and social theoretic foundations of model-based explanations. Central to my claim is the proposition that it is possible to reject both positivism and a vision of society based on neoclassical economics without rejecting a modeling approach to economic geography.

Model Design

Modeling appears to be central to almost all aspects of contemporary scientific activity. Modelers represent the world to themselves and others through the use of their models. In contemporary discourse, it is conventional to argue that models are "idealizations," "abstractions," or "simplifications" designed to account for some aspect of (social) reality (Casti, 1998; Clark, 1998). For any actual geographical system, it seems plausible that the location of economic activities will depend on the space of flows connecting and helping to define those locations, and vice versa. Further, it seems likely that changes in the location of economic activities may depend on the existing space of flows, and changes in the space of flows may depend on the existing location of economic activities. The complexity of the relationship between spatial processes, spatial interaction, and spatial structure suggests that it is difficult, if not impossible, to provide an explanation for all aspects of the evolving economic landscape. Indeed, it is because the geographical world is complex that we need to build relatively simple models to understand how these systems operate. Historically, modelers have attempted to control for the complexity of the evolving economic landscape by developing: (i) models of location for a given space of flows;

(ii) models of spatial interaction between a fixed set of locations; and (iii) models of simultaneous location and allocation of economic activities (Bennett and Wilson, 1985). However, recent research has shifted towards developing models that both account for the mutual temporal adjustment between spatial structure and spatial interaction, and involve some underlying dynamic spatial process or mechanism (Bennett et al., 1985).

Questions regarding which aspects of an actual geographical system should be included in a model are part of *model design*. As part of model design, it is common to use analog models to compare and describe an unfamiliar system in terms of the properties of a familiar model. The objective in using analog models is to establish relationships of similarity between two different model systems rather than between a model and social reality. In economic geography, examples include: the use of Newtonian mechanics as a model for the flow of people, goods, and information (Wilson, 1970); the use of electrical circuits as a model for spatial price competition (Sheppard and Curry, 1982); and the use of crossword puzzles as models for the evolution of spatial economic systems (Curry, 1989).

In contrast, and despite differences in the form of representation, material and logical models share the common feature that they are employed to establish relations of similarity between a model and an actual system. Material (or experimental) models are some of the first models that we encounter in economic geography. They consist of physical (hardware) or scaled (iconic) representations of actual systems. For example, in industrial location theory, there is the Varignon frame that makes use of weights and pulleys as a mechanical model to determine the optimal location for an industrialist pulled between various raw materials sources and markets. In contrast, logical models employ a more formal language of representation that, typically, consists of a mathematical or computational language (e.g. Aristotelian logic, differential calculus, FORTRAN). The essential feature of formal languages is that they consist of a set of abstract symbols assumed to be true without proof (axioms) and a set of rules of logical inference for combining those symbols (grammar, syntax) to generate new symbols and symbol strings (theorems). Using the rules of logical inference, it is possible to explore the properties of the stipulated model. However, the rules of logical inference do not establish the degree to which the model system corresponds to the relevant aspects of the modeler's world.

To ground our discussion, consider one of the simplest models we encounter in economic geography: the (Keynesian) economic base model. Despite its shortcomings, this model continues to form the basis of many contemporary approaches that seek to determine the level of regional economic activity (Armstrong and Taylor, 1993). In its simplest form, an economic base model consists of an accounting identity, relating regional income and expenditures, and behavioral postulates, linking consumption and investment expenditures to regional income (output). The accounting identity (or budget equation) states that regional income (Y) is equal to aggregate consumption expenditures (C) added to aggregate regional investment expenditures (I). This relationship is necessarily true, being determined by income and expenditure accounting conventions. A minimal requirement for a mathematical model to account for some aspect of the evolving economic landscape is that such a model should embody a process or mechanism of change. In other words, our model needs to be dynamic. The simplest way to introduce dynamics into the regional economic base model is to assume that consumption at time t (C_t) is a proportion (β) of income in the previous time period (Y_{t-1}). In addition, aggregate investment expenditures at time t $(I_t = I)$ are assumed to be determined by factors outside the model. That is, both aggregate income and aggregate consumption are assumed to be endogenously determined in the model while aggregate investment is determined exogenously. Mathematically, this model constitutes an accounting identity, a function linking income and consumption (consumption function), and a constant determining the level of investment (investment function):

$$Y_t = C_t + I_t \quad \text{(Accounting identity)} \tag{3.1}$$

$$C_t = \beta Y_{t-1} \quad \text{(Consumption function)} \tag{3.2}$$

$$I_t = I \quad \text{(Investment function)} \tag{3.3}$$

Substituting the consumption function (3.2) and investment function (3.3) into the accounting identity (3.1) and solving for regional income Y_t, yields an equation linking current and previous aggregate regional income:

$$Y_t = \beta Y_{t-1} + I \tag{3.4}$$

In order to determine the trajectory of aggregate regional income, it is necessary to have information on the level of regional investment (I), the responsiveness of consumption to income (β), and the level of aggregate income at the beginning of the dynamic adjustment process, $t = 0$ (Y_0). To illustrate the derived properties of this model, consider a simple numerical example in which $I = \$10,000$ and that $Y_0 = \$20,000$. If the consumption function (equation 3.2) is to be economically mean-ingful it is necessary that consumption expenditures are positive but less than income $(0 < \beta < 1)$. Accordingly, arbitrarily set $\beta = 0.6$. An equilibrium level of regional income (Y_E) is defined by the condition that current income is equal to income in the previous period $(Y_E = Y_t = Y_{t-1})$. Setting the current income level equal to the immediately previous income level and solving equation (3.4) in terms of the equilibrium level of income yields $Y_E = I/(1 - \beta)$. In this numerical example, the equilibrium level of regional income is $\$25,000$. Figure 3.1 illustrates the result-ing dynamic adjustment of regional income with respect to this equilibrium by plotting the trajectory of income against time. Note that, in this numerical example, the dynamics of regional income are *stable* in the sense that, from an initial position of disequilibrium, the model converges monotonically to an equilibrium. Provided the responsiveness of consumption to income is positive but less than unity, it can be demonstrated that the trajectory of regional income will converge monotonically to an equilibrium regardless of the initial level of income (Shone, 1997).

To understand the process of model design in more detail, it is necessary to unpack the way in which this model has been constructed. As with any model, the economic base model is formulated using a set of assumptions about the way in which the actual system operates. In arriving at a simplified representation of an evolving economic landscape, a decision has been made on which elements to include in the model. While the model will include variables and relationships that are hypothe-sized to hold in the actual system being modeled, it will also include assumptions that are false. For example, the economic base model usually assumes that: (i) there

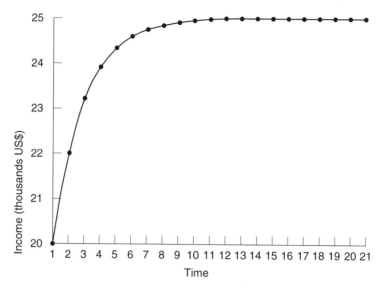

Figure 3.1 Monotonic convergence in a simple linear model
Source: Author

is no trade between regions; (ii) spatial economic behavior can be represented by aggregate variables; (iii) there is a constant relationship between consumption and income; and (iv) aggregate investment does not depend on either income or consumption.

Employing simplified assumptions, which are known to be unrealistic, is justified when it is believed that the omission of more complex variables has only a negligible impact on the operation of the actual system. For example, trade with other regions may be negligible. However, false assumptions can also be employed as heuristic devices when exploring the logical properties of a model. These assumptions can be relaxed in subsequent investigations. The use of heuristic assumptions is an important part of model design in economic geography. For example, an isotropic surface, in which space is treated as a homogeneous and unbounded surface with the cost of movement on the surface being determined by a "distance friction" parameter, represents a powerful *ceteris paribus* or "other things being equal" assumption that is used to isolate the impact of transportation costs on location patterns in models of agricultural land use, central place models, and models of industrial location (Dicken and Lloyd, 1990).

Model Validation

While the selection of appropriate models is part of model design, evaluating the degree to which such a model captures some aspect of an actual geographical system is part of *model validation*. This entails establishing the grounds, if any, on which we are justified in the claim that our model is similar to the object being modeled. Depending on our philosophical orientation, model validation may be guided by the degree of correspondence between our model and social reality, the degree to which our model coheres with our belief system, and/or the utility of the model in solving problems (Laudan, 1990). In practice, the decision to accept a model as an adequate explanation can involve a wide array of sometimes conflicting methodological rules including model elegance, relevance, simplicity, theoretical plausibility, explanatory power, and predictive ability.

From the perspective of a modeling approach, empirical evidence provides a necessary but not sufficient condition for model validation. If the aim of modeling is to establish the degree of correspondence between a model and an actual geographical system, then this entails the assumption that empirical evidence reflects the actual state of some aspect of the social world. Alternatively, the link between a mathematical model and empirical evidence may constitute one way of evaluating either the coherence of our overall belief system or of corroborating the degree to which our model represents a useful problem-solving device. Regardless, model validation entails employing a suitably formulated modeling methodology in such a way as to bring empirical evidence to bear on the claims formulated in our mathematical model.

Broadly speaking, empirical evidence involves quantification and measurement. Accordingly, empirical evidence encompasses both direct sensory data and observations from sensory data that have been measured, however indirectly, using instruments and measurement tools. For example, in our simple economic base model, we would need to be able to quantify regional income and regional investment over a specified time period. How we choose to measure variables such as income and investment will depend, at least in part, on the prevailing conventions and measurement practices underlying income accounting (Shaikh and Tonak, 1994). Empirical evidence that can be successfully replicated using prevailing measurement conventions and practices may become accepted as a "fact" by members of an intellectual community. However, there is no foundational empirical basis from which we are able to construct knowledge of the evolving economic landscape. Rather, empirical evidence is subject to revision and, perhaps, later rejection in the light of changes in the prevailing set of conventions and measurement practices. It is in this sense that we think of "facts" as being theory-laden. The theory-ladenness of observation undermines the positivist ideal of grounding scientific explanation in a theory-neutral observation language (Chalmers, 1987).

In addition to the conceptual constraints that prevent the constructing of a secure empirical basis for knowledge, modeling faces practical limitations regarding the availability of relevant spatial economic information. Typically, we are trying to evaluate complex dynamic models of interdependent spatial economic systems using relatively aggregated cross-sectional data or relatively short spatial time series. In addition, important variables are either not measured or are unobservable. This includes measures of expectations, regional profitability, regional labor values, and regional input–output coefficients (Webber, 1987b; Dewhurst and Hewings, 1991). While there have been many innovative developments in measurement theory, it seems likely that the limitations of empirical spatial information are likely to remain – at least in the near future. Thus even if it is possible to design meaningful empirical tests to establish the validity of mathematical models, it may not be possible to implement such tests due to the limited quantity and quality of the empirical information that can be collected for spatial economic systems.

To confront a mathematical model with empirical evidence, we need a method for linking the "theory" to the "facts." In practice, this involves specifying an empirically estimable model derived from the mathematical model. Given the complexity of the evolving economic landscape, it does not seem plausible to aim for empirically estimable models that are capable of accounting completely for all of the features of our empirical information. The discrepancy between "theory" and "fact" is assumed to result from the set of non-systematic measurement errors, omitted variables, and approximations that are made when formulating the mathematical model. For example, an empirical specification of the dynamic economic base relationship is:

$$Y_t = \beta Y_{t-1} + I + \varepsilon \tag{3.5}$$

Where ε is a random error term that is intended to capture the innumerable, but individually unimportant factors that have not been included in the model (Haining, 1990).

Once an empirical specification has been estimated, the resulting estimates can be used to evaluate the empirical claims of the mathematical model. This includes evaluating the economic behavior contained in the model, making forecasts about the future, and analyzing economic policy alternatives. For example, in the case of the dynamic economic base model, we can use empirical evidence to interpret and evaluate the properties of the dynamics of regional income. Specifically, we can estimate the equilibrium level of regional economic activity ($Y_E = I/(1 - \beta)$) and test the stability properties of the dynamic adjustment process ($0 < \beta < 1$). In simple linear systems, we can make forecasts about the future and analyze alternative policy scenarios based on the rule that small changes in input (investment) will produce correspondingly small changes in the level of output (economic activity).

The dynamic economic base model (equation 3.5) assumes that the parameters β and I are constant over time and space, and independent of the level of consumption, investment, and income. That is, the economic base model represents a closed system in the sense that: (i) the stipulated relationships between inputs and outputs are invariant over space and time; and (ii) the parameters of the system are structurally invariant with respect to changes in the variables in the system (Sayer, 1992). Assuming that a dynamic system can be treated "as if" it is closed is to ignore both geography and history. Fortunately, recent developments in modeling approaches have taken their first steps towards treating space and time seriously. Places are treated as interdependent in the sense that the economic activities in one region may, at least potentially, influence the behavior of all other regions (Hepple, 1996). In models of "fast" and "slow" dynamics, the parameters of a model are allowed to vary over time and space, and in response to changes in the system variables (Bennett, 1979; Bennett and Haining, 1985). In addition, multilevel models are now being employed to account for both contextual differences and place heterogeneity (Jones and Duncan, 1996).

To recapitulate, the objective of employing empirically estimable models is to determine the validity of our mathematical models, make forecasts about the future, and evaluate policy alternatives. This involves confronting the conclusions that are deduced from the mathematical model with empirical evidence, using a suitably formulated empirical modeling methodology. Empirical modeling is not a simple process of confronting "theory" with the "facts." Rather, empirical modeling is a complex and iterative process involving the confrontation, and subsequent revision, of different types of theoretical information. In any given modeling situation we are attempting to evaluate our mathematical model, in conjunction with background knowledge, defined by the set of approximations and omissions that we make in specifying an empirical model. We cannot know for certain whether a lack of correspondence between our mathematical model and empirical evidence is due to the falsity of our mathematical model or the falsity of our background information. As a consequence, we cannot establish for certain whether our model is either confirmed or rejected by the available empirical evidence. The absence of a secure foundation for empirical evidence and our methods for establishing the truth or falsity of mathematical models undermines a positivist conception of inquiry. However, our inability to establish certain and complete explanations of the evolving geographical landscape does not imply either that mathematical modeling is impossible or that some modeling methodologies are no more useful than others. Rather, we can aim for conjectural knowledge, in which our modeling methodologies, mathematical models, and background information are always fallible and subject to continual revision (Musgrave, 1993).

Competing Model Designs

In economic geography, the development of mathematical models owes much to an historical legacy that is grounded in the classical models of agricultural land use (von Thünen), industrial location (Weber), and central place organization (Lösch, Christaller). In the early days of the so-called "quantitative revolution," the objective of accounting for observed patterns of spatial economic activity became translated into a search for geometrical patterns and morphological laws (Bunge, 1966). Typically, these geometric configurations were derived from the simplistic assumption that space can be treated "as if" it is an isotropic surface (Haggett, 1963). The search for spatial patterns and geometric laws lay mathematical modeling open to the charge of "spatial fetishism." That is, spatial outcomes are treated "as if" they are independent of the society within which they are embedded (Smith, 1981). While the assumption of an isotropic surface was undoubtedly fetishized by some researchers, it is more appropriate to consider it as an heuristic assumption, employed as part of model development, rather than a representation of social reality. In fact, relaxing the assumption of spatial homogeneity tends to destroy the elegant geometric patterns of classical location models.

Much of what passes for mathematical modeling in contemporary economic geography is played according to the rules of neoclassical economics. These rules stipulate that a mathematical model can be considered adequate if, and only if, it is possible to derive an equilibrium spatial configuration of economic activity from a foundation based upon individual optimization, constrained by social and spatial structures of production and consumption. In terms of the operation of individual spatial markets, in which firms pursue location and pricing strategies, the focus is on determining the configuration of prices, outputs, and profits that is optimal under conditions of either oligopolistic competition, monopolistic competition, or perfect competition (Nagurney, 1993). At the macro (regional) level, it is now common to

Neoclassical geography

employ models of monopolistic competition to determine optimal configurations of specialization and trade, urban and regional agglomeration, and regional economic growth (Krugman, 1991).

The assumption that the characteristics of spatial markets and regional economies must be deduced from the behavior of individual agents entails *methodological individualism*. In accordance with the postulates of neoclassical economics, economic geographers have extended methodological individualism by stipulating that agents must behave rationally, attempting to maximize a specified objective, subject to constraints. For example, individual producers are assumed to maximize their total profits, subject to resource constraints, and individual consumers are assumed to maximize their utility in accordance with their given preferences. If there exists a competitive mechanism coordinating the behavior of all producers and consumers, then the aim is to derive spatial configurations in which all producers and consumers are maximizing their specified objective. Importantly, such a derived configuration is considered to be optimal in the sense that, simultaneously, collective economic well-being is maximized.

The notion that an equilibrium spatial configuration represents an optimal condition provides justification for a vision of capitalism as a self-regulating system that promotes rationality, stability, and the equitable distribution of resources amongst members of society (Sheppard and Barnes, 1990). However, such an equilibrium is only of theoretical interest if it is stable, in the sense that it can be reached from a position of disequilibrium as time increases (cf. figure 3.1). Neoclassically oriented mathematical models tend to treat the dynamics of competition as an adjustment process in which the rational response of producers and consumers will tend to drive an economic system towards an equilibrium configuration. If a spatial economic system is converging towards an equilibrium configuration, then the dynamics of competition are unimportant relative to the equilibrium configuration of a spatial economy. This provides a rationale for the view that equilibrium analysis is a plausible theoretical entry point for modeling a spatial economic system.

It is well known that simple linear systems, such as equation (3.4), possess only a limited set of possible trajectories with respect to an equilibrium. Specifically, out of equilibrium the time path either diverges from an equilibrium or converges towards that equilibrium. In addition, out of equilibrium dynamics are limited to either monotonic or periodic trajectories with respect to an equilibrium level (compare the two upper panels of figure 3.2). In contrast, if the relationship linking inputs to outputs is nonlinear, then the long-run behavior of the system can be more complicated. For example, if we assume that current regional income depends nonlinearly on regional income in the previous time period (e.g. $Y_t = \beta Y_{t-1}(1 - Y_{t-1})$) then the trajectory of income is no longer limited to either converging to, or diverging from, an equilibrium. Figure 3.2 illustrates that, depending on the value of β, it is possible that the model either: (a) converges monotonically to an equilibrium level of income; (b) converges periodically to an equilibrium; (c) displays sustained periodic fluctuations of income; or (d) displays the type of aperiodic fluctuations that are characteristic of "chaotic" systems (Gandolfo, 1996).

If nonlinear relationships are plausible representations of the relationship between the inputs and outputs of spatial economic systems, then this raises the possibility that such systems are permanently out of equilibrium. In turn, this implies that

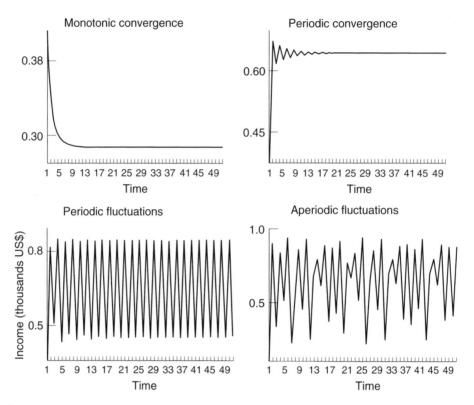

Figure 3.2 The qualitative properties of nonlinear systems
Source: Author

equilibrium analysis should no longer occupy the focus of analytical attention. Rather, equilibria should be relegated to theoretical reference points from which to triangulate the out-of-equilibrium dynamics of spatial economic systems. As a corollary, if the dynamics-of-competition do not tend to drive a spatial economic system towards an equilibrium configuration, then out-of-equilibrium dynamics may well be more important than the equilibrium configuration of a spatial economy.

All too often, the dominance of neoclassical economics has led to the erroneous conclusion in economic geography that all mathematical modeling is necessarily based upon equilibrium theorizing that is grounded in methodological individualism. Nothing could be further from the truth. In contemporary economic geography, there exists an alternative way of playing the game of mathematical modeling: regional political economy (Plummer et al., 1998). According to this approach, the capitalist space economy should be conceived of as evolving out of equilibrium, as a complex and conflict ridden spatio-temporal system. Employing assumptions derived from classical and Marxian political economy, regional political economists have constructed mathematical models to account for differential urban, regional, and national growth (Sheppard and Barnes, 1990; Webber and Rigby, 1996), the dynamics of regional capital and labor markets (Webber 1987a; Clark et al., 1986), and the location and pricing strategies of firms (Plummer, 1996).

According to this alternative vision of the economy, the evolving economic land-scape consists of sets of heterogeneous and interdependent agents, making decisions out of equilibrium and under conditions of economic uncertainty. In contrast to the rationality postulate of neoclassical economics, agents are assumed to possess lim-ited information and computational ability. In addition, their participation in the economic system is conditioned by membership of one or more economic classes. Similarly, space is conceptualized both as heterogeneous and as endogenously cre-ated as a result of changes in transport technology and the location of production and consumption (Sheppard, 1990a). The way in which economic agents adjust their strategies and decision rules depends on how those agents are embedded in evolving social and geographical structures. In turn, the evolution of the spatial economy depends on how these agents behave within a given context (Sheppard, 1990b). That is, agents construct, but are also embedded in, society and space.

A model design based upon these alternative assumptions about the nature of social and geographical reality has two broad implications for the ways in which we understand the evolving economic landscape. First, the neoclassical vision of capit-alism as an economic system that promotes the rational, harmonious, and equitable distribution of resources in society no longer holds. Rather, the distribution of resources amongst members of society depends, in part, on the social and political power of members of economic classes. Furthermore, there exist no equilibria that simultaneously maximize the interests of all economic actors. More generally, individual decisions can result in unintended consequences that oppose the interests of other classes, and those of other members of the same class (Sheppard and Barnes, 1986). This undermines the neoclassical notion that capitalism promotes the rational and equitable distribution of resources in society.

Second, the evolving economic landscape is understood as a fundamentally non-equilibrium system, subject to both equilibrating and disequilibrating forces (Web-ber and Rigby, 1999; Plummer, 1999). Whilst the existence of an equilibrium configuration of prices, outputs, and profits can be shown to exist in theory, the dynamics of spatial competition are such that this equilibrium is only "weakly" stable, and easily destabilized. In turn, if geographical reality does not approximate an equilibrium configuration, then this undermines the methodological justification for searching for predictable relationships between observed patterns. Rather, the orientation of mathematical modeling shifts from a concern with describing and explaining spatial patterns to a concern for the dynamics that drive the spatio-temporal trajectories of capitalist economies. By now, the message should be clear. A modeling approach does not presuppose that our model design should be based upon a particular set of assumptions about the way in which society and space are organized. However, the ways in which both society and space enter our model design are critical to the properties that can be deduced from our mathematical models.

Modeling in the Contemporary Context

It is now commonplace to accept a dualism between an analytical and a contextual approach to understanding the evolving economic landscape. The analytical approach tends to be characterized by quantitative reasoning and a search for spatial patterns that is, at best, grounded in mathematical models derived from neoclassical economics. For many contemporary economic geographers, this form of modeling is too constraining to capture the complexity of actually evolving economies. In place of mathematical models, it is considered more profitable to employ discursive models and qualitative reasoning in an attempt to understand the complexity and contingency of societal processes. In this chapter, I have argued that proclamations about the death of mathematical modeling are premature. To justify my claim, I have attempted to demonstrate that it is possible to engage in mathematical modeling without recourse to either a positivist vision of geographical ways of knowing, or a social ontology that is grounded in neoclassical economics.

Within the contemporary context, if the modeling approach is to provide a plausible framework for explaining the evolution of the economic landscape, then we need to move beyond traditional location theory. This will entail a shift in the orientation of mathematical modeling away from the search for spatial patterns and equilibrium-oriented theoretical models towards a modeling approach that reflects the complex dynamics that operate on and through the economic landscape. If the space economy is viewed as a complex construction of nonlinearly related societal processes, then this has profound implications for what we can expect models to tell us about the social world. The goal of taking seriously the spatial and temporal aspects of capitalist economies may well challenge the limits of what may be possible with mathematical models. In this regard, however, the modeling approach is no different from any other approach to economic geography. Whether we are employing "quantitative" or "qualitative" research methodologies, we all face limitations and possibilities. The challenges of research are to confront the limitations of our approach and to explore the possibilities.

Bibliography

Armstrong, H. and Taylor, J. 1993. *Regional Economics and Policy*. Hemel Hempstead: Harvester Wheatsheaf.

Bennett, R. J. 1979. *Spatial Time Series: Analysis-Control-Forecast*. London: Pion.

Bennett, R. J., Haining, R. P., and Wilson, A. G. 1985. Spatial structure, spatial interaction, and their integration: a review of alternative models. *Environment and Planning A*, 17, 625–45.

Bennett, R. J. and Haining R. P. 1985. Spatial structure and spatial interaction: modelling approaches to the statistical analysis of geographical data. *Journal of the Royal Statistical Society*, 148, 1–36.

Bennett, R. J. and Wilson, A. G. 1985. *Mathematical Models in Human Geography*. New York: Wiley.

Bunge, W. 1966. *Theoretical Geography*. Lund: C. W. K. Gleerup.

Casti, J. 1998. *Alternate Realities: Mathematical Models of Nature and Man*. New York: Wiley.

Chalmers, A. F. 1987. *What is this Thing called Science?* New York: University of Queensland Press.

Clark, G. 1998. Stylized facts and close dialogue: methodology in economic geography. *Annals of the Association of American Geographers*, 88, 73–87.

Clark, G., Gertler, M., and Whiteman, J. 1986. *Regional Dynamics: Studies in Adjustment Theory*. Boston: Allen & Unwin.

Cloke, P., Philo, C., and Sadler, D. 1991. *Approaching Human Geography: An Introduction to Contemporary Theoretical Debates*. London: Chapman.

Curry, L. 1989. Towards a theory of endogenous geographical evolution. Research paper 169, Department of Geography, University of Toronto.

Dewhurst, J. H. and Hewings, G. 1991. *Regional Input–Output Modelling: New Developments and Interpretations*. Aldershot: Brookfield.

Dicken, P. and Lloyd, P. 1990. *Location in Space: Theoretical Perspectives in Economic Geography*. New York: HarperCollins.

Gandolfo, G. 1996. *Economic Dynamics*. New York: Springer.

Giere, R. N. 1984. *Understanding Scientific Reasoning*. New York: Holt, Rinehart and Winston.

Gregory, D. 1978. *Ideology, Science, and Human Geography*. London: Hutchinson.

Haggett, P. 1963. *Locational Analysis in Human Geography*. London: Arnold.

Haining, R. P. 1990. *Spatial Data Analysis in the Social and Environmental Sciences*. Cambridge: Cambridge University Press.

Harvey, D. 1969. *Explanation in Geography*. New York: St. Martin's Press.

Hepple, L. 1996. Directions and opportunities in spatial econometrics. In P. Longley and M. Batty (eds). *Spatial Analysis: Modelling in a GIS Environment*. Cambridge: Geoinformation International, 231–46.

Johnston, R. J. 1986. *Philosophy and Human Geography*. London: Edward Arnold.

Jones, K. and Duncan, C. 1996. People and places: The multilevel model as a general framework for the quantitative analysis of geographical data. In P. Longley and M. Batty (eds). *Spatial Analysis: Modelling in a GIS Environment*. Cambridge: Geoinformation International, 79–104.

Krugman, P. 1991. *Geography and Trade*. Cambridge: MIT Press.

Laudan, L. 1990. *Science and Relativism: Some Key Controversies in the Philosophy of Science*. Chicago: University of Chicago Press.

Massey, D. and Meegan, R. 1985. *Politics and Method*. London: Methuen.

Musgrave, A. 1993. *Common Sense, Science and Skepticism: a Historical Introduction to the Theory of Knowledge*. Cambridge: Cambridge University Press.

Nagurney, A. 1993. *Network Economics: A Variational Inequality Approach*. Boston: Kluwer Publishers.

Plummer, P. S. 1996. Spatial competition amongst hierarchically organized corporations: prices, profits, and shipment patterns. *Environment and Planning A*, 28, 199–222.

Plummer, P. S. 1999. Capital accumulation, economic restructuring, and nonequilibrium regional growth dynamics. *Geographical Analysis*, 31, 267–87.

Plummer, P. S., Sheppard, E. S., and Haining, R. P. 1998. Modeling spatial competition: Marxian versus neoclassical approaches. *Annals of the Association of American Geographers*, 84, 575–94.

Sayer, A. 1984. *Method in Social Science: A Realist Approach*. London: Hutchinson.

Sayer, A. 1992. Explanation in economic geography: abstraction versus generalization. *Progress in Human Geography*, 6, 65–85.

Shaikh, A. M. and Tonak, E. A. 1994. *Measuring the Wealth of Nations: The Political Economy of National Accounts*. Cambridge: Cambridge University Press.

Sheppard, E. 1990a. Transportation in a capitalist space economy: Transportation demand, circulation time, and transportation innovations. *Environment and Planning A*, 22, 1007–24.

Sheppard, E. 1990b. Modeling the capitalist space economy: Bringing society and space back. *Economic Geography*, 66, 201–28.

Sheppard, E. and Barnes T. J. 1986. Instabilities in the geography of capitalist production: collective vs. individual profit maximization. *Annals of the Association of American Geographers*, 76, 493–507.

Sheppard, E. and Barnes, T. J. 1990. *The Capitalist Space Economy*. London: Unwin Hyman.

Sheppard, E. and Curry, L. 1982. Spatial price equilibria. *Geographical Analysis*, 14, 279–304.

Shone, R. 1997. *Economic Dynamics: Phase Diagrams and Their Application*. Cambridge: Cambridge University Press.

Smith, N. 1981. Degeneracy in theory and practice: Spatial interactionism and radical eclecticism. *Progress in Human Geography*, 5, 111–18.

Webber, M. J. 1987a. Rates of profit and interregional flows of capital. *Annals of the Association of American Geographers*, 77, 63–75.

Webber, M. J. 1987b. Quantitative measurement of some marxian economic categories. *Environment and Planning A*, 19, 1303–28.

Webber, M. J. and Rigby, D. 1996. *The Golden Age Illusion: Rethinking Postwar Capitalism*. New York: Guilford Press.

Webber, M. J. and Rigby, D. 1999. Accumulation and the rate of profit: Regulating the macroeconomy. *Environment and Planning A*, 31, 141–64.

Wilson, A. G. 1970. *Entropy in Urban and Regional Modelling*. London: Pion.

Chapter 4

The Marxian Alternative: Historical–Geographical Materialism and the Political Economy of Capitalism

Erik Swyngedouw

"Revolution is not about showing life to people but making them live" (Guy Debord, in Bracken, 1997, p. 1)

Geography is an eclectic and fashion-prone discipline. The attention span in the discipline for major theoretical or methodological perspectives is rather short-lived. For many a geographer, it is very hard to keep up with the endless re-formulations of spatial or geographical perspectives and theoretical influences. In fact, the chapters of this reader illustrate this abundantly. Whims and preferences fade in and out in tune with the tumultuous re-ordering of tastes and interests in modern society. The relative absence in geography of a canonical mode of thinking that turns other disciplines into rather arcane and often idiosyncratic pursuits (think of classical economics, for example) is no doubt also a great advantage. It keeps geography alive and kicking and maintains a vibrant intellectual environment. However, it equally leads to often rather superficial dabblings with epistemological and methodological issues of intellectual traditions that are much more complex, variegated, and sophisticated than their customary cursory introduction into geography usually suggests.

Marxism is one such perspective that has infused geographical theory and practice over the past two decades (in the Anglo-Saxon world at least). While it was the most exciting approach around in the 1970s and early 1980s, today it seems that Marxist geographical enquiries have moved to the back burner of the academic agenda. Marxism in geography seems to be relegated to the status of a mere additional chapter in the intellectual history of geography. This partial blindness for Marxism has of course a lot to do with the fall of really existing socialism in Eastern Europe and elsewhere, the "victory" (however Pyrrhic it may turn out to be) over market-Stalinism by capitalism, and the systematic theoretical onslaught on Marxist modes of enquiry. Yet, even a brief glance at today's key social and geographical issues, such as uneven spatial development, socioeconomic polarization and exclusion, economic and financial crises, geopolitical tension and conflict, environmental degradation, and so on, suggests that Marxist analytical and political perspectives may be more relevant than ever in understanding the world and in contributing to the formulation of political strategies for an emancipatory project (Harvey, 1984, 1998).

Marx and Geography

In the turbulent history of Marxist scholarship, geography and geographers have at best played a rather marginal role. In the Anglo-Saxon world, the academic attention to Marxist analysis emerged together with the critical social movements of the 1960s. Yet, by that time, there had already been a long and rich tradition of Marxist scholarship and activism, much of which was acutely aware of and sensitive to space and to the geographical dynamics and conditions of everyday life under capitalism (see Brewer, 1980).

Karl Marx (1818–1883) is of course the founding father of historical materialism. Drawing on a long intellectual tradition, starting with Aristotle and culminating in Hegel, and critically inspired by other great political economists like Malthus, Ricardo, and Adam Smith, Marx set out the beacons for a social theoretical approach and political practice that would infuse both the social theoretical and political course of modern times in a decisive fashion. As a German intellectual, who lived most of his life as an exiled refugee in Brussels, Paris, and London, he was an eminent scholar spending much of his time in the British Library studying classical theory as well as the conditions of everyday life in the capitalist world. But he was also a political activist who became the leader of the First Socialist International and engaged in political work with other socialists and communists of his time.[1] Although his work, and in particular his critical analyses of the dynamics of capitalist societies, is not explicitly geographical, many of his writings are eminently spatial. The Communist Manifesto (Marx and Engels, 1848), for example, a political pamphlet co-authored with his long-term friend and political comrade Friedrich Engels, is littered with deeply geographical insights, many of which offer astute and still very relevant insights into the geographical dynamics of today's global capitalism (see Harvey, 1998). Later Marxists, writing during the first decades of this century, like Lenin (1970), Hilferding (1910), or Rosa Luxemburg (1951), equally offer penetrating geographical understandings of the economic dynamics of capitalism (see Peet, 1991). And, of course, early geographers like Kropotkin (1885) or Elisée Reclus (Dunbar, 1979) were inspired directly or indirectly by the intellectual milieu and political projects of which Marx was part.

Most accounts would situate the introduction of Marxism into geography some time in the late 1960s (Peet, 1998). A growing counter-cultural movement (associated with anti-Vietnam and human rights activism in the USA), a call for democratization and structural social reform in Europe, emerging "Third World" movements contesting continuing postcolonial domination, and an ecological awakening, bred an intellectual environment that sought to find new ways of capturing and scripting processes of exclusion, repression, and exploitation. This new reading of the world would help formulate political strategies for social transformation and for the construction of a more just and inclusive social order. In Anglo-Saxon geography, where quantitative revolutionaries calculated away and tried to erode the still hegemonic framework of a largely idiographic discipline, the calls for a socially more relevant perspective (pioneered by Bill Bunge (1971, 1973), David Harvey (1973, 1977), Doreen Massey (1973), Bob Colenutt (Ambrose and Colenutt, 1975), Keith Buchanan (1966, 1973) or Jim Blaut (1970, 1975), among many other activists, scholars and students) drove a new generation of scholar-activists to explore social theoretical perspectives that apparently had been written out of geography. What linked this generation together were a passion and a commitment to make life and work relevant and meaningful at both the personal and wider social level. Passion and desire, commitment and solidarity, were the emotional driving forces. Exploring Marxism then was an exciting, liberating, and passionate affair. Obscure and less obscure groups of academics and activists strategized away in solidarity to transform a discipline that seemed to have outlived its useful shelf-life, and a society that seemed to generate exploitation, repression, and domination both locally and globally.

I am always slightly amused (but more often infuriated) when I read the proliferating number of textbooks in geography that treat Marxist geography as a pure academic–theoretical perspective or epistemology alongside the other great "-isms" that have infused geography and inspired generations of geographers. The passion of commitment that was an integral part of the lives of those who introduced Marxism in geography, and their drive to produce a truly humanizing geography (in the sense of a lived everyday experience – not as a merely discursive practice), has been largely scripted out of its intellectual history, and practice. Marxism has undoubtedly had a major influence in the discipline and most of the current debates have the specter of Marx haunting them. Ironically, while texts by Marx or Marxists (and notably Harvey's work) have been among the most quoted and referenced, it is remarkable that very little of their research program, both theoretically and empirically, has been pursued actively.

It would of course be impossible in this context to trace comprehensively the origin, contribution, and resonance of Marxism in current geographical practice. I would rather explore how and why Marxism still has a place not just in maintaining geography as a vibrant and exciting discipline, but – perhaps most importantly – in contributing to the production of a truly humanizing geography.

Historical–geographical Materialism

In the social and political environment of the 1960s, in which questions of social justice were high on the agenda, it became abundantly clear that justice (however conceptualized) is a deeply geographical affair, and that emancipatory or empowering politics and strategies are necessarily geographical projects – in much the same way as capitalism is an inherently geographical process (Merrifield and Swyngedouw, 1997). What Marx and Marxism offered was a comprehension of why, where, and how deep and perverse injustices and inequalities persist. It was, in fact, the gut feeling that "space mattered," both in terms of explaining injustice and inequality and of developing strategies for social change, that prompted this feverish search for alternative formulations. For most, it was not just a matter of radicalizing geography, but first and foremost the search for a politics of transformation and the scripting of a "social science" that would support and develop in tandem with a progressive and emancipatory political commitment (Peet, 1977).

Marxism seemed to offer the epistemologically and substantively most coherent attempt at unraveling how the dynamics of political economic relations were both space-dependent and space-forming. Processes of empowerment/disempowerment, domination/subordination, appropriation/exploitation were not only seen as inscribed in space and spatial configurations, but space was theorized as an active and integral moment and arena expressing and embodying the struggles that unfold along the above dialectical pairs (Gregory and Urry, 1985). Space is deeply political (in the widest possible sense of the word) from the very beginning, and politicizing space became the *leitmotiv* of much of Marxist and related radical geographical research.[2] Marx and Engels' writings were littered with implicit and explicit references to the spatiality of capitalism and spatial strategies of resistance (see Lefebvre, 1968, 1989; Smith, 1984; Harvey, 1982, 1998). It would, of course, be Harvey's academic project not only to excavate how space was produced under capitalism and entered as part of an historical–geographical process into the perpetual transformation of capitalism, but also to push the theory further and fill in some of the gaps and inconsistencies that Marx himself had left open or incomplete.[3]

For Marx, the basis of historical geographical materialism resides in the ontological (foundational) view that "production" is the basis of all social life and of history. "Production" has to be understood here in the broadest possible sense. It refers to any human activity of formation and transformation of nature and includes physical, material, and social processes as well as the human ideas, views, and desires through which this transformation takes place. In this sense, human beings produce (change) their own lives as well as their social and physical environment. It is, more exactly, "the production of his own means of existence, an activity at once personal and collective (transindividual) which transforms him at the same time as it irreversibly transforms nature and which, in this way, constitutes 'history'" (Balibar, 1995, p. 35) and – I would add – geography as well. The "economy" for Marx is, therefore, much broader than what classical economists usually understand by the term. While the latter restrict their analysis to the study of the production and exchange of commodities, Marxists would see this narrow activity as an integral part of a much wider social, political, and environmental process (Benton, 1996; O'Connor, 1998).

Under capitalism, which is a historically and geographically specific form of social organization, the individual and collective form of this transformation (metabolism) of social and physical nature is characterized by a fundamental social division between those owning the means of production (capitalists), and those only owning their labor, which they need to sell as labor force to capitalists in order to secure their own short- and medium-term survival. The dynamics of these social class relations take particular geographical and ecological forms and lead to a series of processes, contradictions (tensions), and social struggles that render capitalism geographically and historically dynamic, but inherently unstable. In addition, these capitalist social relations – by virtue of the unequal power relationship embodied in the above class relations – produce systematic conditions of repression, social and ecological exploitation, uneven development, disempowerment, and social exclusion for many, as well as immense wealth, power, and freedom for a few.

In sum, the particular social form through which nature is transformed in order to permit the continuation of individual and collective human life constitutes the field of inquiry of historical–geographical materialism. The tensions, contradictions, and conflicts that arise from this constitute an arena for social action and struggle. In wrestling with a dialectical and historical–geographical materialist understanding of the world, Debord (1967), Lefebvre (1989), Harvey (1973, 1982), Massey (1984, 1994) and Castells (1977), among many others, rediscovered the deep ontological premise of the role of space-as-process in the unfolding of these capitalist social relations. All lived geography is historical and political, contested and contestable. In what follows, I shall explore some of the key insights raised and developed by Marxist geographers, themes that still infuse vanguard debates in the discipline.

The Concept of Space: A Dialectical Perspective

The theoretical and epistemological thinking on space from a historical–geographical materialist perspective is at times arcane and complex, yet surprisingly simple in its basic formulations.[4] Historical–geographical materialism starts from the premise that things (as objects and phenomena) exist, but that these objects or phenomena are the embodiment of (they interiorize) relationships; things become the outcome of processes that have themselves ontological priority. The latter means that an object or phenomenon (think of, for example, coffee, a romantic idea of nature, or a stock market issue) is the end result of a process. The coffee I sip in the morning reflects and embodies relations between peasants and landowners, between merchants and producers, shippers and bankers, wholesalers and retailers, etc. These relations and processes are more important in terms of understanding the objects/phenomena than the characteristics of things/phenomena themselves. "Coffee" is then no longer just coffee, but also a whole host of other things that are part and parcel of what constitutes "coffee" as coffee. The characteristics of coffee as a thing can only give me clues to what is hidden underneath. The excavation of these relational processes is at the heart of historical–geographical materialist inquiry. If I were to reconstruct the myriad social relations through which coffee becomes the liquid I drink, I would uncover a historical geography of the world that would simultaneously provide powerful insights into the many mechanisms of economic exploitation, social domination, profit-making, uneven development, ecological transformation, and the like.

In other words, everything flows; the flow or the process constitutes the phenomenon, subject, or object. Those relationships shape and define flows and processes precisely because they are heterogeneous or dialectical (Levins and Lewontin, 1985; Harvey, 1996). This means simply that relationships are constituted through differences that produce some sort of tension which, in turn, generates a dynamic and becomes as such the "motor" of historical–geographical change. For example, whenever I desire a good for my use (such as a Walkman), I cannot any longer use it as a commodity for profitable sale on the market. Or if I sell my labor on the labor market, I surrender my personal time and hand over control over part of my time to someone else. The necessity to do so for many people, then, brings in the importance of "free" time, the meaning of which can only become transparent in a context in which part of the time of people's lives is bought by others. And, of course, this leads to all manner of conflict and struggle over time, its value, and its use. These are all examples of contradictory relationships. As Ollman (1993, p. 11)[5] puts it:

Dialectics restructures our thinking about reality by replacing the common sense notion of "thing", as something that has a history and has external connections to other things, with a notion of "process", which contains its history and possible futures, and "relation", which contains as part of what it is its ties with other relations.

Dialectics, combined with a materialist "ontology," provides the intellectual ferment for Marxist inquiry. For example, money as a thing can only be understood in terms of the socio-spatial relationships through which it flows. Money does not make sense if abstracted from ownership, value, the production of exchange values, and the social relations through which monetary exchanges takes place (Castree, 1996; Swyngedouw, 1996a). Similarly, space-as-a-thing acquires meaning, significance, resonance, and even a particular geographical form in and through the multiple relations with which it is infused and through which it becomes produced. For example, the making of the built environment or the cutting down of the Amazon rain forest is realized in and through socio-spatial processes of appropriation, capital accumulation, and the imagineering and scripting of people, place, and nature. It has been the staple of Marxist geography to disentangle the socio-spatial processes through which, particularly – but not exclusively – under capitalism, spatial config-urations are produced (in their economic, political, social, cultural, and ideological instances) and transformed. Places or concrete geographies become then a moment in the perpetual dialectical dynamics of socio-spatial processes (Merrifield, 1993).[6] In other words, historical–geographical materialism turns all geography into a historical socio-environmental geography (not just of the past, but also of the present and the future).

The city, uneven geographical development, and the environment were not surprisingly the central processes that were thoroughly examined by Marxist geo-graphers.[7] The geography of uneven development and the tumultuous re-orderings that characterize the contemporary city are indeed arenas around which many social movements crystallize. They are also the central loci where the dialectics and contradictions of capitalism and its associated socioeconomic inequalities are most acutely expressed. The power relations embodied in the things through which capitalism operate reveal systematic differences in positions of social power which, given their socio-spatial constitution, result in an uneven geography of change at all spatial scales. Whether at the scale of the body, the city, the region, the nation or, indeed, worldwide, the dialectics of socio-spatial power geometries produce geogra-phical difference, heterogeneity, and socio-spatial inequalities, characterized by systematic mechanisms of empowerment/disempowerment, oppression/subordina-tion, and appropriation/exploitation. Great debates have developed over the exact formulations of these dialectical twins and over the exact location(s) of the sources of power, but general agreement was reached that the spatiality of capitalism was profoundly heterogeneous and uneven; a differentiation that was historically and geographically – hence socially – produced, and that could be rendered visible and changed.

Marxist Political Economy: The Contradictions of Capitalism

Marxist political economy should be understood from the vantage point of the above dialectical and historical–geographical materialist perspective. Its basic premises are fairly simple. Production, defined as the social metabolization of nature, is the starting point. The dynamics of the particular social relations under which this takes place in a capitalist market economy are associated with a particular temporal and spatial organization of society. This territorial structure of capitalism is contradictory (that is, full of tensions and conflict), and the socio-spatial dynamics of capitalism are inherently unstable. Since this is not the place to give an exhaustive account of these dynamics (for a detailed analysis, see Harvey, 1982, 1985), it will suffice in this context to summarize them briefly (see also Swyngedouw, 1993).

- A capitalist society is based on the circulation of capital, organized as an inter-linked network of production, exchange, and consumption processes with the socially accepted goal of profit-making as its driving force. Accumulation of capital (or economic growth) is the correlative of this circulation process. Put simply, in a market economy, economic agents invest with the intention of appropriating surplus (profit) after the successful completion of a production and marketing process. Such successful circulation of capital is predicated upon the organization in space and the movement over space of money, commodities, and labor. In the process, concrete geographies (of production, consumption, and communication) are actively produced.
- The above suggests that a capitalist market economy is necessarily expansionary or growth-oriented. Zero growth (let alone negative growth) cannot be sustained for other than a very short period of time without threatening the fundamental social and economic order on which capitalism rests.
- The expansion of the system is based on living labor. Without the application of productive and actual labor, no production and consumption system along the lines outlined above can be maintained.
- As surplus is generated by living labor, but appropriated by the owner of capital in the form of profit (or transferred to the state in the form of taxation, to landowners in the form of rent, or to financial institutions in the form of interest), the above condition suggests that accumulation is based on an exploitative relationship. That is (if stripped from its emotive connotation), accumulation is of necessity the result of the application and appropriation of unpaid living labor. Put simply, the labor power put to work in the production process produces more value than the value the worker receives for his or her effort (salary or wage). The difference between the two constitutes surplus value (which goes to the owner of capital) and this is the source of profit.
- Consequently, the circulation of capital through which accumulation takes place implies antagonistic social relationships, i.e. the entrepreneur/capitalist needs to safeguard his/her accumulation condition, while the worker wants to assure his/her reproduction (short and medium-term survival). In other words, there is a fundamental contradiction between the requirement to make profit and the

necessity to invest on the one hand, and wages (the need for the worker to survive) on the other. With a deepening and expanding social and spatial division of labor, socio-spatial fragmentation increases while power relations and tensions multiply (along class, gender, ethnic, territorial, or other fractures).

- This antagonism is expressed in the struggle for the control and appropriation of the surplus produced through the circulation of capital. Given the territorial organization of this circulation process, this struggle too is inscribed in, and unfolds over, space. Social struggles of a variety of kinds alter pre-existing geographical configurations. This struggle over space can be exemplified by conflicts over land use, or over the distribution, allocation, and appropriation of natural or socially produced resources, infrastructure, and geographical arrangements.

- In addition, individual capitalists operate in a competitive context in which they engage in a struggle with each other over the conditions of surplus production, appropriation, and transfer. Consequently, all manner of struggles unfold for the control over spaces of production, and over the flows of commodities, labor, and money.

- The latter two conditions (inter- and intra-class struggle) render the capitalist economy inherently technologically and organizationally (and hence spatially) dynamic. The doubly competitive character of capitalism induces the need for continuous productivity increases, an expanding resource base, diminishing capital circulation times, lowering costs, and/or expanding markets. It demands continuous changes in the geography of production, consumption, and exchange and, hence, the perpetual production of new geographical land-scapes. A capitalist geography is a "restless landscape," subject to perpetual reconfigurations. In short, the sustainability of capitalism is based on a broadening and deepening resource base, and perpetual upheaval of the physical, ecological, economic, social, cultural, and political forms in which space is organized and controlled.

- This instability of the circulation process erupts from time to time in problems of over-accumulation or over-production. That is a situation in which capital in all its forms (commodities, money, productive equipment, and built environment) and labor lie idle, side by side. Dramatic examples of this include: the Great Depression of the 1930s, the stagflation (economic stagnation combined with high inflation) period of the 1970s, and most recently the 1998 financial crisis in Southeast Asia. These conditions of over-accumulation can take the form of high inflation, high unemployment, low or negative economic growth, idle equip-ment, un- or under-utilized infrastructure, over-production of commodities, and over-capacity in certain sectors. It is the moment when the perversions of capitalism are shown in their most brutal form. While capital is desperately seeking ways to maintain profitability, and ransacking the world's spaces in search of accumulation potential, unemployment increases and many human needs, both locally and globally, remain unfulfilled. Such crises become etched into the geographical landscape. Overaccumulated forms of capital are devalorized, and sometimes even physically destroyed. Forms of chronic or instantaneous devaluations of capital are: inflation (devalorization of the money-form of capital), debt defaulting and writing-off of debt, unemployment, stock market crashes, real-estate depreciation, physical destruction of productive, consumptive, or circulating capital, stockpiling of unsold commodities, falling currency exchange rates, and deindustrialization. Such devalorizations are always place-specific and affect different social groups in different ways, but can easily ripple over space and erupt in general regional, national, continental, or global crises.

- The perpetual threat of crisis is contained by means of a continuous restructuring of the capital circulation process. This takes the form of technological and organizational change (with all the related changes in the organization, requirements, qualifications, and the like of the labor force, of companies, of government regulations), and geographical change and relocation (in the form of foreign direct investment, re-location, and the search for new markets). While older socioeconomic and organizational forms of capital are devalued or become obsolete, others are created anew to reinvigorate the capital accumulation process. This can take the form of, for example, the radical transformation of devalorized city-centers (gentrification, large-scale urban re-development) or the mushrooming of new spaces of production (Southern California, the Third Italy). The dialectic of accumulation/devalorization produces a perpetually shifting mosaic of uneven geographical development (Smith, 1984; Storper and Walker, 1989).

The above socioeconomic geographical dynamics are always inscribed in a historically produced institutional, political, ecological, ethnic, and cultural landscape. This produces territorial configurations or coherences that are highly differentiated and give the geographical landscape its sweeping diversity, heterogeneity, and difference. The geographical dynamics of capital accumulation are faced with permanent struggle between capital and labor over the conditions of production and appropriation of the produced value, and between individual capitals, as well as between different forms of labor. In addition, the search for the "new," and for the production of new spaces of production and consumption, finds on its way all sorts of already existing communities, social ecologies, and geographies which are transformed and/or incorporated. All of these struggles are infused by a myriad non-class-based cleavages and conflicts such as ethnic, gender, or territorial conflicts or conflicts outside the realm of production, and take distinctive geographical forms.

While classical Marxist economists were mainly concerned with excavating the crisis-ridden and socially conflicting nature and characteristics of capitalism, and heroic intellectual fights unfolded over the interpretation of these crisis tendencies, capitalism proved remarkably robust as a socioeconomic and cultural system in the face of the alleged inevitability of crises. From the 1960s, a new generation of Marxists began to shift slightly the question from one that focused on the inevitability of the demise of capitalism to one that asked why capitalist social relations were so robust in the face of ongoing marginalization, exploitation, uneven development, and cyclical crises. In other words, Marxists began to concentrate on the "reproduction" of capitalist societies (Lipietz, 1988). They asked how and why capitalist social relations and accompanying institutions reproduced themselves despite the inherently conflicting nature of the system.

The Althusserian Legacy: Emphasizing Reproduction

One of the key interlocutors in this debate was Louis Althusser (Althusser, 1969; Althusser and Balibar, 1970). He was a French philosopher, Marxist and life-long member of the French Communist party who would exercise a major influence on Marxist thinking in the 1960s and 1970s. Much of the early Marxist geography was directly or indirectly inspired by Althusser's thinking (Lipietz, 1974, 1977; Castells, 1977; Massey, 1973; Peet and Lyons, 1981). His intellectual and political project was concerned with theorizing the importance of the non-economic moments of capitalist societies (culture, ideology, and the state) and identifying the mechanisms of social reproduction. He sought to de-economize the dogmatic economistic and teleological Marxism of earlier generations, who had concentrated on highlighting the structural economic crisis tendencies within capitalism to explain the inevitability of communist revolutions.

The question that began to gnaw was that if capitalism is such a crisis- and conflict-ridden system when analyzed from a purely economic perspective, why does it exhibit such a great resilience to fundamental change? For Althusser, the role of ideology and politics is central in explaining this. While ideology maintains some glue and cohesion in legitimizing the system, the political arena is pivotal in maintaining and organizing the state apparatuses (in particular education) that produce the ideologized visions that lead to the acceptance of capitalism as an apparently "natural" or inevitable social order. Moreover, the state plays a key role in containing and "managing" the tensions that exist within capitalism and in mitigating conflict. Althusser's most important insight is that the state does not function purely as an instrument in the hands of the elites (as the executive branch of the ruling class). Since the state in a democratic society embodies a series of different class fractions (including parts of the working class), state intervention is relatively autonomous from the immediate short-term interests of capital (Poulantzas, 1973, 1978). By intervening, the state contributes to maintaining cohesion in societies in which capitalist social relations of production are dominant. For example, state intervention in the form of social welfare, in the form of regional development assistance, or in the form of providing collective means of consumption (social housing, education, health services, policing, child-care, and the like), permits relatively socially peaceful development by providing functions the private capitalist sector cannot easily fulfil, but desperately needs in order to engage in successful accumulation. In addition, the absence of such services might undermine social cohesion in capitalist society and make tensions more acute, thereby rendering the social order more fragile. Particularly, the "regional problem" and the "urban question" would take an Althusserian perspective.

Although Althusser and his followers were often criticized for being structuralist[8] – particularly in the Anglo-Saxon world – the great contribution of his work consists in arguing that capitalism cannot be understood without proper attention to the ideological (what is today often called "discursive") and political instances. The latter matter in important ways, and exercise a powerful influence over the course of development of capitalist societies. Again, the geographical foundation of political and institutional configurations is of central importance to the understanding of political–economic change. Althusser's views would inspire one of the key theoretical developments in economic geographical theorizations in the 1980s, "regulation theory."

A Regulation Perspective: Economy and Institutions

When Western capitalism experienced the most serious economic crisis since the 1930s after the oil-shock of 1973, French Marxists began to ask why capitalism, despite its inherent instability and crisis-tendency, had been so remarkably successful during the boom years of the 1950s and 1960s, and why this virtuous cycle of growth turned into the vicious spiral of decline from the second half of the 1970s onwards. Aglietta (1979) introduced the "regulation approach" (as it became known). This perspective attempted to theorize the social and economic forms that maintained successful accumulation (or economic growth), and contained the contradictions and instabilities of the capitalist mode of production. "Regulationists" showed how and why the very economic and institutional forms that supported growth would eventually become contradictory or unstable themselves, and plunge post-war "Fordist" capitalism into a major socioeconomic crisis (Boyer, 1989; Dunford, 1990; Lipietz, 1986, 1987, 1992; Moulaert and Swyngedouw, 1989).

Fordism, a term coined by Antonio Gramsci (1971), refers broadly to the post-war period of expansion (up to about the mid-1970s). It is a "mode of development" that combines a "regime of accumulation" (the economic characteristics of capitalism during a particular historical period), with a "mode of regulation" (the dominant institutional and regulatory forms of the period). Under Fordism, the "regime of accumulation" took the form of mass production of standardized commodities in large vertically integrated production complexes that exhibited very detailed social, gender, technical, and spatial divisions of labor (Walker, this volume). Productivity increases through permanent technological and organizational innovations, combined with aggressive and costly advertising, became key competitive strategies, while continuously expanding markets provided a sound basis for the growth of the system. Fordist production systems were characterized by the expansion of the spatial scales of production while minimizing capital circulation times (by spatial expansion and/or further automation), and also by strategies of relocating production processes to places with a different regulatory mode (Swyngedouw, 1997b).

The pivot of "Fordist" regulation centered on the national state. During the post-war period the latter became the pre-eminent spatial scale where conflicts and tensions were negotiated (in corporatist states) and compromises settled (Altvater, 1993; Jessop, 1993, 1994a; Peck and Tickell, 1994). At the same time, the Keynesian view of macroeconomic policies, which defended an interventionist role of the state in economic affairs, constructed a precarious but increasingly important bond between the state and private capital. The state actively combined accumulation-supporting investments with a redistributional welfare system. The relative equalization across national spaces of a series of socio-economic aspects (wages, redistributive schemes, state intervention, socio-economic norms, rules, and procedures) was articulated with a highly uneven and differentiated local and regional development process.

In addition, the expansion of international investment and trade was supported by the national and international regulation of the various functions of money (see Swyngedouw, 1996a). The Bretton Woods agreement was such a compromise. It was anchored on the dollar–gold standard, which guaranteed the convertibility of the dollar to gold at a rate of $35 for an ounce of gold. Other major currencies pegged their value to the dollar. This stabilized the international monetary system by providing a relatively secure container of value. However, while regulating the value of money was cemented in the rules of the Bretton Woods agreement and policed by the IMF (see Swyngedouw, 1992a, 1996a; Leyshon and Tickell, 1994), the regulation of credit or the issuing of money, for example, remained firmly at level of the nation-state. In short, different forms and functions of money were regulated at different scales.

Towards A New Global–Local (Dis)Order?

In this section, we suggest ways in which a political–economic analysis from a "regulationist" perspective can elucidate some of the key geographical changes that have happened over the past few decades. Needless to say, these changes are complex, variegated, and still on-going. We delve into this analysis from the vantage point of how the dynamics of change express important changes in the structure and relevance of particular geographical scales (like the regional, the national, or the international scale), and how the current process of profound re-shuffling of the importance of particular geographical scales alters power geometries between social groups. The so-called "crisis of Fordism" was paralleled by a significant geographical re-ordering of economic processes and regulatory practices (see Moulaert, Swyngedouw and Wilson, 1988; Peck and Tickell, 1992, 1994; Jessop, 1994a, 1994b; Swyngedouw, 1997b). It was characterized by the rapid introduction of new technological and organizational patterns, new forms of corporate organization, a de-regulation of social and economic procedures at every geographical scale, and a more market-oriented policy framework. The overall pattern is one that I have termed elsewhere "glocalization" (Swyngedouw, 1992a, 1992b), which refers to: (i) the contested restructuring of the institutional level from the national scale both upwards to supra-national and/or global scales and downwards to the scale of the individual body, the local, the urban, or regional configurations; and (ii) the strategies of global localization of key forms of industrial, service, and financial capital (see Cooke et al., 1992). This, in turn, changes social power geometries and produces rather disturbing effects in terms of democracy, accountability, and citizenship rights.

The "glocalization" of governance

The regulation of capital/labor relations shifted decisively from some kind of national collective bargaining to often highly localized or individualized forms of negotiating wages and working conditions. Of course, depending on particular political configurations, resistance to these movements toward downscaling has been more successful in some countries (such as Sweden and Germany) than in others (such as the UK). The "Schumpeterian Workfare State" (see Jessop, 1994b;

Peck, 1996), which combines a drive towards competitive innovation with an erosion of traditional redistributional social welfare systems, has abolished a series of institutionalized regulatory procedures to leave them organized by the market and, consequently, by the power of money. At the same time, other forms of governmental intervention are replaced by more local institutional and regulatory forms (where "local" can take a variety of spatial scale forms from local constituencies to cities or entire regions). For example, the restructuring of, and often outright attack on, national welfare regimes erodes national schemes of redistribution, while privatization permits a socially highly exclusive form of protection, shielding the bodies of the powerful while leaving the bodies of the poor to their own devices.

The interventionism of the state in the economy is equally restructured with a much greater emphasis on local (urban or regional) forms of governance, where public/private partnerships shape an entrepreneurial practice and ideology needed to successfully engage in an intensified process of inter-urban competition (Harvey, 1989b), or upwards to super-national arenas. The latter is manifested in the – albeit highly contested and still rather limited – attempt to create supranational quasi-state forms at the level of the European Union. In a different way, institutions such as NAFTA, GATT, and others are testimony to similar processes of scaling up of the state. Furthermore, a host of informal quasi-global political arenas has been formed: the G-7 or G-8 meetings, the Group of 77, the Club of Paris, and other "informal" gatherings of "world" leaders are examples of such forums that attempt (without much success) to regulate the global political economy.

In addition to the socially deeply uneven, socio-spatially polarizing, and selectively disempowering effects of this "glocalization" process, it is also characterized by disturbingly undemocratic procedures. The double re-articulation of political scales (downward, upward, and outward to private capital) leads to political exclusion, a narrowing of democratic control, and consequently a re-definition (or rather limitation) of citizenship. Local or regional public/private initiatives often lack democratic control of any sort, while supranational institutions are notoriously autocratic (Swyngedouw, 1996b).

The "glocalization" of the economy

The "glocalization" or re-scaling of regulatory or institutional forms is paralleled by a variety of spatial re-configurations of the circulation of various forms of money and capital. In production, local or regional inter-firm networks, deeply inserted in local/regional institutional, political, and cultural environments, co-operating locally but competing globally, have become central to a reinvigorated – but often very vulnerable and volatile – local, regional, or urban economy (Amin, this volume). A variety of terms have been associated with such territorial economies, including learning regions, intelligent regions, or innovative regions, while new organizational strategies have been identified (the "embedded" firm (Grabher, 1993), vertical disintegration, strategic alliances, and so forth). Surely, such territorial production systems are articulated with national, supra-national, and global processes. In fact, the intensifying competition on an ever-expanding scale is paralleled exactly by the emergence of locally/regionally sensitive production milieus.

Quite clearly, "glocalizing" production cannot be separated from "glocalizing" levels of governance. The re-scaling of the regulation of wage and working conditions or the denationalization of important companies throughout Europe, for example, simultaneously opens up international competition and necessitates a greater sensitivity to sub-national conditions.

Perhaps the most pervasive process of "glocalization" and re-definition of scales operates through the financial system. The hotly contested and recently implemented introduction of the European Common Currency (euro) exemplifies this, but the chaotic and wildly fluctuating financial system that moves erratically from place to place in the global financial market place also illustrates how local and global processes intersect in often very disturbing ways (Swyngedouw, 1997c).

In sum, the regulation approach attempts to theorize and clarify the perpetual re-orderings of capitalist production and social regulation and the emergence of new territorial or geographical configurations as the processes of change unfold. The emphasis here is, of course, on the dynamics of social relations and the regulatory or institutional forms in which they are embodied and expressed. The "regulation" approach, therefore, permits thinking through alternative socioeconomic trajectories, and the development of political–economic strategies that permit the imagining, if not the construction, of alternative trajectories. While Marxist perspectives have often been criticized for their deterministic, economistic, and teleological analysis, the central tenet of the Marxist intellectual agenda remains not just to interpret the world, but to change it.

Post-Marxism, or Imagining Possible Futures

"The philosophers have only interpreted the world, in various ways; the point is to change it" (Marx, *Theses on Feuerbach*).

Whatever the failings of a Marxist critique, few other perspectives seem to be equipped to grasp persistent inequalities and exploitation here and in the rest of the world with the power of insight and the passion of commitment brought by Marx's original formulations and elaborated by a century of dedicated scholars and activists. In particular, Marxism has never shied away from asking Big Questions.

The genocides in Central Africa and in former Yugoslavia are still in full swing; fascist parties and activists rear their ugly heads all around the world; the Southeast Asian economic–financial bubble has imploded, throwing the lives of many women, men, and children into misery and despair; while the social and ecological disintegration of big cities is turning some urban neighborhoods back to the conditions described so effectively by Engels more than 100 years ago (Engels, 1844). Marxist geography, from its crudest early formulations to later more sophisticated perspectives, has always sought to see and to address the "Big Issues," even in the smallest of places or events. In today's fledgling post-postmodernist and post-deconstructionist world, a re-invented modernity that is sensitive to heterogeneity, solidarity, and emancipation in the construction of a pre-socialist vision for genuinely humanizing lived geographies is where the agenda should be. This may be a rather grand agenda, but the current conditions in many parts of the world demand no less from a self-respecting discipline. In particular, the search for possible different futures, for a social economy embedded in a truly humanizing geography of every-day life at the scale of the body, the urban, the region or the globe, requires urgent attention. The forging of strategic political alliances with those who struggle for freedom from repression and for emancipation from domination is where Marxist geography and geographers still have an important contribution to make. This contribution lies not only in a permanent critique of the conditions and dynamics of capitalism, but also in pushing the frontiers of the imagining of geographical trajectories in which difference, heterogeneity, and the unrepressed expression of desire coincide with a just and inclusive socioeconomic and political order (see Harvey, 2000).

Endnotes

1. For easy introductions to the life and work of Karl Marx, see Eagleton (1997), McLellan (1986), Singer (1980).
2. These themes were developed and presented primarily in *Antipode: A radical journal of geography*.
3. See also Brewer (1980), Peet (1991, 1998), Kasinitz (1995).
4. There is also an important strand of non-dialectical Marxism, usually referred to as Analytical Marxism or Post-Sraffian Marxism (Roemer, 1986; Cohen, 1978). In geography, key contributors to the debate from this perspective are, among others, Sheppard and Barnes (1990) or Webber (1989, 1997).
5. For a detailed account, see Ollman (1993), Merrifield (1993), Castree (1995, 1996), Harvey (1996), Balibar (1995).
6. See, for example, Carney et al. (1981), Corbridge (1986), Davis (1991), Hecht and Cockburn (1989), Watts (1983), Dunford and Perrons (1983), Dunford (1988), Hudson (1989), Harvey (1989a), Swyngedouw (1997a, 1999).
7. Mainly in the pages of *Antipode* and the *International Journal of Urban and Regional Research*.
8. Structuralism is a mode of thinking that argues that social structures unfold and develop largely independently from conscious individual action or agency.

Bibliography

Aglietta, M. 1979. *A Theory of Capitalist Regulation – The U.S. Experience*. London: New Left Books (original work published 1976).

Althusser, L. 1969. *For Marx* (B. Brewster, transl.). London: New Left Books (original work published 1965).

Althusser, L. and Balibar, E. 1970. *Reading Capital*. London: New Left Books (original work published 1968).

Altvater, E. 1993. *The Future of the Market – An Essay on the Regulation of Money and Nature after the Collapse of "Actually Existing Socialism"*. London: Verso.

Ambrose, P. and Colenutt, B. 1975. *The Property Machine*. London: Penguin.

Balibar, E. 1995. *The Philosophy of Marx* (C. Turner, transl.). London: Verso (original work published 1993).

Benton, T. (ed) 1996. *The Greening of Marxism*. New York: Guilford Press.

Blaut, J. 1970. Geographic models of imperialism. *Antipode*, 2, 1, 65–85.

Blaut, J. 1975. Imperialism: the Marxist theory and its evolution. *Antipode*, 7, 1, 1–19.

Boyer, R. 1989. *Regulation Theory, a Critical Perspective*. New York: Columbia University Press (original work published 1986).

Bracken, L. 1997. *Guy Debord – Revolutionary*. Venice, CA: Feral House.

Brewer, A. 1980. *Marxist Theories of Imperialism – a Critical Survey*. London: Routledge & Kegan Paul.

Buchanan, K. 1966. *The Chinese People and the Chinese Earth*. London: G. Bell and Sons.

Buchanan, K. 1973. The White North and the population explosion. *Antipode*, 5, 3, 7–15.

Bunge, W. 1971. *Fitzgerald: Geography of a Revolution*. Cambridge, MA: Schenkman.

Bunge, W. 1973. The geography of human survival. *Annals of the Association of American Geographers*, 63, 275–95.

Carney, J., Hudson R., and Lewis, J. 1980. *Regions in Crisis*. London: Croom Helm.

Castells, M. 1977. *The Urban Question*. London: Edward Arnold.

Castree, N. 1995. The nature of produced nature: Materiality and knowledge construction in Marxism. *Antipode*, 27, 12–48.

Castree, N. 1996. Birds, mice and geography: Marxisms and dialectics. *Transactions of the Institute of British Geographers*, NS 21, 342–62.

Cohen, J. 1978. *Karl Marx's Theory of History: a Defence*. Oxford: Clarendon Press.

Cooke, P., Moulaert, F., Swyngedouw, E., Weinstein, O., and Wells, P. 1992. *Towards Global Localization: The Computing and Communications Industries in Britain and France*. London: UCL Press.

Corbridge, S. 1986. *Capitalist World Development*. London: Macmillan.

Davis, M. 1990. *City of Quartz – Excavating the Future of Los Angeles*. London: Verso.

Dear, M. and Scott, A. (eds). 1981. *Urbanization and Urban Planning in Capitalist Society*. London: Methuen.

Debord, G. 1967. *The Society of the Spectacle*. London: Rebel Press.

Dunbar, G. 1979. Elisée Reclus, geographer and anarchist. *Antipode*, 10, 3, 16–21.

Dunford, M. 1988. *Capital, the State and Regional Development*. London: Pion.

Dunford, M. 1990. Theories of regulation. *Environment and Planning D: Society and Space*, 8, 297–322.

Dunford, M. and Perrons, D. 1983. *The Arena of Capital*. London: Macmillan.

Eagleton, T. 1997. *Marx and Freedom*. London: Phoenix.

Engels, F. 1987 (1844). *The Condition of the Working Class in England*. London: Penguin.

Grabher, G. (ed) 1993. *The Embedded Firm: On the Socioeconomics of Industrial Networks*. London: Routledge.

Gramsci, A. 1971 (1928). Americanism and Fordism. In *Selections from the Prison Notebooks*. New York: International Publishers, 277–318.

Gregory, D. and Urry, J. (eds). 1985. *Social Relations and Spatial Structures*. London: Macmillan.

Harvey, D. 1973. *Social Justice and the City*. London: Edward Arnold.

Harvey, D. 1977. The geography of capitalist accumulation: A Reconstruction of the Marxian theory. In R. Peet (ed), *Radical Geography*. Chicago: Maroufa Press, 263–92.

Harvey, D. 1982. *Limits to Capital*. Oxford: Blackwell.

Harvey, D. 1984. On the history and present condition of geography – an historical materialist manifesto. *Professional Geographer*, 36, 1–11.

Harvey, D. 1985. The geo-politics of capitalism. In D. Gregory and J. Urry (eds), *Social Relations and Spatial Structures*. London: Macmillan, 128–63.

Harvey, D. 1989a. *The Urban Experience*. Oxford: Blackwell.

Harvey, D. 1989b. From managerialism to entrepreneuralism: The transformation in urban governance in late capitalism. *Geographiska Annaler* Series B, 71, 3–18.

Harvey, D. 1996. *Justice, Nature and the Geography of Difference*. Oxford: Blackwell.

Harvey, D. 1998. The geography of the manifesto. In L. Panitch and C. Leys (eds), *The Communist Manifesto now – Socialist Register 1998*. Rendlesham: The Merlin Press, 49–74.

Harvey, D. 2000. *Spaces of Hope*, Edinburgh: Edinburgh University Press.

Hecht, S. and Cockburn, A. 1989. *The Fate of the Forest: Developers, Destroyers and Defenders of the Amazon*. London: Verso.

Hilferding, R. 1981. *Finance Capital* (M. Watnick and S. Gordon, transl.). London: Routledge & Kegan Paul (original work published 1910).

Hudson, R. 1989. *Wrecking a Region*. London: Pion.

Jessop, B. 1993. Fordism and post-Fordism: Critique and reformulation. In A. J. Scott and M. Storper (eds), *Pathways to Regionalism and Industrial Development*. London: Routledge, 43–65.

Jessop, B. 1994a. Post-Fordism and the state. In A. Amin (ed), *Post-Fordism: a Reader*. Oxford: Blackwell, 251–79.

Jessop, B. 1994b. The transition to post-Fordism and the Schumpeterian workfare state. In R. Burrows and B. Loader (eds), *Towards a Post-Fordist Welfare State?* London: Routledge, 13–37.

Kasinitz, P. (ed). 1995. *Metropolis – Centre and Symbol of our Times*. London: Macmillan.

Kropotkin, P. 1979 (1885). What Geography ought to be. *Antipode*, 10, 3, 6–15.

Lefebvre, H. 1968. *La Pensée Marxiste et la Ville* [Marxist Thought and the City]. Paris: Casterman.

Lefebvre, H. 1989. *The Production of Space*. Oxford: Blackwell (original work published 1974).

Lenin, V. I. 1970. *Imperialism: The Highest Stage of Capitalism*. Beijing: Foreign Languages Press.

Levins, R. and Lewontin, R. 1985. *The Dialectical Biologist*. Cambridge, MA: Harvard University Press.

Leyshon, A. and Tickell, A. 1994. Money order? The discursive construction of Bretton Woods and the making and breaking of regulatory space. *Environment and Planning A*, 26, 1861–1890.

Lipietz, A. 1974. *Le Tribut Foncier Urbain* [Urban Land Rent]. Paris: Maspéro.

Lipietz, A. 1977. *Le Capital et son Éspace* [Capital and its Space]. Paris: Maspéro.

Lipietz, A. 1986. New tendencies in the international division of labour: Regimes of accumulation and modes of social regulation. In A. J. Scott and M. Storper (eds), *Production, Work, Territory: The Geographical Anatomy of Industrial Capitalism*. Boston: Allen and Unwin, 16–40.

Lipietz, A. 1987. *Mirages and Miracles: The Crisis of Global Fordism* (D. Macey, transl.). London: New Left Books.

Lipietz, A. 1988. Reflections on a tale: The Marxist foundations of the concepts of regulation and accumulation. *Studies in Political Economy*, 26, 7–36.

Lipietz, A. 1992. The regulation approach and capitalist crisis: An alternative compromise for the 1990s. In M. Dunford and G. Kafkalas (eds), *Cities and Regions in the New Europe: The Global–Local Interplay and Spatial Development Strategies*. London: Belhaven, 309–34.

Luxemburg, R. 1951. *The Accumulation of Capital*. London: Routledge & Kegan Paul.

Marx, K. 1997. *Theses on Feuerbach*. CD-Rom eb0002, London: Electric Book Company.

Marx, K. and Engels, F. 1848 (1997). *The Communist Manifesto*. CD-Rom eb0002, London: Electric Book Company.

Massey, D. 1973. Towards a critique of industrial location theory. *Antipode*, 5, 33–9.

Massey, D. 1984. *Spatial Divisions of Labour*. Basingstoke: Macmillan.

Massey, D. 1994. *Space, Place and Gender*. Cambridge: Polity Press.

McLellan, D. 1986. *Marx*. London: Fontana Press.

Merrifield, A. 1993. Place and space: a Lefebvrian reconciliation. *Transactions, Institute of British Geographers*, NS 18, 516–31.

Merrifield, A. and Swyngedouw, E. (eds). 1997. *The Urbanisation of Injustice*. London/New York: Lawrence & Wishart/New York University Press.

Moulaert, F. and Swyngedouw, E. 1989. A regulation approach to the geography of flexible production systems. *Environment and Planning D: Society and Space*, 7, 327–45.

Moulaert, F., Swyngedouw, E., and Wilson, P. 1988. Spatial responses to Fordist and post-Fordist accumulation and regulation. *Papers of the Regional Science Association*, 64, 11–23.

O'Connor, J. 1998. *Natural Causes – Essays in Ecological Marxism*. New York: Guilford Press.

Ollman, B. 1993. *Dialectical Investigations*. London: Routledge.

Peck, J. 1996. *Work-Place – The Social Regulation of Labor Markets*. New York: Guilford Press.

Peck, J. and Tickell, A. 1992. Local modes of social regulation? Regulation theory, Thatcherism and uneven development. *Geoforum*, 23, 347–64.

Peck, J. and Tickell, A. 1994. Searching for a new institutional fix: the after-Fordist crisis and the global-local disorder. In A. Amin (ed), *Post-Fordism: a Reader*. Oxford: Blackwell, 280–315.

Peet, R. 1977. *Radical Geography*. Chicago: Maroufa Press.

Peet, R. 1991. *Global Capitalism: Theories of Societal Development*. London: Routledge.

Peet, R. 1998. *Modern Geographical Thought*. Oxford: Blackwell.

Peet, R. and Lyons, J. V. 1981. Marxism: Dialectical materialism, social formation and geographic relations. In M. E. Harvey and B. P. Holly (eds), *Themes in Geographic Thought*. London: Croom Helm, 187–202.

Poulantzas, N. 1973. *Political Power and Social Classes*. London: New Left Books.

Poulantzas, N. 1978. *State, Power, Socialism*. London: New Left Books.

Roemer, J. 1986. *Analytical Marxism*. Cambridge: Cambridge University Press.

Scott, A. 1988. *New Industrial Spaces*. London: Pion.

Sheppard, E. and Barnes, T. 1990. *The Capitalist Space Economy: Geographical Analysis after Ricardo, Marx and Sraffa*. London: Unwin Hyman.

Singer, P. 1980. *Marx*. Oxford: Oxford University Press.

Smith, N. 1984. *Uneven Development*. Oxford: Blackwell.

Storper, M. and Walker, R. 1989. *The Capitalist Imperative*. Oxford: Blackwell.

Swyngedouw, E. 1992a. The mammon quest. "Glocalization", interspatial competition and the monetary order: The construction of new scales. In M. Dunford and G. Kafkalas (eds), *Cities and Regions in the New Europe*. London: Belhaven Press, 39–67.

Swyngedouw, E. 1992b. Territorial organization and the space/technology nexus. *Transactions, the Institute of British Geographers*, NS 17, 417–33.

Swyngedouw, E. 1993. Communication, mobility and the struggle for power over space. In G. Giannopoulos and A. Gillespie (eds), *Transport and Communications Innovation in Europe*. London: Belhaven Press, 305–25.

Swyngedouw, E. 1996a. Producing futures: Global finance as a geographical project. In P. W. Daniels and W. F. Lever (eds), *The Global Economy in Transition*. Oxford and London: Longman, 135–63.

Swyngedouw, E. 1996b. Reconstructing citizenship, the re-scaling of the state and the new authoritarianism: Closing the Belgian mines. *Urban Studies*, 33, 1499–1521.

Swyngedouw, E. 1997a. Power, nature and the city. The conquest of water and the political ecology of urbanisation in Guayaquil, Ecuador: 1880–1980. *Environment and Planning A*, 29, 311–32.

Swyngedouw, E. 1997b. Neither global nor local: "Glocalization" and the politics of scale. In K. Cox (ed), *Spaces of Globalization: Reasserting the Power of the Local*. New York: Guilford Press, 137–66.

Swyngedouw, E. 1997c. Excluding the other: The contested production of a new "Gestalt of scale" and the politics of marginalisation. In R. Lee and J. Wills (eds), *Geographies of Economies*. London: Edward Arnold, 167–76.

Swyngedouw, E. 1999. Modernity and hybridity: Nature, "regeneracionismo," and the production of the Spanish waterscape, 1890–1930. *Annals of the Association of American Geographers*, 89, 3, 443–65.

Watts, M. 1983. *Silent Violence: Food, Famine and Peasantry in Northern Nigeria*. Berkeley, CA: University of California Press.

Webber, M. 1989. Capital flows and rates of profit. *Review of Radical Political Economy*, 21, 113–35.

Webber, M. 1997. Profitability and growth in a multiregion system: Theory and a model. *Economic Geography*, 72, 335–52.

Chapter 5

Feminism and Economic Geography: Gendering Work and Working Gender

Ann M. Oberhauser

Two decades ago, a chapter on feminism in an economic geography reader would have been unlikely. The role of gender in labor migration, commuting patterns, and the globalization of capital, however, is increasingly recognized as an important dimension of economic processes. This chapter examines the relationship of feminism to economic geography, from its initial roots in political economy to its contemporary forms turning on poststructural theories of embodiment and identity in the workplace. It is argued that feminism has increased our understanding of economic processes through its analyses of how gender and work are socially constructed at multiple scales and in diverse geographical contexts.

There are three compelling arguments for the relevance of feminism to economic geography. First, feminist theory forces us to confront the heterogeneity of the workforce. Disaggregating economic data by gender reveals two separate patterns and trends in labor force participation, unemployment, and occupational status, one for men and a very different one for women. Identifying and exploring these differences in men's and women's labor contributes to our understanding of not only the complexity of the workforce, but also the role of gender in economic processes.

Second, feminist geography helps us to realize that the economy and its geography are defined by multiple social relations. While class was previously viewed as the dominant social relation under capitalism, feminism has helped to expand economic analyses to include gender as well as race, ethnicity, age, sexuality, and other sets of relations that are linked to economic power and identity. These social relations, in turn, play a role in numerous economic processes such as the location of low-wage factories in developing regions, the unionization of blue-collar workers, and household income-generating strategies.

Finally, feminism contributes to economic geography through its methodological approaches which raise gender-sensitive questions and allow for more inclusive and relevant research. Questions traditionally ignored in conventional economic geography, but highlighted in feminist research include: How do the expenditure patterns of men and women vary? Why is there a prevalence of informal sector work among women in the global South? What explains the continued segregation of women and men in certain occupations? These questions lead to different methods of collecting data to obtain more accurate information on gender relations and divisions of labor in the workforce and the household. Feminist research demonstrates that qualitative approaches such as intensive interviews, participant observation, and personal narratives can be effectively combined with quantitative methods that describe and probe the measurable aspects of women's lives, analyze spatial associations, and document spatial and temporal inequalities (*Professional Geographer*, 1995).

This chapter addresses these themes in five sections. The next section focuses on the emergence of feminist perspectives in economic geography, beginning with its roots in political economy and Marxist geography. Early empirical analyses in feminist geography that examined the spatial distribution of women's work on national and global scales will also be discussed. Section three reviews works that build on these earlier feminist approaches to economic processes and addresses the complexity of gendered labor in diverse geographic and socioeconomic contexts. In particular, I will draw upon a body of work called postcolonialism, which questions dominant notions of gender and work in order to include social categories such as race and ethnicity. Gendered livelihood strategies in the global South, such as informal activities and immigrant labor, are linked to these diverse social categories.

More recent feminist theories that analyze the embodied nature of work and socially constructed identities in the workplace are highlighted in the fourth section. Increasing emphasis on cultural dimensions of human interaction in economic geography raises issues concerning the construction of masculinity and femininity in the workplace. Through case studies of firms in the service sector, the discussion illustrates how gender is embodied in the social and cultural processes that are intrinsic to work and organizational restructuring.

The conclusion summarizes the development of economic perspectives within feminism. I argue that women, people of color, and other disempowered members of society remain marginalized in the workforce because of socially constructed norms and dominant power relations. This chapter demonstrates how various theoretical and methodological approaches in feminism identify and explore the positions and institutional contexts of these power relations at various scales of economic activity. Additionally, comparative research on gender and work illustrates the experiences of workers (especially women) in diverse cultural and political economic contexts. Analyses such as these will hopefully raise awareness about the need to emancipate workers and encourage economic institutions that are sensitive to gender as well as other social differences.

The Emergence of Feminism in Economic Geography

Some of the first topics of study by feminists in the discipline of geography included analyses of women's economic activity. These studies initially criticized conventional approaches for neglecting women's roles and divisions of labor in their analyses of economic processes. Research in this area was mostly descriptive and aimed to make women's work visible. The seemingly unexamined category of woman was soon replaced by the concept of gender which focused on the social construction of male and female roles and behavior in society. This shift provided a more complex analysis and comparison of men's and women's economic activities. In turn, socialist feminism highlighted the role of gender and class in economic processes (Mackenzie, 1989). Geography was central to this early feminist work because it incorporated multiple scales of analysis, and recognized the importance of place-based social relations.

Counting women and work

In the mid-1970s, geography underwent a significant transition as the quantitative methods and spatial science traditions were criticized for their normative assumptions of a single universal truth and what has been cited as totalizing tendencies or "grand narratives" of rational science (Barnes, this volume). Feminists were among these critics and challenged the discipline for its male bias in constructing knowledge and theory. These constructions extended dualisms constructed around masculinity and femininity to the work sphere, where men were associated with paid labor in the public workplace and women with unpaid labor in the private household. The overall impact of this approach in the discipline of geography was to marginalize gender and neglect women's experiences in geographic inquiry (Mazey and Lee, 1983; Monk and Hanson, 1982).

During the 1970s, gender and work were important themes in feminist geography for reasons that coincided with societal concerns about women's inequality in the home, at work, and in society as a whole (Bowlby et al., 1989; Monk and Hanson, 1982). Much of this concern stemmed from the feminist movement which claimed women were materially disadvantaged and marginalized because of their position in lower status jobs and their relative absence from positions of power. Efforts to empower women during this period included pay equity, expanded childcare, and equal opportunities in the workforce (Mackenzie, 1989). Empiricist approaches that focused on case studies and empirical analyses dominated the research during this period.

Not surprisingly, early geographic literature on gender and work also emphasized the spatial dimensions of gender divisions of labor and occupational segregation. This research attempted to explain how individuals reproduce social phenomena in space through everyday activities. According to this approach, the "paths" taken by individuals in their daily lives are subject to certain constraints that include physical limits to movements, the need to gather in schools, workplaces or the like, and social rules that control where people can or cannot go (Rose, 1993). Accordingly, journey-to-work studies document how gender roles influenced unequal access to transportation and constraints stemming from domestic responsibilities (Hanson and Hanson, 1981). Results from these studies indicated that women commute shorter distances to work, travel less frequently, and use different means of transport than men (Hanson and Johnston, 1985), which, in turn, contributes to, and stems from, unequal material conditions of men and women. Hanson and Pratt (1995), for example, argue that women's tendency to travel shorter distances to work is linked to their lower wages.

Early research in feminist geography, however, tended to homogenize women and overlook differences in their economic activity and commuting patterns. Recent studies include social categories other than gender, such as race, ethnicity, and class, to reveal the diversity of social and economic factors that impact the journey to work. For example, extensive research in the Buffalo and New York metropolitan areas demonstrates that the commuting experiences of African American and Latina women are often different from those of white women (Johnston-Anumonwo, 1995; Preston et al., 1993). The claim that women work closer to home than men because of their dual roles as mother and waged worker does not hold for Black and Hispanic women. These women commute as far as Black and Hispanic men, and spend more time commuting than white males and females (McLafferty and Preston, 1991).

Accompanying this focus on the spatial dimensions of individuals' daily activities was a growing literature within feminist geography on suburbanization, and household social relations and divisions of labor. Suburbanization contributed to the social construction of domesticity and its physical as well as psychological separation from paid labor. Miller (1982), for example, examined how suburban women's roles were constructed through advertising and the media in the early twentieth century. Through this and the spatial constraints that structured women's lives, he draws conclusions about the impact of patriarchy on women's domestic roles. Other studies note how the economic sphere of the workplace was socially and spatially separated from the household and, by extension, the private realm of women (Mackenzie and Rose, 1983; Dyck, 1989). This separation of women into the private sphere and care-taking activities is related to the dominant perception of women's roles, and therefore their mobility.

The social relations and spatial dynamics of the domestic sphere and workplace are closely associated with occupational segregation of the labor market. Numerous studies have focused on the way in which labor supply and demand processes play out differently in various locations, but reveal the connections between the supposedly separate spheres of private domesticity and public labor (Hanson and Pratt, 1995; Massey, 1984; Rose, 1993). Figure 5.1 depicts the number of women employed in the top ten occupations of women in the USA. It illustrates how women's segregation in the workforce is related to the perception of women's roles in our society. In the USA, jobs associated with care-taking, nurturing, and other so-called feminine traits are among the most common forms of female employment. Secretaries are the most numerous with nearly 3 million women in 1997, followed by cashiers with 2.3 million women (US Dept. of Labor, 1998). Additionally, most of these occupations are highly segregated: females represent over two-thirds of the workforce in nearly all of the occupations listed. Secretaries as an occupation have the highest proportion of females at 98 percent (US Dept. of Labor, 1998).

Finally, the wages of people in these occupations indicate the relatively low value placed on this type of work. Earnings range from $705 per week among registered nurses to $248 among cashiers. In 1997, the average weekly earning of all full-time women workers was $431 compared to $579 for men. These data demonstrate that occupational segregation is linked to socially embedded divisions of labor and contributes to economic inequalities between men and women. As outlined above, feminist geographers approach this phenomenon from a geographical perspective by drawing connections between relations within the home and the workplace in both social and spatial terms (Hanson and Pratt, 1995).

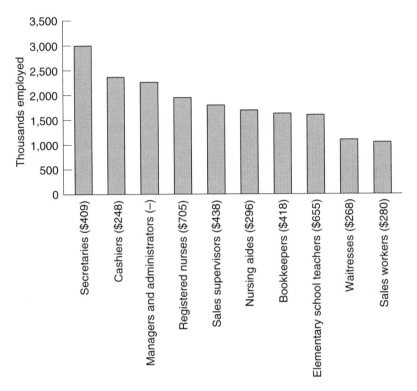

Figure 5.1 The top ten occupations of employed women in the USA, 1997
Figures in parentheses are women's median weekly earnings for that occupation.
Source: US Department of Labor (1998)

In sum, the early phase of feminist economic geography offered empirical analyses of spatial patterns in women's (and men's) everyday experiences, especially work-related activities such as travel to work, employment patterns, and the relation between the domestic sphere and the workplace. The subsequent phase extended the analysis of gender inequality to include broader historical and material structures in society.

Geographies of socialist feminism

Research on gender, women, and work during the emergence of feminist geography was also influenced by scholarship in socialist feminism and Marxism. Marxist geographers analyzed the economy from an historical materialist approach (Swyngedouw, this volume) which emphasized the contradictions of capitalist accumulation (Peet and Thrift, 1989). Some of the earliest pieces on women and geography appeared during the late 1970s and early 1980s, and were couched solidly within this political economy framework (Hayford, 1974; Mackenzie and Rose, 1983; Tivers, 1978). This critical approach to conventional geography resonated with feminists because they were concerned with addressing issues of social justice and individual equality.

Additionally, many socialist feminists felt that answers to questions about women's lower-than-average incomes, unequal travel patterns, and segregation in lower status jobs were not completely addressed by the kind of empirical analyses discussed above. Instead, they tended to emphasize broader structures in society such as patriarchy, where gender relations are characterized by the domination of men over women, and capitalism, to explain social and economic inequality (Mackenzie, 1989; Bowlby et al., 1989). Socialist feminists, however, were skeptical of the somewhat rigid categories and universal laws that were central to orthodox Marxism, because they tended to ignore gender divisions of labor and regard class as the dominant social relationship under capitalism.

An important aspect of socialist feminism was its analysis of the spatial and social dimensions of the domestic sphere, and its link to capitalist production. Socialist feminists drew from Marxist geography in their argument that the domestic work carried out by women was essential for the maintenance and reproduction of the labor force and hence for capitalism more generally (Mackenzie and Rose, 1983; Walby and Bagguley, 1989). They also maintained that capitalist patriarchy created a gender division of labor in which men primarily engaged in waged, productive labor, and women in unwaged, reproductive labor. Massey (1997, p. 113) critically examined these dualistic approaches to work and home in her discussion of the way in which men and women negotiate the boundary between productive and reproductive labor:

Dualistic thinking leads to the closing-off of options, and to the structuring of the world in terms of either/or... Moreover, even when at first sight they may seem to have little to do with gender, a wide range of such dualisms are in fact thoroughly imbued with gender connotations, one side being socially characterized as masculine, the other as feminine, and the former being accordingly socially valorized.

Thus, in this complex and historically dynamic capitalist system, women's reproductive labor was rendered invisible, but was fundamentally connected to productive labor by providing support for capital such as clothing, feeding, and ensuring the reproduction of labor.

During the 1980s, feminist perspectives in economic geography began to explore the intersection of uneven development and the gendering of work. It was argued that unequal social relations are largely derived from capitalism, and are manifest through uneven regional development (Lewis, 1983; Massey, 1984; McDowell and Massey, 1984). For example, capital was invested in particular types of industry in regions where female labor was available and non-unionized. Feminist analyses of social and spatial inequality, however, were not limited to the scale of region, but also incorporated multiple scales such as the household, local, and international levels.

Overall, socialist feminism introduced new theoretical and methodological analyses to economic geography while borrowing from some of the social and economic categories of Marxism. It challenged the binary division of spaces and roles in the domestic sphere and workplace, particularly as they related to livelihood strategies and the material conditions of everyday life. While the voices in this literature were not always in unison, they nonetheless advanced the discipline in its appreciation of the critical role of gender relations in understanding social, economic, and spatial inequality.

Economic restructuring and shifting employment patterns

Feminist analyses of economic processes contributed significantly to our understanding of the economic restructuring that developed during the late 1970s. These analyses focused on the dramatic shift in employment from manufacturing to the service sector. For example, Hanson and Pratt's widely cited study of Worcester, MA (1995) examines the importance of gender and economic restructuring in an industrialized area that has experienced a shift from relatively high-paid, unionized, and male-dominated jobs to lower paying, less secure, and female-dominated jobs. Studies such as this demonstrate that economic restructuring formed the basis for analyses in feminist geography that sought to explain social and spatial divisions of labor under capitalism.

One of the implications of this shift was the steady increase in many Western countries of economically active women, both in absolute numbers and as a proportion of the female population. In the United States, the number of working women doubled between 1970 and 1997, from 30 million to 60 million (US Department of Labor, 1998). In addition, during the post-WWII period, labor force participation – the percentage of working age persons actually employed or looking for work – rose significantly for women, while men's rates slowly declined (table 5.1). In 1950, 29.0 percent of all women in the USA were economically active compared to 78.9 percent of men. By 1995, female participation rates had more than doubled while male participation rates had declined nearly 4 percent. These figures have significant implications for gender roles and identity in our society as women have become more active in the paid labor force.

Aggregate analyses of employment, however, often neglect the diversity that appears in the workforce when consideration is given to race and ethnicity, as well as gender. Research on the impacts of economic restructuring on labor force participation has increasingly addressed differences among ethnic and racial groups. Table 5.1 compares the participation rates of men and women in European American, African American, and Hispanic populations in the USA during the post-war period.

Table 5.1 Labor force participation rates by gender and race, 1950–1995

	Total (%)		European American		African American		Hispanic	
	Male	Female	Male	Female	Male	Female	Male	Female
1950	78.9	29.0	85.6	33.3	–	–	–	–
1960	82.4	37.1	83.4	36.5	–	–	–	–
1970	79.2	42.8	80.0	42.6	73.6	48.7	–	–
1980	–	–	–	–	–	–	81.4	47.4
1985	76.3	54.5	77.0	54.1	70.8	56.5	80.3	49.3
1990	76.1	57.5	77.1	57.4	71.0	58.3	81.4	53.1
1995	75.0	58.8	75.1	59.0	69.0	59.5	79.1	52.6

Source: Jacobs (1997)

Although accurate employment data disaggregated by race was not available until relatively recently, studies indicate that in the early part of the century, African American women were more than five times more likely to be economically active than white women and are more likely to be employed than women of any other racial group (US Dept. of Labor, 1998). Women of European descent, however, have the fastest growth rates and are expected to have a higher participation rate than Black women at the turn of the century. Overall, women's economic activity rates have risen since 1950 except for a minor decline among Hispanic women from 1990 to 1995. (See Amott and Mathaei, 1991, for in-depth analyses of work patterns of women from diverse racial-ethnic groups.) In contrast, male participation rates of European Americans and Hispanics are higher than those of African Americans, but have slowly declined since the 1960s. These statistics reveal the important intersection of gender and race in empirical analyses of how work has changed over time. Feminist theory's engagement with gender, race, and ethnicity underlines the significance of social differentiation in economic practices.

Looking back on the initial phase of feminist geography, analyses of women's work and gender relations dominated much of the research within this subfield. These analyses took place in the context of widespread industrial restructuring, uneven regional development, and the rapid increase of women in the labor force. While some research was based on empiricist approaches, political economy, and particularly socialist feminism, was the dominant framework for much of this work (Christopherson, 1989; Mackenzie, 1989). During this period, feminist geography examined increasing economic activity rates among women and the changing gender composition of the labor force in the context of economic restructuring (Massey, 1984; McDowell, 1991). The following section shifts focus from analyses of gender and work in the context of developed regions to developing regions where Third-World feminism has focused on diversity among women and critical theorization of global capitalist domination.

Globalization of Gender and Work: Incorporating Voices from the South

This section examines analyses of gender and work in diverse geographical and socioeconomic contexts. Globalization and the increasing incorporation of women into the economies of developing regions highlight the fluid boundaries and over-lapping spheres of work and home, formal and informal work. The theoretically and empirically rich literature that has developed in feminist geography since the late 1970s offers critical perspectives on the contested role of gender as it is mediated by race, ethnicity, sexuality, class, history, and geographical context. This discussion draws primarily on recent work in postcolonial feminism that addresses social diversity and promotes strategies of resistance and struggle to overcome forces of oppression.

Regional and global approaches to women and gender in the South[1]

During the 1970s and early 1980s, analyses of Third World women in feminist geography mirrored the descriptive and empiricist trends discussed above. These analyses focused on the material aspects of women's lives in developing countries, and attempted to map social and economic indicators that revealed the overall status of women in the global South (Brydon and Chant, 1989; Momsen and Townsend, 1987). An important contribution to these geographical analyses of women's status is Joni Seager's *The State of Women in the World Atlas* (1997), mapping a variety of social and economic conditions such as women in the labor force, literacy rates, birth and death rates, and poverty. This innovative atlas compares international data on the status of women and unmasks the commonalities and differences that exist among women in diverse cultures.

This graphic representation of women's status remains incomplete, however, because their activities are largely invisible and undervalued in many national statistics. Moreover, descriptive analyses of women's status can reinforce somewhat negative stereotypes of women in the South as impoverished and uneducated, and with high rates of fertility. According to Mohanty (1991), the depiction of women in developing countries as a homogenous population, measured against a Western norm, is misleading and Eurocentric.

In general, gender studies have made important contributions to globalization and have changed the way feminist geographers think about gender and work. In the late 1970s and early 1980s, analyses of the new international division of labor examined how the reorganization of production contributed to global shifts in production from First to Third World locations (Fernandez-Kelly, 1983). This period witnessed significant increases in women's employment in the textile, electronic, garment, and other labor-intensive industries in these regions. Research demonstrated how women have been incorporated into multinational factory work largely because they were low paid, unskilled, and non-unionized (Beneria and Feldman, 1992; Fernandez-Kelly, 1983, 1990; Tinker, 1990). Several projects exposed the highly exploitative nature of this phenomenon. According to Elson and Pearson (1989), the contradictions between the globalization of capital and community values are highlighted in the poor working conditions of women in multinational companies. They argue that women form a significant proportion of those employed in multinational corporations due to their perceived subordination and ability to perform highly repetitive, menial labor. Fernandez-Kelly (1983) also provides an insightful analysis of the oppressive conditions experienced by women in the Mexican *maquiladora* sector who are living in substandard conditions with low-paying jobs.

Finally, the link between women's income-generating strategies and state policies has been a focus of Third World research in feminist geography. Many studies have shown that state industrialization policies and economic reform are closely connected to the globalization of capital. For example, Cravey's (1997) analysis of household dynamics in the context of Mexican industrial transition demonstrates that the shift in industrialization strategies from import substitution to export-based production has affected household gender dynamics. Her research concludes that different production regimes and household relations result in diverse coping strategies. In addition, numerous authors cite the negative impact of structural adjustment policies on women and households in many developing countries and regions, as state spending on education, welfare, and other social services has been cut (Beneria and Feldman, 1992; Elson, 1992). This literature suggests that Western-led economic policies put severe strains on households and women in developing regions.

Overlapping spaces and places of work

Feminist analyses of globalization also examine how the overlapping spheres of production and reproduction, formal and informal sector linkages, and household and workplace highlighted above relate to international economic processes. These analyses link women's economic strategies in an era of global restructuring with the social construction of gender identity. Feminist research on corporate labor practices and the recruitment of women for their "nimble fingers" and nonmilitant character-istics reveal how social roles and norms have been constructed in ways that support certain economic strategies (Elson and Pearson, 1989). For example, Wright (1997) addresses the importance of gender identity in her compelling analysis of Mexican women working in the *maquiladoras*. The representation of the Mexican "Woman" as a docile, submissive, and tradition-bound worker stems from the dominant Western discourse of women in the context of development. This discourse is challenged by the story of Gloria, a worker who subverts this ideological representa-tion of Woman, and resists the *maquiladora* division of labor to become a social agent of change (Wright, 1997). Her experiences as a manager in a firm involved strategies of resistance and compliance that impacted her role as an active particip-ant in the labor process rather than a passive victim of exploitation.

The overlapping spheres of production and reproduction, and household and workplace, are especially relevant when women's labor in the domestic sphere generates income or becomes part of household economic strategies. Often this labor takes place in the household or is intertwined with domestic responsibilities such as childcare, food processing and preparation, or any of the multitude of domestic tasks for which women are primarily responsible (Hays-Mitchell, 1993; Oberhauser, 1995). According to Gibson-Graham (1996), feminist attempts to retheorize and displace "the economy" have far-reaching implications that empha-size the diversity of household forms of economy and exploitation, while opening up the possibility of theorizing class diversity in the non-household sector. Their work draws upon feminist literature that:

...portrays the household as a site of production and distribution as well as consumption, in order to problematize the singular representation of "the economy" as a preeminently capit-alist formation located in the non-domestic sphere and unified by "the market."...[W]e are specifically (re)incorporating the feminized sphere of the household into the masculinized modern economy, acknowledging the household as an economic site rather than simply as a condition of existence of "the economy" more commonly understood (Gibson-Graham, 1996, p. 207).

Feminist research thus challenges the hegemony of capitalism in analyses of the global economy and instead focuses on the heterogeneity and plurality of economic forms.

Much of the research exploring economic activities that occur outside formal workplaces focuses on informal sector labor. Feminist analyses of this form of work have attempted to explain how gender and family ideologies underpin infor-mal labor (Beneria and Feldman, 1992; Brydon and Chant, 1989; Wilson, 1993). What is increasingly evident from this research is the crucial role of women's economic activities, both formal and informal, to households in developing countries, especially with economic reforms being implemented to reduce government spending on basic services and infrastructure that benefit low-income households (Hays-Mitchell, 1993; Faulkner and Lawson, 1991).

The final theme in this discussion of feminism in the global South addresses strategies of resistance and struggle among women, to empower themselves in the face of tremendous hardship and marginalization. Postcolonial feminism criticizes Western feminism for its ethnocentric biases and representation of women in the South as impoverished victims of oppression. This approach forwards a more political and geographically diverse conceptualization of women who oppose sexist, racist, and imperialist structures (Alexander and Mohanty, 1997; McClintock et al., 1997). Alexander and Mohanty (1997) emphasize the importance of different strategies of resistance in their call for a "feminist democracy," in which global processes clearly require global alliances. In this model, the common context of struggle among women of color leads to alliances against specific exploitative structures and systems. These are powerful images that defy modernist and conventional stereotypes and are pervasive themes in the gender and development literature.

Specific examples of both individual and collective struggles are evident in Third World feminist research. Ong's (1987) *Spirits of Resistance and Capitalist Discipline* outlines the struggles of rural Malaysian Muslim women who migrate to urban areas to work in the electronics industry. Struggles and resistance to neo-imperialist and Western structures also occur in rural, agricultural settings. Fieldwork by Carney and Watts (1991) in Senegambia revealed that intensification of agriculture during the last 50 years negatively impacted women's work and the internal structure of households. Consequently, gender-based struggles over property, labor, and conditions of work have become a critical part of agrarian change. Strategies by women in the South to resist oppression and to gain power also include informal income-generating activities that occur at the individual and household level.

Methodologies used in this research often entail the investigation of issues that are related to specific political objectives and are sensitive to uneven power relations among the subjects of research. Studies in the global South are often informed by goals and techniques that engender critical forms of engagement that directly involve the researcher. Consequently, intensive fieldwork and qualitative methods are common means of gathering information. For example, in an excellent study of female-headed households in remote Bolivian villages, Sage (1993) engages in intensive fieldwork to better understand women's involvement in multiple economic activities. Studies such as these reinforce the need to examine critically the internal structure and social relations of the domestic sphere in analyzing the gendered nature of economic strategies.

In sum, international perspectives have been integral to feminist analyses of the economy because they highlight cultural differences and geographical context. This discussion reinforces feminist thinking about redefining social and spatial categories, and exploring qualitative methodologies to research the social and spatial construction of gender and work. While the simultaneous tasks of reproductive and productive labor have been addressed in advanced industrial contexts, research on women in the South advances our understanding of the overlapping spheres of household and workplace. Finally, recent feminism has broadened the scope of understanding to recognize that women's work is a form of political action and struggle. Whether it is in the factory, the fields, or the urban market, women engage in collective and individual struggles for more control over their own and their households' well-being.

Gender and Identity in the Social and Cultural Practice of Work

In the early 1990s, feminist analyses in economic geography shifted from approaches emphasizing economic restructuring and gender divisions of labor to more detailed and nuanced examinations of the practices and cultural attributes that take place in economic interactions (McDowell, 1997a). Increased attention to social and cultural aspects of economic activities has coincided with expansion of the service-based economy and with the "cultural turn" in economic geography that emphasizes how institutions are influenced by societal cultural attributes and values. Research on the relationship between economic behavior and culturally constructed contexts is having a significant impact on contemporary economic geography (Lee and Wills, 1997). An important contributor to this recent body of literature is Linda McDowell (1997a) who explains this approach as a rejection of the concept of economic rationality, and a recognition that economic action is embedded in the social context and specific institutions within which it takes place.

The literature that examines culturally created economic processes is interdisciplinary, borrowing from economic sociology, cultural anthropology, geography, and the sociology of organizations. Geography brings an understanding of the ways in which spatial processes influence these practices, as well as linking culturally embedded economic practices to multiple scales – from individual bodies through to the workplace, the city, the region, and the globe. Consequently, it is not just firms or individuals that are involved in these cultural practices, but a myriad of local, regional, and international players and institutions. Several differentiated levels of embeddedness can, in turn, be mapped onto a hierarchy of places, spatial scales, or levels of analysis.

Embodied gender identities and economic practices

Contemporary feminist geography also analyzes how culturally embedded economic practices, workers, and organizations are embodied in dominant notions of masculinity and femininity (Halford and Savage, 1997; Leidner, 1993; McDowell, 1997a; McDowell and Court, 1994). This research highlights how workers and occupations are mutually constituted and extend to the physical appearance and makeup of the body. In other words, one cannot separate the cultural dimensions of economic behavior from the socially inscribed attitudes and behavior of men and women. Perceptions about workers' respective roles in society affect nearly all facets of their working lives, including the way they are recruited, the occupations they fill, and the promotions they are given. The concept of "performance" is often used in this literature to analyze how gender and sexuality are acted out in the workplace. McDowell, for example, links institutional interests and power relations to microscale social practices in her analysis of capital culture. She argues that:

socially sanctioned gendered identities and ways of behaving are reinforced and policed through a set of structures that keep in place dominant and subordinate social relations.

These structures or mechanisms include not only institutional force and sanctions from above but also self-surveillance (1997a, p. 31).

Thus economic behavior is influenced by dominant norms and culturally embedded social practices in society.

Conceptual arguments such as those outlined above are illustrated by several empirical studies on the socially embedded and embodied nature of economic behavior. Robin Leidner (1993) examines the importance of gender identity in what she calls routinized interactive work in the service sector. Here, employees' tasks are controlled and regulated to not only maximize efficiency, but also please clients. This routinization of work in interactive services challenges the cultural autonomy of the workers. Leidner's study, of the fast food chain McDonalds and an insurance company, addresses the somewhat contradictory position of workers who are expected to be amiable yet coercive with their customers. The gender identity of the employees in these companies is critical to how they approach the customers, their language, appearance, and attitude. For example, working at the take-out window at McDonalds is usually assigned to women because that job often requires accepting abuse calmly, whereas men are assigned to the grill because they are less afraid of getting burned (Leidner, 1993). This research illustrates the role of gender identity in workplace interactions and tasks that reflect and maintain dominant cultural norms of masculinity and femininity.

In another study, McDowell and Court (1994) examine merchant banking in London where the Big Bang of 1987 has shaped and impacted the gender and cultural characteristics of the workforce and workplace. Part of their argument is that the attitudes and behavior of workers are gendered in ways that are influenced by performance in the workplace. According to McDowell (1997a, p. 208),

... gendered identities and interactions in the working environment are, within bounds, fluid and negotiable. Men, as well as women, are able to construct differential performances in the workplace, while continuing to interpret them in ways congruent with hegemonic notions of masculinity.

Her analysis reinforces the notion that these behaviors are neither essentialized nor considered "natural," but are culturally embedded in particular social and economic contexts.

Another dimension of gender identity in the culture of firms and dynamics within the workplace is reflected in geographical research on sexuality. In a series of in-depth studies, Gill Valentine (1993, 1995) notes the predominance of heterosexuality in workplaces and public places in general. Consequently, many gays and lesbians are hesitant to reveal their sexual identities in workplaces for fear of retribution or harassment (Valentine, 1993). What might be considered benign practices and topics of conversation, such as displaying personal pictures, talking about weekend activities, and participating in social gatherings with work colleagues, tend to marginalize gays and lesbians within a dominantly heterosexual workplace. Thus, culturally embedded practices are constructed according to dominant sexual norms and behaviors in society.

In sum, analyses of gender identity in the workplace contribute to, and maintain, the cultural turn in economic geography by focusing on the embedded and embodied nature of economic processes in cultural contexts and institutions. These analyses coincide with economic restructuring and the transition in contemporary capitalism to service-based economies where the workplace environment and attitudes interact with social norms. Gender identity is an integral aspect of this cultural order of the economy, and enhances our understanding of economic restructuring in general.

Conclusion

This chapter demonstrates how feminism contributes to, and draws from, geographical analyses of the spatial dimensions of economic behavior. The discussion outlined feminist approaches to geography that emphasize the complexity of women's economic activities and the gendered nature of work. The socio-spatial dimensions of gender and economic strategies are important aspects of feminist research, and have affected economic geography in a variety of ways. Initially, discussions of gender relations and women's roles in society emphasized the position of women in the labor force and their experiences in the workplace. In addition, feminist perspectives on economic processes have moved beyond industrialized regions and now also include livelihood strategies from the South. Finally, more recent research on gender and work examines how economic behavior is embedded in the social construction of gender identity and dominant discourses of culture and place.

In general, feminism challenges conventional economic geography for neglecting women's experiences and gender roles in the economy. This is especially relevant to empiricist approaches to gender and work, in which careful attention to gender differences and comparison of work experiences yields important insights about ways in which relations of production and divisions of labor are gendered. In addition, feminism emphasizes the need to incorporate a variety of social constructs such as race, ethnicity, sexuality, and class in economic geographical processes. Postcolonialism contributes to our understanding of the diverse experiences of women by emphasizing the historical relations of power in Third-World contexts. This approach also provides a critical view of Western hegemonic discourses pertaining to work, capitalism, and the economy in general.

Finally, the chapter discussed important aspects of feminist methodology for formulating relevant questions and techniques in analyzing often discounted, undervalued, or informal aspects of economic activity. Analyses of household gender relations, heterosexual norms in the workplace, and collective economic strategies among women in the South require critical and innovative approaches that expose the social construction of gender and work. As McDowell (1997b, p. 119) states, "a new set of questions" is needed to "conceptualize both organizations and employees as actors with sets of cultural attributes which are constituted in, affected by and affect the huge range of (economic) interactions." This new conceptualization of the economic landscape requires a reconsideration of methodology used in conventional economic research. In sum, the tasks laid out by feminist economic geography are numerous and far-reaching. Although feminist geography did not emerge as a viable body of literature until the 1970s, it has contributed significantly to our understanding of the geography of economic processes.

Endnote

1. Many scholars and activists question the terminology used in analyses of women, gender, and feminism in developing regions. The terms "Third World" and "lesser developed" imply a hierarchy that is embedded in structures of Western thought and have certainly been adopted in conventional geography. In Mohanty's (1991) discussion about language and power, she uses women of color and Third World women interchangeably as a political constituency versus a geographical or biological common group. In this chapter I use several terms, but favor the South as a geographical reference to developing regions.

Bibliography

Alexander, M. J. and Mohanty, C. T. (eds) 1997. *Feminist Genealogies, Colonial Legacies, Democratic Futures*. New York: Routledge.

Amott, T. L. and Matthaei, J. A. 1991. *Race, Gender and Work: A Multicultural Economic History of Women in the United States*. Boston, MA: South End Press.

Beneria, L. and Feldman, S. (eds) 1992. *Unequal Burden: Economic Crises, Persistent Poverty, and Women's Work*. Boulder, CO: Westview Press.

Bowlby, S., Lewis, J., and McDowell, L. 1989. The geography of gender. In R. Peet and N. Thrift (eds), *New Models in Geography* (Vol. 2). London: Unwin Hyman, 157–75.

Brydon, L. and Chant, S. 1989. *Women in the Third World: Gender Issues in Rural and Urban Areas*. London: Edward Elgar.

Carney, J. and Watts, M. 1991. Disciplining women? Rice, mechanization, and the evolution of Mandinka gender relations in Senegambia. *Signs: Journal of Women in Culture and Society*, 16, 651–81.

Christopherson, S. 1989. Flexibility in the US service economy and the emerging spatial division of labor. *Transactions of the Institute of British Geographers*, 14, 131–43.

Cravey, A. 1997. The politics of reproduction: Households in the Mexican industrial transition. *Economic Geography*, 73, 166–86.

Dyck, I. 1989. Integrating home and wage workplace: Women's daily lives in a Canadian suburb. *The Canadian Geographer*, 33, 329–41.

Elson, D. 1992. From survival strategies to transformation strategies: Women's needs and structural adjustment. In L. Beneria and S. Feldman (eds), *Unequal Burden: Economic Crises, Persistent Poverty, and Women's Work*. Boulder, CO: Westview Press, 26–48.

Elson, D. and Pearson, R. 1989. *Women's Employment and Multinationals in Europe*. New York: Macmillan Press.

Faulkner, A. and Lawson, V. 1991. Employment versus empowerment: A case study of the nature of women's work in Ecuador. *The Journal of Development Studies*, 27, 4, 16–47.

Fernandez-Kelly, M. P. 1983. *For We Are Sold, I and My People: Women and Industry in Mexico's Frontier*. Albany: State University of New York Press.

Fernandez-Kelly, M. P. 1990. International development and industrial restructuring: The case of garment and electronics in Southern California. In W. Tabb and A. McEwan (eds), *Instability and Change in the World Economy*. New York: Monthly Review Press, 147–65.

Gibson-Graham, J. K. 1996. *The End of Capitalism (As We Knew It): A Feminist Critique of Political Economy*. Oxford: Blackwell.

Halford, S. and Savage, M. 1997. Rethinking restructuring: embodiment, agency and identity in organizational change. In R. Lee and J. Wills (eds), *Geographies of Economies*. London: Arnold, 108–17.

Hanson, S. and Hanson, P. 1981. The impacts of married women's employment on household travel patterns: A Swedish example. *Transportation*, 10, 165–83.

Hanson, S. and Johnston, I. 1985. Gender differences in work-trip length. *Urban Geography*, 3, 193–219.

Hanson, S. and Pratt, G. 1995. *Gender, Work, and Space*. London: Routledge.

Hayford, A. 1974. The geography of women: An historical introduction. *Antipode*, 6, 1–18.

Hays-Mitchell, M. 1993. The ties that bind: Informal and formal sector linkages in street vending: the case of Peru's *ambulantes*. *Environment and Planning A*, 25, 1085–1102.

Jacobs, E. (ed) 1997. *Handbook of U.S. Labor Statistics: Employment, Earnings, Prices, Productivity, and Other Labor Data*. Lanham, MD: Bernan Press.

Johnston-Anumonwo, I. 1995. Racial differences in the commuting behavior of women in Buffalo, 1980–1990. *Urban Geography*, 16, 23–45.

Lee, R. and Wills, J. (eds) 1997. *Geographies of Economies*. London: Arnold.

Leidner, R. 1993. *Fast Food, Fast Talk: Service Work and the Routinization of Everyday Life*. Berkeley, CA: University of California Press.

Lewis, J. 1983. Women, work, and regional development. *Northern Economic Review*, 7, 10–24.

Mackenzie, S. 1989. Women in the city. In R. Peet and N. Thrift (eds), *New Models in Geography* (Vol. 2). Boston: Unwin Hyman, 109–26.

Mackenzie, S. and Rose, D. 1983. Industrial change, the domestic economy and home life. In J. Anderson, S. Duncan, and R. Hudson (eds), *Redundant Spaces in Cities and Regions? Studies in Industrial Decline and Social Change*. London: Academic Press, 155–99.

Massey, D. 1984. *Spatial Divisions of Labour*. London: Macmillan.

Massey, D. 1997. Economic/non-economic. In J. Wills and R. Lee (eds), *Geographies of Economies*. London: Arnold, 27–36.

Mazey, M. E. and Lee, D. R. 1983. *Her Space, Her Place: A Geography of Women*. Washington, D.C.: Association of American Geographers.

McClintock, A., Mufti, A., and Shohat, E. (eds) 1997. *Dangerous Liaisons: Gender, Nation, and Postcolonial Perspectives*. Minneapolis, MN: University of Minnesota Press.

McDowell, L. 1991. Life without father and Ford: The new gender order of post-Fordism. *Transactions of the Institute of British Geographers*, 16, 400–19.

McDowell, L. 1997a. *Capital Culture: Gender at Work in the City*. Oxford: Blackwell.

McDowell, L. 1997b. A tale of two cities? Embedded organizations and embodied workers in the city of London. In J. Wills and R. Lee (eds), *Geographies of Economies*. London: Arnold, 118–29.

McDowell, L. and Court, G. 1994. Missing subjects: Gender, power and sexuality in merchant banking. *Economic Geography*, 70, 229–51.

McDowell, L. and Massey, D. 1984. A woman's place? In D. Massey and J. Allen (eds), *Geography Matters! A Reader*. Cambridge: Cambridge University Press, 128–47.

McLafferty, S. L. and Preston, V. 1991. Gender, race and commuting among service sector workers. *The Professional Geographer*, 43, 1–15.

Miller, R. 1982. Household activity patterns in nineteenth-century suburbs: A time-geographic exploration. *Annals of the Association of American Geographers*, 72, 355–71.

Mohanty, C. 1991. Cartographies of struggle: Third world women and the politics of feminism. In C. J. Mohanty, A. Russo, and L. Torres (eds), *Third World Women and the Politics of Feminism*. Bloomington, IN: Indiana University Press, 1–47.

Momsen, J. H. and Townsend, J. (eds) 1987. *The Geography of Gender in the Third World*. Albany: State University of New York Press.

Monk, J. and Hanson, S. 1982. On not excluding half of the human in human geography. *The Professional Geographer*, 34, 11–23.

Oberhauser, A. M. 1995. Gender and household economic strategies in rural Appalachia. *Gender, Place and Culture* 2, 51–70.

Ong, A. 1987. *Spirits of Resistance and Capitalist Discipline: Factory Women in Malaysia.* Albany: State University of New York Press.

Peet, R. and Thrift, N. (eds) 1989. *New Models in Geography: The Political Economy Perspective.* London: Unwin Hyman.

Preston, V., McLafferty, S., and Hamilton, E. 1993. The impact of family status on black, white, and Hispanic women's commuting. *Urban Geography,* 14, 228–50.

Professional Geographer. 1995. Should women count? The role of quantitative methodology in feminist geographic research, 47, 4, 426–66.

Rose, G. 1993. *Feminism and Geography: The Limits of Geographical Knowledge.* Cambridge: Polity Press.

Sage, C. 1993. Deconstructing the household: Women's roles under commodity relations in highland Bolivia. In J. H. Momsen and V. Kinnaird (eds), *Different Places, Different Voices: Gender and Development in Africa, Asia, and Latin America.* London: Routledge, 243–55.

Seager, J. 1997. *The State of Women in the World Atlas.* New York: Penguin.

Tinker, I. (ed) 1990. *Persistent Inequalities: Women and World Development.* New York: Oxford University Press.

Tivers, J. 1978. How the other half lives: The geographical study of women. *Area,* 10, 302–6.

US Department of Labor 1998. Bureau of Labor Statistics, Women's Bureau. Washington D.C.: U.S. Department of Labor.

Valentine, G. 1993. (Hetero)sexing space: Lesbian perceptions and experiences of everyday spaces. *Environment and Planning D: Society and Space,* 11, 395–413.

Valentine, G. 1995. Out and about: Geographies of lesbian landscapes. *International Journal of Urban and Regional Research,* 19, 96–111.

Walby, S. and Bagguley, P. 1989. Gender restructuring: Five labor-markets compared. *Environment and Planning D: Society and Space,* 7, 277–92.

Wilson, F. 1993. Workshops as domestic domains: Reflections on small-scale industry in Mexico. *World Development,* 21, 67–80.

Wright, M. 1997. Crossing the factory frontier: Gender, place, and power in the Mexican maquiladora. *Antipode,* 29, 278–302.

Chapter 6

Institutional Approaches in Economic Geography

Ron Martin

Introduction: The "New Institutionalism" in Economic Geography

Over the past decade, economic geography has undergone something of a renaissance, expanding its theoretical foundations, methodologies, and empirical reach considerably in the process (see, for example, Lee and Wills, 1997; Bryson et al., 1999; Clark et al., 2000). One of the key elements in this expansion and re-orientation has been what might be called the "institutional turn," the recognition that the form and evolution of the economic landscape cannot be fully understood without giving due attention to the various social institutions on which economic activity depends and through which it is shaped.

This "institutional turn" derives from various sources. In part, it has spiraled out of the widespread adoption by economic geographers of French regulation theory, especially the latter's emphasis on the "mode of social regulation;" the ensemble of rules, customs, norms, conventions, and interventions which mediate and support economic production, accumulation, and consumption.[1] Despite the fact that regulation theory itself fails to provide any detailed conceptual account of the nature and evolution of those social frameworks (beyond some broad discussions of the role of the state), it has nevertheless focused geographers' attention on the nature of the socio-institutional structure as an "essential underpinning of efficient capitalist production" (Storper and Walker, 1989, p. 5).

A second catalyst has been the growing recognition of the "socio-cultural" within economic geography. Traditionally, economic geographers have tended to define their subject matter in terms of what they see as the "purely economic," and to exclude social and cultural factors from their analyses, either on grounds of relative unimportance or because of the desire to demarcate their subfield from other specialisms within human geography. But it is now fast becoming recognized that, however artfully one carves the intra-disciplinary joints, as it were, the "purely economic" is a contrived construct. The economic process is also a socio-cultural process (see Thrift and Olds, 1996; Crang, 1997), and institutions are central to the socio-cultural construction of the economic.

A third influence stems from developments elsewhere in the social sciences, especially in economics, sociology, and political science. Over the past two decades, these disciplines have themselves experienced a growth in "institutionalism." In economics, it is now acknowledged that institutional factors play a key role at all levels in the economy, from the structure and functions of the firm, through the operation of markets, to the form of state intervention (see Williamson, 1985; Hodgson, 1988; North, 1990). At the same time, by reinterpreting economic action as social action, sociologists have similarly highlighted the pervasiveness and importance of social institutions in economic life (for example, Granovetter, 1985; Zukin and DiMaggio, 1990; Granovetter and Swedberg, 1992). And for their part, political scientists are increasingly emphasizing the institutional organization of the polity, and how national differences in institutional organization influence political, social, and economic outcomes (see March and Olsen, 1989; Steinmo et al., 1992). In all three instances, institutionalist perspectives have developed as a reaction to the behavioral approaches that became prominent in the 1960s and early 1970s. Given the increasing "blurring of the boundaries" between the social sciences, it is perhaps not surprising that this wave of institutionalism should have begun to permeate economic geography.

Fourthly, without doubt the rise of the institutional turn in economic geography has also been stimulated by the upheavals that have actually been underway over the past two decades in the institutions of capitalism. In regulation theory parlance, the institutional forms that underpinned and regulated the post-World War II "Fordist" regime of economic accumulation have become increasingly ineffectual in the face of the shift to a new regime of "post-Fordist" economic accumulation. In every capitalist nation, the old institutional frameworks are being abandoned as economic organizations, social groups, and states themselves search for new institutional configurations more congruent with the markedly different, and still rapidly changing, economic conditions of "post-Fordism." The institutional landscapes of capitalism are being redrawn and, rightly, geographers have become closely interested in the nature and implications of this process.

For all these reasons, then, "institutions" are firmly on the research agenda in economic geography. This is not to suggest, however, that a fully articulated "institutionalist economic geography" has yet emerged. To the contrary, the field is still very much in its infancy. And, as with any new approach or perspective, the novelty of new ideas, even when set against the omissions and weaknesses of old approaches, is not itself sufficient to proclaim an advance in understanding. Institutionalist terminology may have become widespread within economic geography, but many of the terms and concepts used, such as "institution," "institutional thickness," "social embeddedness," "networks," and "governance," have yet to acquire commonly agreed definitions.[2] This is not to devalue the significance of the "institutionalist turn" in economic geography, but to point up the fact that it is still in its early stages. It is, however, a movement that seems certain to grow in importance over the coming years. My aim in this chapter is to provide a (necessarily selective) map of this emerging terrain of institutionalist economic geography, bearing in mind that it is a terrain that is still being explored.

Delimiting Institutionalist Economic Geography

What, then, do we mean by an institutional approach to economic geography? Conventionally, economic geography has tended to abstract economic action from its social, political, and cultural contexts (Martin, 1994; Sunley, 1996). In the standard location-theoretic models of industrial geography, for example, economic action is atomistic, rational, and maximizing; socio-political context is "held constant," that is, ignored. Even though Marxist economic geography emphasizes the class relations inherent in capitalist accumulation, here too the theoretical framework remains firmly economistic, in that socio-political and cultural structures are assumed to be determined by, rather than being determinants of or reflexively interrelated with, the economic process. The fact is, however, that all economic action is a form of social action and cannot be separated from questions of status, sociability, and power (see also Mitchell, this volume). In other words, economic activity is *socially and institutionally situated*: it cannot be explained by reference to atomistic individual motives alone, but has to be understood as enmeshed in wider *structures of social, economic, and political rules, procedures and conventions*. It is the role of these systems of rules, procedures, and conventions, both of a formal and informal nature, that is the focus of an institutionalist approach to economic geography.[3]

More specifically, we may define an institutionalist approach to economic geography as an attempt to illuminate the following basic question: *to what extent and in what ways are the processes of geographically uneven capitalist economic development shaped and mediated by the institutional structures in and through which those processes take place?* Geographers have long highlighted the ways in which the forces of capitalist economic development – competition, the drive to accumulate profits, the evolution of technology and labor processes, the tendency towards the concentration and centralization of capital – generate spatially differentiated outcomes, in terms of growth, prosperity, and employment. Those forces in turn both forge and are forged by a complex matrix of institutions. Thus to understand the capitalist space economy, geographers need to understand the role, impact, and evolution of the institutions of capitalism. The implicit assumption is that although institutions are unlikely to be the sole "cause" of geographically uneven development, they enable, constrain, and refract economic development in spatially differentiated ways. An institutional approach to economic geography does not, therefore, deny the forces and mechanisms that animate our various theories of geographically uneven development. Instead it seeks to uncover the ways in which institutions shape these forces from place to place, and in so doing influence their outcomes in different places.

Institutional economic geography, as it is currently emerging, has several key themes. First, it seeks to identify *the role of the different sorts of institution* at work in shaping the space economy. Economists often distinguish between the "institutional environment" and "institutional arrangements" (see North, 1990; Rutherford, 1994). The "institutional environment" refers to both the systems of informal conventions, customs, norms, and social routines (such as habitual forms of corporate behavior, consumption cultures, socialized work practices, transaction norms, and so on), and the formal (usually legally enforced) structures of rules and regulations (for example, laws relating to competition, employment, contract, trade, money flows, corporate governance, welfare provision) which constrain and control socioeconomic behavior. The term "institutional arrangements" is used to denote the particular organizational forms (such as markets, firms, labor unions, city councils, regulatory agencies, the welfare state) which arise as a consequence of, and whose constitution and operation are governed by, the institutional environment. Institutional economic geography is concerned with both of these aspects of the "institutional regime" of the economy, and especially the interaction between them. What economic organizations come into existence, and how they function and evolve, are fundamentally influenced by the institutional environment. But equally, in the course of their operation, institutional arrangements (economic-political organizations) not only reproduce but also modify the institutional environment. How this interaction varies across space, and how it shapes local economic outcomes, are central issues of concern in institutionalist approaches to economic geography.

Second, an institutional approach to economic geography emphasizes the *evolution of the economic landscape*. By their very nature institutions are characterized by inertia and durability – these are precisely the features that provide the structured frameworks necessary for the coordination and continuity of economic activity, and which make complex exchange possible across both time and space. Institutions are characterized by "path dependence," that is they tend to evolve incrementally in a self-reproducing and continuity-preserving way (see North, 1990; David, 1994; Setterfield, 1997). As such, institutions are therefore important "carriers of history," in that they serve to impart path dependence to the process of economic development, for example by functioning as a source of increasing returns and positive externalities.[4] As the economic historian North expresses it,

Institutions . . . connect the past with the present and the future, so that history is a largely incremental story of institutional evolution in which the historical performance of economies can only be understood as a part of a sequential story (1990, p. 118).

It is at the regional and local levels that the effects of institutional path dependence are particularly significant. Institutions are important "carriers" of *local* economic histories. Different specific institutional regimes develop in different places, and these then interact with local economic activity in a mutually reinforcing way. If institutional path dependence matters, it matters in different ways in different places: institutional–economic path dependence is itself place-dependent.

Allied to this theme, institutional economic geography also puts great stress on the *role of technological innovation* in shaping the changing fortunes of regional and local economies. Drawing on elements of evolutionary economics and Schumpeterian economics (Magnusson and Ottoson, 1997; Reijnders, 1997), economic geographers increasingly point to, and seek to explain the factors underpinning, the locally based nature of technological change and learning (Simmie, 1997). Technological change is not, as in mainstream economic models, some exogenous disembodied process, but an inherently socio-cultural activity dependent on the institutional setting within which it takes place. Geographers have begun to show

how technological innovation appears to be more easily promoted by certain local institutional environments and arrangements than by others. Certain forms of local institutional regime or "milieu" (such as the presence of a well-developed enterprise culture, supportive regulatory and promotional agencies, research-orientated universities, and locally committed financial structures) appear especially to facilitate the emergence and development of clusters of technologically based activity (Camagni, 1991; *Regional Studies*, 1999). Once established, such local technological clusters in turn generate further specialized local institutional systems, which in their turn foster further high-technology-based economic development and spillover amongst local firms. Uncovering the spatio-institutional foundations of technological innovation and diffusion is currently one of the most prominent research foci in contemporary economic geography.

Equally, an institutionalist approach to economic geography also encompasses consideration of the *cultural foundations* of the space economy. Economic geographers point to the role of cultural processes in the formation of social structures and individual identities, consumption norms, and lifestyles, all of which may influence the formation and nature of informal conventions, constraints, and norms which impact on local and regional economic development. At the same time, cultural processes, to the extent that they serve to transmit knowledge, attitudes, and values from one generation to the next, are a key factor in determining the path-dependent nature of institutional development. Cultural processes operate at various geographical levels, from the global level through the national to the local, all of which have spatially-differentiated economic effects. The rise of (Fordist) mass production–mass consumption industrial society, for example, created its own set of geographically differentiated social institutions, not least among them region-specific work traditions, union structures, and political cultures. Similarly, contemporary high-technology regions and districts are characterized by distinctive socio-cultural identities, lifestyles, and attitudes.

Finally, and critically, institutionalist economic geography focuses on the *social regulation and governance* of regional and local economies. In regulation theory, the term "regulation" refers to a historically-specific institutional regime for coordinating, stabilizing, and reproducing socioeconomic relations. Three main macro-institutional forms are singled out as being of fundamental importance: the form of monetary management, the configuration of the wage relation (that is, the socio-technical division of labor and the social reproduction of the labor force), and the form of competition (see Boyer, 1990). Geographers have extended this framework to include not only aspects of the legal–economic nexus, the apparatus and agencies of the state (both central and local), business institutions (such as employers' associations, chambers of commerce, and so on), and labor organizations (especially unions), but also the plethora of locally and regionally specific institutions of an informal nature, including local social networks, cultures, and traditions. Complex national–local institutional arrangements such as the welfare system, the industrial relations system, and the financial system, are regarded as particularly significant elements of the institutional regulatory framework of the space economy. Further, whereas under regulation theory the geographical unit of regulation is the national economic space, economic geographers also focus on local and regional aspects of regulation, both because national (and supra-national) mechanisms of socioeconomic regulation have local – and locally varying – impacts, and because regulation has a locally-specific micro-level dimension, reflecting the spatial diversity and specificity of local institutional structures.

These themes highlight the emerging richness and diversity of institutionalist economic geography. For institutionalist economic geographers, the economic landscape is more than the "market." Likewise, it is more than some "inner logic" of capitalist accumulation. It is an ongoing socially and legally instituted process with elements that co-evolve through complex forces of political intervention, cultural development, technological innovation, and path-dependence. Institutionalist economic geography is not just about the study of particular institutions and their role in shaping and regulating the spatial dynamics of capitalism, it is also a distinctive *way of thinking about* the space economy and its evolution. But herein lies the challenge: How should such an institutionalist perspective be conceptualized?

Conceptual Frameworks for Institutional Economic Geography

As Samuels (1995) points out in his survey of institutional economics, a distinctive feature – indeed, as he sees it, a key advantage – of institutionalism is its multi-perspectival, multidisciplinary character. Three main conceptual approaches can be distinguished: rational choice institutionalism, sociological institutionalism, and historical (evolutionary) institutionalism (see table 6.1).[5] These provide somewhat different interpretations of institutions, their functions and mode of change.

In the rational choice perspective, the focus is on how particular institutional environments give rise to particular institutional arrangements, that is, organizational forms, and how far and in what ways institutions serve to reduce transaction costs and increase economic efficiency. Institutions are seen as the outcome of market behavior, constantly changing through a process of "competitive selection" in response to shifts in relative prices and transactions costs (Eggertson, 1990; North, 1990). By contrast, the sociological model seeks to understand the economy as a socio-institutionally "embedded" system (Granovetter, 1985; Zukin and DiMaggio, 1990; Swedberg, 1997). Institutions are interpreted as culturally based social repertoires, routines, and networks of trust, cooperation, obligation, and authority. As such, they provide cognitive frameworks or templates of meaning through which economic identities and action are legitimized. Sociological institutionalists see institutional evolution as arising out of collective processes of interpretation, and emphasis is put on the ways in which existing institutions structure and circumscribe the range of institutional change and creation. In its turn, the historical institutionalist approach focuses on understanding how institutional structures evolve over time, and how this evolution impacts on and relates to the historical dynamics of the capitalist economy. Institutions are seen as the products of historically-situated interactions, conflicts, and negotiations amongst different socioeconomic actors and groups. Stress is put on the asymmetric power relations associated with institutions, and on major periods of transformation in the institutions of economic regulation and governance.

Economic geographers have drawn on all three conceptual frameworks. Thus the idea that institutions function to reduce transaction costs has been used to help explain the emergence and development of successful local and regional economies.

Table 6.1 Alternative approaches to institutional analysis and their application in economic geography

Perspective	Main focus	View of institutions	Theoretical basis	Account of institutional change	Geographical applications
Rational choice institutionalism	Understanding how institutions generate particular organizational forms under capitalism.	Institutions structure individual actions through constraint, information, or enforcement. Institutions judged according to whether they reduce transactions costs and increase economic efficiency.	Transaction costs economics, agency theory, contract theory, property rights.	Constantly changing as outcome of market behavior (relative price changes and changes in transaction costs). Evolutionary trajectory determined by competitive selection.	Spatial agglomeration and localization of economic activity creates specialized institutions which lower transaction costs.
Sociological institutionalism	Understanding the economy as a socio-institutionally embedded system.	Institutions as culturally specific social networks of trust, reflexive co-operation and obligation which underpin economic behavior and relationships.	Network theory (institutions as congealed networks), organization theory, group theory and cultural theory.	Institutional change as process of social construction around new logics of social legitimacy or new shared cognitive maps.	The role of locally specific formal and informal networks of trust, cooperation, and knowledge transfer ("untraded interdependencies") in fostering the local embeddedness of firms.
Historical (evolutionary) institutionalism	Understanding the role of institutional evolution in the historical dynamics of the capitalist economy.	Institutions as systems of social, economic, and political power relations, which frame the regulation and coordination of economic activity.	Eclectic, drawing on a range of heterodox frameworks, including post-Keynesian, and evolutionary economics, regulation theory, long-wave theory, and comparative politics.	Durable over long periods, built up through slow accretion, and subject to hysteretic path dependence and lock-in. Long-run evolution is episodic as result of interaction with economic development.	The nature and evolution of local institutional regimes and their role in the social regulation and governance of local economies.

Access to information is the key to the costs of transacting (the costs of locating suppliers and customers, of serving markets, of determining values and prices, and of protecting property rights and enforcing contracts and agreements). By providing sources of information about markets, prices, finance, technologies, labor, etc., by codifying the multifarious histories of past transactions and business practices, and by providing a normative framework for conducting exchange and trade, institutions help reduce costs, and hence improve efficiency, competitiveness, and profitability. Whilst many institutions are national (and even international) in scope, geographical variations in the range and type of local institutions appear to be especially important in shaping the relative performance of local and regional economies. In particular, the spatial localization of an industry or of interrelated industries in an area tends to promote the formation of specialized local institutional structures (such as local business associations and consortia, research bodies, marketing agencies, employment and training agencies, and so on), all of which reduce transaction costs and increase the competitiveness of the local economy.

However, while economic geographers now attribute considerable importance to local institutions as a source of regional competitive advantage and dynamism, they do not necessarily subscribe to the rational choice theoretical framework that underpins transactions cost institutionalist economics.[6] Instead, they have increasingly drawn on the idea of "embeddedness" from sociological institutionalism. According to Granovetter (1985, 1993), a leading exponent of the concept of embeddedness, economic institutions are constructed through the mobilization of resources through social networks, conducted against a background of constraints given by the previous historical development of society, polity, and technology. In effect, institutions are "congealed" social networks. Economic activity is said to be "embedded" in these ongoing social institutions or networks to the extent that it depends on interaction with other agents in those networks. The socio-institutional embeddedness of the capitalist economy permits actors to circumvent the limits of pure rationality and the interactions of anonymous markets. Thus, for example, product and credit markets can exist because they are based on trust in the fulfillment of future transactions. But trust is not easily manufactured. When and where economic activity is tightly connected with dense social relationships and networks based on family, tradition, skills, political culture, religion, educational ties, etc., trust can be used to build useful and efficient economic institutions. Considerable evidence now exists demonstrating that many organizations, public and private, operate effectively precisely because they are embedded in, and incorporate, these social networks.

Within economic geography, the "embeddedness hypothesis" argues that trust, reciprocity, cooperation, and convention have a key role to play in successful regional development (see Storper and Salais, 1997). These social relations require community-like structures, associations, and networks for their existence. Suitably institutionalized, that is subsumed within local business and work cultures and routines, these social relationships serve to reduce uncertainty amongst local economic actors by providing tacit or collective knowledge, by disseminating strategic information, by sharing risks, and by establishing acceptable norms of competition. Yet, although most economic geographers would agree on the importance of such localized interactions among economic actors for successful economic development, the concept of embeddedness nevertheless remains under-theorized. In sociology, it is now recognized that Granovetter's original notion of embeddedness in social networks is only one type of embeddedness (Swedberg, 1997). There is also "political embeddedness" or the fact that economic action is always set within a specific context of political structures and conflict. Then there is "cognitive embeddedness," which has to do with the systems of meaning within which economic activity is conducted. And, finally, there is "cultural embeddedness," the embeddedness of economic action within specific cultural systems of ideological and normative beliefs. Likewise, different types of regional economic development tend to foster and depend upon different types of local institutional embeddedness. Thus the sort of socio-institutional embeddedness that develops in an area based on highly specialized, large-employer-dominated activity (say, a coal or steel community) will be very different from the pattern that is likely to emerge in an area of small-firm-based, flexible specialization (of the sort found in many of the Italian industrial districts, for example; Amin, this volume). At present we know little about how and why the process of embeddedness differs between different regional and local economies.

Furthermore, while the idea of the embeddedness of economic activity in local institutional structures provides some important insights into the differential patterns of economic activity and performance across the space economy, it says relatively little about the nature of institutional change and how that change influences the evolution of the economic landscape. How are local institutional structures built up, and how do they change? How are new institutional structures created? What happens if institutions prove resistant to change? What are the implications of these processes for the regulation and embeddedness of local economic activity? Historical institutionalism provides some concepts and ideas for addressing these questions.

Two of the weaknesses of old historical institutionalism (of the original Veblenian tradition) were its "structuralism," that is, its prioritization of institutional structures over action (human agency), and the consequential tendency to attribute the existence of institutions to their history, as if this in and of itself explains their origins and persistence. Recently, however, some new conceptualizations have appeared which go beyond this limitation, concerning in particular the causes and consequences of periodic transformations in institutional structures, and the tendency for some institutions to persist, despite being no longer conformable with or conducive to economic development. Incremental evolution is only one way in which institutional change occurs. Much more significant is the tendency for institutional structures to undergo major episodic reconfigurations. These periodic upheavals can arise because institutions themselves ossify and become "sclerotic," thereby hindering economic growth and development (see Hodgson, 1989).[7] Eventually, pressure will accumulate to reform or abandon particular elements of the institutional environment or particular institutional arrangements. Alternatively, economic developments may "outgrow" particular institutions, rendering them obsolete or inefficient, again stimulating the search for new institutional forms and structures more appropriate to the new economic conditions (or at least as these are perceived). Institutional evolution thus consists of periods of relative stability (or slow change), punctuated historically by phases of major transformation.

Setterfield (1997) has combined both of these aspects of institutional evolution – slow, incremental path-dependent change, and occasional historic transformations –

in his theory of "institutional hysteresis." This approach focuses on the complex interaction between institutions and economic activity in a way that recognizes the importance of current behavior in shaping future institutions, but which at the same time takes account of the extent to which this behavior is constrained by pre-existing institutional structures. The distribution of power amongst socioeconomic actors and groups is central to the nature and outcome of this interaction. Economic pressures for institutional change arise continuously. Provided these pressures are not major or crisis-making in nature, they are counterbalanced by the forces of institutional durability and inertia. But, eventually, when economic trends build up to the point where existing institutional structures prove dysfunctional to economic growth and development, then those structures will come under intense scrutiny, and economic and political agents will search for a new institutional fix. The process of transformation and its particular outcome may be consensual in nature, but is much more likely to involve conflict between different power groups (especially capital and labor) and between old institutions and new. Some old, inefficient institutions may persist, and even new institutional structures are likely to be based – at least to some extent – on pre-existing ones. By institutional hysteresis, then, is meant a process in which current institutions influence the nature of current economic activity, which in turn influences subsequent institutional forms. Integral to this model is a view of institutional evolution that foresees periods of institutional stability (deriving from the short-run exogeneity of institutions) punctuated by phases of substantial institutional change (reflecting the long-run endogeneity of institutional structures to the economic system).

The task for economic geographers is to conceptualize the spatial dimensions of this hysteretic process, not simply in order to determine whether and in what ways institutional change may have different effects on different regional and local economies, but also to determine how far and for what reasons the process of institutional change itself is likely to vary geographically. In this context, one of the notions that economic geographers have sought to explore is that of "regional lock-in," the situation in which a local institutional regime gets "stuck in a groove." Because of a particularly strong degree of "interrelatedness" (mutual interdependency, entrenched vested interests, etc.) within and between its constituent elements, a local institutional regime may resist change, and in consequence may hinder economic development. As Grabher (1993) highlights in the case of old industrial regions (he cites the example of the Ruhr in Germany), strong local ties to once favorable but now outmoded institutions can become a major structural weakness, contributing to economic sclerosis and decline, and holding back the much-needed process of economic restructuring and renewal. In other areas, by contrast, typically those less dependent on old, heavy, specialized industry, local institutional structures may be much more "flexible," and hence adapt more readily and painlessly to new economic circumstances. In other words, different forms of regional development may give rise to regional differences in institutional hysteresis. Institutional transformation thus has a specifically local dimension.

As yet, however, local processes of institutional transformation remain poorly theorized. One of the key agents of institutional change, of course, is the state. State-led institutional change can take various forms, involving the reform of the legal–regulatory environment of economic activity (for example, changing the nature of competition law, employment law, etc.); major shifts in policy programs (such as changes in industrial policy, or monetary policy); changing the legislative framework governing the form and operation of institutional arrangements and other economic organizations (such as corporations, and labor unions); and reconfiguring the regulatory structures and apparatus of the state itself (for example, setting up new regulatory institutions, and changing the division of policy responsibilities between central and local government).

Unraveling the specific effects of these different forms of state-led institutional change on individual regions and localities is a complex issue. The pressure for state-instigated institutional change may itself have particular geographical origins. Thus, in some instances the impetus may be the lobbying by business or labor in particular regions, especially those where political support for the government is concentrated. In other instances, it may derive from an ideological assault by the government on what are seen as outmoded or dysfunctional institutional structures in certain regions, structures that are deemed to be impeding not only the economic prospects of the regions in question but also the economy as a whole. What is almost certain is that, because of the regionally differentiated nature of the national economy, state attempts to change institutional structures are likely to generate conflict between regions and the political center. In the UK during the 1980s, for example, the Thatcher government's strident recasting of employment law, curbing of union power, and privatization of public sector activities found favor in the more prosperous, less unionized and pro-Tory south and east areas, closely linked into and benefiting from the dominant economic institutions centered in London, but were greeted with hostility in the less prosperous, highly unionized, and pro-Labour north. Such conflicts between central state and regions over institutional reform may influence the eventual path that such reforms take.

"Institutional Thickness" and Regional Development

In the course of applying these various theoretical perspectives on institutions, economic geographers have sought to construct their own, specifically spatial concepts. Two of these stand out, namely the notions of "institutional spaces" and "institutional thickness." By "institutional space" is meant the specific geographical area over which a given institution is constituted and has effective reach or influence. On the one hand, we can define a hierarchy of institutional spaces ranging from supra-national institutional spaces (such as internationally agreed rules of competition, trade, and monetary relations), through national institutional spaces (such as each nation's welfare system, its financial system, its employment laws, property laws, and national labor unions), some of which may incorporate sub-nationally differentiated structures, to regional and local institutional spaces (such as local state structures, locally specific legal arrangements, local employers' associations, local labor union cultures, and other local social or economic traditions and conventions). The "nestedness" – that is, the combination, interaction and mode of articulation – of these institutional spaces varies from place to place.[8] Thus, as we move from one area to another within a national economic space, not only may the specifics of institutional nestedness change but also the interactions within that ensemble. It is in this sense that we may speak of different "local institutional regimes."

National-level differences in institutional regimes are now widely acknowledged to be a key factor in explaining national differences in economic organization, development, and growth dynamics (see, for example, Berger and Dore, 1996; Rogers Hollingsworth and Boyer, 1997). There is not one model of "capitalism," but many national models. In each nation, capitalism is embedded in different nationally specific institutional structures – differences that make "US capitalism" different from "UK capitalism," and these in turn different from "Japanese capitalism," and so on.[9] Differences in national institutional regimes underpin the distinctions economic geographers make between different national versions of post-war Fordism (see, for example, Tickell and Peck, 1995). But, equally, important differences in institutional regimes also exist between regions and even localities within national economies. Economic geographers have attempted to capture these differences in their use of the concept of "institutional thickness."

Amin and Thrift (1995) define "institutional thickness" in terms of four key constitutive elements. The first is a strong institutional presence, in the form of institutional arrangements (firms, local authorities, chambers of commerce and other business associations, financial institutions, development agencies, labor unions, research and innovation centers, and various voluntary bodies). The second is a high level of interaction amongst these institutions so as to facilitate mutual and reflexive networking, cooperation, and exchange, thereby producing a significant degree of mutual isomorphism amongst the ensemble of local institutional arrangements. Thirdly, institutional thickness depends on there being well-defined structures of domination, coalition-building, and collective representation in order to minimize sectionalism and inter-institutional conflict. This leads, fourthly, to the notion of inclusiveness and collective mobilization, that is, the emergence of a common sense of purpose around a widely-held agenda, or project, of regional or local socioeconomic development. The specific local combinations of these elements define the degree and nature of local "institutional thickness" or "integrity"

...which both establishes legitimacy and nourishes relations of trust...[and] which continues to stimulate entrepreneurship and consolidate the local embeddedness of industry... [W]hat is of most significance here is not the presence of a network of institutions per se, but rather the process of institutionalization; that is, the institutionalizing processes that both underpin and stimulate a diffused entrepreneurship – a recognized set of codes of conduct, supports and practices which certain individuals can dip into with relative ease (Amin and Thrift, 1995, pp. 102–3).

In this way, by seeking to define regions and localities in terms of situated institutional capacities, the concept of "local institutional thickness" provides a framework for underpinning the idea of "locality as agent" (Cox and Mair, 1991).

However, although now widely invoked by economic geographers, the concept of "local institutional thickness" is not without problems and limitations. Although highly suggestive, the term still lacks definitional and theoretical precision (even in the work of Amin and Thrift). Neither case-by-case examples nor somewhat tautological definitions are substitutes for a general conceptualization of how "institutional thickness" emerges at the regional and local level and what precise role it plays in regional economic development. If the key process is the mobilization and alliance of local institutions around a common agenda, what stimulates this mobilization, and who sets the agenda? Does local institutional thickness depend on consensus building and negotiation amongst diverse local interest groups and institutions, or is it most likely to develop and work best where there is a dominant elite or leading interest group? In particular, why does "institutional thickness" develop in some areas but not others? And how, exactly, does local institutional thickness promote local economic development and prosperity? Further, how do we explain the fact that some very successful regions do not appear to be underpinned by a highly integrated and coordinated matrix of institutions? An additional issue is that by focusing on specifically local institutions, the concept tends to downplay the significant role of the central state, both as a multi-scaled set of institutional forms in its own right and as a central influence on the formation and functions of other, national and local, institutions. As noted above, no matter how committed local institutions are to a particular regional agenda, if the latter runs counter to national-level economic policies and objectives its chances of being implemented and succeeding are very much reduced. There is a real need to address these various issues, since the notion of "institutional thickness" has rapidly moved from being an analytical concept to a prescriptive one, in the sense of being seen as a necessary prerequisite to regional restructuring and regeneration, and particularly as a policy tool for stimulating new regions of high technology and innovation-based industrial development.

Contemporary Institutional Transformations and the Space Economy

This increasing focus on regional and local institution-building is especially apt at the present time. At all geographical scales, from the international to the national and the local, formal and informal institutions are being challenged, abandoned, and reconfigured. We are witnessing an historical wave of institutional change and experimentation, and, as part of this process, a spatial rescaling of institutions and their role in economic coordination, regulation, and governance. While regulation theorists see this process as marking a shift from a Fordist institutional framework to a new, post-Fordist institutional system, American "social structure of accumulation" theorists see it more as marking the passage from one institutional "long wave" (life cycle) to another. Still others see it as part of an historic move to a new "reflexive modernization" phase of capitalism (Beck et al., 1994). Whatever the interpretation, it is clear that by the late 1970s the institutional forms and arrangements that had supported the post-war boom in capitalist countries had become incompatible with, and no longer capable of providing a stable environment for, economic accumulation and competition. For many neoliberals, the institutions of post-war capitalism had ceased to be efficient, and were themselves largely to blame for the economic slowdown that had swept through world capitalism during the 1970s. The solution, therefore, was to "de-institutionalize" and socially "dis-embed" the ailing economy, so as to release a new wave of "free enterprise." Correspondingly, over the past two decades states have restructured their welfare systems, privatized public sector activities, de-regulated financial markets and labor markets, and sought to promote a new culture of self-interested individualism. The impact on the economic landscape of this assault on post-war institutions has been profound, leading to a marked increase in social and economic inequality within and between places.

But even though the neoliberal assault of the 1980s and 1990s has contributed to the contemporary transformation of institutional forms,[10] there are also fundamental forces at work within capitalism itself which are driving institutional change, particularly globalization and the expansion of the post-industrial information economy. Both of these are undermining the relevance and significance of old (modernist) institutional structures, and fueling the search for new ones. Globalization poses real challenges for institutions at all spatial scales (Webber, this volume). At one level, globalization is argued to be eroding the authority and economic impact of the nation-state and other national-level institutions. At another, it is seen as a "de-localizing" force, disconnecting and dis-embedding economic activities from their local socio-institutional contexts and frameworks, whilst simultaneously exposing local economies to the increasing uncertainties associated with global competition and globally mobile capital. And alongside and intersecting with globalization, the advent of a new "information age," a new post-industrial mode of economic development and social organization, signals the end of industrial capitalism as we have known it, and of the various institutions and traditions that historically have formed part of industrial society.

Thus, by virtue of its own inherent dynamic, industrial capitalism appears to be refracting back on and undercutting the very basic social structures, traditions, class identities, and local contexts that it had created and on which it depended. These traditions and institutions have now become the focus of reflexive interrogation and reconfiguration in the face of the increased autonomy and fluidity of the new globalized, information-based economy. There is no longer a "fixed landscape" of traditions and institutions structuring economic and social relations.

This does not mean the end of traditions and institutions, however, but rather a new phase of institution-building. Globalization may be emptying economic spaces of their pre-existing industrial activities and accompanying social institutions, but it is simultaneously relocalizing economic activity and thereby creating opportunities for new institutional forms and structures. Economic geographers have drawn attention to the "rebirth" and "re-emergence" of regional and local economies as nodes within this new globalized capitalism (see, for example, Storper, 1997; Scott, 1998). Many of these new local and regional economies are beginning to see themselves as socioeconomic and institutional entities. In fact many nation-states are themselves decentralizing and devolving their institutional and political structures down to regions and localities, partly in an attempt to "flexibilize" their space economies, partly in recognition of the locally variable nature of economic development, and partly to curb state expenditures. The localization of workfare in the USA, and the creation of local Training and Enterprise Councils and, more recently, Regional Development Agencies, in the UK, are examples of this institutional decentralization.

At the same time, "bottom-up" institution-building, networking, and experimentation is gathering pace at the regional and local scales, involving, for example, local economic development consortia and partnerships, strengthened local chambers of commerce, the emergence of local enterprise unionism, and a myriad of community development and employment organizations. The growing focus on learning and reflexivity (Storper, 1997) and the building of new forms of trust and cooperation (Sabel, 1993) as the foundations for successful regional development has also highlighted the emergence and policy significance of new institutional models of regional economic regulation and governance, including the so-called "associational economy" (Cooke and Morgan, 1998), the formation of a dense integrated network of collaborative social, economic, political, and civic institutions (see also Amin, 1995, 1999; Hirst, 1997). Boyer and Rogers Hollingsworth (1997) see these various local experiments and developments as marking a shift from a post-war Fordist institutional structure dominated by national-level forms to a much more complex, and spatially dispersed structure in which regional and local institutions are assuming an increasingly formative role in shaping economic activity. Similarly, Regini (1991) identifies the rise of this new "local institutionalism" as a central element of what he sees as a shift away from macroeconomic regulation to a more decentralized regime of micro-socio-institutional regulation.

How these new local institutional forms intermesh and interact with the changing structure of national and supra-national institutions is a point of some contention. Likewise, whether the increasing complexity of institutional spaces and local institutional nestedness – and the new "local policy heterodoxy" that is emerging as a consequence (Storper, 1997) – will enhance or impede the ability of regional and local economies to adapt to and compete in the new globalized capitalism is also a critical question. Institutional economic geographers, therefore, not only have a unique opportunity to observe a major wave of institution-building and experimentation actually underway, but also, hopefully, to contribute to the debates and discussions informing this process.

Acknowledgment

In writing this chapter I benefited from useful discussions with Geof Hodgson and Tony Lawson. Trevor Barnes made useful comments on the original draft. None are responsible, of course, for what the chapter says, or leaves unsaid.

Endnotes

1. See Boyer (1990) for an exposition of regulation theory. In US academic circles, the "social structure of accumulation" approach (see Kotz, 1994) bears a close similarity to the "mode of regulation" concept used by regulation theorists. The central idea of both is that crucial features of the long-run development of the capitalist economy are the product of the supporting and coordinating role played by social institutions.
2. I am certainly not suggesting that only institutional economic geography is guilty of definitional and conceptual "fuzziness." Unfortunately, the same problem is characteristic of much of contemporary economic and human geography as a whole.
3. Within economics, the term "institution" has some ambiguity. A typical definition is that given by Rutherford (1994, p. 182): "An institution is a regularity of behavior or a rule that is generally accepted by members of a social group, that specifies behavior in specific situations, and that is either self-policed or policed by external authority". There are numerous variants and inflections of this definition. Thus, according to Hodgson (1988, p. 10), an institution is "a social organization which, through the operation of tradition, custom or legal constraint tends to create durable and routinized patterns of behavior". Similarly, the "new" economic historian North (1990, p. 3) refers to institutions as the "self-imposed constraints that shape human interaction ... Institutions reduce uncertainty by providing a structure to everyday life".

4. In recent years, the notion of "path dependence" has attracted considerable attention within economics. There are several mechanisms making for path persistence in the economic landscape over time. Economists working in the "new geographical economics" (such as Krugman, 1991; Arthur, 1994) stress the role of increasing returns to industrial localization, agglomeration economies which generate a self-reinforcing tendency for particular activities to concentrate in particular areas (see Martin, 1999 for a survey of this literature). As North (1990), Setterfield (1997) and others have shown, institutional frameworks are another important source of the externalities that shape local economic path dependence.

5. For an excellent review of these three forms of institutionalism, and their respective strengths and weaknesses, albeit from a political science vantage point, see Hall and Taylor (1996).

6. For a critique of the rational choice foundations of the "new" (transactions costs) institutionalism in economics, see Hodgson (1988) and Rutherford (1994)

7. Like other aspects of the economy, institutions may be viewed as having "life-cycles," in the sense that as they "age" they develop features that render them increasingly rigid and less efficient.

8. For examples of the idea of institutional spaces and their local significance see Martin, Sunley and Wills (1996); Jones (1999). The term "spatial nestedness" of institutions comes from Boyer and Rogers Hollngsworth (1997).

9. For this reason, theories about economic development, and about regional and local economic dynamics, do not necessarily travel well between different national contexts. At the same time, different cultural forms of capitalism are not necessarily confined to their respective national territorial spaces. One thinks here, for example, of overseas Chinese business communities with their own distinctive sets of institutional practices, customs, and conventions.

10. A process that continues, with the current Anglo-American shift to so-called "Third Way" politics aimed at creating a new institutional landscape that avoids the over-centralized state system of the post-war period on the one hand, and the excessive ideological pre-occupation with free markets and individualism of the 1980s and 1990s on the other.

Bibliography

Amin, A. 1995. Beyond associative democracy. *New Political Economy*, 1, 309–33.

Amin, A. 1999. An Institutionalist Perspective on Regional Economic Development. *International Journal of Urban and Regional Research*, 23, 2, 365–78.

Amin, A. and Thrift, N. J. 1994. Living in the global. In A. Amin, and N. Thrift (eds) *Globalization, Institutions and Regional Development in Europe*. Oxford: Oxford University Press, 1–22.

Amin, A. and Thrift, N. J. 1995. Globalization, institutional "thickness" and the local economy. In P. Healey, S. Cameron, S. Davoudi, S. Graham, and A. Madinpour (eds) *Managing Cities: The New Urban Context*, Chichester: Wiley, 91–108.

Arthur, W. B. 1994. *Increasing Returns and Path Dependence in the Economy*. Ann Arbor: University of Michigan Press.

Beck, U., Giddens, A., and Lash, S. 1994. *Reflexive Modernization: Politics, Tradition and Aesthetics in the Modern Social Order*. Cambridge: Polity Press.

Berger, S. and Dore, R. (eds) 1996. *National Diversity and Global Capitalism*. Ithaca: Cornell University Press.

Boyer, R. 1990. *The Regulation School: A Critical Introduction*. New York: Columbia University Press.

Boyer, R. and Rogers Hollingsworth, J. 1997. From national embeddedness to spatial and institutional nestedness. Chapter 14 in J. Rogers Hollingsworth, and R. Boyer (eds) *Contemporary Capitalism: The Embeddedness of Institutions*. Cambridge: Cambridge University Press, 433–84.

Bryson, J., Henry, N., Keeble, D., and Martin, R. L. (eds) 1999. *The Economic Geography Reader*. London: Wiley.

Camagni, R. (ed) 1991. *Innovation Networks: Spatial Perspectives*. London: Belhaven Press.

Clark, G. L., Gertler, M., and Feldman, M. (eds) 2000. *The Handbook of Economic Geography*. Oxford: Oxford University Press.

Cooke, P. and Morgan, K. 1998. *The Associational Economy: Firms, Regions and Innovation*. Oxford: Oxford University Press.

Cox, K. and Mair, A. 1991. From localized social structures to localities as agents. *Environment and Planning, A*, 23, 197–213.

Crang, P. 1997. Cultural turns and the reconstitution of the economic. In R. Lee, and J. Wills (eds) *Geographies of Economies*. London: Arnold, 3–15.

David, P. 1994. Why are institutions the "carriers of history"? Path dependence and the evolution of conventions, organizations and institutions. *Structural Change and Economic Dynamics*, 5, 2, 205–200.

Eggertson, T. 1990. *Economic Behaviour and Institutions*. Cambridge: Cambridge University Press.

Grabher, G. 1993. The weakness of strong ties: The lock-in of regional development in the Ruhr area. In G. Grabher, (ed) *The Embedded Firm: On the Socio-Economics of Industrial Networks*. London: Routledge, 255–77.

Granovetter, M. 1985. Economic action and social structures: The problem of embeddedness. *American Journal of Sociology*, 91, 481–510.

Granovetter, M. 1993. Economic institutions as social constructions: A framework of analysis. *Acta Sociologica*, 35, 3–12.

Granovetter, M. and Swedberg, R. 1992. *The Sociology of Economic Life*. Boulder: Westview Press.

Hall, P. and Taylor, R.C.R. 1996. Political science and the three new institutionalisms. *Political Studies*, 44, 5, 936–57.

Hirst, P. 1997. *From Statism to Pluralism: Democracy, Civil Politics and Global Politics*. London: University College London Press.

Hodgson, G. M. 1988. *Economics and Institutions*. Cambridge: Polity.

Hodgson, G. M. 1989. Institutional rigidities and economic growth. *Cambridge Journal of Economics*, 13, 79–101.

Jones, M. 1999. *New Institutional Spaces: Training and Enterprise Councils and the Remaking of Economic Governance*. London: Jessica Kingsley.

Kotz, D. M. 1994. Regulation theory and social structures of accumulation. In D. M. Kotz, T. McDonough, and M. Reich (eds) *Social Structures of Accumulation: The Political Economy of Growth and Crisis*. Cambridge: Cambridge University Press, 85–98.

Krugman, P. 1991. Increasing returns and economic geography. *Journal of Political Economy*, 99, 483–99.

Lee, R. and Wills, J. (eds) 1997. *Geographies of Economies*. London: Arnold.

Liebowitz, S. J. and Margolis, S. E. 1995. Path dependence, lock-in and history. *Journal of Law, Economics and Organization*, 11, 1, 205–26.

Magnusson, L. and Ottosson, J. (eds) 1997. *Evolutionary Economics and Path Dependence*. Cheltenham: Edward Elgar.

March, J. and Olsen, J. 1989. *Rediscovering Institutions: The Organizational Basis of Politics*. New York: The Free Press.

Martin, R. L. 1994. Economic theory and human geography. In D. Gregory, R. L. Martin and G. E. Smith (eds) *Human Geography: Society, Space and Social Science*. Basingstoke: Macmillan, 21–53.

Martin, R. L. 1999. The new "geographical turn" in economics: Some critical reflections. *Cambridge Journal of Economics*, 23, 65–91.

Martin, R. L., Sunley, P., and Wills, J. 1996. *Union Retreat and the Regions: The Shrinking Landscape of Organized Labour*. London: Jessica Kingsley.

North, D. 1990. *Institutions, Institutional Change and Economic Performance*. Cambridge: Cambridge University Press.

Regini, M. 1991. *Uncertain Boundaries: The Social and Political Construction of European Economies*. Cambridge: Cambridge University Press.

Regional Studies 1999. Special issue on regional networking, collective learning and innovation in high technology SMEs in Europe. *Regional Studies*, 33, 4.

Reijnders, J. (ed) 1997. *Economics and Evolution*. Cheltenham: Edward Elgar.

Rogers Hollingsworth, J. and Boyer, R. (eds) 1997. *Contemporary Capitalism: The Embeddedness of Institutions*. Cambridge: Cambridge University Press.

Rutherford, M. 1994. *Institutions in Economics: The Old and the New Institutionalism*. Cambridge: Cambridge University Press.

Sabel, C. 1993. Studied trust: Building new forms of cooperation in a volatile economy. In D. Foray, and C. Freeman (eds) *Technology and the Wealth of Nations*. London: Pinter, 332–52.

Samuels, W. J. 1995. The present state of institutional economics. *Cambridge Journal of Economics*, 19, 569–90.

Scott, A. J. 1998. *Regions and the World Economy*. Oxford: Oxford University Press.

Setterfield, M. 1997. *Rapid Growth and Relative Decline: Modelling Macroeconomic Dynamics with Hysteresis*. London: Macmillan.

Simmie, J. (ed) 1997. *Innovation, Networks and Learning Regions*. London: Jessica Kingsley.

Steinmo, S., Thelen, K., and Longstreth, F. (eds) 1992. *Structuring Politics: Historical Institutionalism in Comparative Perspective*. Cambridge: Cambridge University Press.

Storper, M. 1997. *The Regional World: Territorial Development in a Global Economy*. London: Harvard University Press.

Storper, M. and Walker, R. 1989. *The Capitalist Imperative: Territory, Technology, and Industrial Growth*. Oxford: Blackwell.

Storper, M. and Salais, R. 1997. *Worlds of Production: The Action Frameworks of the Economy*. Cambridge, MA: Harvard University Press.

Sunley, P. 1996. Context in economic geography: The relevance of pragmatism. *Progress in Human Geography*, 20, 338–55.

Swedberg, R. 1997. New economic sociology: What has been accomplished, what is ahead? *Acta Sociologica*, 40, 161–82.

Thrift, N. and Olds, K. 1996. Refiguring the economic in economic geography. *Progress in Human Geography*, 20, 311–37.

Tickell, A. and Peck, J. 1995. Social regulation *after* Fordism: Regulation theory, neo-liberalism and the Global-Local Nexus. *Economy and Society*, 24, 357–86.

Williamson, O. E. 1985. *The Economic Institutions of Capitalism: Firms, Markets, Relational Contracting*. New York: The Free Press.

Zukin, S. and DiMaggio, P. (eds) 1990. *Structures of Capital: The Social Organization of the Economy*. Cambridge: Cambridge University Press.

Chapter 7

Poststructural Interventions

J. K. Gibson-Graham

Poststructuralism is a theoretical approach to knowledge and society that embraces the ultimate undecidability of meaning, the constitutive power of discourse, and the political effectivity of theory and research. Beginning in the 1960s as a movement within French philosophy, poststructural theory soon migrated into the English-speaking world where it has had a transformative impact within philosophy as well as literary and cultural studies. More recently, poststructuralism has gained ground within human geography and the other social sciences. The goals of this chapter are to offer a brief overview of the key insights and concerns of this relatively new theoretical tradition and to chart its nascent development within economic geography, giving a sense of its powers and potentials.

Poststructuralism is sometimes equated with "postmodernism," a term that tends to be loosely defined. In geography, for example, postmodernism variously refers to an historical epoch (Harvey, 1989) characterized by a particular set of socioeconomic practices and ideological "conditions;" an aesthetic style in film, architecture, and other cultural forms; or a theoretical approach to knowledge and society (Gibson and Watson, 1995, p. 1). The latter is what we are calling poststructuralism – a philosophically informed and theoretically distinctive approach to knowing and the world (Amariglio, 2000). What is postmodern about poststructuralism is its rejection of certain readily identifiable modernist conceptions of knowing, the knower, and the known. While knowledge is understood within a modernist frame as singular, cumulative, and neutral, from a poststructural perspective knowledge is multiple, contradictory, and powerful. The implications of this difference for economic geography are what we wish to explore.

Poststructuralist Antecedents and Origins

It was within and against the modernist tradition of structuralism that poststructuralism emerged. Perhaps its most direct antecedent was the linguistic structuralism of Ferdinand de Saussure, who rejected the view of traditional linguistics that words are symbols standing in for objects in the world. Saussure (1966) argued instead that words could be seen as "signs" constituted by a relation between two parts, the signifier – the acoustic or visual image – and the signified – the concept invoked by this image. In any language meaning emerges not from the connection of words to their extralinguistic referents (which is entirely arbitrary) but from the socially constructed relations of difference between signs (Yapa, 1999a).

Consider the word "factory." In the pre-Saussurian view this word is a symbol that represents or stands in for a building where production takes place – its real world referent. By contrast, in Saussure's structuralist view, the word "factory" is a sign that encompasses the spoken or written word FACTORY (the signifier) and the IDEA (the signified) of a building that is differentiated from other things because of the culturally encoded practices that take place within and around it. The word "factory" takes on meaning through the distinctiveness of its sign in relation to all other signs in the language, for example, in its differentiation from "office" or "house" or "field" or "playground." What was revolutionary about this formulation was the view that meaning is created within a complex social structure of relation and difference rather than through words substituting for objects, representing them in their absence (Hewitson, 2000).

While structuralism can be seen as destabilizing to the modernist presumption that language is a mirror of reality, in other ways it was a quintessentially modernist project. For structuralist thinkers, underlying the flux and contingencies of the social world lay unified formations that gave shape to social life. The linguistic theories of Saussure, the cultural theories of Claude Levi-Strauss, the economic theories of Karl Marx and the psychoanalytic theories of Sigmund Freud each, in their various ways, traced the origin and organization of complex social phenomena to deeper underlying structures. Uncovering or discovering those structures was the task of structuralist science (Amariglio, 2000).

The philosophers who were to become known as "poststructuralists" confronted the structuralist project with a skeptical attitude toward determination by "underlying" structures and attempts to grasp the ultimate "truth" of language, culture, society, and psyche. But perhaps their most salient move was to call into question the fixed relationship between signifier and signified that characterized Saussurian linguistics. From a poststructuralist perspective, language does not exist as a system of differences among a single set of signs. Rather the signifier–signified relations that generate meaning are continually being created and revised as words are recontextualized in the endless production of texts. The creation of meaning is an unfinished process, a site of (political) struggle where alternative meanings are generated and only temporarily fixed. Thus the meaning of the word "woman" in the context of "husband," "home," and "family" is very different from its meaning in the context of "lesbian," "work," and "politics." Political struggles undertaken by feminists can be seen as multiplying the contextualizations and significations of "woman" and, in the process, destabilizing the fixities of meaning associated with a patriarchal order (Daly, 1991).

So far in this discussion we have encountered both the anti-foundationalism and the anti-essentialism that characterize poststructuralist theory. An anti-foundationalist epistemology refuses a vision of knowledge as "grounded in reality" or as charged with the task of mirroring or "reflecting" the world (Rorty, 1979). Rather than being a dependent reflection of an independent real, poststructural knowledge is a social process in its own right, interacting fully with other social and natural processes in the constitution of social life. This anti-foundationalist epistemology is directly linked to an anti-essentialist ontology. If the word "woman" does not "correspond" to a generic female human being, its meaning cannot be understood as fixed by an essential sameness that unifies all women. Instead "woman" is resignified every time the word is recontextualized. There is no essential, fundamental, or invariant concept of "woman" to anchor the word, but rather an infinity of contextualizations that provide multiple and contradictory readings of what woman is or could be.

Despite their antagonism to foundations, poststructuralist thinkers do not see meaning and knowledge as unconnected to other aspects of social life. Meaning is understood as produced under specific social and intellectual conditions, and knowledge is not a "true reflection" but a productive and constitutive force. Although knowledges cannot be differentiated according to their greater or lesser accuracy (their success or failure in reflecting the world), they can be distinguished by their effects – the different subjects they empower, the institutions and practices they enable, and those they exclude or suppress. Thus is it a matter of *consequence* rather than a matter of *indifference* what kind of knowledge you produce (despite the familiar criticism that, for poststructuralists, "anything goes").

Poststructuralist Strategies

Poststructuralism offers a number of strategies for calling into question received ideas and dominant practices, making visible their power, and creating openings for alternative forms of practice and power to emerge. The ones that we explore here are deconstruction, genealogy and discourse analysis, and the theory of performativity.

Deconstruction

Deconstruction is a reading practice originating with the contemporary French philosopher Jacques Derrida (1967). Working against what he calls the "metaphysics of presence," or "logocentrism," Derrida calls into question certain fundamental axioms of Western thought. These include:

- the law of identity and self-presence (if a building is a factory, it is factory);
- the law of non-contradiction that establishes identity in relation to its "other" (if a thing is a factory, it cannot also be a not-factory); and
- the law of the excluded middle (factory and not-factory contain all the possibilities in a given situation) (Jay, 1991, pp. 92–3; Gross, 1986, pp. 26–8; Hewitson, 2000, p. 4).

Taken together these "laws" give us objects/identities that are stable, bounded, and constituted via a negation (of all that "factory" is not).

What Derrida identified as logocentrism is the Western pattern of producing meaning through a binary structure of positive and negative (A/not A, factory/not factory). Within the frame of logocentrism, the first term in a binary is endowed with positivity at the expense of the other: presence and value are attached to factory, while not-factory is absent and devalued. The binary structure establishes a relation of opposition and exclusion rather than similarity and mixture between the two terms – thus if the factory is a site of production, then not-factory, such as the household, is not (or if production exists within the household, it is inferior to that taking place in the factory). This structure of opposition is associated with a metaphysic of valuation that may be subtle but is impossible to avoid – the pervasive tendency to take seriously presence, positivity, and being, and to discount absence, negation, and non-being/becoming. It is easy to see from this example how logo-centrism might give rise to a concept of "economy" as bounded, stable, and inherently more important than the non-economic.

Feminist poststructuralists, among others, have observed that attempts to (re)value the absent or subordinated term within a binary hierarchy are easily undermined. This reveals the presence of what Saussure identified as a "master signifier," that operates to stabilize relations of difference. Poststructuralist feminists have renamed logocentrism "phallogocentrism" to highlight the way in which the figure of the masculine (the phallus) fixes meaning and endows presence and positivity to one side of the binary, producing an aligned chain of dominant terms within Western Enlightenment thought, as in the following example:

Man / woman
Mind / body
Reason / emotion
Objectivity / subjectivity
Self / other
Economy / society
Production / reproduction
Factory / household

We can detect the role of the phallus as master signifier in the regularity of the association of the first term with masculinity, dominance, and importance. To come back to our factory example, the very identity and positivity of factory is gained within a socio-linguistic structure that associates what goes on in a factory with reason, objectivity, mind, man, and economy. These dominant terms reinforce each other, differentiating the kind of production that takes place in the factory from the kinds of production taking place in households, backyards, streets, and fields, endowing it with greater "reality," independence, and consequence.

Derrida's deconstructive strategy is interested in rethinking difference, outside a binary and hierarchical structure, as part of an ultimately political project of creating spaces of "radical heterogeneity." A Derridean deconstruction of the sign FACT-ORY might begin by revaluing the subordinate term in the factory/not-factory binary. Feminist economic theorists, for example, have attempted to reverse the flow of cultural valuation by pointing out how many hours are spent in unpaid domestic labor in the household and how this contribution to Gross Domestic Product, if measured, would outweigh the labor performed in factory-based production (Waring, 1988). To take this a bit further, in economic geography the household is traditionally seen as the site of social reproduction – a dependent, less determining, often devalued set of (what are presented as non-economic) practices such as housework and child-rearing. Reversing the production–reproduction binary, we might represent reproduction as the activity that engages everyone at all times; it is the entire process of creating the conditions for society to continue to exist. Reproduction is thus a more embracing process than production. It is the general case of which production is the special case, the whole of which production is a part.

One of the problems with reversal is that it leaves the binary structure intact, with the hierarchy of valuation simply switched around. Another and more potent deconstructive strategy is to blur the boundaries between the terms, highlighting similarities on both sides of the divide, undermining the solidity and fixity of identity/presence, showing how the excluded "other" is so embedded within the primary Identity that its distinctiveness is ultimately unsustainable. Thus we might represent the household as also a site of production – of various domestic goods and services – and the factory as also a place of reproduction. So-called "non-economic" activities said to take place in the domestic realm – the display of emotions, the performance of sexual and gendered identities, socialization, training, and caring – are not only also practiced in the public realm of the factory, but can be seen to undermine the integrity of the factory as a site of efficient production, rational calculation, and profit maximization. The presence of the excluded other "within" renders the Identity unfamiliar and hollows out its meaning (Doel, 1994). Suddenly the stability of what we understood as "factory" begins to crumble.

Deconstruction highlights moments of contradiction and undecidability in what appears to be a neatly conceived structure or text (Ruccio, 2000). It points to the endless deferral of meaning within a system of differentiation, and foregrounds the inability of any sign to totally embody an essential meaning. Meaning is created and recreated within specific texts and contexts. Since ultimately no master term exists to fix concepts to particular signifiers, meaning is always in process and incomplete.

Genealogy and discourse analysis

Whereas Derrida highlights the unfixity and contestability of meaning (albeit within a frame in which meaning is constrained by the "metaphysics of presence"), Michel Foucault's project was to examine how certain knowledges and meanings become normalized and accepted as Truth. Foucault's work highlights the ways in which the construction of meaning is an enactment of power that is not only traced within language but also etched upon the body and continually re-enacted in social life. In his use of the term "discourse," Foucault (1991) refers to a rule-governed practice that includes meanings set within a knowledge system as well as institutions and social practices that produce and maintain these meanings. To return to our factory example, a shed with a saw-toothed roof in which people take one set of materials and, using various kinds of machines, transform them into other materials becomes known as a "factory" both because of the differentiation of FACTORY within a system of linguistic signs and also because of its place in an even wider system of signs made up of social conventions, routinized bodily movements, rules of behavior, institutional actors, and so on. This collection of metal, glass, bodies, energy, and produced materials takes on meaning and is endowed with positive value only within discourse – in this case, perhaps, a discourse of industrialization.

Foucault (1981) challenges the universality and "truth" of meaning by developing a distinctive method of discourse analysis. This involves: (1) a *critical* analysis of the violences enacted by any theory or system of meaning (what it excludes, prohibits, and denies); and (2) a *genealogical* analysis of the processes, continuities, and discontinuities by which a discourse comes to be formed. His work directs our attention to the way knowledges exercise and produce power – through apparatuses of regulation (for example, institutions like schools, prisons, workhouses, and factories where techniques of bodily discipline and surveillance prevail) and through the development and application of technologies of self-management that help to organize the daily business of living (for example, budgets, diets, maps, sexual advice columns, advertisements for personal hygiene products).

A critical analysis of the discourse of industrialization might highlight how the bodies and material production taking place in households are devalued within the disciplinary knowledge systems of economics or economic geography. And a genea-logical analysis might trace the formation of these disciplined understandings of "industry" and "economy," focusing upon the ruptures and discontinuities as well as the regularities and correspondences associated with key words of industrialization discourse. Such a genealogy might start, for example, with a Physiocratic vision of the economy in which an agricultural surplus makes non-agricultural activities possible (thus establishing the dependency and secondary status of industry). It might then move to the centrality and originary status of industry in the discourse of industrialization, with its vision of the economy and economic growth – including demand for agricultural products and services – being driven by productivity increases in manufacturing; and finally to the literature on the postindustrial eco-nomy, in which high-level financial institutions and transactions establish once again the subordination of industry, this time to financial speculation and the vagaries of the international finance sector. What emerges from this project of tracing changing knowledges of industry is a relatively stable conception of the economy as an integrated totality, centered on a determinant site that constrains, drives, or dictates to other sites/activities. Through the process of genealogy this notion of economy is "denaturalized" made visible as a discursive construction.

Foucault's influence on poststructuralism has produced a focus upon how differ-ent forms of power intersect with knowledge production to create certain valorized conceptions of the subject in any historical period. While much of Foucault's work appears to emphasize the construction and consolidation of dominant discourses that "subject" the individual to powerful forces beyond her control, his intervention also opens the way for examining the proliferation and multiplicity of discourses that can create subjects able to resist and reconstitute power in different ways. Here the work of Judith Butler and other queer theorists on the performativity of dis-course conveys some of the incompleteness and openness of processes of "subjec-tion."

Performativity

Performativity for Butler is "the reiterative and citational practice through which discourse produces the effects that it names" (1993, p. 2). In *Gender Trouble* (1990) Butler explores performativity specifically with respect to gender. Gender is not a stable characteristic of the subject that emanates from some biologically determined or culturally inscribed binary structure. Rather, gender identity is practiced through the repetitive performance of certain acts. Gender must be continually re-enacted in order to secure its seeming fixity. This notion of iterative performance as constitutive of what is taken to be a stable reality opens up some interesting insights into the politics of knowledge. Butler is drawn to emphasize the inevitable differences between performances, the slippages between iterations, suggesting that gender (or other) identities are always unfinished and open to subversion.

The concept of performativity opens a pathway through the sometimes disconcerting ungroundedness of the poststructuralist project and points toward engaged interventions that challenge the hegemonic knowledge/power systems so clearly outlined by Foucault. What intrigues Butler are the openings through which queer subjects can be seen to emerge, outside an established heteronormative order. Butler is committed to destabilizing the binary gender categories that function to buttress compulsory heterosexuality. By emphasizing the uncertainties and discontinuities inherent in gender performance, she brings to light existing possibilities of disruption and invention in the cultural process of gendering. In this way she opens a space for "agency" and unpredictability, in a mode of subjection that is often seen as biologically or culturally given.

For Butler and the other theorists we have discussed here, poststructural interventions are not equated with a retreat into theory and a disengagement from the world, politics, ethics, and social change. Rather, poststructuralism opens up a new role for theory as a political intervention. Poststructural knowledge actively shapes "reality" rather than passively reflecting it. The production of new knowledges is a world-changing activity, repositioning other knowledges and validating new subjects, practices, policies, and institutions.

Poststructuralist Moments in Economic Geography

In economic geography, the poststructural reconfiguration of the relation between knowledge and action, research and reality, has inspired a number of new directions in research. It has also increased our responsibility to ask "What kind of knowledge do we want to participate in creating? What are the effects of the knowledge we construct? What possibilities does our research enable?" In the rest of this chapter, we briefly address some of the economic geographic work that is animated by the strategies of deconstruction, genealogy and discourse analysis, the theory of performativity, and a vision of the researcher as an agent/intervener.

Deconstruction

J. K. Gibson-Graham (1996) employs a deconstructive approach in *The End of Capitalism (As We Knew It)*, highlighting the ways that a capitalism/noncapitalism binary operates in economic discourse to constitute capitalism as a necessarily and naturally dominant form of economy. Noncapitalist economic practices (whether in the household, the formal market sector, the informal economy, alternative economic experiments, or socialist or cooperative undertakings) are generally understood with respect to capitalism – as either the same as, the opposite of, the complement to, or contained within capitalism. They are seen as weaker and less able to reproduce themselves. Noncapitalism is represented in the interstices, in experimental enclaves, or as scattered and fragmented in the landscape. Capitalism, by contrast, is represented as systemic, naturally expansive, and coextensive with the national or world economy. As a consequence, many studies of noncapitalist activities focus upon their imminent destruction, their proto-capitalist qualities, their weak or determined position within a local economy. These activities are rarely represented as resilient, ubiquitous, capable of generative growth, or driving economic change. In the face of "globalization," which is represented as the march of an ever more powerful capitalism across the economic landscape, noncapitalist activities are rendered powerless or subsumed.

Drawing on the insights of feminist economic theorists and theorists of the informal sector, Gibson-Graham has attempted to undermine the "capitalocentrism" of economic discourse, both popular and academic, pointing out that nonmarket and thus by definition noncapitalist activities produce at least half of the total product worldwide. Yet in the dominant economic discourse of both left and right, these activities, as well as noncapitalist production for a market, subsist in the shadows of capitalism, relatively invisible and subordinated to its presumed dominance and importance.

In addressing the capitalism/noncapitalism binary, Gibson-Graham's principal deconstructive move is to bring the subordinate and negative term into independent existence, representing "noncapitalism" positively as an array of different economic forms rather than simply as an absence, insufficiency, or dependency. Using a language of economic difference derived from Marx, she identifies in the economic landscape a variety of economic relations in addition to capitalist ones, including feudal, slave, independent, and communal relations in both the market and nonmarket sectors. As the black boxes of the formal, informal, alternative, and domestic economies are opened up for scrutiny, "the economy" emerges as a complex interdependency of different economic relations within the variously constituted household, volunteer, self-employed, family business, prison, and illegal, as well as industrial market, sectors.

In a simultaneous deconstructive move, the unity on the other side of the binary can be dissolved and represented as a contradictory multiplicity; capitalism can be deprived of its solid and coherent identity, becoming different from itself and difficult to generalize. One approach to this project has been through reconceptualizing the enterprise. Enterprises are traditionally assumed to share a common structure, be driven by a common imperative – the logic of growth, accumulation, or profitability – and encounter a fundamentally similar external terrain, the business world or market. Yet there is a growing literature emerging from economic geography, as well as economic sociology, anthropology, accounting, organization theory, and Marxian political economy, that highlights the heterogeneity of companies. Erica Schoenberger's (1997) work, for example, on the social embeddedness of firms traces the effects of culture, tradition, and affinity upon enterprise behavior, and demonstrates the ways in which personal values and relationships within management undermine many of the assumed corporate goals of efficiency and profit maximization.

O'Neill and Gibson-Graham (2001) explore the role of competing discourses of management in shaping the fluid entity that is unproblematically represented as "the capitalist firm." Examining an Australian minerals and steel multinational, they produce a disruptive reading that emphasizes the decentered and disorganized actions taken in response to multiple logics circulating within and without the corporation. Their analysis represents the enterprise as an unpredictable and potentially open site, rather than as a set of practices unified by a predictable logic of profit maximization or capital accumulation. Untethered from a preordained economic logic, the enterprise becomes recognizable as an ordinary social institution, one that often fails to enact its will or realize its goals or even fails to come to a coherent conception of what these might be. In the context of such a representation, the conflation of capitalism with power, and the related vision of noncapitalism as excluded from power, becomes more difficult to sustain.

By conceptualizing the economy as diverse and heterogeneous, not centered upon any privileged set of activities, or ordered by any core dynamic, but always in the process of (discursive) construction, Gibson-Graham (1996) has embarked on a project of reconstructing the political terrain on which alternative economies are imagined and enacted. Using deconstruction to open up concepts of economy to include unpaid, nonmarket, and noncapitalist activities, and to envision economic difference outside the binary frame, she is producing alternative economic representations as a contribution to innovative forms of economic policy and activism.

Genealogy and discourse analysis

Poststructuralist research on "the economy" highlights the social and discursive construction of local, regional, national, and global economies, and works against dominant representations of these entities as real containers and determinants of social life. The literatures on post-Fordism, globalization, and development, to name but a few, are not taken as sophisticated descriptions of economic reality but as discourses that actively constitute economic possibility, shaping and constraining the actions of economic agents and policymakers.

One of the prevalent discourses that organizes knowledge of economic change is that of "development" – the story of growth along a universal social trajectory in which regions or nations characterized by "backwardness" are seen to progress towards modernity, maturity, and the full realization of their potential. This organic model of change has structured theoretical work in a large number of disparate fields but is now under scrutiny, not least because of its Eurocentrism and its devaluing and disabling effects on the "less developed." In anthropology, Arturo Escobar (1995) has begun the task of critically analyzing and producing a genealogy of development discourse. His work traces the historical production of the "Third World" – that collection of countries whose populations came to be represented as poor, illiterate, malnourished, underemployed, requiring aid, and in need of Western models of development. The Third World was the problem for which "development" provided the solution – through the establishment of a range of institutions, practices, and experts that were empowered to exercise domination in the name of the scientifically justified development project. Escobar's close reading reveals how the practice of identifying barriers to growth and prescribing development pathways has in effect violently "subjected" individuals, regions, and entire countries to the powers and agencies of the development apparatus. The subjects produced within and by this discourse are ill-equipped to think outside the presumed Order and Truth of the economic development story and to reject a vision of the "good society" emanating from the West. Escobar's Foucauldian approach to development discourse has opened the way towards an "unmaking" of the Third World, by highlighting its constructedness and the possibility of alternative constructions. Importantly, his work points the way toward a repositioning of subjects outside a discourse that produces subservience, victimhood, and economic impotence (Gibson-Graham and Ruccio, 2001).

Less entrenched but perhaps more ubiquitous is the discourse of globalization that circulates today in popular, academic, and policy discussions. In a recent paper on the integration of the New Zealand domestic economy into global networks of trade, finance, and production, Wendy Larner (1998) takes on the "globalization" story. She emphasizes that globalization is not, as many economic geographers have claimed, a "new reality" that forces nation-states and their citizens into new roles (p. 600), but a discourse that powerfully poses a different conception of the relationship between the national and international economies. In New Zealand the globalization "imperative" has been constituted out of specific political ambitions, rationalities, and social practices. The current representation of New Zealand's economy as a "node in the flows and networks of the Pacific Rim" (p. 607) rather than as a self-contained entity has ushered in new forms of economic governance that privilege the market over the state as a mechanism for social provision, and situate the individual within a global rather than a national spatial imaginary. That these changes are the product of currently dominant but not entirely robust discourses Larner takes as encouragement to begin to identify different political strategies and alternative spatial imaginaries that might have more progressive (or at least different) effects. While not ignoring the very real impacts of job loss, privatization, capital redirection, and the replacement of old systems of governance by new, Larner challenges economic geographers to intervene in the constitution of the world around them by refusing to reify globalization as an extradiscursive *fait accompli*.

Performativity

To recognize the performativity of discourse is to recognize its power – its ability to produce "the effects that it names" (Butler, 1993, p. 2). But the process of repetition by which discourse produces its effects is characterized by hesitancies and interruptions. This performative dimension of discourse is highlighted in recent geographic research on economic subjects. Unlike the coherent and rational modernist subject, the poststructuralist economic subject is incompletely "subjected." Her identity is always under construction, constituted in part through daily and discontinuous practices that leave openings for (re)invention and "perversion."

In the work of Linda McDowell on gender in the City of London, for example, masculinity, femininity, and the gender division of labor are represented as arising within a particular geographical and temporal milieu rather than as manifestations, in a new economic environment, of a patriarchal system of oppression. Through observations of behavior, clothing, and demeanor on the trading floor, and interviews with participants in the merchant banking business, McDowell (1997) explores the ways in which gender is produced and transformed in the very process of buying and selling money, futures, and shares. Gender is not constructed outside the City and "taken to work" but is constituted in and through participation in "economic" practices. Its limits and possibilities are not ordained by a sexed body, a patriarchal structure, or culturally transmitted gender norms but are constructed and reconstructed in the moment of performance. "By making it clear that what seems an inalienable part of a sexed self is actually a temporally and spatially specific performance" (p. 165), McDowell highlights the possibility of transforming and multiplying "acceptable fleshy styles and corporeal practices" and enabling new forms of gendered power in the workplace.

Jenny Cameron (1996/97, 1998) takes up the issue of gender performativity in the context of a site often ignored by economic geographers – the households of "middle-class" women in Australia. In these domestic economies Cameron finds the familiar gender inequality in the division of domestic labor, but chooses not to interpret the unequal allocation of housework as evidence of the resiliency of patriarchal structures of male domination and female exploitation. Her close reading of interview texts reveals instead a complex process through which the performance of particular household tasks constructs heterosexual identity for both women and men. Emphasizing the precarious and shifting boundary between what lines up as masculine and feminine in Australian society, Cameron shows how subjects actively work to sustain the fiction of a natural and stable gender identity which their lives and social contexts continually undermine. Moreover, the inevitable slippages and contradictions between performances suggest the possibility of alternative genderings and the different sexualities they might signal or permit.

Poststructuralism and the Politics of Research

This discussion so far, with its emphasis on the constructive and disruptive powers of discourse, suggests that poststructuralism has the potential to offer a new model of geographic research. If knowledge is not assigned the task of providing an accurate reflection of reality (Rorty, 1979), then research doesn't simply reveal "what's out there" in the world. Recognizing the effectivity of knowledge creates an important role for research as an activity of producing and transforming discourses, creating new subject positions and imaginative possibilities that can animate political projects and desires (Gibson-Graham, 1994). In this concluding section of the chapter, we focus on three projects of poststructural research that actively engage in producing or destabilizing discursive formations, thereby participating in the constitution of power, subjectivity, and social possibility.

In a study of Filipina domestic workers and labor market segmentation in Vancouver, British Columbia, Geraldine Pratt (1999) traces the ways various discourses operate to devalue and disempower Filipina immigrants. Pratt's critical and genealogical project reveals how Filipina women are assigned to a variety of subject positions and defined in relation to "what they are inferior to." As Live-in Caregivers, they are identified as non-Canadians, taking jobs no Canadian would want and deprived until recently of rights accorded to Canadian workers. As "housekeepers," often with a university education or professional qualifications (e.g., registered nurse) from the Philippines, they are positioned as less qualified than European "nannies." And as Filipinas within their community, they are seen as "nannies" rather than regular "immigrants" and stigmatized as promiscuous "husband stealers."

Pratt collaborated with organizers at the Philippine Women Centre, running focus groups with Filipina domestic workers in which these discourses with their devalued subject positions emerged. What Pratt emphasizes, however, is that they emerged in the context of silences and gaps, contradictory discourses, and alternative subject positions, signifying that discourse is never fully capable of exercising power and producing "subjection." Pratt sees the possibility for "agency, and for the creative redirection and redefinition of subject positions" (p. 35), in the coexistence of contradictory or incommensurable discourses. When domestic workers begin to represent themselves to themselves and to their employers as workers rather than "family members," for example, they move from a "racial" subject position within a discourse of immigration and citizenship to a "class" subject position in a discourse of employment and employee rights.

Pratt understands her critical discourse analysis as an intervention that is an "important element of disrupting...oppressive institutional practices" (p. 35). Discourse analysis denaturalizes subjectivities and social practices, exoticizing them and making them "remarkable" as elements of a particular formation. The process of analyzing a discourse highlights the contingency of its alignments and reveals it as an attempt at stabilization. It thus simultaneously suggests its vulnerability to destabilization and reconstruction.

In the Philadelphia Field Project, Lakshman Yapa (1999b) is attempting to integrate the Pennsylvania State University's mission of teaching, research, and service with a poststructural intervention into urban poverty. The initial and critical step in this project is to problematize the dominant discourse of poverty, which Yapa sees as having three prominent characteristics: it assumes that knowledge *reflects* meanings already existing in objects; it is founded on a subject/object duality in which poverty is the object/problem and the social scientist is the subject/non-problem; and it embodies an essentialist ontology in which problems like poverty have root causes and solutions. Yapa proposes, by contrast, that meaning is *constructed* in discourse, that social science is part of the problem of poverty, and that the causes of poverty are "multiple, diffused, and reside in thoroughly overdetermined systems" (p. 11).

The dominant discourse of poverty treats it as a problem with lack of income at its root. Social scientists collect information on the poor to understand who and where they are, what levels of poverty they experience, and why they have insufficient income. Depending on the political context, policymakers may then devise strategies to improve access to income among the impoverished. Yapa offers a different, anti-essentialist understanding of poverty, seeing it as a multidimensional social construction of scarcity that has an infinity of origins or causes and therefore an infinite number of points of intervention. Reducing poverty to a problem of inadequate income not only suggests that increased income will alleviate the problems of poverty – including poor nutrition, lack of access to health care, lack of transportation, inadequate educational resources, substandard housing – but closes off (by making invisible) other avenues for addressing these forms of deprivation.

Yapa's vision is that poverty involves scarcities of basic goods like food, clothing, housing, health, and transportation which are complexly constructed by the entire set of conditions constituting a society. Since none of these conditions is the root cause of poverty (Yapa, 1998), none of them constitutes the privileged approach to addressing it: "Poverty is created at a large number of sites spread throughout the larger society," so "no one grand project of economic development, jobs, income or even affirmative action can provide a solution to poverty" (p. 8). Yapa's argument is that to deal with poverty we can "start where we are" and focus on available agents, their substantive competencies, and a range of feasible forms of intervention. Thus a nutritionist and food activist who has no knowledge of economic development theory does not have to create enterprise zones to solve the general problem of poverty, but can directly address a specific substantive problem (for example, inadequate food supplies and nutrition) through nutrition education, developing cooperatives for purchasing food, organizing classes for the intergenerational transfer of cooking knowledge, etc.

Students in the Philadelphia Field Project do not speak with inner city residents about poverty or lack of income but instead ask substantive questions about why they have "problems meeting their daily needs for food, housing, transport, etc." This accords with the view that there is no general logic of poverty but rather specific conditions constructing scarcity in each case. The students also do not offer themselves as "general change agents" but as individuals with certain substantive competencies and the ability to create knowledge that may be relevant to the needs of the community. Recognizing the ways that academia has been implicated in poverty, the students are attempting to create a knowledge and practice of anti-poverty that is not based on a subject/object hierarchy but rather on the specificity of scarcities, the substantive competencies of communities, and the availability of agents.

In both the United States and Australia, J. K. Gibson-Graham is undertaking community-based research in a regional setting, attempting to generate a discourse of economic diversity as a contribution to a politics of economic innovation (Byrne et al., 1999). The Australian phase of the project is situated in a region where privatization of a large state energy authority has produced widespread unemployment and a sense of regional victimhood and hopelessness. Early on in the project, researchers conducted focus groups with a range of economic actors (planners, business people, unionists, media workers) and social and community actors (welfare workers, clergy, artists, teachers) to explore narratives of regional change. In both groups the story of economic restructuring that emerged identified the regional economy with mining and energy production, and positioned local actors as powerless in the face of global and national forces. When the groups were asked to think of the strengths and successes of their communities, a different but halting story of economic diversity and cultural and social innovation began to emerge (Gibson et al., 1999). In conversation with some of the participants, an action research project was planned that would begin to develop this alternative but subordinated discourse of regional identity and economic possibility.

Participatory action research (PAR) is a practice that has traditionally been associated with politically motivated modernist research projects. Designed to empower communities that are marginalized or oppressed, this research methodology actively involves community members in the initiation, design, conduct, and evaluation of research (Fals Borda and Rahman, 1991). In its original mode, PAR is seen as bringing oppressed people to a recognition of their common interests, based upon their shared humanity or their structurally determined experiences of oppression. In its poststructuralist guise (Gibson-Graham, 1994; Reinharz, 1992), PAR involves enabling conversations across different identities, building partial and temporary communities among a "complex and diverse 'we'" (Brown, 1991, p. 81). Interactions between academic and nonacademic researchers in the process of research generate new languages and social representations that can become constituents of alternative social visions and practices.

During the early stages of the Australian project, both nonacademics and academics recognized the importance of language and representation. In the absence of a language of economic diversity, the mainstream discourse of "development" had positioned the region as entirely dependent on investment by capitalist firms, which might or might not be attracted by various blandishments. This vision, especially as it was reiterated by the press, had assumed the status of grim "reality," making it nearly impossible to talk hopefully about regional capacities and potentials. Using community researchers, the action research project is designed to produce an alternative "accounting" of the regional economy, one that recognizes a wide variety of economic relations and economic subjects, along with the particular gifts and capacities the latter possess (especially those of three economically marginalized groups – retrenched workers, unemployed youth, and single parents). This alternative accounting will serve to destabilize the identity of the area as primarily a "resource region." Its collective compilation will be a first step in creating opportunities for new regional actors and different economic subjects to come together to talk about and construct alternative regional futures.

In this poststructuralist research project, research is a process of engendering "conversations" through which new languages, identities, communities, and social possibilities emerge. Focus groups, individual interviews, interactions between academic and community researchers, community conferences, and other "conversations" are both the sites and products of research. One goal of the project is to produce new models of regional development that exceed the theory and practice of capitalist industrialization. But, just as importantly, the research aims to validate a different set of economic subjects, energizing them to intervene in the ongoing conversation that is economic development. In these ways the project will enable new political identities and initiatives.

Bibliography

Amariglio, J. 2000. Poststructuralism. In J. Davis, D. Wade Hands, and Uskali Maki (eds), *Handbook of Economic Methodology*. Cheltenham: Edward Elgar (forthcoming).

Brown, W. 1991. Feminist hesitations, postmodern exposures. *Differences*, 3, 63–84.

Butler, J. 1990. *Gender Trouble: Feminism and the Subversion of Identity*. New York: Routledge.

Butler, J. 1993. *Bodies that Matter: On the Discursive Limits of "Sex"*. London: Routledge.

Byrne, K., Forest, R., Gibson-Graham, J. K., Healy, S., and Horvath, G. 1999. Imagining and enacting noncapitalist futures. Rethinking Economy Project, Working Paper No. 1, Dept. of Geosciences, University of Massachusetts, Amherst.

Cameron, J. 1996/97. Throwing a dishcloth into the works: Troubling theories of domestic labor. *Rethinking Marxism*, 9, 2, 24–44.

Cameron, J. 1998. The practice of politics: Transforming subjectivities in the domestic domain and the public sphere. *Australian Geographer*, 29, 293–307.

Daly, G. 1991. The discursive construction of economic space: Logics of organization and disorganization. *Economy and Society*, 20, 79–102.

Derrida, J. 1967. *Of Grammatology* (G. Spivak, Trans.). Baltimore: Johns Hopkins University Press.

Doel, M. 1994. Deconstruction on the move: from libidinal economy to liminal materialism. *Environment and Planning A*, 26, 1041–59.

Escobar, A. 1995. *Encountering Development: The Making and Unmaking of the Third World*. Princeton, NJ: Princeton University Press.

Fals Borda, O. and Rahman, M. A. 1991. *Action and Knowledge: Breaking the Monopoly with Participatory Action Research*. New York: Apex.

Foucault, M. 1981. The order of discourse. Inaugural lecture at the College de France, given 2 December 1970. In R. Young (ed), *Untying the Text: A Poststructuralist Reader*. Boston, London and Henley: Routledge & Kegan Paul, 48–78.

Foucault, M. 1991. Politics and the study of discourse. In G. Burchell, C. Gordon and P. Miller (eds), *The Foucault Effect: Studies in Governmentality with Two Lectures and an Interview with Michel Foucault*. Chicago: University of Chicago Press, 53–72.

Gibson, K. and Watson, S. 1995. Postmodern spaces, cities and politics: An introduction. In S. Watson and K. Gibson (eds), *Postmodern Cities and Spaces*. Oxford: Blackwell, 1–12.

Gibson, K., Cameron, J., and Veno, A. 1999. Negotiating restructuring: A study of regional communities experiencing rapid social and economic change. Working Paper No. 11, Australian Housing and Urban Research Institute.

Gibson-Graham, J. K. 1994. "Stuffed if I know!" Reflections on post-modern feminist social research. *Gender, Place and Culture*, 1, 205–24.

Gibson-Graham, J. K. 1996. *The End of Capitalism (As We Knew It): A Feminist Critique of Political Economy*. Oxford: Blackwell.

Gibson-Graham, J. K. 1997. Re-placing class in economic geography: Possibilities for a new class politics. In R. Lee and J. Wills (eds), *Geographies of Economies*. New York: Arnold, 87–97.

Gibson-Graham, J. K. and Ruccio, D. 2001. "After" development: Negotiating the place of class. In J. K. Gibson-Graham, S. Resnick and R. Wolff (eds), *Re-presenting Class: Essays in Postmodern Political Economy*. Durham, NC: Duke University Press (forthcoming).

Gibson-Graham, J. K., Resnick, S., and Wolff, R. (eds) 2000. *Class and its Others*. Minneapolis: University of Minnesota Press (forthcoming).

Gross, E. 1986. Derrida and the limits of philosophy. *Thesis Eleven*, 14, 26–43.

Harvey, D. 1989. *The Condition of Postmodernity*. Oxford: Blackwell.

Hewitson, G. 2000. The body of economic theory: A feminist poststructuralist investigation. In J. Amariglio, S. Cullenberg, and D. Ruccio (eds), *Postmodernism, Knowledge, and Economics*. New York: Routledge (forthcoming).

Jay, N. 1991. Gender and dichotomy. In S. Gunew (ed), *A Reader in Feminist Knowledge*. London: Routledge, 89–106.

Larner, W. 1998. Hitching a ride on the tiger's back: Globalization and spatial imaginaries in New Zealand. *Environment and Planning D: Society and Space*, 16, 599–614.

McDowell, L. 1997. *Capital Culture: Gender at Work in the City*. Oxford: Blackwell.

O'Neill, P. and Gibson-Graham, J. K. 1999. Enterprise discourse and executive talk: Stories that destabilize the company. *Transactions: An International Journal of Geographical Research*, 24, 11–22.

Pratt, G. 1999. From registered nurse to registered nanny: Discursive geographies of Filipina domestic workers in Vancouver, B.C. *Economic Geography*, 75, 215–36.

Reinharz, S. 1992. *Feminist Methods in Social Research*. New York: Oxford University Press.

Rorty, R. 1979. *Philosophy and the Mirror of Nature*. Princeton, NJ: Princeton University Press.

Ruccio, D. 2000. Deconstruction. In J. Davis, D. Wade Hands, and Uskali Maki (eds), *Handbook of Economic Methodology*. Cheltenham: Edward Elgar (forthcoming).

Saussure, F. de 1966 (1915). *Course in General Linguistics*. In C. Bally and A. Sechehaye (eds), (W. Baskin, transl.). New York: McGraw-Hill.

Schoenberger, E. 1997. *The Cultural Crisis of the Firm*. Oxford: Blackwell.

Waring, M. 1988. *Counting for Nothing: What Men Value and What Women are Worth*. Sydney: Allen & Unwin.

Yapa, L. 1998. The poverty discourse and the poor in Sri Lanka. *Transactions of the Institute of British Geographers*, 23, 95–115.

Yapa, L. 1999a. A primer on postmodernism with a view to understanding poverty. *www.geog.psu.edu/~yapa/Discorse.htm*

Yapa, L. 1999b. Integrating teaching, research, and service – a Philadelphia Field Project. Unpublished paper, Dept. of Geography, The Pennsylvania State University, State College, PA.

Part II Realms of Production

Chapter 8

The Geography of Production

Richard A. Walker

The heart of every economy is industrial production, and the heart of economic geography lies in the spatial patterns and physical landscapes industry creates. Industry location may seem a limited provenance, at first. One starts off comfortably enough with the localization of different factories and sectors, such as steel mills and automobile plants, but all too quickly the division of labor outruns traditional notions of what constitutes an "industry" and the spatial division of labor disperses into a thousand pieces cast hither and yon – tire factories here, engine plants there, electronic ignitions and engineering plants somewhere else. These all have to be knit together into discrete units called factories, offices, or design houses, and into much looser connections called firms, sectors, or networks. To make matters worse, these webs of production overlap and interconnect in surprising ways that can never be entirely untangled, and they stretch far and wide across boundaries and to the far reaches of the global economy. The modestly ambitious industrial geographer must pull on seven-league boots as we try to mark off the immense tapestry of localization and globalization woven by contemporary production and trade.

This immense geography of production is in constant motion, rendering moot all fixed ideas about industry location patterns. Industrialization drives sectors and places along divergent paths of growth, and disrupts all established geographic habits. That divergence and instability is essential to the uneven development of the industrialized world. But movement alone does not capture the creative (and destructive) powers of modern industry. Successive industrial revolutions have built up the great cities, transport systems, and landscapes of production that surround us; industry does not locate in a known world so much as it produces the places it inhabits. This jagged process of industrial development repeatedly outruns prediction and liquidates the geographies of the past, generating the endless novelty that makes economic geography such a lively and challenging area of inquiry.

What is Industrial Production?

Production is making something. That something may be as rock solid as a car, as passing as a meal, or as shadowy as a program flickering on a TV screen. As the world moves toward more information-rich forms of production and products, like computer software and video games, there are more products that come in small packages, like CDs, and fewer bulky objects like steel girders. But production is, in all cases, an act of human labor; it involves work, plain and simple. In one case there is a definite product, something you can kick with your foot, a good like a television or a hamburger. In the other, there is only a change of condition, a labor-service such as a pierced ear, clean floor, or plane ride to Jamaica.

In traditional usage, "industrial location" referred to manufacture, or the making of goods. There is little reason for economic geographers to be so restrictive today, when manufacturing has dwindled down to one-fifth of employment, on average, in the advanced industrial economies. We have to come to grips with the larger compass of modern industrial activity and its spatial distribution. Nonetheless, we still inhabit an industrial economy with the production of goods at its core. There is a good deal of confusion on this point because official censuses call almost every-thing today a service.

A great many things called services are, in fact, work on goods. Some kinds of goods take unusual forms, like french fries to go, downtown skyscrapers, or the water coming out the tap. Some come in modest physical carriers with lots of critical information, like newpapers or software disks. Some very large goods are used by many people through lease and rental arrangements, as with hotels, airplanes, and cranes. Long-lived goods require repeated maintenance and repair, which is why there are so many auto shops, janitors, and house-painters (Sayer and Walker, 1992).

Nor should one forget the important arenas of agriculture and resource extrac-tion, which no longer take up much of the workforce in the advanced industrial nations but share most of the characteristics of modern industries (Hanink, this volume; Page, this volume). Then there are the crucial functions of distribution and sale of all goods that involve extensive distribution networks of warehouses and stores. Together, all such labor involved in the production, distribution, and repro-duction of goods of every kind more than doubles the usual estimates of manufac-turing, bringing the total employment in basic industry up to well over half the private sector. Most of the remainder is taken up by three major labor-using activities: health care, education, and personal service. Their output unquestionably takes the form of labor-services, not goods.

Whatever category these segments of goods and services production fall under, most are organized today on a large-scale industrial basis – ask any Las Vegas maid if she works in the "hotel industry" or Miami janitor if he works in the "cleaning industry." All such industries play a role in shaping the landscape of modern cities and regions, whether the factory districts of Kuala Lumpur, the malls surrounding Minneapolis, the hotels of Cancun, or the sea of warehouses around O'Hare Air-port. They are all a part of modern economic geography, even if its heart remains manufacturing, from Manchester to Silicon Valley (Daniels, 1993).

In saying that production is an act of human labor, we mean that people are centrally involved in organizing, orchestrating, and carrying out the tasks of modern industry. This means that securing a labor force is a prime task of any industrial operation, and critical to its locational calculus. Firms must recruit labor either by locating near to where the workers live or by attracting them from long distances; this matching of labor demand and supply is a base point for economic geography.

Different kinds of work demand different kinds of capacities from workers and provide varying levels of wages and other rewards, and here lies an elemental force for spatial differentiation of industrial activities, or spatial divisions of labor (Peck, this volume).

Workers and capitalists actively shape each other in the process of production. They do so in the workplace through labor training, managerial adjustment, and conflict – from everyday jostling for autonomy and control in the office or on the shopfloor, to full-blown strikes, lockouts, or violent confrontations. They do so outside the workplace in the way workers settle into residential communities with their own ways of life, and the way industrialists try to influence local development patterns and politics. And they do so by the way institutions and expectations evolve around well-established employment centers and communities, giving a strong element of inertia to industrial and worker location (Hanson and Pratt, 1995).

Human effort is augmented in three vital ways, however. The first is by tools and machines, from the simplest axe to the most complicated computer. The second is by nature and natural processes, such as bacteria turning milk into yogurt, chemical reactions in a steel furnace, or the electrons racing along a microcircuit. Workers initiate, guide, and oversee such gifts of nature, and we all benefit from them. Third, every worker benefits from accumulated knowledge, passed along from other people and embodied in plans, blueprints, books, machinery, and software. Technology is frequently imagined to be the prime mover of modern industry in commentaries on "computerization," "high tech," or "the information age," but production is basically a labor process to which machines and technical wonders are allied. The contemporary fascination with technology too often overshadows the central place of ordinary work and workers in economic geography (Herod, 1998).

Spatial Divisions of Labor: The Geography of Industries

Today's market economies produce millions of different commodities for sale, and employ hundreds of millions of people. They are immensely complex systems of production, made up of an extraordinarily large number of pieces. Those bits and pieces constitute "the division of labor," and are the basic building blocks of the industrial system and of economic geography. Without a division of labor, there would be no differentiation of economic activities, no factories to site, no localization of industries. The pied and dappled geography of modern economies comes about precisely because of the wide variety of work being done at different places. Yet, the division of labor is not infinite: individualized work, where the person labors alone on a project like making a pair of shoes, is rare in advanced economies. Work today is mostly collective labor, where each person is responsible for a part of the whole. These collectivities range from small groups, such as fabric cutters, to whole garment factories, to entire commodity chains. A basic concept for the study of industrial geography is, therefore, the social division of labor.

The term industrial location was largely replaced by spatial divisions of labor in the lexicon of economic geographers during the 1980s. The former had come to mean the optimal siting of production units according to their specific needs for inputs, in the tradition of Alfred Weber (1909), or the optimal spatial allocation of sellers according to a highly abstract calculus of access to customers, in Walter

Christaller's (1935) and August Lösch's (1944) central place theory. Industry was assumed to conform to pre-existing patterns of people and resources on the land. Doreen Massey (1984) turned this around. For her, spatial divisions of labor signified a view of industrial patterns that recognized powerful forces for spatial differentiation coming out of industry itself and projected onto the landscape.

To speak of "industries" and their locations is to carve out specific arenas of social labor. Yet there are many ways of slicing up the industrial economy into observable units (Sayer and Walker, 1992). The most common definition of industries is on the basis of the product division of labor, or the distinction among types of output, such as the toy or shoe industry. Similarly, the most basic meaning of industrial location is the distribution of such sectors across places, as in case of the automobile industry of Detroit, the steel complex of Pyongyang, or the old cutlery district of Sheffield.

A difficulty immediately arises in that these sectors are really clusters of products, and can be further differentiated. The garment industry, for example, can be broken down into segments such as men's suits, sporting outfits, or blue jeans. These are likely to have surprisingly different locational patterns, as illustrated by the sports garments district of Los Angeles, the fashion industry of Paris, and the blue jean belt along the US–Mexico border. Conversely, one can lump industries into larger groupings, such as consumer goods and producer goods. Instead, studies of industry have gone in the other direction, paying closer attention to the fine grain of the social division of labor and its spatial manifestations.

Almost all products today are sufficiently complex that they require many parts, and several stages of production, before they are complete. This is rarely done under one roof. Whole hosts of intermediate inputs arrive at the typical factory to be assembled into a car or airplane (or at separate factories for car bodies or engines or axles before final assembly). A different sort of production occurs where the same material requires several stages of processing, as in the making of thread into cloth or refining of crude oil into petrochemicals. Here, too, the products are likely to move between factories before they are finished and ready for sale.

Such production systems are frequently called "commodity chains" (Gareffi and Korzeniewicz, 1994), but only some look like chains; most spread out like tree roots. Economic geographers find the linkages binding such production systems to be more significant than the bits and pieces considered separately (a car engine is a marvelous product, but who cares apart from the final automobile?). The geography of commodity chains may be regional, as in the case of Toyota's amazing subcontracting network in the Nagoya area (Tabb, 1995). Or their geography may be global, as is famously the case for Nike's network of subcontractors for sports shoes (Dongahu and Barff, 1991). Your personal computer, for example, may be assembled in Texas from a casing made in Taiwan, a liquid-crystal screen made in Japan, a disk drive put together in Silicon Valley, a CPU chip fabricated in Phoenix, and memory chips made in Malaysia (Dedrick and Kramer, 1998).

The definition of an industry and its geography is clouded by the way commodity chains branch into other commodity chains, becoming a thicket of connections. The formal name for this is "input–output systems." If we map out where every product goes and every part, piece, and machine comes from, we get a network of product flows criss-crossing almost every sector. Ponder the complexity of national input–output accounts and one is likely to despair of isolating finite industries at all. Yet businesses and financiers speak of the biotechnology or wood-products sector all the time, and most corporations know very well which sectors they operate in.

On closer inspection, then, every economy has practical ways of drawing boundaries around industries, even if those lines are fuzzy at the edges. For the economic geographer, this implies that the boundaries of industries are social conventions. These often have a strong national or regional character, and are rooted in past business practices as much as in the technical logic of production (Tabb, 1995; Herrigel, 1996). This is why seed companies in Germany are part of the chemical industry, whereas in the United States they have traditionally been independent – although they are now being transforming into biotech companies, some of which are attached to the chemicals industry and some to agribusiness companies. Similarly, computer production in Japan grew out of the domain of consumer electronics under expansive companies such as NEC and Hitachi, whereas in the United States it developed out of business machines (IBM, NCR) and new firms (Apple, Sun) (Henderson, 1989). Research into the location of biotech or electronics must be aware of national differences and changing sectoral boundaries.

Spatial Divisions of Labor: Geographies of Skill and Management

A quite different way of slicing up social labor is by occupations, or what individual workers do in their jobs. An occupational division of labor can be found within any workplace and every industrial sector, regardless of the final output, and it can be aggregated across sectors. The number of occupations certainly runs into the tens of thousands, with some highly specific to their sector, like cinematographers or meat-cutters. But the same general occupations can be found across widely different product industries: engineers, computer operators, machinists, secretaries, managers, and so on. These can, in turn, be lumped into highly aggregated categories such as "skilled," "semi-skilled," or "unskilled," the definition of which is both technical and social, and forever open to question.

Within any industry there are different workplaces with different locational patterns. Economic geographers have become keenly aware of two divisions of labor of vital importance for the functioning of all production (Sayer and Walker, 1992). The first is the expanding provenance of management, or the administration of industry. Managers fill up a substantial proportion of the occupational positions in what can be called a "hierarchical division of labor," accompanied by armies of accessory workers in clerical and personal assistance, computing and communications, billing and accounting, copyrights and litigation.

A second area of expanding industrial division of labor has been lateral – extending production in time. This includes several critical types of work, mostly of an intellectual and creative sort, that necessarily take place before a new product or process is introduced. One is research and development on new technologies. Big corporations have long set aside divisions for R&D, often far from manufacturing facilities: a good example was the Bell Labs of AT&T (now spun off as Lucent Technologies). Another is product design, as in the contrasting fortunes of Apple's fashionable iMac computers and its failed "Newton" think pad. All automobile corporations have design teams who come up with new makes and models every year.

While management, research, and design used to be part of the internal division of labor within the large firm, these functions have spilled out into the open market. This has led to rapid growth in enterprises and employment in the "business services" sector. Corporations have long sought outside help, of course; advertising has traditionally been an independent sector centered on New York's Madison Avenue. But all this has vastly expanded. Today, managerial functions are commonly turned over to consultants such as McKinsey & Company, engineering firms such as Bechtel provide technical design input, telemarketing companies handle phone sales, billing companies take care of consumer accounts, leasing companies supply office furniture, and food service firms staff the lunchrooms (on "financial services," see Leyshon, this volume).

Headquarters, R&D, and business services each have their own geographical calculus, and these have become a staple of economic geography (Daniels, 1993). Corporate managers and their high-level advisors cluster in prestigious locales, traditionally in major urban centers like New York. More recently, many have decamped to salubrious suburban estates not too far from the airports on the edge of big cities. But they are likely to consign their accessory workers to back offices in distant suburbs or cities, like the billing offices of Chevron in the outer reaches of the San Francisco Bay Area.

Sites of Production: Factories and Other Workplaces

In conjuring up an image of industrial production, most people think of the factory, which has been virtually synonymous with industrialization for most of the last two centuries. Factories are a highly visible and striking part of the landscape of industrialized countries, so shocking to early observers that they earned William Blake's epithet "dark satanic mills." But why do factories exist at all? We take them for granted, little considering that they are a way of coping with the division of labor and orchestrating industrial work.

Factories were a spectacularly successful innovation in business organization, crucial to the success of the industrial revolution and long pre-dating the modern corporation. The factory is a way of organizing social labor by bringing together many workers and tasks in one place, rationalizing the allocation of work and flow of materials over a limited space, allowing close oversight of workers by bosses, and driving machinery from a single power source (Nelson, 1995). Factories soon grew to embrace the work of hundreds or thousands of people. Some nineteenth-century New England textile mills were a mile long, Douglas's Long Beach aircraft plant once employed 100,000, and the steel mill complex at Tienshin, China, held over 200,000 workers.

Large factories include so much of the relevant social labor that they are more independent of linkages to other workplaces than smaller facilities. They have, therefore, tended to disperse toward the edges of cities or into rural sites where they dominate the landscape (Scott 1982). Pittsburgh's many steel mills spread out like stars in a spiral galaxy. Yet large factories often obscure the vital role of smaller workshops and plants in many industries, which have long clustered in industrial districts embedded in large and small cities (Scranton, 1997). Most such districts are rich combinations of large and small facilities, as in Silicon Valley today.

The most striking workplace of the twentieth century has been the office building, particularly the skyscraper. Offices are like factories in that they bring together large numbers of workers doing related tasks under direct supervision, with the provision of suitable equipment, from desks to photocopiers. But the occupations are different, "white collar" versus "blue collar," because the work is different: management, design, or marketing, not the hands-on labor of manufacturing. From humble beginnings off to the side of factories, they were set off on their own and driven ever skyward as they clustered together in the centers of big cities. The visual impact of the skyscraper at the turn of the twentieth century was as great as that of the satanic mills of a century before (Markus, 1993). But, like factories, many offices have shifted toward suburban office parks.

There are many kinds of workplaces besides factories and offices. Hotels, restaurants, airports, convention centers, and warehouses are common workplaces, and often as large as the biggest factories (Las Vegas hotels run to over 5,000 workers). Some workplaces are single open-air locales, like wheat fields, golf courses, and building sites. Others are extensive: cable networks, highway systems, or gas lines. Some workplaces are inside moving machines, such as ships and airplanes. Some workers move about from one designated site to another, as in theater or musical performance. And some roam freely, as in the case of truck-drivers or mail deliverers. Many sales representatives, designers, engineers, and repairers, not to mention gardeners and janitors, work in someone else's workplace instead that of their own employer.

The workplace is an essential building block in the organization of industrial production, and the invention of more effective workplaces is an important part of the evolution of business and management (Sayer and Walker, 1992). Workplace form has been neglected in contemporary industrial geography, which focuses on markets, firms, and networks. Yet it has been a key term (under such names as "plant" or "establishment") in the lexicon of economic geographers going back to Weber. Part of the logic of workplaces is dictated by technical considerations that give rise to economies of scale and scope in the shared use of machinery, bring together a set of tasks under one roof, or drive powerlines across hill and dale. But the social component in deciding workplace function, boundaries, scale, and location is sizeable. The early factory, with its high walls and clock tower, was a strict reminder of the new industrial order. Office towers and urban skylines are equally bold statements about the power of business. But the kind of megalomania entailed in Henry Ford's River Rouge plant, which did everything from make steel to assemble Model As, is rarer today. The importance of the large factory as a business strategy has been diminishing for the last half century, in favor of more dispersed, flexible, and externalized forms of organization (Noble, 1986; Nelson, 1995).

The Location of Firms and Corporate Geography

The other major "container" for production is the firm, and firms are key players in the geography of industry. We take the firm for granted, but the modern firm evolved in tandem with capitalism and the industrial revolution. The big advances in the nineteenth century were "limited liability" (which kept failed businessmen out of debtor's prison), joint stock holding (which made large investments by outsiders easy), and unlimited charters (which freed business from special legislation and state monopolies). Impersonal, professional management arose alongside these, and became generalized in the modern corporation of the twentieth century.

The modern corporation notwithstanding, other types of firm are still commonplace. Private companies and partnerships are typically small enterprises, although some are huge, like clothing manufacturer Levi Strauss. Syndicates and holding companies are a passive form of ownership without direct management, more popular in Britain than in the USA. State-owned enterprises have been a major component of most economies, though they are currently out of favor; they vary from wholly owned to quasi-public companies with minority government holdings, like the US postal service.

Size is a key dimension of the firm. Economists long ignored the question of size, following Alfred Marshall's (1890) assumption of a world of small firms that guaranteed price competition. Yet large corporations only became bigger as the twentieth century wore on, driven by economies of scale and throughput (as explained by historian Alfred Chandler, 1977). Only Ronald Coase (1937) saw clearly that in some cases small firms are the most efficient way of organizing production, and in other cases large firms. The comparative cost of large versus small units depends on the characteristics of product and commodity chains, technical competence and patent control, and size and reliability of markets. This is usually put, following Oliver Williamson (1985), in terms of "economies of scale and scope" in production (the advantages of combining or separating production systems) versus "transactions costs" of market exchange or administrative hierarchies. In practice, firms decide to take on new functions or divest themselves of peripheral activities they are not capable of managing.

Another way of putting this puzzle of company size and activity mix is to go back to the division of labor. The general question is where to draw boundaries around clusters of social labor. Only the smallest firms still specialize in only one product: most quickly diversify into several related products and many diversify into a wide range of product groups. Some, like the gigantic *chaebol* of Korea, orchestrate multisectoral industrial empires (Amsden, 1989). Management may handle such complexity by setting up autonomous divisions or profit centers. But firms today have become remarkably cautious about branching into domains outside their core competency.

Complexity and managerial strategy raise three key boundary problems in industrial organization and economic geography. First, an industry may consist of firms of varying sizes (and internal arrangements). Second, companies cross over industry lines. Third, the same sector may be organized differently in different countries (or regions). In short, there is no single best solution, and no a priori unit of production to be located. Not only do sectoral technologies and economics diverge, but local history, politics, and institutions also make a real difference. German industry never consolidated into giant firms to the degree that American industry did. British industry remained a province of family firms and holding companies longer than either. France and Taiwan relied more heavily on large state-owned companies in key sectors, yet both have many small family firms as well (Chandler 1990; Herrigel, 1996; Hsing, 1998).

Recently, firms of all nations, especially in high-tech sectors, have been rushing to form agreements, alliances, and joint-ventures of all kinds, which blur company, industry, and national boundaries. This presents another sharp challenge to management: how best to operate in a matrix of global relations. The hot-button idea of the last decade has been the "network firm" and how to build up and manage a complex set of contracts and alliances with complementary firms across an effective range of products and technologies (Castells 1996; Cooke and Morgan, 1998).

Changes in business practice have leveled off the percentage of the economy controlled by large corporations, with the proliferation of small and medium-sized enterprises in many sectors, especially those with rapid innovation like electronics and biotechnology. But disaggregation is effective in sectors such as fast-foods, where McDonald's and Burger King expand by franchise and keep the core firm focused on managing and marketing the whole assemblage. An extreme trend is toward the "virtual firm," in which all non-core functions, including production, supply, and sales, are externalized (Davidow and Malone, 1992). This has had a significant effect on industrial geography.

In classical location theory, the firm and the workplace were identical and focused on a single product. This reductionist approach was later countered by "corporate geography," which took as its object the sprawling patterns of operations within the large enterprise. At their best, studies of corporate strategy can be indicative of broad trends in the spatial economy (Schoenberger, 1997). But this approach, too, has run into limits. Economic geographers have recently directed their attention to clusters of smaller firms and factories, rediscovering insights of Weber on "agglomeration effects" and Marshall on "industrial districts" that had been neglected by mainstream economics and location theory. This research demonstrates that industrial districts have mutually beneficial effects, or external economies, that make them viable alternatives to the large factory and giant corporation (Storper and Salais, 1997; Scott, 1988b). Indeed, they have undergone a substantial revival in our times (Amin, this volume).

Geography Unbound: International Divisions of Labor

In a globalizing economy, industrial location can no longer be treated at the national level alone. There is a "new international division of labor" that demands attention from economic geographers. This has replaced, in part, the old trading system between those nations that exported manufactures and those consigned to the export of natural resources. There are several aspects of this global geography that grow out of what has already been said about sites, industries, firms, and goods/services. These have to be supplemented by considerations of national and continental scale boundaries.

With globalization, local sites of production become specialized within a worldwide division of labor, whether in tourism, steel, or textiles. A handful of factories may supply the demands of the entire world, such as watch makers in Hong Kong and Switzerland. The auto industry counts perhaps one hundred assembly plants in all. Studies show that the degree of geographic specialization is increasing with global integration (Storper, 1997). Such specialization is usually attributed to the lowering of barriers to trade and improvements in communications and transport.

Often forgotten, however, is the power of production at the workplace or industrial district. If a locale is to have global reach, it must be able to serve an immense number of customers, which requires a high level of productivity (and acceptance of mass-produced goods), as in tire manufacture. Or, local producers must offer a specific but widely desired product (technically sophisticated or highly fashionable), such as machines to etch silicon wafers from Applied Materials in Silicon Valley. In addition, large multinational corporations have been major carriers of investment and industry across national boundaries, helping to expand the industrialized world by leaps and bounds. In so doing, they imprint their internal hierarchies of labor and management on the global arena (Dicken, 1992).

Most big industries operate at a variety of territorial scales, from the local to the global (Storper, 1997; Scott, 1998). This can make their worldwide locational pattern seem "all over the place," and it takes closer analysis to reveal a geography of large and small clusters, networks of linkages between nodes, and sub-specialization and hierarchy. Aircraft, for instance, have a global geography that includes not only focal points such as Boeing in Seattle, but components in Los Angeles, electronics in Orange County and Silicon Valley, assembly in Alabama and Missouri, jet engines in Connecticut, or in Europe (around Toulouse, in southwest England, Silicon Glen, southern Germany, and central Italy) (cf. Dedrick and Kraemer, 1998).

A different geography prevails where products don't travel well. Manufactured goods are often contrasted with so-called services in this regard. Yet some services, like finance, consulting, and tourism, travel exceedingly well. Conversely, some basic activities, such as retailing, electric power, and prepared foods, are relatively immobile. Such local-serving industries tend not to have a strong spatial division of labor; rather, they are replicated from locale to locale. A way around this limit to size is to clone, either by merging regional firms (as in paint manufacturing) or creating a chain of identical facilities from place to place (as in fast foods). A false contrast is often set up between localized and globalized industries, when Taco Bell, for example, can be both.

There are, of course, boundaries that limit the flow of goods and services, not to mention labor and capital. Industrial production systems are not all internationalized, much less globalized. They still flourish within certain bounded arenas. National borders are the most obvious of these bounded spaces. Modern economic development cannot be understood apart from the emergence of nation-states since the eighteenth century (Hobsbawm, 1990). Consolidation of national markets and trade barriers has confined many industries within national boundaries, with distinctive patterns of production, business organization, labor relations, and consumer preferences (Chandler, 1990; Biernacki, 1994). For example, almost all large industrialized nations have their own automobile, weapons, garment, and machine-tool industries. They also have their own favorite specialities and competitive advantages that continue even as trade barriers break down, from fashion shoes from Italy to modernist wood furniture from Scandinavia, based in long-nurtured national skills, well honed tastes, or government favoritism (Best, 1990; Porter, 1990; Walton, 1992). The reported death of the nation-state has been greatly exaggerated.

One of the most striking shifts in global geography has been the independent development (endogenous growth) of newly industrialized countries. This is particularly true of East Asia's amazing rise to near parity with North America and Europe over the last fifty years. First Japan, then the "Four Tigers" (Taiwan, Korea, Hong Kong, and Singapore), and now China and Southeast Asia have industrialized at a breakneck pace (Amsden, 1989; Wade, 1990; Tabb, 1995). Their expansion has depended less on foreign direct investment than on high rates of internal saving and investment, under hothouse conditions forged by strong national states (Wade and Veneroso, 1998). They have often specialized in labor-intensive and technically unsophisticated products, dramatically altering the global pattern of industries like toys and garments in the process. This pattern depends a little on corporate hierarchies, to some extent on technical change (see below), and mostly on international differences in skills and wages (Webber, this volume).

Rising productivity in these places creates new patterns of international trade and takes a competitive toll on many industries in the advanced capitalist nations (Walker, 1999). This has forced a reconsideration of the dynamics of international competitiveness. Students of comparative development, such as Michael Porter (1990) and Peter Evans (1995), have come to realize how much of the "competitive advantage" of one country over another is not due to pre-given conditions, but to the way that labor productivity, product quality, and business capability can be improved through innovation, education, state assistance, and industrial strategy (Sheppard, this volume).

Another limit to globalization is the embrace of the three great continental trading blocs of the developed world, which have been growing faster than the global economy. The European Union (EU) is the leading exemplar, followed by the North American Free Trade Alliance (NAFTA) and the East Asian sphere. There has been a dramatic shift in Europe toward production for the EU and away from national markets, accompanied by increased competition and consolidation across national borders, whereas trade beyond the bloc has diminished (Urwin, 1995). The North American Free Trade Alliance consolidates American economic integration under US domination; here, too, Canada, the USA and Mexico are each other's largest trading partners (Orme, 1996). The countries of the East Asian region lack an equivalent to the EU and NAFTA, but Japan's overseas investment has fueled explosive growth in Southeast Asia, while investment from Hong Kong and Taiwan has done the same for China (Hsing, 1998; Brenner, 1998).

Industrialization: The Rising Tide of Production and Technology

The idea of a spatial division of labor is a powerful one, but remains static. Economic geographers must capture the relentless dynamism of modern industry. The great flaw in classical location theory is the belief that one already knows beforehand what industry looks like, and that industrial location consists of finding the best location for a known product and technique of production. Location, in this view, is like arranging furniture. One is searching for the optimal (least cost, highest profit) assignment of chairs and sofas, given the shape of the rooms. Once that arrangement is found, the furnishings are said to be at equilibrium. Location theory becomes a sort of economic *feng shui*, putting things in harmony with the gods of commerce.

This misses the dynamism of industrialization and industrial geography. As economic geographers discovered in the 1980s, industrial restructuring repeatedly breaks up the furniture in one corner of the house while adding new pieces in another. Accepted assumptions about industry are forever thrown into disarray along with the landscape. One has to rethink fundamentals, reject equilibrium and optimization analysis, and go back to classical political economy with its emphasis on the sources of growth and development (Storper and Walker, 1989; Swyngedouw, this volume).

The term industrialization has a somewhat antiquated ring today; one speaks of "technical innovation" instead. The technology of production has been rapidly and continuously improved over the last two centuries. "High tech" is the popular term for the leading edge of technology in our times, including such things as electronics, genetics, and aeronautics. But the age of High Tech is nothing new: flour and spinning mills were the high tech of the early nineteenth century, slaughterhouses and steel furnaces the wonders of the late nineteenth century, and electricity and chemicals the breakthrough domains of the early twentieth century. The Industrial Revolution is not a single event, but an upheaval that still goes on today (Landes, 1970; Mokyr, 1990; Pursell, 1995).

Industrialization is, first, a process of improving production methods, raising the productivity of labor and other inputs. Such improvements include the use of better tools, rationalization of tasks, application of machinery, automatic control of machines, moving assembly lines, savings on materials, and the like. Rising productivity means falling costs per unit of output, which translates on the consumer side into declining prices. This, in turn, brings more customers into the market and sharpens competition between best practice and lagging producers (Rigby, this volume).

Just as important are improvements to products and the creation of entirely new goods and services. From early on in the Industrial Revolution factories have sent forth a stream of commodities that never existed before, such as the locomotive, metal-cutting lathe, or analine dyes. That stream has become a veritable Mississippi of invention, making product proliferation an essential strategy of every industry from soft-drinks to software (now called "niche marketing"). Moreover, one of the clearest lessons of the new global competition has been the importance of quality control, design, and performance of goods. Not that this is a wholly new idea: consider the design of Wedgwood pottery or the reliability of Ford's Model T (Forty, 1986; Hounshell, 1984).

Changes in technique alter the input mix of manufacturers, which changes their locational calculus (in the terms of Weber). Improvements in meat-packing, for example, shifted the beef industry from East Coast butchers to Chicago's industrial packers at the turn of the last century. At the same time, shifts in competitive advantage and market share of firms and workplaces due to production advances will rearrange the overall geography of an industry. The industrial chicken-house has made the poultry districts of the South into major agribusiness centers (Boyd and Watts, 1997; Page, this volume). Finally, wholly original products give rise to new sectors of industry. The invention of the movie camera made a motion picture industry possible, leading to Hollywood, while a shift from mini- to micro-computers in the 1980s favored Silicon Valley over Boston's Route 128 electronics complex (Saxenian, 1994).

Organizational forms and management practices have come to the fore as an arena in economic geography. It used to be thought that business organization was rather static and foreordained; technological innovation referred to machines and chemicals. Firms were either small or large, competitive or oligopolistic. Little attention was paid to progress in management practices or the possibility of competitive advantage through organizational change, except to make the one-time jump from small to big (Chandler, 1977; but see Pollard, 1968; Beniger, 1986). This view has passed with the recognition not only of international differences in organizational methods, but also the rapid learning and innovation taking place in the face of just-in-time supply, global networking, small firm spin-offs, and flattening hierarchies of control. The Japanese system of mass production through better integration within and across firms shifted the geography of automobiles and consumer electronics dramatically toward East Asia during the 1980s, for example (Kenney and Florida, 1993). Today, the Silicon Valley networking model, which has given the place such a decisive advantage in global electronics, is all the rage with managers and governments around the world (Davidow and Malone, 1992).

Technology means human knowledge as much as machines, circuitry, or genetically altered organisms. The latter are the embodiment of a state of knowledge, involving scientific research, skilled labor, and practical know-how. This means that the education, training, and technical competence of the labor force is an essential consideration to industries and their locational calculations. A premium on technologically sophisticated labor keeps many industries operating within the compass of the advanced countries and has long been a barrier to the spread of industrialization to the backward areas of the world economy.

Moreover, knowledge acquisition, application, and improvement are part and parcel of technical innovation. That is why economic geography has taken a keen interest in "learning by doing" in the process of technical innovation. Early research on the geography of innovation concentrated on the location of corporate research and development labs, showing that these units demand the skilled scientific labor best found in favored regions such as New England, southeast England, or California (Malecki, 1980). But attention has shifted toward "learning regions", or the way collective interaction, practice, and creativity contribute to innovation and production improvements throughout an industrial territory (Storper, 1997; Malecki, 1997).

For all the R&D and learning involved, technology would not progress very far without the investment of large sums of money, which is the province of capitalists and governments. Nor would it proceed with such fearsome, manic energy without the spur of capitalist competition and lust for accumulation. Rates of technical change and productivity improvement track rates of investment rather closely, and the latter track rates of profit (Brenner, 1998). High rates of investment have been vital to such rapid growth places as Japan or California over many decades (Japan took advantage of high savings and capital controls; California benefited from massive stocks of gold, silver and oil; Walker, 2000). Nor should we forget the unfortunate spur that military competition and warfare have given to technical innovation, from the invention of interchangeable parts in the early nineteenth century in France and New England, to the creation of computers in the mid-twentieth century in Britain and the United States (Hounshell, 1984).

Uneven Development: Many Paths of Industrialization

Industrialization does not proceed evenly across the whole front of the division of labor or the whole world of industrial places, despite the rapid circulation and diffusion of new ideas. Different industrial sectors and segments have their peculiar conditions of production to cope with, including the nature of the product, type of demand, level of competition, and labor traditions. The result is that the same abstract forces compelling technical change are applied differently from one sphere to another (Storper and Walker, 1989).

Thus, only some kinds of industries are fruitful areas of mass (quantity) production, while others work in batches, and some by special order. Cars can be made en masse, but work stations are made in large batches, and specialized machinery for steel mills must be crafted on a custom basis. Big mass production factories tend to locate in more dispersed places than batch and custom workplaces, where unstandardized inputs, skilled labor, and continual interaction with customers and suppliers are the rule. The latter types of production generate strong agglomeration economies, as pointed out by observers of industrial districts (Scott, 1988a; Amin, this volume). Custom and batch production have also generated some of the most dramatic innovations of our time (Saxenian, 1994; Scranton, 1997), while many producers have backed off from the fetish of sheer volume in favor of a more careful coordination of supply with demand along the whole chain of production.

Different industries follow their own peculiar paths of industrialization. Most elementally, every sector has a sort of "technological backbone" that aligns the body of work around certain characteristics. Agriculture, for example, has never been assimilated into theories of automation based on manufacture, and health care does not conform well to conventional theses about mechanization. Ideas derived from automobile assembly do not carry over well to pharmaceuticals or textiles, where products and methods of production are so different. This means that sectoral studies are essential to economic geography and that abstracting from many sectors, rather than generalizing from only a few, is the only way to grasp the essential tendencies of geographical industrialization (Storper and Walker, 1989; Page, this volume).

Beyond the technical spine of industrialization lies the "soft body" of social development, which is anything but formless. One of most exciting ideas in contemporary economic geography is that industrial history is literally embodied in the present. That is, choices made in the past – technologies embodied in machinery and product design, firm assets gained as patents or specific competencies, or labor skills acquired through learning – influence subsequent choices of methods, designs, and practices. This is usually called "path dependence" or "industrial trajectories" (Dosi, 1984). It does not mean a rigid sequence determined by technology and the past, but a road map in which an established direction leads more easily one way than another – and wholesale reversals are difficult. This logic applies to industrial location, as well: a Silicon Valley, once established, takes on a life of its own, driving the electronics industry and building on the accumulated advantages of the past (Saxenian, 1994).

Technical and institutional practices of industry vary across regions, countries, and continents, as well as sectors. Economic geographers no longer make simple presumptions about the homogeneity of capitalism and industrialism, or a common trajectory for all industrialization. Japanese industry is not a replicant of American industry, which did not replicate British industry. When Japanese managers applied the lessons of American Fordism to their situation, they came up with a new hybrid, "Toyotism;" and when American car companies tried to learn from the Japanese, in turn, they came up with an altered system of "lean production" (Schonberger, 1982; Womack et al., 1990).

One can therefore speak of "national systems of innovation," "national patterns of business organization," and even "national capitalisms" because habits and styles of industrialization diverge between countries (Freeman, 1995). There are also substantial regional variations in technologies and industrial practices within supposedly homogeneous countries such as Canada and the United States (Rigby, this volume), and continental scale differences, as in comparing North American labor markets (flexible hire and fire, high youth employment, low entry wages) with those in Europe (greater job security, high entry wages, high youth unemployment) (Freeman, 1994). The same may be said of the contrast between East Asian statist development and American liberalism with its dread of state intervention (Wade, 1990).

The Production of Place and New Industrial Spaces

Recent upheavals in the world economy and its geography have forced geographers to come up with a more forceful way of imagining how the industrial map has been put together over time. In a dynamic world, production is not just the production of goods and services; industrialization is the production of industry itself and of the industrial world. Before the Industrial Revolution there were no corporations, no factories, no captains of industry, no high tech; there was nothing to locate. The whole apparatus of industrial production had to be created out of whole cloth, and with it the landscape of modern industry.

The act of industrial creation has not been once-and-for-all but a thing repeated over and over – industrialization itself. Before the invention of open-hearth furnaces, there was hardly a steel industry and no Ruhr Valley metals complex. Before the internal combustion engine, there was only a small oil industry, and no Houston petroleum equipment and refining district. Before the automobile, no Detroit or Coventry. Today's economic geographers may delve into the secrets of Silicon Valley, but California was no more than a province of Spain when the iron furnaces of Derbyshire and cotton mills of Lancashire were creating the first industrial landscapes of Britain. Or they go in search of the miracles of Chinese and Malaysian industrialization, hardly thought possible a generation ago.

Industries and their locales have to be built from scratch. Studies of industrial districts have been illuminating in this regard, because they call attention to the way industrial growth creates new pieces of itself continuously, whether through product innovation, firm spinoffs, or new machinery. Industrial district growth does more: it generates collective learning, new labor skills, machines never before seen, and new organizational forms. An expanding division of labor and proliferating external economies give a new twist to the theory of industrialization, which has traditionally focused on rationalization and mechanization within the factory. But the power of the district lies in the way the place produces more of itself (Scott, 1988a).

This "production of place" connotes something even more general and powerful than external economies and entrepreneurial start-ups. It means the creation of whole urban landscapes, from the factories to the houses of the workers, from the infrastructure to the commercial life of cities. It begins with capital investments, building workplaces and installing machines. It grows as the workforce and firms expand and proliferate. It draws in new migrants, who in turn support a mass of secondary industries and ancillary activities. It pays taxes to governments which build supporting infrastructure and develop their own dense bureaucracies. It generates profits to support further investment, in an upward spiral (Storper and Walker, 1989).

Large urban agglomerations like Tokyo mix a variety of industries that make metropolitan economies much more than a set of industrial districts, and send the scale of modern cities off the charts with each passing generation (the largest are now on the order of 25 million people) (Tabb, 1995; Hall, 1998). Industrialization has, in time, built up the entirety of the modern urban system of advanced capitalist countries and the great production heartlands of the English Midlands, the Midwest, the Rhine, or Lombardia. Very simply, it has created the industrialized world as we know it.

Industrialization does not unroll its wonder like a red carpet to the future, however. It jigs and jumps and explodes from one epoch to another, one place to another. New factories open up, while others close down. New firms rise like the phoenix, while others go bankrupt. Entirely new industries, never before imagined, march across the face of the earth setting up outposts like the Assyrians of old, bedecked in silicon and gold. Newly industrialized countries sprout on the margins of the world. One of the most important topics of investigation in economic geography is the development of such "new industrial spaces" – the impact craters of the meteor showers of industrialization (Scott, 1988b).

These spaces open up in the course of every sector's evolution. Industries normally relocate and rebuild several times over the long term. Steel in the USA began as iron-making in the ferrous bogs of New Jersey in the seventeenth century, then shifted west in the eighteenth century to eastern Pennsylvania's iron deposits and woods where charcoal could be made. When cast iron began to be produced with coal in the nineteenth century, the industry clustered around the anthracite regions. But with the shift to bituminous coal and Bessemer furnaces after the Civil War, Pittsburgh became the transcendent steel center. That pattern was altered as the iron deposits of Lake Superior came into use and the Midwest's boisterous demand for steel drew factories to the Great Lakes in the twentieth century. The sharp decline of American steel in the 1980s left the playing field open to a wave of Japanese and Korean mills in new sites across the Midwest.

New industrial spaces are not confined to the expansion of single industries. Industrialization periodically hurls huge new landscapes onto the map, as can be seen in the expansion of industrial Europe outward from Britain in the nineteenth century (Pollard, 1981), or in today's "Edge City" developments of north Dallas, the Francelline arc of southern Paris, and Orange County, California (Scott, 1998b; Garreau, 1991). They can revolutionize the geography of nations with remarkable speed. Mexico, for example, has been turned upside down geographically twice in the last half century: the tornado of Mexico City sucked in millions of people during the national industrialization era of the 1940s to 1960s; then came border industrialization under US auspices that put the North back atop Mexican development (Cravey, 1998). The way vigorous national economies enter the global scene is another case of new industrial spaces, as seen in places like Italy, Thailand, and Brazil. Indeed, the whole continental zone of East Asia swept onto the field of play of the industrialized world in our time, in the same way North America rose to challenge Europe a century before.

Nor should we forget the virtual deindustrialization of the former Soviet Union. One should never imagine that the unstable and uneven development of industrialization only moves in one direction, the upward curve of progress. Deindustrialization, which smote the northern cities of the USA, Britain, and France with such ferocity from the 1970s to the 1990s, is the frowning face of the geography of capitalism (Bluestone and Harrison, 1982). This has been a repeated threat since the dawn of industrial revolution, as can be seen in the ruins of the once-glorious iron district of Shropshire or the silk-weaving region of Lyon.

Conclusion

Our tour of the geography of production has no real conclusion. The restless hand of industry writes and then moves on, and the book of industrial revolutions has yet to be closed. Thus, economic geography is not a stale field of study, but partakes of the most startling developments of political economy. It forces history upon a reluctant social science and futurism upon recalcitrant historians. And it demands spatial imagination and a geographical turn of mind among every economist, technologist, or sociologist studying capitalism and the immense powers of production it has unleashed. As Marx and Engels (1848) put it eloquently in the year of the California Gold Rush, "All that is solid melts into air." Melts, rather, into space – a geography full of the sound and fury of modern industrial production.

Bibliography

Amsden, A. 1989. *Asia's Next Giant: South Korea and Late Industrialization*. New York: Oxford University Press.

Barnes, W. and Ledebur, L. 1998. *The New Regional Economies: The US Common Market and the Global Economy*. Thousand Oaks: Sage Publications.

Beniger, J. 1986. *The Control Revolution: Technological and Economic Origins of the Information Society*. Cambridge: Harvard University Press.

Best, M. 1990. *The New Competition: Institutions of Industrial Restructuring*. Cambridge: Harvard University Press.

Biernacki, R. 1994. *The Fabrication of Labor: Germany and Britain, 1640–1914*. Berkeley: University of California Press.

Bluestone, B. and Harrison, B. 1982. *The Deindustrialization of America*. New York: Basic Books.

Boyd, W. and Watts, M. 1997. Agro-industrial just-in-time: the chicken industry of postwar American Capitalism. In Michael Watts and David Goodman (eds). *Globalising Food: Agrarian Questions and Global Restructuring*. London: Routledge, 192–225.

Brenner, R. 1998. Uneven development and the long downturn: the advanced capitalist economies from boom to stagnation, 1950–1998. *New Left Review*, 229, 1–264.

Castells, M. 1996. *The Rise of the Network Society*. Oxford: Blackwell.

Chandler, A. 1977. *The Visible Hand*. Cambridge: Harvard University Press.

Chandler, A. 1990. *Scale and Scope*. Cambridge: Harvard University Press

Christaller, W. 1935. *Central Places in Southern Germany*. Engl. edition, 1966. Englewood Cliffs, NJ: Prentice-Hall.

Coase, R. 1937. The nature of the firm. *Economica*, 4, 386–405.

Cooke, P. and Morgan, K. 1998. *The Associational Economy: Firms, Regions and Innovation*. Oxford: Oxford University Press.

Cravey, A. 1998. *Women and Work in Mexico's Maquiladoras*. Lanham, MD: Rowman & Littlefield.

Daniels, P. 1993. *Service Industries in the World Economy*. Oxford: Blackwell.

Davidow, W. and Malone, M. 1992. *The Virtual Corporation*. New York: HarperCollins.

Dedrick, J. and Kraemer, K. 1998. *Asia's Computer Challenge: Threat or Opportunity for the United States and the World?* New York: Oxford University Press.

Dicken, P. 1992. *Global Shift*, second edition. London: Guilford Press.

Donaghu, M. and Barff, R. 1991. Nike just did it: international subcontracting and flexibility in atheletic footwear production. *Regional Studies*, 24, 537–52.

Dosi, G. 1984. *Technical Change and Industrial Transformation*. New York: St. Martin's Press.

Evans, P. 1995. *Embedded Autonomy: States and Industrial Transformation*. New York: Cambridge University Press.

Forty, A. 1986. *Objects of Desire: Design and Society, 1750–1980*. London: Thames and Hudson/Cameron.

Freeman, C. 1995. The national system of innovation in historical perspective. *Cambridge Journal of Economics*, 19, 5–24.

Freeman, R. (ed) 1994. *Working Under Different Rules*. New York: Russell Sage Foundation.

Garreau, J. 1991. *Edge City: Life on the New Frontier*. New York: Doubleday.

Gereffi, G. and Korzeniewicz, M. (eds) 1994. *Commodity Chains and Global Capitalism*. Westport: Praeger.

Hall, P. 1998. *Cities in Civilization*. New York: Pantheon.

Hanson, S. and Pratt, G. 1995. *Gender, Work and Space*. London: Routledge.

Henderson, J. 1998. *The Globalization of High Technology Production*. London: Routledge.

Herod, A. (ed) 1998. *Organizing the Landscape: Geographical Perspectives on Labor Unionism*. Minneapolis: University of Minnesota Press.

Herrigel, G. 1996. *Industrial Constructions: The Sources of German Industrial Power*. Cambridge University Press.

Hobsbawm, E. 1990. *Nations and Nationalism since 1780*. Cambridge: Cambridge University Press.

Hounshell, D. 1984. *From the American System to Mass Production, 1800–1932*. Baltimore: Johns Hopkins University Press.

Hsing, Y. 1998. *Making Capitalism in China: The Taiwan Connection*. New York: Oxford University University Press.

Kenney, M. and Florida, R. 1993. *Beyond Mass Production: The Japanese System and its Transfer to the United States*. New York: Oxford University Press.

Landes, D. 1970. *The Unbound Prometheus*. London: Cambridge University Press.

Lösch, A. 1944. *Die Räumliche Ordnung der Wirtschaft*. Jena. English transl. 1954.

Malecki, E. 1980. Corporate organization of R&D and the location of technological activities. *Regional Studies*, 14, 219–34.

Malecki, E. 1997. *Technology and Economic Development: The Dynamics of Local, Regional, and National Competitiveness*. Harlow, Essex: Longman.

Markus, T. 1993. *Buildings and Power: Freedom and Control in the Origin of Modern Building Types*. London: Routledge.

Marshall, A. 1890. *Principles of Economics*. London: Macmillan.

Marx, K. and Engels, F. 1848. *Manifesto of the Communist Party*. Moscow: Progress Publishers. 1952 edition.

Massey, D. 1984. *Spatial Divisions of Labor: Social Structures and the Geography of Production*. London: Macmillan.

Mokyr, J. 1990. *The Lever of Riches: Technological Creativity and Economic Progress*. New York: Oxford University Press.

Nelson, D. 1995. *Workers and Managers: Origins of the New Factory System in the United States, 1880–1920*, second edition. Madison: University of Wisconsin Press.

Noble, D. 1986. *Forces of Production: A Social History of Industrial Automation*. New York: Oxford University Press.

Orme, W. 1996. *Understanding NAFTA: Mexico, Free Trade, and the New North America*. Austin: University of Texas Press.

Pollard, S. 1968. *The Genesis of Modern Management*. Harmondsworth: Penguin.

Pollard, S. 1981. *Peaceful Conquest: The Industrialization of Europe, 1760–1970*. New York: Oxford University Press.

Porter, M. 1990. *The Competitive Advantage of Nations*. New York: The Free Press.

Pursell, C. 1995. *The Machine in America: A Social History of Technology*. Baltimore: Johns Hopkins University Press.

Saxenian, A. 1994. *Regional Advantage: Culture and Competition in Silicon Valley and Route 128*. Cambridge: Harvard University Press.

Sayer, A. and Walker, R. 1992. *The New Social Economy: Reworking the Division of Labor*. Oxford: Blackwell.

Schoenberger, E. 1997. *The Cultural Crisis of the Firm*. Cambridge: Blackwell.

Schonberger, R. 1982. *Japanese Manufacturing Techniques*. New York: The Free Press.

Scott, A. 1982. Production system dynamics and metropolitan development. *Annals of the Association of American Geographers*, 72, 185–200.

Scott, A. 1988a. *Metropolis: From the Division of Labor to Urban Form*. Berkeley and Los Angeles: University of California Press.

Scott, A. 1988b. *New Industrial Spaces*. London: Pion.

Scott, A. 1998. *Regions and the World Economy: The Coming Shape of Global Production, Competition, and Political Order*. New York: Oxford University Press.

Scranton, P. 1997. *Endless Novelty: Specialty Production and American Industrialization, 1865–1925*. Princeton: Princeton University Press.

Storper, M. 1997. *The Regional World: Territorial Development in a Global Economy*. New York: Guilford Press.

Storper, M. and Walker, R. 1989. *The Capitalist Imperative: Territory, Technology and Industrial Growth*. New York: Blackwell.

Storper, M. and Salais, R. 1997. *Worlds of Production: The Action Frameworks of the Economy*. Cambridge: Harvard University Press.

Tabb, W. 1995. *The Postwar Japanese System*. New York: Oxford University Press.

Urwin, D. 1995. *The Community of Europe: A History of European Integration Since 1945*, second edition. London: Longman.

Wade, R. 1990. *Governing the Market: Economic Theory and the Role of Government in East Asian Industrialization*. Princeton: Princeton University Press.

Wade, R. and Veneroso, F. 1998. The Asian crisis: the high debt model versus the Wall Street–treasury–IMF complex. *New Left Review*, 228, 3–24.

Walker, R. 1999. Putting capital in its place: globalization and the prospects for labor. *Geoforum*, 30, 3, 263–84.

Walker, R. 2000. California's golden road to growth, 1848–1940. *Annals of the Association of American Geographers* (forthcoming).

Walton, W. 1992. *France at the Crystal Palace: Bourgeois Taste and Artisan Manufacture in the 19th Century*. Berkeley: University of California Press.

Weber, A. 1909. *Über den Standort der Industrien*. Tübingen. English trans. C. J. Friedrich, University of Chicago Press, 1929.

Williamson, O. 1985. *The Economic Institutions of Capitalism: Firms, Markets, Relational Contracting*. London: Macmillan.

Womack, J., Jones, D., and Roos, D. 1990. *The Machine That Changed the World*. New York: Rawson Associates.

Chapter 9

Places of Work

Jamie Peck

Labor is at the same time "the most fundamental and the most inherently problematic of all economic categories" (Block, 1990, p. 75). Contemporary economic geography now takes it as axiomatic that work and workplace restructuring are inherently social processes, that labor markets are structured by power relations and institutional forces, and that above all these are *social-spatial* phenomena. But it was not always so. Neoclassical economics maintained that labor, which after all is bought and sold on labor "markets," should be regarded like any other commodity. Along with capital and raw materials it is simply another "factor input" to the production process, whose price (wages) will fall when supply exceeds demand and rise when there is a shortage of workers, and so forth. When economic geography was in thrall to neoclassical economics, back in the 1960s and 1970s, this was how labor was conceived. As just another economic variable, labor represented little more than an accounting line in the calculus of industrial location, its price and availability routinely quantified and mapped. In these ostensibly apolitical accounts, labor was rendered an unproblematic category. Workers did not seem to strike, raise children, or make creative inputs into the production process. And while the economic geographers of the time might study journey-to-work patterns, they rarely delved into what the workers were *doing*, either at home or at work.

Some time in the 1970s, this began to change. A new generation of economic geographers, taking their cues from Marxian political economy rather than orthodox economics, sought to place labor–capital relations and workplace reorganization at the very center of their accounts of industrial restructuring and regional economic change. This new concern with "labor geography" represented more than a shift in academic fashion, for it reflected profound changes that were underway in the world outside. Waves of job losses and plant closures had begun to sweep through the old industrial heartlands of North America and Western Europe, triggering defensive responses on the part of communities and labor unions. As Bluestone and Harrison (1982) saw it, this new process of deindustrialization and disinvestment was pitching "capital against community" in ways that were profoundly destructive but also potentially radicalizing. Many of the old certainties, indeed the economic "rules of the game" themselves, were being brought into question, as large-scale labor-shedding and unemployment became persistent realities. Few could be certain about the future course of events, but most agreed that neoclassical notions of "equilibrium" and seamless economic adjustment were now singularly inappropriate.

Over the 25 years that have passed since the emergence of "labor geography," this vibrant subfield of economic geography has continued to thrive and develop. In the early days, the focus was on the ways in which the changing organization of work and production systems in manufacturing industry reflected the deep dynamics of capitalist accumulation and restructuring. More recently, issues of race and gender, political agency, and worker identities have received attention, along with a consideration of the institutional and social "embeddedness" of labor markets. Empirically, the field of vision has been broadened to encompass domestic and service work as well as factory labor, while at a theoretical level there is a greater concern now with complex processes of governance and regulation as well as with the underlying dynamics of capitalist accumulation. Issues around labor, and its roles in production and reproduction, open up these questions in ways that are often challenging but never less than revealing. Indeed, for Harvey (1989, p. 19) "[t]he history of the urbanization of capital is at least in part a history of its evolving labor market geography." The purpose of this chapter is to highlight some of the key developments, directions, and debates in labor geography, starting at the beginning and with some of the fundamentals – the process of industrial restructuring.

Restructuring Industrial Regions

In the early 1960s, the manufacturing regions of North America and Western Europe had never had it so good. Employment was plentiful, wages had been growing year-to-year, and the conspicuous benefits of the "consumer economy" had filtered down to the men and women who *made* the automobiles, the washing machines, and the electric toasters. The big factory and the suburban home came to epitomize the "modern" economy. But some time in the late 1960s, the trend began to bend, silently at first. The onset of "deindustrialization," signaled by absolute declines in manufacturing employment and output, represented a key moment for what all had grown accustomed to calling the "advanced industrial nations." In the 1970s and 1980s, many communities – which in an earlier era had literally been built by industrialization, and had evolved distinctive working-class cultures and patterns of political organization – found themselves abruptly abandoned, or held to ransom, by industrial capital. As the "rustbelt" regions of North America and Western Europe went into apparently terminal decline, and as factories and jobs relocated to low-cost, non-unionized locations, critical attention was focused on the underlying *causes* of these changes in the restructuring of production. The concerns of economic geography assumed an unprecedented relevance – and indeed a political potency.

The main challenge was to provide an explanation for a phenomenon that was clear to everyone – that the shifting geography of industry and employment was somehow related to processes of restructuring. Were the problems of capital flight and economic decline problems *affecting* the region or problems *in* the region?

Perhaps the most compelling response to these questions came in the form of Doreen Massey's notion of the "spatial division of labor," which linked economic development and restructuring processes at the regional and local scales with wider, national and international changes in the organization of production. Massey explored the relationship between the changing imperatives of production and the emerging geographies of employment, arguing that these shifts in the "spatial structures of production" should be understood as:

...changes in class relations, economic and political, national and international. Their development is a social and conflictual process; the geography of industry is an object of struggle... [So] new spatial divisions of labor are more than just patterns, a kind of geographical re-shuffling of the same old pack of cards. They represent whole new sets of relations between activities in different places, new spatial patterns of social organization, new dimensions of inequality and new relations of dominance and dependence. Each new spatial division of labor represents a real, and thorough, spatial restructuring. There is more than one kind of "regional problem" (Massey, 1984, pp. 7–8).

Massey's account rested heavily on the "labor factor" in the process of industrial restructuring – an analytical privilege also assigned by Storper and Walker (1983; Walker and Storper, 1981) – but she did not *reduce* the industrial location decision to this single dimension. Rather, she stressed how changes in the workplace and in the wider division of labor are deeply implicated in processes of economic restructuring. So it is not just the labor process (the social relations of work at the point of production) that "determines industrial location, but the search after profit and the fluctuating conflict between capital and labor" (Massey, 1984, p. 25). Decisions about labor and location are inescapably entangled, but one does not determine the other.

Massey's account reveals how geographical differentiation itself is exploited by capital in the competitive search for profit. Crucially, geography is part of the calculus of profitable production here: changes in production generate new demands for labor and therefore new locational imperatives; in turn, these reorganized landscapes of labor trigger further restructuring opportunities/threats. Her central contention is that production is organized systematically – not randomly – over space, the underlying dynamic of the system deriving from the competitive search for profit. Massey's analysis of the changing geography of production is embedded within a Marxian framework in which the interactions of economic, social, and political factors are afforded significant roles – an approach which marked a fundamental break with the neoclassical tradition. Hers is a complex and nuanced account, but at root it can be summarized in six points:

- Imperatives of global competition are driving processes of industrial restructuring.
- Restructuring is in turn entailing the reorganization and "breaking up" of established labor processes and production chains, as managerial functions ("control") are increasingly separated from those of manual labor ("execution").
- The separation of control from execution has a geographical manifestation, as different functions are located in accordance with the geography of their preferred labor supply (typically, with "deskilled" execution and assembly functions

locating in peripheral areas, whereas control and R&D functions concentrate in dominant cities and their satellites).

- This process is conceptualized as one in which "rounds of accumulation" unfold across the economic landscape, producing in their wake new geographies of production (Walker, this volume) and new sets of relationship between places (which may be gaining or losing functions in new rounds of accumulation). Consequently, the "layering" of successive rounds of accumulation results in changes to both the pattern of regional inequality and the form of inter-regional relationships, as regions gain or lose control functions, such as headquarters operations, for example.
- New spatial divisions of labor are forged through the complex process by which new rounds of accumulation (say, in financial services or biotechnology) interact with, and remake, pre-existing geographies of production and social relations, resulting in a range of locally contingent "combination effects."
- A reciprocal relationship is therefore envisaged between local economic, social, and political conditions and unfolding rounds of accumulation, one conceptualized through the notion of the spatial division of labor.

The spatial division of labor framework consequently permits analysis along two dimensions. First, processes of restructuring in companies and sectors can be conceived in terms of unfolding *rounds of accumulation*, "a product of the interaction between, on the one hand, the existing characteristics of spatial differentiation and, on the other hand, the requirements at that time of the particular process of production" (Massey, 1979, p. 234). From this sectoral perspective, the fundamentally uneven development of production (and its associated social relations) can be conceptualized within a framework that foregrounds the strategic calculations of profit-seeking capital confronted with constantly shifting landscapes of labor qualities and (therefore) of opportunities for profitable production. Secondly, the changing economic fortunes of individual *localities* (both "internally" and relative to other places) can be understood as a "complex result of the combination of [their] succession of roles within the series of wider, national and international, spatial divisions of labor" (ibid., p. 235). Local economic and social structures can consequently be seen to derive, in large part, from the distinctive role played by localities in successive spatial divisions of labor, understood here as a cumulative, local "layering" of rounds of accumulation, mediated and shaped by specifically local interactions and responses, such as those associated with local political systems or labor-union traditions ("combination effects").

This notion of layering has led some to suggest that Massey's framework rests fundamentally on a geological metaphor, in which:

...successive rounds of accumulation deposit layers of industrial sediment in geographical space. That sediment comprises both plant and persons, the qualities of the latter, deposited in one round, being of primary importance at the beginning of the next round....The chain of reasoning about the political effects of industrial change [is] a direct one, from industrial structure, through occupational structure, to regional class structure. This is not necessarily an objectionable initial procedure, but it renders all spatial effects as class effects (Warde, 1985, pp. 196–7).

Massey herself has never used this simple metaphorical device, and has since laid stress on the complex interactions between class relations and non-class relations. Nevertheless, Warde's critique is revealing as it draws attention to the way in which her account of regional economic change is unambiguously anchored in the dynamics of *accumulation* (see Herod, 1997). Although the imperatives of profitable production and labor control may be relatively unambiguous in the abstract, the particular forms these take locally are in fact quite variable. Moreover, the profit motive itself does not determine which labor control strategy is adopted. As a result, the outcomes of restructuring have been rather more variegated than was apparent in Massey's early work, which at least empirically was a product of both its time and its place – the onset of deindustrialization in the UK (Lovering, 1989).

Together with the parallel work of Storper and Walker (1983) and Bluestone and Harrison (1982) in the USA, Massey's spatial division of labor approach helped to establish a new orthodoxy in economic geographical research in the 1980s. Issues relating to the "labor factor" were foregrounded as researchers sought to chart the restructuring strategies associated with different industries/sectors and different regions/localities (see Clark et al., 1986; Cooke, 1986; Martin and Rowthorn, 1986; Scott and Storper, 1986). Reflecting the material conditions and political concerns of the time, much of this work focused on the interrelated processes of deindustrialization and job loss, the reorganization of corporate hierarchies and work systems, and – perhaps above all – regional economic decline.

In retrospect, we can now see that this "restructuring approach" had been forged through an analysis of what would later be characterized as the "Fordist crisis." The 1970s had witnessed an historic downturn in the mass production economies of Western Europe and North America, coinciding with political and institutional attacks on the Keynesian welfare state and organized labor. What is striking is that, prior to the mid-1980s, comparatively little attention had been given to *growing* regions or *growing* industries, the other side of the coin of "restructuring." In a remarkably prescient article, Sayer (1985) observed that much "radical" research of the time seemed unduly preoccupied with mass production or assembly industries, employing male manual workers, and typically based in working-class regions with traditions of combative industrial relations. But this was about to change, as the focus shifted from the old industrial regions of Fordism to the new economic spaces of "flexible accumulation."

Flexibilizing economic spaces

By the late 1980s, the buzzword "flexibility" had become a powerful political–economic signifier for a host of changes in production and labor relations. Accounts of economic change increasingly began to deploy "transition models" that contrasted the (old) mass-production regime of Fordism with an ascendant regime of flexible accumulation or "post-Fordism" (cf. Amin, 1994). Both conservatives and radicals began to point to what many saw as structural shifts in the organization of labor markets and production systems. On the one hand, a range of managerially orientated, prescriptive studies of the "new flexibility" not only described, but actively advocated and even celebrated, moves towards flexible employment arrangements and lean corporate structures as "solutions" to the problems of low profitability and productivity. On the other hand, writers drawing on the regulation approach, and working at the macro rather than the micro level, underlined the unstable and contradictory nature of "after-Fordist" experiments in the production system and labor market. Somewhere between these two positions, another influential group of researchers drew attention to a series of meso-level shifts in industrial organization and regional growth systems, which they saw as indicative of an emerging paradigm of "flexible specialization."

In pronounced contrast with much of the previous work on industrial restructuring, with its focus on the (slowly moving) deep dynamics of capitalist profit-seeking and labor control, the transition models focused almost exclusively on *change*. They drew attention specifically to *new* forms of production organization, *new* methods of labor management, and *new* geographical dynamics, purposefully contrasting these with the previous practices of Fordism. Table 9.1 provides a stylized account of the key differences that have been identified between Fordist and contemporary, "flexible" labor markets. The far-reaching "flexibilization" of labor markets is an indisputably significant development, but there is continuing debate over what it means causally or historically. Specifically, does this represent a new "regime" of post-Fordist labor relations, or the continued unraveling and restructuring of the Fordist system? Transition models, of course, tend to privilege the former interpretation, though there is clearly a danger that, in focusing on change and discontinuity, a host of loosely-related contemporary developments are bundled together indiscriminately, when all they share is the (causally trivial) characteristic of novelty. Moving beyond such simplified "binary histories" (Sayer, 1989), of an institutionalized, rigid past and a market-orientated, flexible future, calls for a careful analysis of the logic, sustainability, and incipient contradictions of contemporary developments.

It must be acknowledged that real-time analysis of such potentially "systemic" shifts in economic structures and regulatory conventions is never going to be an easy task. Theoretically informed analyses of suggestive empirical developments, however, may open up new and productive lines of inquiry or fields of debate. Take the geography of post-Fordism. It has been argued that flexible production systems are associated with distinctively new spatial dynamics, underlined by the shift away from the sectorally specialized, old industrial regions of Fordism to what Scott has called the "new industrial spaces" of flexible accumulation. The locational logic of flexible accumulation consists, according to Scott (1988c, pp. 11–15), of a dual tendency for, first, the evasion of Fordist labor pools (with their politicized working class, institutionalized labor processes and high cost-structures), and second, the selective re-agglomeration of production in locations socially or geographically insulated from the core regions of Fordist industrialization. Empirical examples of new industrial spaces include regions of conspicuous growth and dynamism, such as Emilia-Romagna in Italy, Baden-Württemberg in Germany, Cambridge in the UK, and Silicon Valley in the United States, but nodes of flexible growth have also been identified in the central business districts and suburban extensions of cities like Los Angeles. Significantly, flexible production/labor-market norms established in these new industrial spaces are also seen to be imported back into the old industrial regions, through a sort of "backwash effect," as the new regime of accumulation becomes established. Through this process, the logic of flexibility assumes the role

Table 9.1 Ideal–typical forms of work organization in Fordist and after-Fordist labor markets

	Fordist-Keynesian labor markets	After-Fordist labor markets
Production organization	Mass production, typified by the large, integrated, capital-intensive factory producing standardized outputs	Flexible production, typified by the networked, vertically-disintegrated firm producing specialized goods and services
Labor process	Deskilled and Taylorized, based on a detailed division of labor and routinized tasks; socially alienating work regimes	Flexible, based on increased use of contingent labor and cross-skilling/adaptability amongst core workforce
Industrial relations	High union densities; worker rights strongly embedded in law and convention; centralized bargaining	Disorganization of trade unions; individualization of employment relations; decentralized bargaining
Labor segmentation	Institutionalized; rigidly structured work hierarchies according to skill and status; large, internal labor markets	Fluid; deepening segmentation between core and periphery; break-down of internal labor markets
Employment norms	Privileging male, full-time workers; occupational stability and job security	Privileging "adaptable" workers; normalization of employment insecurity
Skills and training	Occupationally specific; focused on young workers, typified by the trade apprentice	Generic, broad-based training; lifelong learning and reskilling, typified by the part-time student
Income distribution	Rising real incomes and declining pay inequality	Polarization of incomes and growing pay inequality
Domestic sphere	Unacknowledged appendage to the "mainstream" economy; presumption of women's unpaid domestic labor, reinforced by gendered welfare regime	Commodification of domestic economy; rising female participation in wage labor; cuts in welfare; displacement of job-market risks to households
Macroeconomic strategy	Maintenance of aggregate demand; smoothing of business cycles; securing rising productivity and incomes	Maintenance of low inflation; deference to international financial markets; enforcement of labor flexibility
Labor market policy	Full employment: secure and high level of male employment	Full employability: ensuring workforce adaptability
Welfare regime	Expansive welfare state, based on (legally embedded) principles of entitlement and universality; welfare as a wage "floor"	Localized workfare systems, based on mandatory participation in work or simulated work; enforcement of contingent work
Local characteristics	Distinctive, specialized and relatively stable regional employment systems	Flexibly localized and volatile employment systems; ubiquitous rise of service employment
Scale characteristics	Privileging of national economy for macroeconomic management and labor regulation	De-privileging of national scale; deference to global economic imperatives; decentralization and devolution of labor regulation
Geographical tendencies	Dispersal of employment and production	Agglomeration of employment and production

of the dominant mode of organization in both production systems and labor markets.

Scott's (1988c) theorization of new industrial spaces begins with the reorganization of the production system, but is closely attentive also to the parallel ways in which labor markets are being restructured. In particular, he highlights a trend towards dynamic vertical disintegration in the production system, as companies seek to enhance their flexibility and responsiveness by "externalizing" many of the functions previously performed within the firm. This amounts to more than a "breaking up" of established production chains, because flexible production systems are portrayed as expansionist and innovation-rich: as the system expands, new and independent forms of specialist production emerge. In turn, this creates a deep and complex social division of labor, as individual producers become locked into "networks of extremely malleable external linkages and labor market relations" (Scott, 1988a, p. 174). The growing importance of small, specialized producers (of services as well as goods) is associated with heightened interdependence in the production system, as firms become deeply embedded within complex webs of inter-organizational transactions. This is where new geographical dynamics become evident, as the dual imperatives of minimizing external (inter-firm) transaction costs and of establishing an appropriate set of labor-market relations bring about a marked agglomeration of economic activity. New industrial spaces, then, are seen as densely networked centers of intensive innovation in production and labor practices (Amin, this volume). Empirically, they were also the "hot-spots" of growth in the late 1980s, the mirror image in many ways of the declining industrial regions of the 1970s.

Contrary to those, such as Hudson (1989), who see the trend towards "flexible" labor markets as a plundering strategy on the part of capital (exploiting labor's vulnerability under conditions of high unemployment and political disorganization), Storper and Scott (1990) seek to attach these developments to the flexible reorganization of production. They point to three important restructuring strategies that are currently being used by firms in their attempts to "flexibilize" labor relations:

- There is an attempt to individualize the employment relation, moving away from institutionalized collective bargaining and negotiation systems in spheres such as wage-setting and training.
- Firms are seeking to achieve enhanced internal flexibility through labor process changes such as multi-skilling and reduced job demarcation.
- External flexibility is being sought through strategies that enable rapid quantitative adjustments to the labor intake to be made in accordance with fluctuating production needs (such as the deployment of part-time and temporary workers).

Scott argues that the search for labor-market flexibility serves to intensify agglomeration tendencies. "Flexible workers" tend to be drawn to large, volatile job markets, where their chances of finding (and re-finding) work are highest. In turn, employers of such workers gravitate toward the spatial core of their preferred labor supply. Scott suggests that processes of local socialization reinforce these agglomeration tendencies, as workers become acclimatized to particular work rhythms and as they develop appropriate job-market coping strategies. Over time, patterns and processes of labor-market behavior thus become locally embedded and, to a degree, self-perpetuating.

Scott's account of the locational dynamics of new industrial spaces is attractive, in part, because it is linear and logical: first, uncertainty and fragmentation in markets leads to dynamic vertical disintegration in the production system; second, this increases the reliance of firms on external transactions, both with other firms and in the labor market; and third, spatial agglomeration enables these external transaction costs to be minimized, while also embedding flexible employment norms as the dominant ones in local labor markets. Lovering (1990) has argued, however, that the effect here is to privilege a particular form of restructuring strategy, when in fact firms tend to mix and match different strategies in a wide variety of ways. Lovering maintains that the outcomes of restructuring are more complex, contingent, and contradictory than Scott allows for.

The tendency to prioritize certain restructuring strategies, and then to seek to link these in a quasi-functional way with particular spatial outcomes, is in many ways a recurrent theme in economic-geographical research. Scott's stylized account links flexible production and the flexible utilization of labor with the emergence of new industrial spaces: flexible production ⇒ flexible labor utilization ⇒ agglomeration. In some senses, this represents the flip-side of the restructuring strategies which were emphasized in the early 1980s, when the drive to maximize profits was equated with the intensification of work and the de-skilling of labor, and subsequently with the creation of "branch-plant" economies: dynamics of accumulation ⇒ labor control ⇒ decentralization. In reality, of course, firms are confronted with a wide range of possible restructuring strategies; some involve spatial relocation, whereas others do not. Likewise, while the reorganization of labor relations invariably plays an integral part in the restructuring process, it is not (pre)determined by this process.

Thus there is no one-to-one correspondence between restructuring imperatives, labor strategies, and geographical outcomes, but a picture of considerable variety and contingency, within which the "contextual" circumstances of different local labor markets and regulatory regimes tend to assume particular significance. For example, the imperative to drive down production costs while weakening the influence of organized labor might result, under one set of local circumstances, in a firm relocating to a greenfield site, complete with its typically "green" labor supply, while in other local circumstances a firm might utilize *in situ* restructuring strategies such as subcontracting or homeworking (Peck, 1996). The cost and viability of alternative strategies tends to be strongly conditioned by local labor-market factors – such as the pattern of inter-sectoral competition for labor, the local industrial relations climate, and the social organization of the labor supply. Thus local labor markets are not simply by-products of employers' restructuring strategies, or the "background scenery" of the restructuring process, but are deeply implicated in shaping and structuring these strategies.

Localizing Labor Markets

"Unlike other commodities," Harvey (1989, p. 19) has pointed out, "labor power has to go home every night." Labor markets should not be conceived as narrowly "economic" systems; they are also "lived," social spaces. The capacity to work

– labor power – is socially produced and reproduced; processes which tend themselves to be culturally embedded, institutionalized, and locally specific. Local labor
markets develop distinctive characters, structures, and dynamics, in part by virtue of
the way that the institutions of labor reproduction (education systems, family
support networks, social service regimes, and so forth) evolve in a geographically
uneven fashion, but also because the daily mobilization of labor for waged employment is essentially a local matter. Storper and Walker note that,

> . . . labor's relative day-by-day immobility . . . gives an irreducible role to place-bound homes
> and communities. . . . It takes time and spatial propinquity for the central institutions of daily
> life – family, church, clubs, schools, sports teams, union locals, etc. – to take shape. . . . Once
> established, these outlive individual participants to benefit, and be sustained by, generations of
> workers. The result is a fabric of distinctive, lasting local communities and cultures woven
> into the landscape of labor (1989, p. 157).

Local labor markets do not exclusively mirror the (unmediated) imperatives of
capital accumulation and labor control, but reflect the strategies – individual and
collective – of workers, at home, at work, and in between (Hanson and Pratt, 1995;
Wright, 1997; McDowell, 1999). They are also deeply structured by race and gender
relations. These are not simply "secondary" influences, but profoundly shape the
ways that labor markets work. For example, McDowell (1991) examined how
changes in the prevailing "gender order" are deeply implicated in patterns of economic restructuring. The crisis of Fordism was not only associated with the destruction of manufacturing jobs, it also signaled the demise of certain forms of
patriarchal, working-class community and of a particular pattern of industrial
masculinity. Working-class young men in deindustrialized areas, in particular, were
denied access both to factory jobs and to one of the principal means of affirming
their masculinity and independence, with social alienation being the predictable
consequence. On the other hand, the "post-Fordist" labor market is, in some
respects, a feminized one. Not only do women comprise a large share of the workforce, but the design of jobs themselves has also been feminized in the sense that job
characteristics like flexible work scheduling, service cultures, and multi-tasking –
established features of "women's work" – are fast becoming normalized. The decline
of the old gender order of Fordism – based on full-time male employment and a
masculine factory culture, nuclear families reliant on the unpaid domestic labor of
women but headed by male "breadwinners," and underpinned by a Keynesian
welfare state which presumed (and reproduced) these gender relations – illustrates
how processes of economic change do not just "result in" new gender divisions but
are themselves constituted in part through gender relations.

Gendered identities are not simply "taken to work," but are also reproduced
through workplace cultures (see McDowell, 1999). Moreover, "work" and
"home" are not the hermetically sealed, separate spheres they were once assumed
to be, but interpenetrate in a variety of ways, strongly conditioning the job-market
behavior of both men and women. Hanson and Pratt (1995) reveal how both
landscapes of employment opportunity and channels of access to jobs are spatially
uneven, creating a situation of "labor-market containment" in which women's
marginality is reinforced. They demonstrate, for example, how labor-market

participation is profoundly shaped by the networks of spatial–temporal routines which women must construct (and repeatedly renegotiate) between waged work on the one hand, and various domestic and caring tasks on the other. This has the effect of "trapping" women spatially, as well as occupationally, within the labor market. Hence, daily processes of getting by and making do are linked with structures of gender segregation and with the very functioning of localized labor markets. In this way, the routines of daily life are connected to the distinctive *local and social* form of labor markets: "the friction of distance not only helps generate and sustain gender divisions of labor but also is central to the construction of different places...[as] labor market segmentation is literally mapped onto the ground" (Hanson and Pratt, 1995, p. 224).

The spatial organization of labor markets and local patterns of labor segmentation consequently reflect and reproduce gender relations. This is one of the ways in which labor markets are socially embedded in a range of (locally variable) institutional, political, and cultural structures. These are not rigid and unyielding structures, but have their own dynamics that are not crudely reducible to an "economic logic." Importantly, they also bear the imprint of labor's *active* struggles – notably, the collective agency of workers – the focus of an important strand of contemporary labor geography (Herod, this volume). Through explicit means such as union strategies and disputes, through the daily routines of the workplace, through historically embedded custom and practice, and through formal channels such as legislation and administrative regulation, labor exerts an influence on work practices and on the functioning of labor markets. Much of this activity is defensive rather than proactive, but it is also historically cumulative. For example, many of labor's incremental gains around issues like working time, pay rates and systems, welfare entitlements, and apprenticeship arrangements have become enshrined in legislation and/or normalized as an element of taken-for-granted practice, such that they become difficult for capital (or the state) to reverse. Once enshrined, these historically embedded norms, conventions, and laws in turn shape the subsequent strategies and practices of firms, in ways that are sometimes defensive (e.g. capital flight to low-wage, "deregulated" locations) and sometimes creative and productive (e.g. the adoption of "high-road" competition strategies, based on highly-skilled labor processes and quality outputs). This means that the labor–capital relationship is not a zero-sum game, in which labor's gain equals capital's loss. Rather, *qualitatively different* developmental trajectories and restructuring strategies become feasible/desirable under different local conditions (with their distinctive regulatory environments, characteristic gender and race relations, cultural conventions, patterns of power relations, modes of institutionalization, and so forth).

This underscores the fact that taking labor seriously does not simply mean highlighting the capacity of workers to *resist* or *ameliorate* the restless imperatives of capital, though this is certainly part of the picture. In a more complex fashion, the institutions of social reproduction and the (past and current) active strategies of labor serve to shape the social form of the accumulation process itself. To take one example, once welfare provisions were established on a generalized basis in the Atlantic Fordist countries of North America and Western Europe (themselves the outcomes of prevailing gender orders, race politics, and workers' struggles, as well as "concessions" from the state and capital), then the nature of labor-market relations and rules were altered in a fundamental way: competitive pressures were somewhat ameliorated and channeled, thereby shaping the design of jobs, gender relations, wage differentials, rates of exploitation, and so on (see Martin and Sunley, 1997). Geography, institutions, and politics matter here. As Dunford (1996) demonstrates, each national welfare regime was associated with a particular pattern of labor-market regulation, exclusion, and inclusion.

More generally, Rogers Hollingsworth (1997, p. 265) maintains that modes of economic regulation and governance in general, and the strengths/weaknesses of different "national capitalisms" in particular, are themselves "embedded in social systems of production distinctive to their particular society." This means that phenomena like Japanese work practices, American models of work-based welfare reform, or German labor union strategies cannot simply be transferred from one spatial context to another, because they are in a sense rooted in historically and geographically embedded regulatory frameworks, political structures, social norms, and so forth (Martin, this volume). This leads some to conceive of the process of international competition (and capitalist development more generally) as an ongoing political and economic struggle between different "varieties" of capitalism (see Albert, 1993; Rogers Hollingsworth and Boyer, 1997).

More often than not, these varieties of capitalism are substantially defined by reference to their distinctive labor practices and systems of social reproduction. Yet while social and institutional specificities may be most boldly expressed at the national scale – reflecting national legislative systems, policy regimes, and political struggles – they are not, either in principle or in practice, restricted to this scale. Distinctive regulatory configurations have indeed been identified at the local and regional scales, particularly with respect to social conventions concerning (paid and unpaid) work, and the governance of labor markets and inter-firm relations (see Jonas, 1996; Peck, 1996; Storper, 1997; Martin, 2000). This has potentially significant implications for arguments that the putative regime of flexible accumulation is rooted in a set of "common causal dynamics" (Scott, 1988b). If the new industrial spaces of Emilia-Romagna, Silicon Valley, or Baden-Württemberg can be shown to be regulated in locally specific ways (DiGiovanna, 1996), or indeed if they can be shown to be embedded in distinctive ways within their respective national regulatory frameworks (Gertler, 1992), then it follows that the underlying sources of competitive flexibility may not actually be "common" at all. In reality, there are many (local) forms of "labor market flexibility."

The process of labor regulation plays a key role here. This tends to be especially "locally embedded" due to the fact that labor itself – reflecting the social nature of its production and reproduction – is the most place-bound and geographically variable of the "factors of production." The production and reproduction of labor-power is dependent upon the supportive effects of certain key social institutions and, as a consequence, requires a substantial degree of stability and longevity. Following this logic, the social institutions that underpin the regulation of labor markets tend themselves to take on distinctively local forms. Labor markets can therefore be seen as one of the most socially – and in this sense also *locally* – embedded of economic systems, a theoretical contention with empirical echoes in the distinctive, "subnational" geographies of union militancy, systems of skill formation, gender relations, welfare regimes, employment norms, pay bargaining arrangements, and so on (Hanson and Pratt, 1995; Peck, 1996; Herod, 1998; Martin, 2000).

This leads Walker and Storper (1981, p. 497) to conclude that "a strong argument can still be made for the primacy of labor over all other market factors in influencing industrial location." But this is not simply a case of capital getting a once-and-for-all "lock" on the map of labor qualities. The process is a dynamic one, in part because the landscape of labor is itself constantly on the move. The "slash and burn" logic of uneven development is reflected in capital's periodic abandonment of regions in which socialized labor practices are seen to have fallen out of step with the current requirements of profitable production (Webber, 1982). Places that have evolved traditions of union militancy are often most susceptible to disinvestment: castigated for their "rigid" working practices and oppositional politics, such areas will often find themselves deserted by capital in search of greener pastures – *and "greener" labor*. Peet (1983) accounted for the "Rustbelt–Sunbelt" shift in the United States in just such terms, demonstrating how the pattern of new investment tended to favor "low class struggle" states in the south and west.

Relocation, of course, is not always an option for capital. Some organizations have high "sunk costs" in the form of long-term fixed investments, some are tied to localized markets or suppliers, others are heavily dependent on scarce labor supplies or the local (physical or social) infrastructure (see Harvey, 1982; Cox and Mair, 1988; Martin et al., 1993). Yet while globalization discourses tend to exaggerate the mobility of capital, the comparatively place-bound nature of labor (and social reproduction), combined with the *potential* for capital flight, mean that capital has the advantage of *relative* mobility. Even if capital does not move, the threat of movement can be used as a potent weapon against labor and as a lever in workplace negotiations, the practice of "whipsaw bargaining" (see Hayter and Harvey, 1994; Herod, 1994). This is another dimension of the asymmetrical power relation between capital and labor (see Offe, 1985), one which is increasingly tipping in favor of capital. Burawoy (1985, p. 150) sees this as a new form of capitalist despotism, observing that whereas:

...labor used to be *granted* concessions on the basis of the expansion of profits, it now *makes* concessions on the basis of the relative profitability of one capitalist vis-à-vis another – that is,

Figure 9.1 "The Directors"
Source: Private Eye, No. 974, 16 April 1999

the opportunity costs of capital. . . . The new despotism is not the resurgence of the old; it is not the arbitrary tyranny of the overseer over *individual* workers (although this happens too). The new despotism is the "rational" tyranny of capital mobility over the *collective* worker. The reproduction of labor power is bound anew to the production process, but, rather than via the individual, the binding occurs at the level of the firm, region or even nation state. The fear of being fired is replaced by the fear of capital flight, plant closure, transfer of operations, and plant disinvestment.

The tyranny of (potential) capital mobility, coupled with the erosion of welfare systems which previously afforded workers some measure of protection from the vagaries of the labor market, threatens to unleash a "race to the bottom" in labor standards (see Peck, 1996). One consequence of this is that workers' class positions are complicated by geography, as they become engaged, for example, in efforts to attract or retain investment (Sadler, this volume). Herod (1997, p. 16) argues that they do this "not as cultural or class dupes but as active economic and geographical agents." This is one of the many ways in which, in Herod's terms, labor seeks to assert its own "spatial fix" – pursuing investment not for its own sake but in the defense of jobs, livelihood, and community. Thus while capital exerts a profound and pervasive influence over the way that local labor markets work – configuring the design of jobs and attendant patterns of labor demand, establishing working time and payment systems, shaping through hiring and selection procedures the distribution of waged work (and therefore unemployment) between individuals and social groups – it does so under conditions that are not entirely of its own making.

Conclusion

This chapter has charted some of the changing material concerns and evolving analytical strategies in the field of labor geography. There have been continuities as well as changes. In some of the early research on industrial restructuring there was a curious echo of the neoclassical preoccupation with managerial decisionmaking: while orthodox analysis tended to reduce labor to a *statistical variable* (calibrated according to wages, skills, etc.) in the locational calculus of profit-maximizing firms, radical researchers tended often to reduce labor to *variable capital* (subject to commodification and control) in the profit-driven accumulation process. In contrast to these mechanistic treatments of labor, in recent years there has been a growing acknowledgment of the (pro)active role of labor in shaping the landscape of capitalism and the geographically distinctive ways in which workers themselves are produced and reproduced. Particular attention has been paid to the gendered nature of work and to the process of economic restructuring itself. Along the way, contemporary economic geography has become rather less concerned with asserting the pivotal importance of the "labor factor" in industrial location, focusing more on the variegated and culturally inflected ways in which work regimes and labor markets are *socially* constructed, restructured, and lived.

The preoccupation with transition models, in which the complexities of labor regulation and restructuring were collapsed into a simplified "binary history" of mass production/flexible accumulation, seems to have passed. But one of the useful legacies of the flexibility debate has been a deeper understanding of the "light sides" and the "dark sides" of the restructuring process, and their connectedness. It is now widely accepted that the evolution of capitalism is not a unilinear process, towards de-skilling and decentralization; neither is there a straightforward trend towards flexible re-agglomeration. Rather, each time the geography of labor is remade, contradictions and counter-tendencies are set in motion. Regimes of labor regulation – themselves subject to perpetual reform and restructuring – variously seek to sustain productivity and competitiveness, and to ensure the reproduction of labor-power, but do so in ways that are inchoate and imperfect. Maps of labor, just like the more restless geographies of capital accumulation with which they intersect, are therefore in a persistent state of remaking.

Acknowledgment

Thanks to Andy Herod and Eric Sheppard for helpful comments on a previous version of this chapter.

Bibliography

Albert, M. 1993. *Capitalism vs. Capitalism*. New York: Four Walls Eight Windows.

Amin, A. (ed). 1994. *Post-Fordism*. Oxford: Blackwell.

Block, F. 1990. *Postindustrial Possibilities*. Berkeley: University of California Press.

Bluestone, B. and Harrison, B. 1982. *The Deindustrialization of America*. New York: Basic Books.

Burawoy, M. 1985. *The Politics of Production*. London: Verso.

Clark, G. L., Gertler, M. S, and Whiteman, J. 1986. *Regional Dynamics*. Boston: Allen & Unwin.

Cooke, P. 1986. The changing urban and regional system in the United Kingdom. *Regional Studies*, 20, 243–51.

Cox, K. and Mair, A. 1988. Locality and community in the politics of local economic development. *Annals of the Association of American Geographers*, 78, 307–25.

DiGiovanna, S. 1996. Industrial districts and regional economic development. *Regional Studies*, 30, 373–86.

Dunford, M. 1996. Disparities in employment, productivity and output in the EU: the roles of labor market governance and welfare regimes. *Regional Studies*, 30, 339–58.

Gertler, M. 1992. Flexibility revisited: districts, nation-states, and the forces of production. *Transactions of the Institute of British Geographers*, 17, 259–78.

Hanson, S. and Pratt, G. 1995. *Gender, Work, and Space*. London: Routledge.

Harvey, D. 1982. *The Limits to Capital*. Oxford: Blackwell.

Harvey, D. 1989. *The Urban Experience*. Oxford: Blackwell.

Hayter, T. and Harvey, D. (eds). 1994. *The Factory and the City*. London: Mansell.

Herod, A. 1994. Further reflections on organized labor and deindustrialization in the United States. *Antipode*, 26, 77–95.

Herod, A. 1997. From a geography of labor to a labor geography: labor's spatial fix and the geography of capitalism. *Antipode*, 29, 1–31.

Herod, A. (ed). 1998. *Organizing the Landscape*. Minneapolis: University of Minnesota Press.

Hudson, R. 1989. Labour market changes and new forms of work in old industrial regions: maybe flexibility for some but not flexible accumulation. *Society and Space*, 7, 5–30.

Jonas, A. E. G. 1996. Local labor control regimes: uneven development and the social regulation of production. *Regional Studies*, 30, 323–38.

Lovering, J. 1989. The restructuring debate. In R. Peet and N. Thrift (eds) *New Models in Geography*, Volume one. London: Unwin Hyman, 198–223.

Lovering, J. 1990. Fordism's unknown successor: a comment on Scott's theory of flexible accumulation and the re-emergence of regional economies. *International Journal of Urban and Regional Research*, 14, 159–74.

Martin, R. L. 2000. Local labor markets: their nature, performance and regulation. In G. Clark, M. Gertler, and M. Feldman (eds) *A Handbook of Economic Geography*. Oxford: Oxford University Press, forthcoming.

Martin, R. and Rowthorn, B. (eds). 1986. *The Geography of De-industrialization*. London: Macmillan.

Martin, R. and Sunley, P. 1997. The Post-Keynesian state and the space economy. In R. Lee and J. Wills (eds) *Geographies of Economies*. London: Arnold, 278–89.

Martin, R., Sunley, P., and Wills, J. 1993. Unions and the politics of deindustrialization: some comments on how geography complicates class analysis. *Antipode*, 26, 59–76.

Massey, D. 1979. In what sense a regional problem? *Regional Studies*, 13, 233–43.

Massey, D. 1984. *Spatial Divisions of Labour*. London: Macmillan.

McDowell, L. 1991. Life without Father and Ford: the new gender order of postfordism. *Transactions of the Institute of British Geographers*, 16, 400–19.

McDowell, L. 1999. *Gender, Identity and Place*. Minneapolis: University of Minnesota Press.

Offe, C. 1985. *Disorganized Capitalism*. Cambridge: Polity.

Peck, J. 1996. *Work-Place*. New York: Guilford Press.

Peet, R. 1983. The geography of class struggle and the relocation of United States manufacturing. *Economic Geography*, 59, 112–43.

Rogers Hollingsworth, J. 1997. Continuities and changes in social systems of production: the cases of Japan, Germany, and the United States. In J. Rogers Hollingsworth and R. Boyer (eds) *Contemporary Capitalism*. Cambridge: Cambridge University Press, 265–310.

Rogers Hollingsworth, J. and Boyer, R. (eds). 1997. *Contemporary Capitalism*. Cambridge: Cambridge University Press.

Sayer, R. A. 1985. Industry and space: a sympathetic critique of radical research. *Society and Space*, 3, 3–29.

Sayer, A. 1989. Postfordism in question. *International Journal of Urban and Regional Research*, 13, 666–96.

Scott, A. J. 1988a. Flexible production systems and regional development: the rise of new industrial spaces in North America and Western Europe. *International Journal of Urban and Regional Research*, 12, 171–86.

Scott, A. J. 1988b. *Metropolis*. Berkeley: University of California Press.

Scott, A. J. 1988c. *New Industrial Spaces*. London: Pion.

Scott, A. J. and Storper, M. (eds). 1986. *Production, Work, Territory*. Winchester, MA: Allen & Unwin.

Storper, M. 1997. *The Regional World*. New York: Guilford Press.

Storper, M. and Scott, A. J. 1990. Work organization and local labor markets in an era of flexible production. *International Labour Review*, 129, 573–91.

Storper, M. and Walker, R. 1983. The theory of labor and the theory of location. *International Journal of Urban and Regional Research*, 7, 1–41.

Storper, M. and Walker, R. 1989. *The Capitalist Imperative*. Oxford: Blackwell.

Walker, R. and Storper, M. 1981. Capital and industrial location. *Progress in Human Geography*, 5, 473–509.

Warde, A. 1985. Spatial change, politics, and the division of labor. In D. Gregory and J. Urry (eds) *Social Relations and Spatial Structures*. London: Macmillan, 190–212.

Webber, M. J. 1982. Agglomeration and the regional question. *Antipode*, 14, 1–11.

Wright, M. 1997. Crossing the factory frontier: gender, place, and power in the Mexican *maquiladora*. *Antipode*, 29, 278–302.

Chapter 10

Industrial Districts

Ash Amin

Around eighty years ago, the eminent English economist Alfred Marshall (1919) noted the possibility of organizing manufacturing industry along two lines: either under the one roof of a big enterprise; or by small enterprises within localities specializing in a particular industry, which he called industrial districts. These industrial districts were the dominant form at the time. Gradually, in the course of the twentieth century, and notably after WWII, mass-production technologies developed and patterns of demand converged towards the mass consumption of relatively standardized products, to shift the balance towards the large enterprise. The growth of the giant corporation reaching out to all parts of the world for its inputs, factories, and markets seemed inevitable and unstoppable.

By the 1960s, in the heyday of growing global demand for cheap mass-produced commodities such as Coca-Cola, Levi jeans, Hoovers, and Fiat Cinquecentos, the industrial district seemed to have had its day – forgotten and relegated to the pre-factory phase of industrialization. The norm had become the vertically integrated corporation drawing on internally generated scale economies to produce standardized goods for a predictable market. But, this was not the end. In 1984, sociologists Michael Piore and Charles Sabel published a seminal book in which they argued that the late twentieth century was the "second industrial divide," a turning point that could reverse the order, from giant corporations back towards regional economies organized around networks of small firms in the same industry. Their claim was that new patterns of demand and the availability of new production technologies and techniques were re-enabling the resurgence of small firms, notably those locked into reciprocal relations with other firms. They noted that affluence and rising incomes in certain parts of the world were increasing the demand for design-intensive and customized products. They argued that these new, quality-based and volatile consumption patterns stretched the capabilities of enterprises geared up for large volumes of the same product for a predictable market. They also noted the rise of re-programmable technologies, often numerically controlled, which made it possible for smaller firms to respond to such demand by allowing flexible usage across both tasks and volumes of output. Large-scale usage of expensive and inflexible technologies was no longer seen to be essential for cost savings in the production process.

Piore and Sabel argued that the new market and technological circumstances provided a unique historical opportunity to reverse the industrial order from large-scale production in impersonal corporations to more decentralized forms drawing upon skills, flexible technologies, and small-scale production units. They labeled the new order as the age of flexible specialization, to mark a production system based on the utilization of flexible technologies and flexible work arrangements within task-specialist units. This allows the final product to be put together in different combinations without loss of efficiency at the level of both the individual unit and the system as a whole.

Sabel (1989), like geographers Michael Storper and Allen Scott (1989), went on to claim that there was substantial and growing evidence for the resurgence of flexibly specialized, decentralized business systems, all geographically agglomerated. In particular, three forms at the cutting edge of competitiveness in quality-driven markets were identified: high-technology agglomerations such as Silicon Valley; craft-based networks, including industrial districts and clusters of small firms in urban areas, specializing in quality-based niche markets of traditional consumer goods such as clothing and footwear; and the business networks of high performance large corporations such as Daimler Benz, drawing on the competencies of relatively autonomous branch units and their local supply chains.

For these commentators it was clear that decentralized production, including industrial districts, could replace mass production, centered around the hierarchically controlled large corporation with its branches scattered around the world, as the dominant industrial paradigm. Flexible specialization would be very different from the preceding paradigm.

Why the Interest?

In truth, the interest in industrial districts far exceeds their empirical significance. Its explanation has to be placed in the broader context of fascination in the revolutionary changes promised by flexible specialization. Ten to fifteen years ago, there was scarcely a mention of industrial districts, while now, few publications in economic geography, industrial sociology, or business studies fail to mention their importance. Yet, beyond the notable examples in advanced economies such as Italy, Spain, France, Japan, and Denmark, where craft traditions remained preserved in the age of mass, science-driven production, the evidence for industrial districts is relatively scant. No doubt small artisan enterprises are to be found everywhere, especially in the developing countries, but rarely are they organized into industrial districts, which possess a distinctive set of characteristics (see next section). The interest in industrial districts has to be explained by other, symbolic or conceptual, reasons. At least four stand out. At first, they symbolized the possibility of small firms and craft democracy in a world of skill-reducing and impersonal big firms (Brusco, 1982). Novel combinations of advanced flexible technologies (such as mini-robots and mini-computers) and the craft skills and ingenuity of small workforces, made it possible again for small to be economically viable. At the same time, evidence of close worker–management co-operation, informality, and mutual reliance in such firms promised the return of human-centered and democratic industrial relations. After decades of worker alienation, oppression, and de-skilling under the tight rules of the hierarchically organized large enterprise, a new industrial democracy came as an unexpected and welcome relief.

Second, the gathering number of case histories of flexible specialization – from regional examples such as artisan districts and high-tech agglomerations to examples of organizational decentralization in high-performance companies – served to reinforce important claims stressing the collective institutional and social foundations of economic life. We learnt that economic success had far less to do with the entrepreneurial virtues of rational economic man as theorized by neoclassical economics, than with collective foundations such as interdependence among economic agents, and the presence of local business support systems, conventions of dialogue, trust and reciprocity, and, in some localized cases, a culture of social and civic solidarity (see, for example, Aoki, 1988 and Sabel, 1994 on Japan; Trigilia, 1986 and Putnam, 1993 on Italy; Saxenian, 1994 on Silicon Valley; and Herrigel, 1995 and Staber, 1996 on Baden Württemberg). These were seen as essential supports for (smaller) firms, facilitating the sharing of risks, costs, information, knowledge, and expertise, and easing competition with larger firms with access to a greater level of internal resources.

Most recently, and coinciding with the rise of evolutionary economics (Hodgson, 1999; Metcalfe, 1998; Storper, 1997), the interest in these local stories of success has begun to turn towards what they can tell us about mechanisms and sources of learning and adaptation. It is becoming increasingly clear, in today's context of rapid technological change, heightened product obsolescence, and intensively contested markets, that an essential condition for economic survival and growth is the ability to keep ahead of the game by learning new tricks and adapting to, or shaping, ever-changing circumstances. Until the late 1970s, during the heyday of the large firm, questions of innovation were either largely ignored, or reduced to technological capability, narrowly defined as the ability of firms to generate or harness the fruits of science and technology through product and process innovation. Now, the discussion has broadened considerably to situate innovation in the broader context of learning and adaptation, acknowledging the importance of both formal (e.g. science and education) and informal (e.g. grounded skills, craft cultures) sources of innovation. In addition, it has come to recognize that innovation – or better learning – is not a sufficient condition for economic success, as it does not automatically secure adaptation, because of entrenched organizational habits and cultures (Cohen and Sproull, 1996). The knowledge that industrial districts rely on informally constituted learning (e.g. learning-by-doing) and are adaptable due to flexible specialization (e.g. ability to mix products or humans and tools in varying ways) has reinforced the conceptual stress on evolution and path-dependency.

Finally, for geographers in particular, the rediscovery of decentralized production systems has renewed hope in the powers of place and the locality or region as a unit of self-sustaining economic development. The age of mass production represented the erosion of local linkages as large multi-locational companies embarked upon fragmenting the production process to seek out cost-efficient regions around the world for their branch units. These units came to be tightly locked into a global intra-corporate division of labor, undermining local affiliations and prospects for local development. Development in a region no longer secured the development of the region. In contrast, all the examples of flexible specialization cited above point to the resurgence of regions as self-contained units of economic development. We are told that producers draw on local supply chains and are solidly locked into the local labor market, knowledge fabric, industrial conventions, and business support institutions, acting as genuine development poles. The regional production complex has become a feasible option among a variety of "worlds of production" (Storper and Salais, 1997).

This rediscovery has helped to rekindle the hopes of the urban and regional policy community, after years of despair over the problem of global integration without local self-sustaining growth, and has forged a "new regionalism" (Amin, 1999a; Lovering, 1999) informed by a radically new theorization of regional development. With the help of insights from recent examples of growth based on the mobilization of "endogenous" resources, regional policy has begun to move away from its traditional emphasis on universally applicable instruments such as support for technological innovation and training, promotion of entrepreneurship, attraction of inward investment, and upgrading of the transport and communications infrastructure. It has become more sensitive to local contexts, and recognizes the broadly defined social and institutional conditions conducive for sustainable development (Storper, 1997; Cooke and Morgan, 1998).

In short, the interest in industrial districts draws on a much wider fascination with a new phase of capitalism that is human-centered, democratic, and regionally oriented. It is also part of a new theoretical project: understanding the socio-institutional foundations and evolutionary processes of economic life.

This chapter begins with a definition and typology of industrial districts, followed by examples drawn notably from central and northeastern Italy – the cradle of contemporary industrial districts. It then examines the various theories that have sought to explain their success. The chapter ends with a discussion of the future of industrial districts in the face of contemporary challenges to their classical form.

Placing Industrial Districts

Definitions

Marshall (1919), drawing on his turn of the century observations in complexes such as the Sheffield cutlery industry and various wool textile areas in West Yorkshire, saw industrial districts as rivals to large-scale industry. For him, it was the concentration of small firms in the same industry and the indivisibility of the local industrial system from local society that marked the industrial district. As the noted Marshallian economist Giacomo Becattini (1991, p. 84) remarks:

Marshall proved in his early writings that most of the advantages of a large scale of production can be achieved also by a population of small-sized firms concentrated in some area, which are specialized in different phases of production and find their labor supply in a single local market. In order for [the industrial district] to develop, it is necessary that such a population of small firms merge with the people who live in the same territory, and who, in turn, possess the social and cultural features (social values and institutions) appropriate for a bottom-up industrialization process.

Marshall famously explained the economic advantages of industrial districts in terms of the localization economies resulting from the geographical agglomeration of firms in the same industry. Agglomeration offered small firms a series of cost savings and economic opportunities normally denied to the isolated small firm. First, there were advantages associated with proximity, such as reduced transaction and transport costs, and ease of access to inputs such as specialized labor, services, and know-how. Second, there were economies resulting from specialization, both by the locality in a given product, and by its firms in a particular task. The "division of labor" between firms allowed the individual firm to specialize in a given task or phase and sell its product to a variety of customers. In other words, the industrial district benefited from economies of variety resulting from the possibility of making up the final product in different ways without loss of productive efficiency (Bellandi, 1996a), and from the benefits of scale economies through task specialization. Third, the specialization of an area in the same industry continuously stimulated spin-off and new entrepreneurship, cushioned to a degree by the incorporation of firms into an interdependent local production system providing the necessary market opportunities.

Importantly, however, Marshall also stressed the indivisibility of industry from local society, which generated the social norms and values he considered to be critical for innovation and economic co-ordination. One aspect was mutual knowledge and trust – the product of economic interdependency, social familiarity, and face-to-face contact – which helped firms to reduce the cost of their transactions (from transport to information costs), facilitate the flow of information and knowledge, control the behavior of those firms trespassing local conventions, and strike a delicate balance between competition and co-operation between economic agents. Another aspect, famously stated in Marshall's words, was a particular "industrial atmosphere" resulting from the involvement of the whole local society in a common industrial project. For Becattini (1991), this atmosphere includes a life ethic based on self-help, entrepreneurship, and a sense of local belonging; a regular flow of bottom-up innovations generated by the industrial atmosphere; a culture of emulation resulting from the mobility of labor between firms; and an area reputation (e.g. "made in Sheffield") that attracts consumers and traders in a given niche market (e.g. cutlery).

Contemporary definitions of industrial districts are remarkably close to Marshall's original definition. Becattini (1990, p. 38), for example, summarizes:

I define the industrial district as a socio-territorial entity which is characterized by the active presence of both a community of people and a population of firms in one naturally and historically bounded area. In the district, unlike in other environments, such as manufacturing towns, community and firms tend to merge.

Even observers not wedded to the Marshallian tradition stray not too far, e.g., Oinas and Malecki (1999, p. 11):

Industrial districts...embody the interaction and dense network of linkages that comprise a local production system, usually around a single or highly related industries.

In the non-Marshallian definition, what is considered central is the division of labor among task-specialist units within the locality (hence the term "local production system"). Thus industrial districts might include, as implied by Storper and Scott (1989), large-firm-dominated regions such as Baden Württemberg, and high-tech regions such as Silicon Valley which combine networks of small and large firms, as well as Marshallian industrial districts and urban centers housing specialized producer or consumer services firms trading with each other. Local interdependencies are the common feature across these production systems. Other commentators (e.g. Markusen, 1996) have sought to broaden the definition still further by emphasizing the agglomeration of firms in the same or related industries in the same locality or region. In my view, this loses the central feature of industrial districts, further blurring the distinction between a production system and the co-presence of firms in a locality. Agglomeration is not the same as interdependence.

For Marshallians, the distinctive feature of an industrial district is not only inter-firm dependency, but also the weaving of economy and society into a local "communitarian market" (dei Ottati, 1994). The business system, cultural values, social structure, and local institutions are mutually reinforcing. Such an emphasis is partly based on analysis of the dynamics of craft areas that have reappeared in the countryside or small towns of such advanced economies as Italy, France, Japan, Denmark, and Spain in the last two to three decades.

The most researched and celebrated districts – reflecting their success in international markets and their numerical importance – are those of central and north-east Italy, scattered across the regions of Tuscany, Emilia-Romagna, and Veneto (see figure 10.1). All three regions have become highly prosperous, with the lion's share of their prosperity accounted for by the dynamism of small firms employing fewer than 15 workers, operating in the specialized niche markets of traditional consumer industries characterized by volatile and design-intensive demand patterns. The regions contain many districts, many of a Marshallian nature, as listed in table 10.1, and include internationally famous areas such as Prato (textiles), Modena (machine tools), Santa Croce (leather tanning), Carpi (knitwear), and Sassuolo (ceramics).

A Marshallian example

Let us focus on one example to get a feel for a typical contemporary Marshallian industrial district: Santa Croce sull' Arno, a leather tanning district in Tuscany. Santa Croce is a small town, 40 kilometers east of Pisa, which specializes in the production of medium- to high-quality cured bovine leather for predominantly the "fashion" end of the shoe and bag industries. There are only two other major leather tanning areas in Italy: Arzignano in the Veneto, which is dominated by a small number of large, vertically integrated, and highly mechanized tanneries, orientated towards the furnishing and upholstery industry; and Solofra in the South (Campania), which specializes in less refined, non-bovine, cured leather for the clothing industry. The Arno Valley accounts for about 25 per cent of national employment in the leather and hide tanning industry.

Remarkably, in an area no larger than 10 square kilometers, are clustered 300 artisan firms employing 4,500 workers and 200 subcontractors employing 1,700 workers. The real figures are probably much higher as the subcontractors' figures cover only those firms officially registered with the Santa Croce Association of

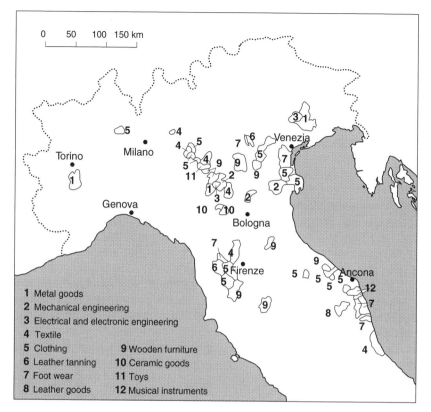

Figure 10.1 Industrial districts in Italy
Source: Sforzi (1990)

Leather Tanners or the Association of Subcontractors. On average, the area derives 15–20 percent of its sales revenue from exports, almost 80 percent of which are destined for the European Union (EU). Although the share of exports has been growing, the industry is still heavily dependent on the Italian market, particularly upon buyers in Tuscany, who account for 30–40 percent of the domestic market.

Twenty years ago, Santa Croce was not a Marshallian industrial district. There were many fewer firms, production was more vertically integrated, the product was more standardized, and the balance of power was very much in favor of older and larger tanneries. Today, Santa Croce derives its competitive strength from specializing in the seasonally-based fashion-wear niche of the industry. Typically, market conditions in this sector, such as product volatility, short product life cycles, design-intensity, and flexibility of volume, demand an innovative excellence and organizational flexibility that Santa Croce has been able to develop and consolidate over the last two decades by building upon its early artisan strengths.

The boom in demand for Italian leather fashion-wear in the 1970s and 1980s provided the opportunity for area-wide specialization and growth in the output of cured leather. That such growth was to occur through a multiplication of independent small firms supported by a myriad task-specialist subcontractors, was perhaps

Table 10.1 The regional distribution and sectoral specialization of localities (communes) with industrial districts in Italy

Region Sector	Piemonte	Lombardia	Emilia-Romagna	Veneto	Fruili Venezia Giulia	Toscana	Marche	Abruzzi
Metal goods	Carmagnolo	Rivarolo Mantovano						
Mechanical engineering		Suzzara	Novellara Cento Copparo Guastalia	Conegliano				
Electrical and electronic engineering								
Textiles		Asola Urgnano Quinzano d'Oglio	Carpi			Prato		
Clothing	Oleggio	Manerbio Pontevico Verolanuova Ostiano		Noventa Vicentina Piazzola sul Brenta Adria Porto Tolle		Castelfiorentino Empoli	Mondolfo Urbania Corinaldo Filottrano	Rosetto degli Abruzzi
Leather tanning				Arzignano		Santa Croce sull'Arno		
Leather goods							Tolentino	
Footwear				San Giovanni Ilarione Piove di Sacco		Lamporecchio Montecatini Terme	Civitanova Marche Fermo Grottazzolina Montefiore dell'Aso Montegranaro Monte San Pietrangeli Torre San Patrizio	

	Viadana	Modigliana	Bovolone	Sacile	Poggibonsi	Saltara
Wooden furniture	Viadana	Modigliana	Bovolone Cerea Nogara Motta di Livenza Oderno Montagnana	Sacile	Poggibonsi Sinalunga	Saltara
Ceramic goods		Sassuolo Casalgrande				
Toys	Canneto sull'Oglio					
Musical instruments						Potenza Picena Recanati

Source: Sforzi (1989)

more a result of specific local peculiarities than an outcome of the new market conditions. Opposed to the highly polluting effects of the tanning process – Santa Croce is one of those places where you can recognize the "industrial atmosphere" by its smell – the local administration was unsympathetic to factory expansion applications and also refused, until very recently, to redraw the local land-use plan to allow for more and better factory space. This, together with the strong tradition of self-employment and small-scale entrepreneurship in rural Tuscany, encouraged the proliferation of independently owned firms, scattered in small units all over Santa Croce. Two further stimuli for fragmented entrepreneurship were the preference of local rural savings banks to spread their portfolio of loans widely, but thinly, to a large number of applicants as a risk-minimization strategy, and the availability of a variety of fiscal and other financial incentives offered by the Italian state to firms with fewer than 15 employees.

This initial response to rapidly expanding demand was gradually turned into an organizational strength capable of responding with the minimum of effort and cost to new and rapidly changing market signals. The tanners – many calling themselves "artists" – became more and more specialized, combining their innate "designer" skills with the latest in chemical and organic treatment techniques to turn out leathers of different thickness, composition, coloration, and design for a wide variety of markets. The advantage for buyers, of course, was the knowledge that any manner of product could be made at the drop of a hat in Santa Croce. The small firms were also able to keep costs down without any loss of productive efficiency, in part through different forms of co-operation. One example is the joint purchase of raw materials in order to minimize price. Another is the pooling of resources to employ export consultants.

The main device for cost flexibility, however, has been the consolidation of an elaborate system of putting-out between tanners and independent subcontractors (often ex-workers). The production cycle in leather tanning is composed of 15–20 phases, of which at least half are subcontracted to task-specialist firms (e.g. removal of hair and fat from the uncured skins, splitting of the skins, flattening and drying). Constantly in work, and specializing in operations that are most easily mechanized, the subcontractors have been able to reduce drastically the cost of individual tasks at the same time as providing the tanners with the fluctuations in order size and specification demanded by the market. This articulate division of labor among and between locally based tanners and subcontractors – combining the advantages of complementarily between specialists and competition between the numerous firms operating in identical market niches – is perhaps the key factor of success.

Product specialization and agglomeration have also played a major role. Santa Croce, like other past and present industrial districts, is a one-product town which offers the full range of external or localization economies associated with local excellence along the entire chain of activities associated with leather tanning. In the locality, there are warehouses of major international traders of raw and semi-finished leather as well as the offices of independent import agents, brokers, and customs specialists. There are depots of the major multinational chemical giants as well as locally owned companies selling paints, dyes, chemicals, and customer-specific treatment formulae to the tanners. There are at least three savings banks that have consistently provided easy and informal access to finance. There are several manufacturers of plant and machinery, tailor-made for the leather tanning industry, and there is a ready supply-base for second-hand equipment and maintenance services. There are several scores of independent sales representatives, export agents, and buyers of finished leather in the area. The local Association of Leather Tanners, the Mayor's office, the bigger local entrepreneurs, and the Pisa offices of the Ministry of Industry and Trade also act as collective agents to further local interests at national and international trade fairs. There are several international haulage companies and shipping agents capable of rapidly transporting goods to any part of the world. There is, at the end of the value-added chain, a company that makes glue from the fat extracted from the hides and skins. No opportunity is missed in Santa Croce.

The entire community in Santa Croce, through its enterprises, families, institutions, associations, clubs, restaurants, shops, and piazza gatherings, is associated in one way or another with leather tanning. This provides new opportunities, through spin-off into new specialized tasks, thus guaranteeing the local supply of virtually all of the ingredients necessary for entrepreneurial success in quality-based and volatile markets. It also provides specialized skills and artisan capability, and a continual supply of industry-specific information, ideas, and knowledge – in short, Marshall's "industrial atmosphere" – geared towards supporting innovation and learning.

A typology

As suggested earlier, not all contemporary industrial districts are Marshallian. Artisan districts such as Santa Croce and many others in Italy, but also elsewhere (e.g. Gnosjö in Sweden, Sakaki in Japan, Alcoy in Spain) draw on craft excellence, multi-use technologies, and the Marshallian social tradition of self-reliance and co-operation. They are classical craft districts, centered around a very large number of small firms locked into an elaborate division of labor, and bound by strong informal traditions and craft institutions supporting the needs of firms (from artisan associations and rural banks to technical schools and trade centers).

There are other small-firm districts, however, many in Emilia-Romagna (see table 10.1) which Sebastiano Brusco (1992) has described as Mark II industrial districts, in which the small firms have come to be surrounded by more formalized institutional support, as well as increased capacity for technological innovation among some firms. Institutional support includes service centers (public or private, and usually located in or near the districts), providing industry-specific expertise to individual firms (from market information and legal or financial services to technological and managerial know-how). This means that a search for these services is not limited to the opportunities provided through inter-firm dependencies (Cooke and Morgan, 1998; Mistri, 1998). Innovative firms include so-called network leader firms within the districts, usually medium-sized companies, which have emerged in technology- or research-intensive industries with high levels of customized demand (e.g. agro-machinery, biomedical instruments). They subcontract products and tasks to other much smaller specialized firms, but provide the managerial, commercial, and technological expertise that takes the district forward into international markets driven by advancement in science and technology. Thus, Mark II industrial districts are less dependent on informal Marshallian traditions and craft institutions.

Table 10.2   Typology of industrial districts with respect to innovative capability

		Strong local co-operative environment	
		Industrial district Mark I	Industrial district Mark II
SMEs Internal resources and competence	Low	I Local production systems with *low* potential for technological capability-building (e.g. Gnosjö, Sweden)	II Local production systems with *some* potential for technological capability-building (e.g. Carpi and Reggio-Emilia in Emilia-Romagna)
	High	III Local production systems with *good* potential for technological capability-building (e.g. Jaeren, Norway; Sassuolo, Emilia-Romagna)	IV Local production systems with *high* potential for technological capability-building (e.g. Modena, Emilia-Romagna; Baden-Württemberg, Germany)

Source: Asheim (1997).

With these differences in mind, Bjorn Asheim (1997) has argued the case for a fourfold typology of industrial districts based upon differences in the potential for technological capacity-building and innovation (see table 10.2). In all four cases, the assumption is that the districts are more than agglomerations, that is, they constitute a local production system with strong links among firms. Asheim explains his typology (p. 151):

Square I represents the original Marshallian model of the industrial district. However, the problem with these industrial districts is their relatively low potential for endogenous technological capacity-building; i.e. owing to the relatively low level of codified knowledge and technological know-how of SMEs [small and medium-sized enterprises]...they are mainly able to adopt, adapt, and develop incremental innovations. In Square II we find industrial districts with some potential for technological capability-building, due to the collective resources of the district as they belong to the mark II model, which to some extent compensates for the low level of internal resources and competence of the individual firm. Square III represents industrial districts with a good potential for technological capacity-building due to a strong horizontal inter-firm cooperation normally found in these districts between firms with high levels of internal resources and competence [i.e. firms with significant technological competence]. Last, Square IV is characterized by a high potential for technological capacity-building due to the combined effect of the presence...of SMEs with high levels of internal resources and competence together with considerable public intervention.

Asheim's typology is helpful for noting important distinctions between Marshallian and other types of industrial district. In doing so, it raises the question of whether the differences between types of industrial district matter less than the similarities. Take, for example, the inclusion of Baden-Württemberg in Square 4. This is a large region, dominated by large corporations, a range of commanding industries, increasingly internationalized production linkages, and formidably large and multitudinous research organizations (Cooke and Morgan, 1998; Staber, 1996). This region is quite different from a craft industrial district such as Santa Croce. Both areas, of course, conform to the definition offered earlier by Oinas and Malecki (1999), who stress the centrality of product specialization and the localization of the division of labor, but the differences are also significant. Thus, while some may wish to retain the broad definition, it is wise not to lose sight of the very different industrial processes at work within the above typology.

Theorizing (Marshallian) Industrial Districts

With such internal differences, it is not feasible to provide a theory of industrial districts that incorporates all the variety, but without degenerating into bland generalizations. This section therefore focuses on craft industrial districts, acknowledging the risk of skewing the above typology in one direction. It places particular emphasis on the socio-political foundations and learning assets of craft industrial districts, so as to add value to accounts in English that are already well known.

In the early years of rediscovery of industrial districts – the 1980s – two analytical models dominated research published in English: Piore and Sabel's model of flexible specialization, emphasizing the combined advantages of craft traditions, yeoman democracy, multi-purpose technologies, and division of labor; and the neo-Marshallian model advanced by economists such as Becattini, emphasizing the importance of localization economies, the combination of scale and scope (or variety) through product and task specialization, industrial atmosphere, and long local histories of competition and co-operation. There were considerable overlaps between the two models, with differences between them largely a matter of emphasis rather than dispute. Sociologists tended to focus on the production process while economists and geographers focused on the properties of the locality.

Communalism

A parallel strand of literature on the Third Italy, well known in Italy at the time due to the seminal studies of Arnaldo Bagnasco (1988) and Carlo Trigilia (1986), but appreciated abroad only more recently following the work of Robert Putnam (1993) and his research collaborators (Nanetti, 1988; Leonardi and Nanetti, 1990), emphasized the nature of local political subcultures to explain the "long histories of collaboration and competition in industrial districts." This literature helped enormously to explain local social and cultural dispositions towards reciprocity and trust. It noted the decades-old strongly communitarian political culture that cut across class, gender, and institutional divides (socialist in Tuscany and Emilia-Romagna, and Catholic in Veneto). Carlo Trigilia (1991, p. 39) summarizes:

... in these communities there is often the prevalence of a specific political tradition, which generally dates back to the start of the century, and a complex of institutions – parties, interest groups, and cultural and charitable structures – that derive from the same political-ideological matrix.

The social practices and conventions of business in industrial districts, such as reliance on extended family labor, persistence of peasant values, belief in the values of work (over profit), and pride in professionalism and product quality, help to explain self-help and entrepreneurial spirit, but not co-operation with others. This is where communitarian local political subcultures helped. For example, in Emilia-Romagna, the Communist and Socialist parties were both pro-worker and pro-business and gained majority influence among both the unions and the artisan associations and co-operatives to which the small entrepreneurs flocked. In turn, these business organizations became important centers of economic power, serving not only to further the interests of small entrepreneurs, but also to provide training in business formation and management. As a consequence, Capecchi (1990, p. 28) notes:

> . . . a kind of Communist and Socialist "political community" was formed wherein people of the same political leaning came to be in charge of local and regional government, labor unions, small artisan associations and industries, and firms organised as co-operatives.

This political community, first, saw to the business needs of the small firm, but importantly it also inculcated a culture of collective action through interest groups. In some regions, such as Emilia Romagna, the local authorities started to offer business premises and services to small firms into the 1970s. Importantly, labor unions, industry associations, small-firm organizations, and local chambers of commerce developed research intelligence for the use of their members and sponsors, but also contributed, through widely attended and frequent public seminars and conferences, towards constructing a public reservoir of knowledge, opinion, debate, and reputation. In addition, the artisans' federations lobbied for favorable legislation and policies, established sector-specific training programs, provided access to a range of business services (from legal advice to technical information), helped to establish consortia for joint purchasing and sales, organized fairs and market publicity, and secured loans or credit. Finally, they gave legitimacy to craft or co-operative economic values, which elsewhere in the world were being discarded as anachronistic or inefficient.

Second, the political community helped to intermediate between sectional interests, without dampening the advantages of associational independence (i.e. effort and loyalty based on membership of interest associations). The role of long-standing ruling parties wedded to communitarian beliefs was critical in this regard. In an industrial district or region, the dominant party drew together, into a heterogeneous coalition, the urban working class, the peasantry and agricultural workers, an urban middle class won over by administrative efficiency and good public services, and an entrepreneurial class satisfied by the latter as well as the offer of business services. In addition, it was able to exercise considerable "network" influence (Bellini, 1996), through the common set of beliefs and values shared by its voters and activists, newspaper readers, recreation club members, and participants at mass festivals and rallies. This network influence also helped to establish consensus up and down the hierarchies of various local institutions. Inter-personal familiarity, and the frequent mobility of the party elite through senior positions within these organizations played an important role in establishing a common agenda as well as nurturing a culture of consultation and compromise.

It should not be assumed that this culture of intermediation has been simply the product of party alliances. It was also the product of what Robert Putnam (1993) has described as the democratic culture of civic regions, finely balanced between an efficient state and strong associationalist tendencies in civil society. In the Third Italy, at least two of the regions – Emilia-Romagna and Tuscany – are replete with voluntary associations, and with high levels of public participation in all areas of public life, from recreation, sports, and culture to housing, welfare services, and education. This fine balance has served to inculcate, first, a tradition of associative governance in which real authority is placed in the hands of autonomous groups (for example, the empowerment of voluntary organizations and charities in welfare provision). Secondly, it has bred a fiercely republican culture composed of belief in individual and group entitlements, rights, and responsibilities, an inclusive and shared public arena, and consultative and democratic decisionmaking. One effect of this culture has been that the public expects efficiency and accountability from the local authorities, and, in return, the political community has expected public endorsement of the local state's commitment to wealth creation and social solidarity.

Thus, beyond the politics of intermediation and communalism, and the institutions of flexible specialization or Marshallian industrial atmosphere, lies a way of life that cherishes – at least in the most civic regions – regional preservation, progressive values, and active civic life. In the Emilian context, Capecchi and Pesce (1993) have related this way of life to the region's strong tradition of women's autonomy, commitment to collective resolution of problems, appetite for cultural innovation, production and consumption, openness to outsiders, and advanced sense of citizenship.

Trust, tacit knowledge, and incremental learning

More recently, the success of industrial districts has been traced to their ability to draw on the economics of trust. The sources of trust have been sought in the nature of the networks of reciprocity which bind firms together (Sabel, 1992; Lazaric and Lorenz, 1998; dei Ottati, 1994; You and Wilkinson, 1994) and/or in the nature of local subcultures (Putnam, 1993). In industrial districts, firms are highly dependent upon each other and on supporting institutions for supply and markets and, for this reason, bound into ties of reciprocity. This kind of "studied trust" (Sabel, 1992), which is different from culturally enforced loyalty, is said to combine the benefits of competition and co-operation. While interdependence allows firms to establish long-term commitments, mutual regard, common learning patterns, and reduced transaction and search costs, the possibility of selecting partners and customers from an array of local firms is said both to help avoid the formation of strong ties of dependence which might stifle innovation and change, and to preserve the autonomy and independence that is necessary for entrepreneurial excellence and new learning opportunities. The industrial district is seen as a perfect example of the strength of weak ties (Granovetter, 1973; Grabher, 1993).

Equally, the commonality of the economic project in industrial districts, together with the political and institutional features of communalism described above, are also local sources of trust. More precisely, they are sources of conventions of mutuality and social obligation, which play a vital role in legitimating certain forms of economic behavior (e.g. mixing contractual and informal agreements, and tolerance for payment lags) and sanctioning against other forms of behavior which threaten the fine balance of power within districts (e.g. repeated violation of payment schedules, price cutting, and hierarchical business practices which threaten the principle of decentralization). There are tacit rules of the game in the air based, at least in part, on communal solidarities of some sort.

The informal basis upon which conventions such as trust are reproduced at the level of both the production system and local society, also lies behind the emphasis on the industrial district as a particular type of innovation environment, as seen above from the typology suggested by Asheim. The consensus seems to be that within the typical industrial district, it is informal, non-scientific, and interactive knowledge that plays a more significant role, in contrast to the technologically advanced firm which derives its dynamism from access to the fruits of scientific knowledge, technological advances, and strategic leadership (Asheim, 1997; Maskell and Malmberg, 1999; Bellandi, 1996b). Success – at individual and net-work level – is the product of craft knowledge and experience, apprenticeship, imitation, and incremental innovation and adaptation. Learning is achieved through imitating, doing, and using (Braczyk et al., 1998). Industrial districts thus are specific learning environments, equipped for continuous and incremental adaption within given niche-markets through the mobilization of informal ties and tacit knowledge.

Conclusion: Prospects

In the early 1990s, discussion of the future of industrial districts tended to be framed in terms of whether or not they would survive in the face of new global challenges. My own view then (Amin and Robins, 1990) was that they would not survive competition from better equipped big firms starting to move away from mass production towards flexible specialization. Similarly, the late Ben Harrison (1992) argued that they would be incapacitated by the predatory behavior of incoming firms and financial institutions, who would incorporate them into a wider spatial division of labor, destroy local tacit arrangements between firms and banks, and shake out very many small firms by concentrating production and power into their own hands. Optimists – from Marshallians and followers of Piore and Sabel to geographers predicting a decisive shift from vertical integration and global production hierarchies to vertical disintegration and local production net-works (Storper and Scott, 1989) – disagreed. The debate remains polarized and unresolved.

Now, interestingly, the discussion has moved on to speculate less on the survival or death of the classical industrial district, than to ask about the ways in which industrial districts are changing and evolving. This has helped to take the debate out of the cul-de-sac of having to be optimistic or pessimistic about their future. One topic, for example, concerns the innovative strengths of industrial districts. Given their greater disposition towards incremental learning within a given product matrix, industrial districts appear to be less well equipped to cope with path-break-ing changes in product or technological trajectory (Asheim, 1997). The firms have limited R&D capacity, and their tendency to adapt to externally driven changes hinders strategic, path-shaping or environment-changing behavior. Charles Sabel (1995, p. 4), for example, contrasts task-oriented co-ordination in the Italian craft model with goal-oriented co-ordination among Japanese decentralized firms:

...forms of coordination, derived from Japanese experience, that encourage deliberate, experimental revision of the definition and distribution of tasks within and among economic institutions outperform those based on notions of craft or entrepreneurship, that pursue the reintegration of conception and execution of tasks within a division of labor assumed to be natural and beyond reflection. This system of coordination I will call learning by monitoring because of the way it links evaluation of performance to reassessment of goals.

For Sabel the craft system generates skill-based interdependency among constituent units, while the goal-oriented system allows individual parts to experiment and adapt as the "system oscillates between determining the division of labor for itself and reconsidering that determination in light of execution" (Sabel, 1995, p. 9)

To a degree, however, this is to typecast the industrial district and deny the possibility of its mutation towards new forms that might permit path-breaking behavior without violating its defining organizational principles. Other commentators have observed that among those industrial districts that have been forced to confront intensified international competition, rapidly changing industry standards and aggressive market leaders, there is an emerging potential for strategic behavior as well as radical innovation. For example, some have seen the rise (from within) of leader firms, capable of technological and market leadership, and managing complex subcontracting relationships, so that the task-specialist units can remain less experimental (Bellandi, 1996a; Varaldo and Ferrucci, 1996).

Within Emilia-Romagna, and Type II districts in general, there is mounting evidence of the emergence of network leader firms, displaying signs of "learning by monitoring," especially in technology-intensive sectors such as automatic machinery or the agromechanical sector, and in new research-intensive sectors such as biomedical products (Lipparini and Lomi, 1996). These are medium-sized firms (80–100 employees), run by highly qualified or creative entrepreneurs with decades of business experience and leadership in a particular industry, and often commanding considerable influence within the regional business community and related organizations (e.g. technical schools, research centers, local authorities). In contrast to the past, they are like a holding company or network co-ordinator, marketing a range of related goods that are fashioned and assembled through a series of product-specific subcontracting networks, each with its own leader and follower firms. Their role is to provide international market access, strategic leadership, and resources, respectively through their extensive commercial experience and presence, investment in appropriate managerial and technical expertise, and command of financial and other resources. Their own survival is based on developing strategic capability and adaptive capacity, so that markets can not only be anticipated, but also shaped.

The example of network leader firms helps also to mollify the warning by commentators such as Ben Harrison concerning the rise of destructive hierarchies. Gabi dei Ottati (1998) argues that such firms are not interested in internalizing production or exercising central control, but are reliant on complex subcontracting and partnership arrangements for the production of specialized and non-specialized inputs. They are rather like the gathering houses of products and sub-assemblies fashioned by small firms within districts, elsewhere in a region, and possibly also abroad. Their role is to provide managerial leadership, markets, and innovation capability, but they are also reliant on the grounded knowledge, skills, and incremental learning capabilities of the firms that surround them. In this relationship of mutuality, the integrity of the industrial district as a local production system of inter-firm dependencies is somehow preserved.

To see industrial districts in a perspective that emphasizes evolution rather than decline or preservation should not imply any complacency about their future. There are serious threats which also need to be appreciated. First, a frequently voiced complaint by entrepreneurs in the Italian districts is that their sons and daughters are reluctant to enter the family firm, preferring professional careers to the risks and hard work of running a small firm (Mistri, 1998). Second, the solidarity and "democratic associationalism" (Amin, 1999b) that characterized Marshallian industrial districts in their formative years is now waning. Ideologies of individualism, market efficiency, and instrumental allegiance are now all-pervasive, even in places with strong family ties, and communal and civic affiliation. It remains unclear whether the balance between competition and co-operation that has proven so critical for industrial districts can be sustained without such local Marshallian commonalities. Finally, it is well known that in many industrial districts, such as Santa Croce, the division of labor is no longer locally contained, as some firms begin to source cheap raw and semi-finished materials from abroad, and establish market outlets overseas. At what point in its insertion into a wider division of labor does an industrial district cease to be one? Does it simply become a center of design, ideas, and innovation – a "Marshallian node in global networks" (Amin and Thrift, 1992) – or does it have to integrate head and hand, and contain the entire division of labor in an industry, to qualify as an industrial district?

Acknowledgment

This chapter was written during a Fellowship in 1999 at the Swedish Collegium for Advanced Study in the Social Sciences (SCASSS) in Uppsala. I wish to thank SCASSS for its generous support, as well as Eric Sheppard for his comments on the first draft.

Bibliography

Aoki, M. 1988. *Information, Incentives, and Bargaining in the Japanese Economy.* Cambridge: Cambridge University Press.

Amin, A. 1999a. An institutionalist perspective on regional development. *International Journal of Urban and Regional Research*, 23, 2, 365–78.

Amin, A. 1999b. The Emilian model: institutional challenges. *European Planning Studies*, 7, 4, 389–405.

Amin, A. and Robins, K. 1990. Industrial districts and regional development. In F. Pyke, G. Becattini, and W. Sengenberger (eds) *Industrial Districts and Inter-Firm Co-operation in Italy.* Geneva: ILO.

Amin, A. and Thrift, N. 1992. Marshallian nodes in global networks. *International Journal of Urban and Regional Research*, 16, 571–87.

Asheim, B. 1997. "Learning regions" in a globalised world economy: towards a new competitive advantage of industrial districts? In S. Conti and M. Taylor (eds) *Interdependent and Uneven Development: Global–Local Perspectives.* London: Avebury, 143–76.

Bagnasco, A. 1988. *La Costruzione Sociale del Mercato*, Bologna: II Mulino.

Becattini, G. 1990. The Marshallian industrial district as a socioeconomic notion. In F. Pyke, G. Becattini, and W. Sengenberger (eds) *op. cit.*, 37–51.

Becattini, G. 1991. Italian industrial districts: problems and perspectives. *International Studies of Management and Organization*, 21, 83–9.

Bellandi, M. 1996a. La dimensione teorica del distretto industriale. *mimeo*, Department of Economics, University of Florence.

Bellandi, M. 1996b. Innovation and change in the Marshallian industrial district. *European Planning Studies*, 4, 357–68.

Bellini, N. 1996. Regional economic policies and the non-linearity of history. *European Planning Studies*, 4, 63–74.

Braczyk, J. Cooke, P., and Heidenreich, M. (eds). 1998. *Regional Innovation Systems*. London: UCL Press.

Brusco, S. 1982. The Emilian model: production decentralization and social integration. *Cambridge Journal of Economics*, 6, 167–84.

Brusco, S. 1992. Small firms and the provision of real services. In F. Pyke and W. Sengenberger (eds) *Industrial Districts and Local Economic Regeneration*. Geneva: ILO, 177–96.

Capecchi, V. 1990. A history of flexible specialisation and industrial districts in Emilia-Romagna. In F. Pyke, G., Becattini, and W. Sengenberger (eds) *op. cit.* 20–36.

Capecchi, V. and Pesce, A. 1993. L'Émilie Romagne. In V. Scardigli (ed.), *L'Europe de la diversité*, Paris: CNRS.

Cohen, M. and Sproull, L. (eds). 1996. *Organisational Learning*. London: Sage.

Cooke, P. and Morgan, K. 1998. *The Associational Economy: Firms, Regions and Innovation*. Oxford: Oxford University Press.

dei Ottati, G. 1994. Cooperation and competition in the industrial district as an organizational model. *European Planning Studies*, 2, 463–83.

dei Ottati, G. 1998. The remarkable resilience of the industrial districts of Tuscany. In J. Braczyk, P. Cooke, and M. Heidenreich (eds) *Regional Innovation Systems*. London: UCL Press. p. 28–47.

Grabher, G. (ed). 1993. *The Embedded Firm*. London: Routledge.

Granovetter, M. 1973. The strength of weak ties. *American Journal of Sociology*, 78, 6, 1360–80.

Harrison, B. 1992. Industrial districts: old wine in new bottles? *Regional Studies*, 26, 469–83.

Herrigel, G. 1995. *Reconceptualizing the Sources of German Industrial Power*. New York: Cambridge University Press.

Hodgson, G. M. 1999. *Economics and Utopia*. London: Routledge.

Lazaric, N. and Lorenz, E. (eds). 1998. *Trust and Economic Learning*. Cheltenham: Edward Elgar.

Leonardi, R. and Nanetti, R. Y. (eds). 1990. *The Regions and European Integration: The Case of Emilia-Romagna*. London: Pinter.

Lipparini, A. and Lomi, A. 1996. Relational structures and strategies in industrial districts: an empirical study of interorganizational relations in the Modena biomedical industry. *mimeo*, Faculty of Economics, University of Bologna.

Lovering, J. 1999. Theory led by policy: the inadequacies of The New Regionalism. *International Journal of Urban and Regional Research*, 23, 2, 379–95.

Markusen, A. 1996. Sticky places in slippery space: a typology of industrial districts. *Economic Geography*, 7, 293–313.

Marshall, A. 1919. *Industry and Trade*. London: Macmillan.

Maskell, P. and Malmberg, A. 1999. Localised learning and industrial competitiveness. *Cambridge Journal of Economics*, 23, 2, 167–86.

Metcalfe, J. S. 1998. *Evolutionary Economics and Creative Destruction*. London: Routledge.

Mistri, M. 1998. Industrial districts and local governance in the Italian experience. *mimeo*, Department of Economics, University of Padova.

Nanetti, R. Y. 1988. *Growth and Territorial Politics: The Italian Model of Social Capitalism*. London: Pinter.

Oinas, P. and Malecki, E. 1999. Spatial innovation systems. In E. Malecki and P. Oinas (eds) *Making Connections: Technological Learning and Regional Economic Change*. Aldershot: Ashgate, 7–33.

Piore, M. and Sabel, C. F. 1984. *The Second Industrial Divide*. New York: Basic Books.

Putnam, R. 1993. *Making Democracy Work: Civic Traditions in Modern Italy*. Princeton, NJ: Princeton University Press.

Sabel, C. F. 1989. Flexible specialisation and the re-emergence of regional economies. In P. Hirst and J. Zeitlin (eds) *Reversing Industrial Decline? Industrial Structure and Policy in Britain and Her Competitors*. Oxford: Berg, 17–70.

Sabel, C. F. 1992. Studied trust: building new forms of cooperation in a volatile economy. In F. Pyke and W. Sengenberger (eds) *Industrial Districts and Local Economic Regeneration*. Geneva: International Institute for Labour Studies.

Sabel, C. F. 1994. Learning by monitoring: the institutions of economic development. In N. Smelser, and R. Swedberg (eds) *Handbook of Economic Sociology*. Princeton, NJ: Princeton University Press, 137–65.

Sabel, C. F. 1995. Experimental regionalism and the dilemmas of regional economic policy in Europe. *mimeo*. Cambridge, MA: MIT.

Saxenian, A. 1994. *Regional Advantage*. Cambridge, MA: Harvard University Press.

Sforzi, F. 1989. The geography of industrial districts in Italy. In E. Goodman, J. Bamford, and P. Saynor (eds) *Small Firms and Industrial Districts in Italy*. London: Routledge, 153–73.

Sforzi, F. 1990. The quantitative importance of Marshallian industrial districts in the Italian economy. In F. Pyke, G. Becattini, and W. Sengenberger (eds) *op. cit.*, 75–106.

Staber, U. 1996. Accounting for variations in the performance of industrial districts: the case of Baden Württemberg. *International Journal of Urban and Regional Research*, 20, 299–316.

Storper, M. 1997. *The Regional World: Territorial Development in a Global Economy*. New York: Guilford Press.

Storper, M. and Salais, R. 1997. *Worlds of Production*. Harvard, MA: Harvard University Press.

Storper, M. and Scott, A. J. 1989. The geographical foundations and social regulation of flexible production complexes. In J. Wolch and M. Dear (eds) *The Power of Geography: How Territory Shapes Social Life*. Boston: Unwin Hyman, 21–40.

Trigilia, C. 1986. *Grandi Partiti e Piccole Imprese*. Bologna: Il Mulino.

Trigilia, C. 1991. The paradox of the region: economic regulation and the representation of interests. *Economy and Society*, 20, 306–27.

Varaldo, R. and Ferrucci, L. 1996. The evolutionary nature of the firm within industrial districts. *European Planning Studies*, 4, 27–34.

You, J.-I. and Wilkinson, F. 1994. Competition and co-operation: toward understanding industrial districts. *Review of Political Economy*, 6, 259–78.

Chapter 11

Competition in Space and between Places

Eric Sheppard

Competition is all the rage. It is, as Erica Schoenberger (1998) puts it, a hegemonic discourse in economics and economic geography, and has been central to how economic geographers think – at least since they took economic theory on board in the 1960s. Economic geographers do not write many articles explicitly on competition but, like any hegemonic discourse, it percolates our thinking without us realizing it. Discourses are ways of talking about phenomena that frame how we think about them, what we take to be natural or unexceptional, and what we find controversial. Barnes (this volume) defines a discourse as "a network of concepts, statements, and practices that produces a distinct body of knowledge." A hegemonic discourse is one that dominates thinking to the point where we have difficulty conceiving of alternative ways of discussing the phenomenon. Competition is also a discourse of the powerful, both in academia and in the real world. It is broadly believed that unfettered competition is good for society. The World Trade Organization has been created to reduce barriers to international trade. A Multilateral Investment Agreement is currently being forged to eliminate political barriers to the international flow of investment capital in all forms. Structural adjustment agreements signed by countries with the International Monetary Fund and the World Bank, reducing government regulation within a country and at its borders, have become pervasive and accepted as the key to economic growth. Not only firms, but also nations, cities, and localities are enjoined to be more competitive if they wish to prosper. The political and economic elites of places, large and small, rich and poor, accept the legitimacy of this discourse – even those from places that have been hurt historically by competition.

Competition is not only a foundational idea in economic and social theory, but also in biological evolutionary theory. Indeed, over the last century social scientists have frequently appealed to notions of struggle and selection in Darwinian evolutionary theory to justify the centrality of competition in human societies. In this view, just as animals struggle to survive and evolve, so must humans compete to survive and prosper – implying that competition and self-interested behavior simply reflect human nature. In the late nineteenth century, in the form of social Darwinism (the view that competition in society, as in nature, promotes the survival of the fittest and progress to a better future), Darwinian theory was used to justify the competitive ethos of Victorian capitalism (Spencer, 1851; Hofstadter, 1955; Bowler, 1984). It provided a rationale for both colonial expansion ("the white man's burden") and eugenics (selective breeding of humans to eliminate "deviance" and promote "intelligence").[1]

In fact, Darwin borrowed the idea from economics. He was inspired to make struggle and competition central to his evolutionary theory by the economist (as well as population theorist and priest) Thomas Malthus (Stigler, 1987; Livingstone, 1991). Darwinian evolutionary theory remains controversial among biologists. Prominent critics question the view that competition drives evolution, and that the survival of the fittest results in superior life forms. Alternatives can be conceived. The geographer Kropotkin was among the first to argue that cooperation is pervasive among animals (Kropotkin, 1939 [1902]). Stephen Jay Gould (1989) argues that evolution is chance-ridden and does not justify views that animals currently at the top of the food chain, mammals in general and humans in particular, are superior. Such attempts to create alternative discourses have had little impact, however, on our tendency to think of competition as age-old, inevitable, natural, and beneficial.

Within economic thinking, the discourse of competition is that market-driven (capitalist) competition is generally economically and socially beneficial. This has been articulated through two prevalent metaphors expressing how competition works. The first and dominant one is *competition as invisible hand*. Here, competition, unfettered by social or political constraints, is seen as resulting in a stable equilibrium allocation of economic resources among members of society; one that is both efficient and equitable (on equilibrium, see Plummer this volume; on the invisible hand, see below). The second is *competition as evolutionary progress*. In this conception, competition is an ever-changing and unstable dynamic process following endless twists and turns, but generally promoting technological progress, increased productivity, and higher wages. (Nelson and Winter (1982) call this Schumpeterian competition or progressive capitalism, although Schumpeter (1942) was less optimistic that capitalism must progress and survive.) Both metaphors can also be found in economic geography. Location theory is exemplary of the former (Lösch, 1954; Krugman, 1991; Fujita et al., 1999), and recent research on competitive advantage and new industrial spaces draws on the latter (Porter, 1990; Storper, 1997).

In this chapter, we look at how these ideas have structured theories about competition among capitalists in space, examining in turn: single firms competing within the same industry, spatial competition among different economic sectors, and competition among places. In each case we will see how one or both of these metaphors has structured thinking. At the same time, however, we will see how careful attention to the geographically extensive nature of economies can call into question the logical validity of the economic discourse of competition. In pursuing this second theme, I am using competition to make a more general argument: economic geography is much more than simply applying economics to things geographical, even if we restrict our focus to economic processes. A geographical perspective can call into question some time-honored beliefs in economics itself.

Firms Competing in Space

Understanding the behavior of firms competing in space to sell the same product to geographically scattered consumers has been a defining problem of economic geography since the development of central place theory. The German economist August Lösch (1954 [1940]) invoked the invisible hand metaphor as a normative ideal for society, applying economists' theories of perfect competition to firms located in space. He did this both because he saw competition as promoting choice, and thus the human freedom envisioned by the idealist German philosophers he admired (Gould, 1999), and also because it seemed a better alternative to the National Socialist regime in which he lived.

The idea of the perfectly competitive market has shaped economic thinking about competition since Adam Smith (1776), often seen as the world's first economist. Smith argued that when firms compete to sell the same product in the same market, then the more firms there are, the more the competition and the lower the prices that consumers will pay. Later, neoclassical economists refined this claim into a precise, mathematical argument. Perfect competition is defined by: (i) egalitarianism – the presence of so many buyers and sellers in a market that no individual has the power to influence market prices; (ii) free entry – anyone can enter the market and start selling if they wish; (iii) full information – everyone is always fully informed about conditions in the market; and (iv) absence of the state – the market is not subject to state regulation. Under these conditions, prices can be defined that ensure that supply matches demand (and the market "clears"). In this market equilibrium, everyone pays the same price for the same product and capitalists make zero profits (implying that consumers are getting the lowest possible price). Smith's view was that, even when capitalists only pursue their own self-interest, free competition provides an "invisible hand" which ensures that their actions are socially beneficial. The theory of perfect competition exemplifies this notion and provides conditions under which it will work. Its "popularity...in theoretical economics is as great today as it has ever been" (Stigler, 1987, p. 535). Neoclassical economic theory also shows that under perfect competition the market equilibrium is *stable* (cf. Plummer, this volume). This means that once equilibrium is reached no further change will occur, because no-one in the market has any incentive to change his behavior – for example by lowering prices. This means that the second metaphor, competition as evolutionary progress, is irrelevant under perfect competition because no change occurs after equilibrium is reached.

Lösch imagined such a market operating in the simplest possible spatial context: identical retailers, evenly spaced apart in an unlimited uniform plain, selling to identical consumers, also uniformly spaced across the plain, who visit the closest retailer. Applying all the assumptions and logic of perfect competition to this idealized geographical model, he made an interesting discovery: while firms will locate as close to consumers as possible, and will make minimal excess profits, the prices they charge are higher than those predicted by non-geographic theories of perfect competition. This is because when competition occurs in this landscape, firms are able to exert monopolistic influence over nearby customers who have no alternative sellers nearby. This local spatial monopoly enables them to charge higher prices. Empirical studies confirm this theoretical result. For example, supermarket prices are higher in lower-income areas whose residents are less able to travel to more distant competitors (cf. Fik, 1988).

Space, then, calls into question the applicability of the invisible hand metaphor, a core idea in standard economic theory, to economic geography, even in a hypothetical case constructed to look as much like perfect competition as possible. Others have confirmed that when perfect competition is attempted in space, the result is imperfect or monopolistic competition, not perfect competition (Curry and Sheppard, 1982; Greenhut et al., 1987; Ohta, 1988; Mulligan and Fik, 1989). Competition in space thus challenges the claim that perfect competition eliminates profit and minimizes prices.

Things are even more complicated once the real geographies of markets are taken into account. When markets do not operate on the head of a pin, customers cannot know the price in every store they might visit, even in stores with which they are familiar, and would not always go to the cheapest store anyway. Space is also not a uniform plain. In the real world, space is differentiated into central and peripheral locations, and the economic distance between customer and retailer depends less on physical distance than on transportation technologies and the time-geographies of daily life. Customers' uncertainty means that there is no uniform price paid by everyone (unlike economic theories of perfect competition). Spatial differentiation means that some firms, because they occupy advantageous locations, will always do better, making considerable profits while others make none at all. Firms in realistic geographical landscapes are not competing on a level playing field, even when the conditions of perfect competition hold.

Some other widely accepted ideas in economics also seem questionable. In the theory of perfect competition, and in micro-economics in general, it is assumed that self-interested capitalists seek to maximize their total profits (their revenues minus their costs). When the landscape is spatially differentiated, this assumption can be questioned. Firms competing in space can increase their profitability if they seek to maximize the rate of profit on capital advanced, i.e. revenues divided by costs, instead of total profits (Sheppard et al., 1998).[2]

Paying attention to space also challenges the idea that the market equilibrium will be stable. Koopmans (1957) showed long ago that "equilibrating prices cannot form in any spatial location/allocation model" (Harvey, 1999, p. xxvi). Recent work has elaborated on why this is the case in realistic geographical landscapes, suggesting that market equilibria are at best *quasi*-stable in spatially differentiated landscapes. This means that there are incentives for firms to disrupt the competitive equilibrium by engaging, for example, in price wars. By reducing its price substantially below the equilibrium, a firm can increase its profitability – inducing other firms to do the same thing (Sheppard et al., 1992). Such instabilities suggest a spatial economy characterized by fluctuation and change, or evolutionary progress, rather than by an invisible hand.

Of course, economists know that their theory of perfect competition is unrealistic. In reality, some firms always do better than others, and not only because they may occupy more advantageous locations. Michael Porter (1985) dubs this a firm's "competitive advantage:" its ability to make a higher rate of profit and expand its market share. As we abandon the invisible hand for the evolutionary progress metaphor, efficiency is no longer sufficient to guarantee competitive success. In a world of uncertainty and change, capitalists must also be imaginative, flexible, and opportunistic. They must develop more efficient production methods, build better mousetraps and new products, reduce labor costs, and pay attention to developing opportunities such as trends in consumer demand. Capitalists with these qualities are said to be entrepreneurial, and entrepreneurialism is seen as the factor differentiating winners from losers in economic competition. Invoking the idea of entrepreneurialism in this way is much like invoking "survival of the fittest" in biological evolutionary theory. Firms, like organisms, are seen as living on the edge. The challenge is to remain competitive, with survivors being those individuals best able to live off their wits.

Those mainstream economists who do employ the competition as evolutionary progress metaphor stress entrepreneurialism as the key to competitiveness.[3] Implicit in this approach is the assumption that the only important difference between firms is the entrepreneurial skills to be found in the firm itself. Little attention is paid to the broader context within which firms are embedded. In all other ways firms are seen as competing on a level playing field. In this view, competition is promoted by supporting entrepreneurship and eliminating regulations and constraints on capitalists' actions. This is argued to stimulate dynamic competition, to allow anyone who is entrepreneurial enough to succeed, to support innovative and creative behavior, and to benefit society. Thus the shift from the first to the second metaphor does not disrupt the hegemonic discourse that capitalist competition in space is beneficial. The second metaphor becomes a kind of dynamic hidden hand, with the same prescription for economic ills as the first metaphor – unfettered capitalist competition.

This seemingly straightforward conclusion is called into question, however, by economic geographers. They have studied a variety of factors affecting the performance of firms – factors suggesting that other reasons, in addition to entrepreneurialism, are necessary to account for success or failure. First, firms do not occupy favored locations only because they made the right choices. They may simply set up by chance in locations that are close to customers, in industrial districts, or accessible to information (Webber, 1971). Second, firms' successes also depend on their past history. For example, they may face sunk costs – money invested in old production technologies that have gone out of style, or locations that have been abandoned by the gales of economic restructuring (Walker, this volume). Clark and Wrigley (1995) argue that sunk costs vary for idiosyncratic reasons from one firm (and location) to the next. Third, economic processes are always embedded in particular societal contexts (Martin, this volume). Firms found in places where the state favors capitalism, facilitates a well trained and cheap workforce, and/or subsidizes the industry that they belong to, inherit a competitive advantage from the particularities of their embeddedness (Storper, 1997). Fourth, decisionmaking within firms, and its impact on their success, is based on such non-economic factors as corporate culture (Schoenberger, 1997). Finally, firms respond to market uncertainty by bending the rules, completing mergers, lobbying, and colluding, all of which reduce competition in a market (Harrison, 1997). These considerations suggest that luck, history, geography, favoritism, culture, and ruthlessness, and not just entrepreneurialism, affect a firm's competitive success. Research into inter-sectoral and place-based competition clarifies the importance and significance of these considerations.

Competition in the Capitalist Space Economy

Extending the invisible hand metaphor

The metaphor of competition as invisible hand initially was developed by Adam Smith for a very particular situation: firms competing to sell the same product in the same market. In order to support the argument that competition benefits society as a whole, economists had to extend their theory to consider competition among different types of firms; it had to apply to steel firms competing with computer firms, not just to steel firms competing with one another. Economists refer to this as competition between firms in different markets. John Bates Clark (1899) pioneered this extension, helping solidify the so-called "marginalist" revolution in economic theory which came to form the core of neoclassical economics (Plummer, this volume). Clark's approach was to focus on the technologies used by firms, defined by the quantity of capital and labor they use in production rather than on the products they produce. He took as his starting point the conditions of perfect competition listed above. In addition, capital and labor, the inputs constituting a firm's production technologies, were seen as homogeneous inputs, which are not produced within the economy but available in limited quantities to all firms. It was also assumed that labor and capital can be transferred without cost from one firm to another (implying no geographical barriers to capital and labor flows).

Beginning with these assumptions, Clark and subsequent theorists conclude that in a competitive market the wage rate and the rate of profit (i.e. the prices paid to purchase labor and capital as inputs for production) are equal to the *marginal productivity* of labor and capital.[4] This extends the invisible hand metaphor in the desired way, because it suggests that competition makes sure that wages and profits reflect the value of labor and capital to society (i.e. their productivity): "what a social class gets is, under natural law, what it contributes to the general output..." (Clark, 1891, p. 313). This suggests that pure forces of economic competition assure a rational basis for wages and profits. If wages are low and profits high, this is not because workers are exploited by capitalists, but because their value to society (i.e. their marginal productivity) is low. The prices paid by consumers for a good are also shown to equal the desirability to them of its purchase (i.e. its marginal utility): "There is no conflict between the interests of... the producers and those of the consumers" (von Mises, 1949, p. 357).

In essence Clark and subsequent neoclassical analysts argue that competitive markets create a stable and harmonious outcome for capitalist production. Each good is sold at a price reflecting its usefulness; profits and wages represent the social value of capitalists and workers; firms make no excess profits and live on the edge of survival; and no rational economic actor would disturb the equilibrium. The market ensures that competitive and self-interested economic action produces the desirable if unintended consequence of a socially beneficial economic harmony. Thus the state should not intervene unless markets fail.

Two important aspects of real economies were neglected by this argument, however. First, it is assumed that firms never sell to one another, but only directly to consumers. Second, it is assumed that there are no transportation or communication costs. These turn out to be rather critical assumptions: if they are made more realistic, neoclassical conclusions about the general validity of the hidden hand metaphor can break down. They deserve detailed consideration.

Critiquing the invisible hand: interdependent firms

Firms do not in fact purchase capital as a homogeneous input for production, but purchase a variety of raw materials, machinery, and infrastructure, manufactured by other firms producing such "capital goods." They need money capital to pay for these inputs, but money is not itself a production input – it is a means of paying for them. A more complex but also more realistic view is to think of an economy as an input–output system of interdependent firms (Walker, this volume). This means that each firm buys inputs from other producers, and sells its outputs to other producers and consumers; commodities are produced by means of commodities, not by some mythical homogeneous capital input (Sraffa, 1960).

A competitive market equilibrium can be described for such interdependent firms; the critical question for our purposes is whether such an equilibrium is consistent with the invisible hand metaphor. Karl Marx (1972 [1896]) in fact described just such an equilibrium in the third volume of *Capital*, published posthumously in German, at the same time that Clark was writing about perfect competition. In Marx's equilibrium, firms in all sectors of the economy (and all regions) make the same rate of profit on the capital they advance to pay for production. If the rate of profit is higher in steel production and lower in wheat production, for example, then investors are expected to disinvest from wheat production, raising its price and increasing profits as wheat supplies diminish, and transfer funds to steel production, where prices and profits will fall. Marx did not have available to him the tools used in economic theory today, but subsequent analysts have confirmed the logical validity of his insights (Sraffa, 1960; Morishima, 1973; Roemer, 1988).

This equilibrium is quite unlike the invisible hand (Walsh and Gram, 1980; Roemer, 1981). First, wages and profit rates in competitive markets cannot be determined solely by economic considerations, i.e. by their contribution to "the general output." They are always influenced by the social and political power of workers and capitalists as they struggle over how to share the surplus created by capitalist production. Second, equilibrium market prices are not equal to the desirability of a commodity to consumers, but are equal to the cost of production incremented by the general rate of profit.[5] Third, firms do not live on the edge of survival making zero profits, but accumulate profits. The rate of profit a capitalist makes depends on the overall power of capitalists to increase their share of the pie, and on any particular competitive advantages they have, for example, based on monopoly power, on lower production costs, or on better products.

Fourth, the equilibrium that results is neither harmonious nor stable. Workers can always improve themselves by organizing to increase wages (Herod, this volume) – and capitalists can lobby government to lower taxes and wages and reduce state regulation, thus increasing profits. In effect, there is competition between social classes that cannot be resolved by the market, since there is no socially optimal wage or profit rate. When workers or capitalists successfully ally to promote their collective class interests (Sadler, this volume), they can contest and destabilize any market equilibrium, forcing the state to step in (often in support of one interest group over others, cf. Painter, this volume: table 22.1).

In addition to social classes disrupting competitive equilibrium, individual firms are motivated to engage in what Storper and Walker (1989) call strong competition. They act strategically to enhance their market share by many mechanisms other than pricing strategies (Kalecki, 1938; Semmler, 1984; Eichner, 1987; Lee, 1994). As geographers have also documented at length, they use their profits to develop new technologies, invent new products, relocate, and exclude competitors (Walker, and Schoenberger, this volume).

Critiquing the invisible hand: space and geography

Clark's neglect of interdependencies between firms is paralleled by a neglect of the costs of transportation and communication. In essence, the economy is treated as if it exists on the head of a pin: " 'There must be perfect, continuous, costless intercommunication between all individual members of the society' – so Jones in Seattle would know the price of potatoes and be able to ship to Smith in Miami a bushel of potatoes at every moment in time" (Stigler, 1987, p. 534, quoting Knight, 1921, p. 78). Only recently have the full implications of this, for both neoclassical and Marxist economics, been laid out by economic geographers (Harvey, 1982; Smith, 1984; Storper and Walker, 1989; Sheppard and Barnes, 1990; Swyngedouw, 1992; Webber, 1996; Webber and Rigby, 1996). The stuff of economic geography has been the geographical variation in what firms produce, how they produce it (and thus their linkages with other firms), labor relations, and access to finance. As a result of complex spatial divisions of labor (Massey, 1984), commodities are traded between regions, both from firms to consumers and from one firm to another. A unique economic sector also exists – transportation – which produces the necessary commodity of transportation services, so that commodities can be shipped from one place to another.

Economic geographers examining the functioning of a spatially extensive capitalist economy have concluded that incorporating space into our thinking poses further challenges to economic theory. First, the complexities of space mean that the decisions individual capitalists make, about where to locate, how to set prices, what to produce in which quantities, which technology to use, and who to trade with, may well have unintended consequences that undermine the functioning of competitive markets. Even when firms make decisions that seem to be economically beneficial in the short run, once the ramifications of these decisions have concatenated through the geographical economy, the result may be geographies of production that are less profitable than before, not more profitable. Marx referred to the tendency of the rate of profit to fall, even as capitalists work to increase it, as one of the forces undermining capitalist production. Such tendencies seem to be enhanced by the economy having a spatial dimension (Sheppard and Barnes, 1990).

Second, geographically uneven development is not only consistent with but frequently facilitates capital accumulation. When there is uneven development, workers in wealthy regions may find that their interests correspond more with those of local capitalists than with those of workers in poorer regions. This has long been recognized at the international scale. For example, colonialism was heartily supported by workers in eighteenth-century Europe – not only because of racial prejudice but also because their wealth was enhanced by impoverishment in the colonies (Galtung, 1971; Blaut, 1993). It has been less widely recognized that the same differentials apply within nations. It is often assumed, for example, that class interests are rather homogeneous within a nation (Barnes and Sheppard, 1992). Yet, spatiality complicates the standard Marxian argument that an economic actor's interests depend only on the class(es) to which he or she belongs (Sadler, this volume). Not only are class alliances weakened by social differences among workers (in skills, gender, race, etc.) and among capitalists (whose interests depend on the economic sector to which they belong, their firm size, and their social identities, Sayer, 1995), but they are also weakened by differences in location. Place-based alliances arise where capitalists and workers in a place ally to defend it against economic uncertainty, and compete against those in other places (cf. Urry, 1981; Hudson and Sadler, 1986; Sheppard and Barnes, 1990, chapters 10–12).

The added complications that the spatial differentiation of economic processes bring to the potential fractions and alliances that may develop within and between classes qualitatively increase the instability of capitalist competition (Sheppard and Barnes, 1990). Even finance markets, often pointed to as the lubricant for competitive markets, may undermine equilibrium in the space economy (Webber, 1987). Thus, as suggested by Koopmans, space is a destabilizing factor. Economists interested in economic geography keep competitive market equilibrium at the center of their attempts to explain the geography of production (for an overview, see Sheppard, 2000). By contrast, many economic geographers have become skeptical of the usefulness of equilibrium models as a tool for making sense of the economic landscape. Some argue that much systematic economic geographical analysis and theorizing is still possible (Plummer, this volume). Sayer (1995) deduces that such problems beset competitive capitalist and centrally planned socialist societies alike, arguing that the complex socio-spatial divisions of labor and of economic interest found in actual economies require a "third way," market socialism, which combines the flexibility of markets with the egalitarian vision of socialism. Webber and Rigby (1996) develop a disequilibrium approach to national economies in a global context. Others suggest that a more drastic modification to economic geography is necessary. Barnes (1996, p. 250) argues that the equilibrium models consciously or unconsciously employed by economic geographers entail an essentialist way of thinking about economic geography that is just as problematic as the equilibria themselves: "The best we can hope for are shards and fragments."

It may seem that these economic geographers are adopting the evolutionary progress metaphor for competition as being more appropriate to spatially extensive economies. This is not the case, however. While having a similar vision of the processes of competition as an evolutionary out-of-equilibrium process, they typically draw very different conclusions about the merits of competition, arguing that capitalist competition undermines social harmony and enhances social and spatial inequality. This is a discourse of competition as uneven development, requiring state intervention or social action to mitigate its worst consequences, not as a socially beneficial process that society should avoid disrupting.

In light of the difficulties in applying the invisible hand to spatially extensive economies, it is perhaps surprising that the discourse about the social function of "free markets" remains hegemonic today despite its logical flaws. One reason for this may be located in the history of economic thought. The marginalist approach of J. B. Clark et al. developed in the late nineteenth century in response to Marx's relentless criticism of and pessimism about capitalism (Marx, 1867; Harcourt, 1972; Pasinetti, 1981). In economics, too, ideological beliefs about capitalism color the theories and discourses used.

Places in Competition

When competition occurs between firms, and classes, competition also occurs between the different places in which they are located. It is thus important to consider whether place itself makes a significant contribution to competition. Economists tend to argue that competition between places is simply competition between the economic actors in those places, and can be analyzed by the same (aspatial) theory of competition. Certainly, economists' prescriptions for reducing geographical inequalities in development are the same as those for reducing social inequalities: the elimination of barriers to free competition. This belief carries over into the policy arena, where it has long been argued that free trade and unrestricted capital and labor mobility result not only in a harmonious social allocation of economic assets but also a harmonious and appropriate spatial allocation – a geographical version of the invisible hand.

David Ricardo (1951 [1817]) famously argued that when a place specializes in producing commodities for which it possesses a comparative advantage, trading its surplus for other commodities produced more efficiently elsewhere, then the international economy operates more efficiently (Grant, this volume). Subsequent trade theorists refined this approach. They deduced from the assumptions of perfect competition that local capitalists, acting self-interestedly, will produce the commodities that exploit a place's comparative advantage, and that perfect competition operating at an international scale will allocate the benefits of specialization and trade equitably between countries (Ohlin, 1933; Wong, 1995). Trade theory presumes, however, that capital and labor do not move between places, only commodities. This is not true, but neoclassical economics has also considered the opposite case. Suppose each place has access to the same production technologies, but has available different quantities of labor and capital as inputs. Suppose, further, that there are no restrictions to the mobility of capital and labor between places (something that is promoted for capital these days, but certainly not for labor, cf. Leitner, and Mitchell, this volume). In this case, beginning with the assumptions of perfect competition, it is concluded again that the actions of self-interested local capitalists, and of workers and investors seeking to find the best place to sell their labor and capital, will create a harmonious geographical allocation of economic activity with equal growth rates everywhere (Borts and Stein, 1964; Siebert, 1969; Henderson, 1987).

The real world falls somewhere between these two hypothetical extremes, of immobile commodities and immobile capital and labor. Capital, labor, and commodities are all mobile, to some extent. Yet, it tends to be presumed that if the invisible hand metaphor holds for the extreme cases it must also be true for more realistic intermediate situations. There have been significant elaborations of trade theory, in the "modern" and "new" international trade and growth theories (for a review see Wong, 1995), but these tend to reproduce the claim that free trade and mobile production factors reduce geographical inequalities in wealth, *ceteris paribus* (Fujita et al., 1999; Sheppard, 2000). Thus the invisible hand metaphor is argued to apply to competition between places, because in large measure competition is represented as perfect competition between capitalists.

The writings of Michael Porter in recent years have provided an analogous discourse about competition between places, but one that draws on the evolutionary progress metaphor. He begins by making inter-firm competition equivalent to inter-place competition, extending his theory of the competitive advantage of firms (Porter, 1985) to the competitive advantage of nations, regions, and even urban districts (Porter, 1990; Porter, 1995). The distinctiveness of his work stems primarily from his use of the evolutionary progress metaphor instead of the invisible hand. He argues that places can create competitive advantage; that competitive advantages are not all equal; and that the local state can intervene to help identify the right opportunities, selectively supporting those firms that efficiently pursue them. He sees the right opportunities as those that characterize successful industrial districts, where clusters of related firms generate dynamic external economies (Amin, this volume). Finally, he argues that promotion of competitive advantage can achieve a desirable development path for places, characterized by good jobs at high wages in a "green" physical environment.

Yet in the final analysis, Porter's vision remains within the discursive frame of neoclassical theories of competition. First, places are treated as independent actors, like the firms of neoclassical theory, for each of which a desirable competitive advantage can be identified. If this means computing in Silicon Valley or high technology in Paris, in the American inner city it means food distribution, discount retailing, suppliers of trade show exhibits, and courier services (Porter, 1995). Porter locates the distinctive competitive advantages of places in their particular values, culture, economic structures, institutions, and histories. Second, the role of states in facilitating this process is limited to creating "specialized factors" (education, infrastructure, health care); enforcing product, safety, and environmental standards; shaping investment goals; deregulating finance markets; promoting free trade; and enact anti-trust regulation – basically correcting market imperfections.[6] Third, the key to pulling this off is local initiative: competition among forward-looking firms, with favorable state policies, is necessary for innovation and competitive advantage.

Thus, competition is conceived of as occurring among places that begin competing on a level playing field, with fortune favoring the entrepreneurial (Leitner and Sheppard, 1998). While different specific advantages will develop in different places, the result is a positive-sum game in which all places can achieve desirable growth. This vision has spawned innumerable studies by consultants hired by local governments to identify their competitive advantage. Indeed, it is now widely argued that entrepreneurial competition between localities can reduce geographical inequalities in economic welfare and promote national economic growth (Peterson, 1981; Lovering, 1995; Hall and Hubbard, 1998). Yet this analysis ignores the other reasons that can make the difference between success and failure in competition between places: luck, historical geography, favoritism, and ruthlessness.

Thus the recent interest of economists in competition between places deploys the same discourse about competition, whether couched in terms of the hidden hand or of evolutionary progress. By contrast, economic geographers conceive of places as more than just a point on a map, and argue that competition between places cannot be reduced to competition between firms. As for the case above, of competition in the capitalist space economy, they argue that in a geographically differentiated economy the appropriate discourse for capitalist competition is that of uneven development.

In doing so, they contribute to a long-standing literature critical of arguments that all places have the same chances of reaching prosperity. Dependency theorists were very critical of international trade theory for suggesting that it does not matter what a place specializes in or how it is plugged into the world economy. They documented the growing gap between the ability of First World countries predominantly exporting manufactures, and of Third World countries predominantly exporting food and minerals, to gain from international trade. They argue that free trade is not fair trade, because trade enhances inequalities between core and peripheral countries in the world economy (Prebisch, 1959; Frank, 1967; Porter and Sheppard, 1998). Similarly, economic geographers argue that processes of uneven development and periodic restructuring better characterize competition between places than do mainstream theories predicting regional convergence (Harvey, 1982; Smith, 1984; Martin and Sunley, 1998; Sunley, this volume). Regional political economists, elaborating a critique of the neoclassically inspired "invisible hand" arguments summarized above, argue that the claims of free trade theory and neoclassical growth theory are not necessarily correct for a capitalist space economy (Sheppard and Barnes, 1990).

Turning to Porter's approach, places do not, and never will, compete with one another on a level playing field. They also are neither like, nor reducible to, firms. Cities, for example, have distinctive characteristics and histories, and are differently situated within the larger political economy. At least three dimensions of difference can be identified, each of which tilts the playing field to favor some cities over others: embeddedness, historical geographical trajectories, and favoritism (Leitner and Sheppard, 1999). First, every city is *embedded* in a set of national and regional institutions, regulatory systems, traditions, and norms (Martin, this volume). For example, European Union (EU) cities are embedded in a broader context very different from that of US cities – despite some convergence in recent years. In the EU, it is still seen as more legitimate for states to intervene in markets, and the belief that individuals are responsible for their own success or failure is less popular. There are also differences within the EU. The tradition of antagonism between capital and labor in the UK, for example, is very different from the corporatist tradition in Germany or Austria, where labor may be more influential in local economic development policymaking.

Second, each city occupies a *unique geographical trajectory* as a consequence of its historical role and location within the broader evolving political and economic system, a uniqueness that creates differences in the ability of individual cities to respond to economic and political restructuring. Economic restructuring favors locations that are well suited to new growth industries, and hurts those better suited to declining industries. How suitable a place is may have more to do with geo-historical happenstance than initiative; indeed there are many cases of cities whose very success in attracting the previous wave of industrialization creates a built and social environment that the next wave of growth industries finds unattractive (Harvey, 1985). For cities occupying different trajectories, identical strategies may have very different consequences, and different strategies may be necessary to achieve the same goals.

Third, higher levels of the state frequently exercise *political favoritism*, either deliberately through spatially targeted policies or as the unintentional result of national policies with different local impacts (Painter, this volume). Markusen et al. (1991) show, for example, how US Federal defense policies, combined with the geostrategic thinking of the US Joint Chiefs of Staff, systematically encouraged defense-related industries to move from their original Midwestern locations to the "gunbelt" of the southeast and southwest.

An uneven playing field also means that the broader consequences of competition between places are different. Michael Porter argues that all cities can use competitive advantage to create a "high road" to urban development, where growth and prosperity reinforce one another. On an uneven playing field, however, disadvantaged locations frequently feel compelled to lower wages to compensate for their disadvantage. Intensive spatial competition can then drive wages down everywhere, resulting in beggar-thy-neighbor competition (Leitner and Sheppard, 1998, 1999), as the First World learnt to its cost after the mid-1970s.

There are other differences between competing firms and competing places. Cities are fixed in place and must adapt to or seek to alter the particular advantages and disadvantages of that location. They are governed by a more-or-less demo-cratically elected government, which has limited powers and serves at the pleasure of the electorate. The legitimacy of the governance structure depends on a local state's ability to juggle growth agendas and welfare needs. There are also very few controls over who enters the city and who leaves, making it impossible to exclude undesirable residents from the city. Finally, many of the firms in a city sell to urban residents, implying that higher urban wages can increase local capitalists' sales, and profits.

By contrast, firms can relocate their activities and workers, if it suits their purposes, when their current location becomes undesirable. They are governed by a management structure with a simpler goal: to meet external (particularly stock-holders') perceptions of an efficiently run firm. Firm management is autocratic, with absolute power, in principle, over the operation of the firm, subject to the cooperation of its employees. This includes the power to recruit those who can contribute to its efficiency and to exclude those who do not – by firing them or denying them access to its private property. Finally, the employees of most firms are not major customers for its products, meaning that higher wages are seen as a drain on profitability instead of a way to increase revenues. Firms thus have more options and powers than cities in seeking to increase their competitive advantage, and pursue much less complex and tentative goals.

Beyond Competition?

As we have seen, economic thinking about competition draws on two metaphors about competition. In the first, competition is a harmonious process, an "invisible hand" enabling capitalism to achieve an equilibrium that can provide for the wishes of all. In the second, competition is a dynamic process of evolutionary progress, through which strategic firms and places play a game of ever-shifting competitive advantage. These different visions of how capitalist competition works, however, contribute to the same hegemonic discourse: unrestricted capitalist competition is a socially beneficial process facilitating personal freedom and social rationality, which other institutions should not interfere with. Even strong critics of capitalism have come to accept competitiveness as a geo-economic imperative, however, and have sought to identify how places can solidify their position in the competition for mobile finance capital (Evans, 1995; Sayer, 1995; Markusen, 1996; Storper, 1997). The result is the common idiom of the "free market," a phrase so widely used by academics, politicians, and in advertisements that we do not question it. Yet the "free market" is a figure of speech: outside the intellectual utopia of perfect competition, there is no logical link guaranteeing that markets are free, or that freedom implies competition.

Paying serious attention to how geography affects competition helps expose contradictions in the discourse of competitive markets. The emphasis on place-based difference in this discourse, both as the source of a distinctive competitive advantage and as a location for economic strategy, suppresses the equally central role of spatial inequality. When the larger spaces within which spatial competition occurs, and the uneven development that typifies economic differences between places, are reintroduced into the analysis, the dynamics of competition are revealed as fraught with negative rather than positive connotations. Competition is reframed as breeding inequality and constraining the possibilities of places, even as it celebrates difference and the possibilities of places.

Discourses of "good" competition should be expected to dominate those of "bad" competition in this era, when capitalism seems triumphant, making it all the more imperative to engage in the kind of deconstruction attempted here. For example, revealing these contradictions can reveal ways of thinking about how to cope with mobile finance capital other than local entrepreneurialism. Indeed, firms already practice alternative strategies. Mergers, strategic alliances, collusion, and lobbying are standard forms of collective action used by firms to deal with those vicissitudes of competition that they cannot handle separately. Equivalent strategies for places include inter-urban international collaborative networks in Europe; living wage initiatives in cities across the USA seeking to require all firms receiving local government subsidies to pay their workers decently; and international collaboration by labor and grassroots organizations (Leitner and Sheppard, 1999; Schoenberger, 1998; Herod, this volume). Revealing the darker underbelly of competition also creates space to analyze how discourses of competition (rooted in European eighteenth- and nineteenth-century thought) are masculinist, postcolonial, and Eurocentric, and push out of the picture more radical alternatives based on collaboration, cooperation, and emancipation (cf. Gibson-Graham, this volume).

Acknowledgment

I would like to thank Trevor Barnes for his patience in identifying flaws in earlier drafts of this chapter, without blaming him for how it turned out.

Endnotes

1. In 1899, on the occasion of the US takeover of the Philippines from Spain, English novelist Rudyard Kipling penned:

   > Take up the white man's burden –
   > Send forth the best ye breed –
   > Go bind your sons to exile
   > To serve your captives' need...
   > (Kipling, 1917, 215)

2. The difference between total profits and the rate of profit is like the difference between thinking about your bank account in terms of the total interest you are paid, in dollars, on the one hand, and the annual rate of interest you get, on the other.
3. This differs from the world of perfect competition and the hidden hand. When all actors are equally powerless to influence prices, and are fully informed about current and future states of affairs, capitalists need no entrepreneurial vision. They simply need to be efficient and rational.
4. The "marginal productivity" of capital or labor is the amount of extra output that can be obtained by employing one extra unit of capital, or one extra hour of labor, in production.
5. This idea actually dates back to Adam Smith.
6. For example, if markets fail to pay adequate attention to the environmental consequences of production, because the full social and environmental costs are not internalized in market values, then state regulation is necessary to ensure that this occurs.

Bibliography

Barnes, T. 1996. *Logics of Dislocation: Models, Metaphors, and Meanings of Economic Space*. New York: Guilford Press.

Barnes, T. and Sheppard, E. 1992. Is there a place for the rational actor? A geographical critique of the rational choice paradigm. *Economic Geography*, 68, 1–21.

Blaut, J. 1993. *The Colonizer's Model of the World*. New York: Guilford Press.

Borts, G. H. and Stein, J. L. 1964. *Economic Growth in a Free Market*. New York: Columbia University Press.

Bowler, P. J. 1984. *Evolution: The History of an Idea*. Berkeley: University of California Press.

Caves, R. E. 1984. Economic Analysis and the Quest for Competitive Advantage. *American Economic Review*, 74, 127–32.

Clark, G. L. and Wrigley, N. 1995. Sunk costs: A framework for economic geography. *Transactions of the Institute of British Geographers*, 20, 204–23.

Clark, J. B. 1891. Distribution as determined by a law of rent. *Econometrica*, XXXVI, 291–301.

Clark, J. B. 1899. *The Distribution of Wealth*. London: Macmillan.

Curry, L. and Sheppard, E. 1982. Spatial price equilibria. *Geographical Analysis*, 14, 279–304.

Eichner, A. S. 1987. *The Macrodynamics of Advanced Market Economies*. New York: M. E. Sharpe.

Evans, P. 1995. *Embedded Autonomy: State and Industrial Transformation*. Princeton: Princeton University Press.

Fik, T. J. 1988. Spatial competition and price reporting in retail food markets. *Economic Geography*, 64, 29–44.

Frank, A. G. 1967. *Capitalism and Underdevelopment in Latin America*. New York: Monthly Review Press.

Fujita, M., Krugman, P., and Venables, A. J. 1999. *The Spatial Economy: Cities, Regions and International Trade*. Cambridge, MA: MIT Press.

Galtung, J. 1971. A Structural Theory of Imperialism. *Journal of Peace Research*, 2, 81–116.

Gibson-Graham, J. K. 1996. *The End of Capitalism (As We Knew It): A Feminist Critique of Political Economy*. Oxford: Blackwell.

Gould, P. 1999. *Becoming a Geographer*. Syracuse, NY: Syracuse University Press.

Gould, S. J. 1989. *Wonderful Life*. New York: W. W. Norton.

Greenhut, M. L., Norman, G. and Hung, C.-S. 1987. *The Economics of Imperfect Competition: A Spatial Approach*. Cambridge: Cambridge University Press.

Hall, T. and Hubbard, P. (eds). 1998. *The Entrepreneurial City*. London: John Wiley & Sons.

Harcourt, G. C. 1972. *Some Cambridge Controversies in the Theory of Capital*. Cambridge: Cambridge University Press.

Harrison, B. 1997. *Lean and Mean: The Changing Landscape of Corporate Power in the Age of Flexibility*. New York: Guilford Press.

Harvey, D. 1982. *The Limits to Capital*. Oxford: Basil Blackwell.

Harvey, D. 1985. *The Urbanization of Capital*. Oxford: Basil Blackwell.

Harvey, D. 1999. Introduction. In *The Limits to Capital*, second edition. London: Verso.

Henderson, J. V. 1987. Systems of cities and inter-city trade. In P. Hansen, M. Labbé, D. Peeters, J.-F. Thisse, and J. V. Henderson (eds). *Systems of Cities and Facility Location*. London: Harwood, 73–118.

Hofstadter, R. 1955. *Social Darwinism in American Thought*. Boston: Beacon.

Hudson, R. and Sadler, D. 1986. Contesting works closures in Western Europe's old industrial regions: defending place or betraying class? In A. J. Scott and M. Storper (eds). *Production, Work, Territory – the Geographical Anatomy of Industrial Capitalism*. London: Allen & Unwin, 172–93.

Kalecki, M. 1938. The determinants of the distribution of national income. *Econometrica*, 6, 97–112.

Kipling, R. 1917. *Collected Verse of Rudyard Kipling*. Garden City, NY: Doubleday, Page.

Knight, F. H. 1921. *Risk, Uncertainty and Profit*. Boston: Houghton Mifflin.

Koopmans, T. 1957. *Three Essays on the State of Economic Science*. New York: McGraw Hill.

Kropotkin, P. 1939 [1902]. *Mutual Aid: A Factor of Evolution*. Harmondsworth, UK: Penguin.

Krugman, P. 1991. Increasing returns and economic geography. *Journal of Political Economy*, 99, 483–99.

Lee, F. S. 1994. From post-Keynesian to historical price theory, part I: facts, theory and empirically grounded pricing model. *Review of Political Economy*, 6, 303–36.

Leitner, H. and Sheppard, E. 1998. Economic uncertainty, inter-urban competition and the efficacy of entrepreneurialism. In T. Hall and P. Hubbard (eds). *The Entrepreneurial City*. London: John Wiley & Sons, 285–308.

Leitner, H. and Sheppard, E. 1999. Transcending urban individualism: Conceptual issues, and policy alternatives in the European Union. In D. Wilson and A. Jonas (eds). *The Urban Growth Machine: Critical Perspectives Two Decades Later*. Albany, NY: State University of New York Press, 227–43.

Livingstone, D. 1991. *The Geographical Tradition: Episodes in the History of a Contested Enterprise*. Oxford: Blackwell.

Lösch, A. 1954 [1940]. *The Economics of Location*. New Haven: Yale University Press.

Lovering, J. 1995. Creating discourses rather than jobs: The crisis in the cities and the transition fantasies of intellectuals and policy makers. In P. Healey, S. Cameron, S. Davoudi, S. Graham, and A. Madani-Pur (eds). *Managing Cities: The New Urban Context*. New York: Wiley, 109–26.

Markusen, A. 1996. Sticky places in slippery space: A typology of industrial districts. *Economic Geography*, 72, 293–313.

Markusen, A., Hall, P. Campbell, S. and Dietrick, S. 1991. *The Rise of the Gun Belt: The Military Remapping of Industrial America*. New York: Oxford University Press.

Martin, R. and Sunley, P. 1998. Slow convergence? The new endogenous growth theory and regional development. *Economic Geography*, 74, 201–27.

Marx, K. 1867. *Das Kapital*. Vol. 1. Hamburg: Otto Meissner.

Marx, K. 1972 [1896]. *Capital*. Vol. 3. Harmondsworth: Penguin.

Massey, D. 1984. *Spatial Divisions of Labour: Social Structure and the Geography of Production*. London: Methuen.

Morishima, M. 1973. *Marx's Economics: a Dual Theory of Value and Growth*. Cambridge: Cambridge University Press.

Mulligan, G. and Fik, T. 1989. Asymmetrical price conjectural variation in spatial competition models. *Economic Geography*, 65, 19–32.

Nelson, R. R. and Winter, S. G. 1982. *The Evolutionary Theory of Economic Change*. Cambridge, MA: Belknap Press.

Ohlin, B. 1933. *Interregional and International Trade*. Cambridge, MA: Harvard University Press.

Ohta, H. 1988. *Spatial Price Theory of Imperfect Competition*. College Station, TX: Texas A & M University Press.

Pasinetti, L. L. 1981. *Structural Change and Economic Growth*. Cambridge: Cambridge University Press.

Peterson, P. E. 1981. *City Limits*. Chicago: Chicago University Press.

Porter, M. 1985. *Competitive Advantage*. New York: Free Press.

Porter, M. 1990. *The Competitive Advantage of Nations*. New York: Free Press.

Porter, M. 1995. The competitive advantage of the inner city. *Harvard Business Review*, 74 (May–June), 55–71.

Porter, P. W. and Sheppard, E. 1998. *A World of Difference*. New York: Guilford Press.

Prebisch, R. 1959. Commercial policy in the underdeveloped countries. *American Economic Review*, 49, 251–73.

Ricardo, D. 1951 [1817]. *On the Principles of Political Economy and Taxation*. Cambridge: Cambridge University Press.

Roemer, J. 1981. *Analytical Foundations of Marxian Economic Theory*. Cambridge: Cambridge University Press.

Roemer, J. 1988. *Free to Lose*. London: Radius.

Sayer, A. 1995. *Radical Political Economy: A Critique*. Oxford: Blackwell.

Schoenberger, E. 1997. *The Cultural Crisis of the Firm*. Oxford: Blackwell.

Schoenberger, E. 1998. Discourse and practice in human geography. *Progress in Human Geography*, 22, 1–14.

Schumpeter, J. A. 1942. *Capitalism, Socialism and Democracy*. New York: Harper.

Semmler, W. 1984. *Competition, Monopoly and Differential Profit Rates*. New York: Columbia University Press.

Sheppard, E. 2000. Geography or economics? In G. Clark, M. Gertler and M. Feldman (eds). *Handbook of Economic Geography*. Oxford: Oxford University Press, forthcoming.

Sheppard, E. and Barnes, T. J. 1990. *The Capitalist Space Economy: Geographical Analysis after Ricardo, Marx and Sraffa*. London: Unwin Hyman.

Sheppard, E., Haining, R. P. and Plummer, P. 1992. Spatial pricing in interdependent markets. *Journal of Regional Science*, 32, 55–75.

Sheppard, E., Plummer, P. and Haining, R. 1998. Profit rate maximization in interdependent markets. *Journal of Regional Science*, 38, 659–67.

Siebert, H. 1969. *Regional Economic Growth: Theory and Policy*. Scranton, PA: International Textbook Company.

Smith, A. 1776. *An Inquiry into the Nature and Causes of the Wealth of Nations*. London: A. Strahan and T. Cadell.

Smith, N. 1984. *Uneven Development: Nature, Capital and the Production of Space*. Oxford: Basil Blackwell.

Spencer, H. 1851. *Social Statics: or, the Conditions Essential to Human Happiness Specified and the First of them Developed*. London: Chapman.

Sraffa, P. 1960. *The Production of Commodities by Means of Commodities*. Cambridge: Cambridge University Press.

Stigler, G. J. 1987. Competition. In J. Eatwell, M. Milgate, and P. Newman (eds). *The New Palgrave: A Dictionary of Economics*. London: W. W. Norton & Co, 531–35.

Storper, M. 1997. *The Regional World: Territorial Development in a Global Economy*. New York: Guilford Press.

Storper, M. and Walker, R. 1989. *The Capitalist Imperative: Territory, Technology and Industrial Growth*. Oxford: Basil Blackwell.

Swyngedouw, E. A. 1992. Territorial organization and the space/technology nexus. *Transactions of the Institute of British Geographers*, 17, 417–33.

Urry, J. 1981. Localities, regions and class. *International Journal of Urban and Regional Research*, 5, 455–74.

von Mises, L. 1949. *Human Action: A Treatise on Economics*. Chicago: Regnery.

Walsh, V. and Gram H. 1980. *Classical and Neoclassical Theories of General Equilibrium*. Oxford: Oxford University Press.

Webber, M. 1971. The empirical verifiability of central place theory. *Geographical Analysis*, 3, 15–28.

Webber, M. 1996. Profitability and growth in multiregional systems: Theory and a model. *Economic Geography*, 72, 335–52.

Webber, M. and Rigby, D. 1996. *The Golden Age Illusion: Rethinking Postwar Capitalism*. New York: Guilford.

Webber, M. J. 1987. Rates of profit and interregional flows of capital. *Annals of the Association of American Geographers*, 77, 63–75.

Wong, K.-Y. 1995. *International Trade in Goods and Factor Mobility*. Cambridge, MA: MIT Press.

Chapter 12

Urban and Regional Growth

Peter Sunley

The Mysteries of Urban and Regional Growth

In economic terms, 1998 was a hard year in the Scottish Borders. The long decline in the region's woolen textile industry, due in large part to increasing foreign competition, was compounded by the closure of electronic components plants in response to over-capacity in world markets. The region has lost 3,000 manufacturing jobs in three years, and local commentators bemoan the region's poor communications and predict entrenched unemployment and out-migration. Just fifty or so miles to the North however, the economy of the city of Edinburgh provided a striking contrast. Buoyed by the devolution of political power to a new Scottish Parliament, media, legal, and other business services mushroomed, and financial services and tourism continued to thrive, despite rising congestion costs. The aim of this chapter is to review recent work on the causes of such pronounced local differences in economic growth. One of the key themes of recent economic geography is that despite the ever-increasing integration of local economies into global flows of trade and capital, such local economic differentiation remains endemic to capitalism, and may even be intensifying as transport and communication costs fall. Despite the numerous glossy predictions of the death of distance and the end of geography, local and regional differences in growth may be intensifying across the industrialized world. Thus, in the era of slower growth since 1973, convergence between sub-national regions has at most been slow, and authors have suggested a resurgence of regions as meaningful economic units (Scott, 1998). Similarly, widespread urban decline associated with deindustrialization has been replaced by talk of an urban revival (Frey, 1993), and recentralization in some cities (Cheshire, 1995). Above all, the search for simple trends in urban and regional disparities has been confounded by the new complexity and unpredictability of local economic changes. In the developing world too, regional and urban inequalities have reached unprecedented scales. Thus, it seems more important than ever to understand the processes causing local economic growth.

Writing a decade ago, Schoenberger (1989) claimed that while geographers have a good understanding of the factors underlying economic growth, much less is known about precisely how these factors combine in specific places. In retrospect this conclusion seems over-confident, as the certainty about basic factors has been brought into question. Partly this reflects a sense of unease surrounding economic growth, with the recognition that aggregate economic growth is not an end in itself but a possible rather than inevitable means to higher standards of living. It also reflects an increasing awareness of the difficulty of measuring growth when many economic outputs and inputs, as well as the real costs and externalities of growth, are hard to quantify and to price, and remain invisible to official statistics. More specifically, however, this uncertainty reflects a widespread dissatisfaction with conventional theories of economic growth and a renewed search for more realistic insights into its prerequisites and dynamics. As this chapter will show, many of these new approaches attempt to incorporate increasing returns and more intangible factors into accounts of economic growth. There has been a shift, both in economics and geography, towards a focus on "softer" factors such as knowledge, innovation, and learning. Advantages have come to be seen less as natural and pre-given and increasingly as constructed and accumulated endogenously over time through economic activity itself. However, despite these common themes, major methodological and theoretical differences exist in the approaches taken to such issues. While economists and regional scientists have dusted off their parsimonious formal models of abstract locational landscapes, economic geographers have been more inclined to adopt evolutionary approaches to the understanding of growth. Instead of distilling the fundamentals of the growth process into mathematical models, this evolutionary economic geography has highlighted the social and institutional advantages of more successful regional and urban economies. This chapter compares the relative strengths and weaknesses of these two types of approach.

The Rediscovery of Increasing Returns

Increasing returns occur when any defined increase in inputs generates a disproportionately larger increase in quantities of outputs. It has long been known that one means of achieving increasing returns is through agglomeration economies. Two sorts have conventionally been identified. Localization economies arise where firms in the same industry cluster together. Marshall (1919) described the three classic types of localization economies: the formation of a pool of skilled labor, the nearby presence of supplying and supporting industries, and the local circulation of trade knowledge and secrets. Urbanization economies, on the other hand, arise when different industries locate in an urban area and benefit from general infrastructural advantages and common externalities. Cramming different trades and industries together in close quarters may also stimulate innovation so that a diversity of industries may increase growth (Jacobs, 1970). Increasing returns were fundamental to Myrdal's (1957) cumulative causation approach which identified types of centripetal backwash effects.[1] Keynesian and Kaldorian approaches also argued that regions with fast-growing outputs and exports would benefit from increasing returns through faster productivity growth.

In contrast, mainstream economics has conventionally been dominated by the decreasing marginal returns[2] assumed by neoclassical growth theory and by comparative advantage. However, in recent years increasing returns have also entered the economics mainstream for four main reasons. First, there has been mounting theoretical dissatisfaction with the neoclassical growth model. It assumes that there are decreasing returns to the main factors of capital and labor, so that the long-term rate of growth is determined by the growth of technology (Rigby, this volume). Despite this, technological change is left as an unexplained residual. Second, the empirical predictions of this model have looked increasingly untenable. The model predicts long-run convergence in rates of growth across both countries and regions, but there is little corroborating evidence. Studies of the movements of regional incomes per head in North America, Europe, and Japan have found that regional incomes over the long term converge at a rate of about 2 percent per annum (Barro and Sala-i-Martin, 1995), which is much slower than one would expect from the standard neoclassical model. Third, trade theory also looks inadequate, as it sees trade as driven by differences in factors and resource endowments and as allowing mutual gain (Grant, this volume). It cannot account for the growing importance of intra-industry trade between countries with very similar factor endowments. In response, trade theorists propose models of imperfect competition in which increasing returns allow different regions and countries to specialize in different varieties of similar products. Finally, advances in mathematics mean that it is also now possible to incorporate increasing returns in formal mathematical models.

The key point in the new location theory is that increasing returns are primarily realized through agglomeration. As Krugman (1991) argued, the primary means for increasing returns is the concentration of industries in particular localities. Much of this new location theory has offered different models of how it is that agglomeration raises the rate of growth. Local specializations of industry have been modeled using the trio of Marshallian external economies. Some new location models have also drawn on the new growth theory, which tries to make increasing returns endogenous to the growth process (Romer, 1986). There are several main types of endogenous growth theory (Martin and Sunley, 1998). One focuses on the returns to capital investment, another on learning-by-doing and the improvements in knowledge, skills, and human capital that workers accrue as a result of being employed. Another variant is called Schumpeterian and is based on the temporary monopoly rents[3] which companies gain from innovations, which in turn drive the growth process. In all three cases some of the increasing returns generated in human capital and through innovations may be geographically defined.

Drawing on these ideas, new location models depend on three main types of externality[4] at various spatial scales: pecuniary externalities, productivity externalities, and innovation externalities. For example, the formation of cities and industrial regions can be partly explained in terms of the operation of pecuniary external economies, in which firms benefit through having access to larger and more diversified markets. Larger markets allow firms, for example, to increase their output without reducing their prices. Productivity externalities may be realized by having a wider variety of intermediate inputs available locally. Productivity is also raised through external returns in human capital. These may drive city economies as workers learn faster in cities and earn more, thus attracting further migrants (Lucas, 1988). Innovation externalities arise primarily from the local diffusion of technological knowledge, which raises the technological capability of firms. Very little is known at present about the geographies of these technological spillovers, although Jaffe et al. (1993) show that, within the USA, citations of patents, which occur when firms make use of innovations, tend to be concentrated in the local area where the innovation originated. This implies that there is a distance-decay effect in learning about new technologies (Rigby, this volume).

The most applied parts of the new location theory have predicted the regional effects of globalization and trade integration, and considered the future of cities. Krugman and Venables (1990) argue that external market economies become more important as trade costs fall, meaning that regional integration in Europe, for example, will increase the specialization of European regions. Moreover, capital and labor mobility may reinforce the concentration of production. In North America, on the other hand, it is predicted that manufacturing will concentrate on the US border with Mexico in order to provide components to Mexican producers (Hanson, 1996). If regions do become more locally specialized as integration proceeds, then it is imperative to understand the various possible adjustment mechanisms, such as population mobility, capital mobility, and fiscal transfers through tax systems, through which regions cope with demand shocks.

These new location theories formalize ideas of cumulative causation by modeling the ways in which similar regions can endogenously differentiate into cores and peripheries (Ottaviano and Puga, 1997). They have also highlighted the importance of what has been called path dependence – a systems or biological metaphor referring to the way that the evolutionary path of a system depends on its past history (Arthur, 1996). It can be argued, for example, that the initial location of many industrial cities and industries was accidental. Once established, however, industrial cities and sites benefit from cumulative advantages and may dominate their rivals. Such cumulative causation has also been described as a type of "lock-in," which refers to the way in which the interactions between components of a system fix its behavior. The most familiar economic case is where a technology becomes dominant because of user externalities.[5] Although the QWERTY keyboard, for example, is relatively inefficient, the more people use it the harder it is to shift to another layout of the keys (David, 1985). New location theories have at least begun to ask questions about the social capability of different states and regions to absorb innovations introduced elsewhere and thus to catch up, and they have re-opened a debate on whether diversity or specialization are more conducive to growth. Some argue that localization economies are more important, but others insist that employment growth is faster in diversified cities (see Harrison et al., 1996).

These theories and models have been rightly criticized. It has been questioned whether they really contain anything new or whether they are simply building formal models of old and familiar ideas. The benefits of having larger markets in wealthy regions have long been known, and Pred (1966) argued for an evolutionary approach to urban manufacturing growth that highlighted inventiveness and uncertainty. Much of this earlier work, however, did not try to represent these ideas in mathematical and equilibrium-based models but was more interested in social and institutional questions. In contrast, the new location theories are abstract and, in many instances, they tend to substitute spatial units which are independent of scale for geographical places (Martin, 1999). They also tend to treat cities and regions as laboratories for testing model parameters, rather than as real places and objects of enquiry. Much of the new work has become increasingly preoccupied with the relative merits of different statistical measures of convergence rather than the underlying processes.

The new location models begin with restrictive assumptions. As Dymski (1997) argues, Krugman's models are based on rational choice equilibria[6] which depart from the neoclassical methodology by relaxing only one of its assumptions, such as by increasing (or: increasing for decreasing) returns (Plummer, this volume). They therefore remain very close to economic orthodoxy and tend to try to control for other factors by focusing on only one aspect of regional and urban growth at a time. Moreover, most of the models assume monopolistic competition,[7] implying that firm strategy cannot be discussed. Furthermore, in this approach factors that cannot be quantified and statistically manipulated are too quickly dismissed as sociological and impossible to study rigorously (Martin and Sunley, 1996). The consequent lack of realism also means that these models seem unable to explain broad historical trends in rates of convergence. For instance, it has been widely noted that rates of regional convergence are more rapid in periods of economic growth, while divergence and stagnation are more typical during recessions. Yet many geographers have argued that a widespread slowdown in economic convergence since the mid-1970s signals a structural change in economic growth (Dunford and Perrons, 1994). It is therefore unsurprising that this resurgence of location theory has been empirically weak and has resolved few issues. Despite statistical studies of trends in regional growth rates, the key ideas on growth have been subjected to very little empirical testing and evaluation. The major problem is that the new location theory is trying to force increasing returns ideas into equilibrium models, when the ideas resist being flattened and squashed in this manner. This partly explains why economic geographers have been drawn instead to examining local characteristics.

The Evolution of Local Competitiveness

In contrast to this revival of formal modeling, most recent accounts of urban and regional economic growth in economic geography have been more concerned with describing the underlying characteristics of economic activity. Rather than simply looking at aggregate outcomes summarized by economic statistics, there has been much more focus on the form of growth. Within this broad approach, however, there has been a major change in the way in which this form is understood: a movement away from macro-structural theories of transitions in the economy, with predictable and deterministic spatial consequences, towards approaches placing much more emphasis on the specific characteristics of particular cities and regions and how these evolve over time. Macro-structural accounts of changing growth identified outcomes in terms of the spatial pattern of growth. For example, the concept of a new spatial division of labor emphasized that aggregate growth conceals a hieararchy in terms of job quality, and predicted regional functional specialization, as high quality control functions concentrate in core cities and peripheral regions become dependent on branch plants (Massey, 1984; Peck, this volume). Similarly, the notion of a transition away from Fordist mass production to a new style of growth based on networks of small firms foresaw the rise of new industrial spaces, supposedly free from the industrial infrastructures and collective institutions of Fordist mass production (Scott, 1988). Again in a deterministic style, Hall (1985) argued that the appearance of a new long wave of economic expansion (or Kondratieff wave), based on a new set of key innovations, would lead to an inevitable restructuring of regional outcomes, or creative destruction, involving the inexorable decline of old regions and the rise of new technological sites.

None of these approaches fully captures the logics of contemporary economic growth, and they struggle to explain the complexity of urban and regional outcomes. At root, the notion of clean transitions between different forms of growth has been steadily undermined. For example, there is no simple transition from mass production to a more flexible mode of growth, but rather a variety of types of flexible mass (or diversified quality, or lean) production which do not have simple and obvious geographies (Hudson, 1997). Instead, evolutionary approaches have become dominant in economic geography. As Barnes (1997) argues, evolutionary approaches allow for indeterminacy as well as complex and unpredictable results. At the heart of these ideas are analogies between genes and firm capabilities and routines, and between genes and local economic relations and conventions. Those that are selected by the market, or replicated by imitation and diffusion, survive and shape the trajectory of change. Such gradualism allows a better understanding of the types of path and place dependency through which the historical geography of regions and cities shapes their future development.

This change in emphasis reflects the way in which growing awareness of a global economy has stimulated debates on the idea of competitiveness. Competitiveness is a controversial concept originating in notions of economic evolution, which has often been accepted too uncritically (Schoenberger, 1998). Some authors insist that the real meaning of competitiveness is the growth of productivity (although Maskell et al. (1998) suggest that competitiveness means a high return to all factors of production, including labor – which of course is not necessarily synonymous with high productivity). This has been associated with the increasing popularity of the notion of competitive advantage. In contrast to comparative advantage, competitive advantage suggests that gain is not mutual: winning regions which capture "first mover" advantages can benefit from the increasing returns of leading sectors, at the expense of other localities.

Explanations of success have emphasized the local economic capabilities and assets of the cities and regions themselves. Many studies have implied that if local supply-side characteristics are right, then regions can create their own demand by gaining market share from their rivals and by attracting investment.[8] A remarkably consensual theme has been that agglomeration can help to create and sustain these local economic capabilities. Porter's (1990) identification of clusters of related industries, linked by material and knowledge flows, has been highly influential in arguing that geographical proximity reinforces the interchanges between competitors, supplying industries, factor and demand conditions, and thus provides a major spur to continuous improvement and innovation. Such clusters, he argues, should be the focus of urban and regional policy. The claim that agglomeration is a key to competitiveness has become a stylized fact, which is routinely accepted rather than tested. Yet much of the evidence on agglomeration continues to be anecdotal and based on well-known examples, and some skeptics suggest that its significance has been exaggerated (Amin, this volume). There are only a few dissenting voices, however, and it is much more frequently argued that agglomeration is the most distinctive feature of the space economy.

Early attempts to explain this feature used the notion of flexible specialization, which suggests that networks of smaller firms can respond more quickly to differentiated and changing consumer demands, and can therefore be both specialized and adaptable (Piore and Sabel, 1984). In this view, mass production gives way to vertical disintegration, as firms contract out their requirements. The transaction costs of such new flexible production systems can be lowered by agglomeration, and the uncertainties faced by producers reduced. Agglomeration is especially effective in reducing those transaction costs that are spatially dependent, where transactions are small-scale, and where they are also irregular and unpredictable – involving frequent changes of specification, and where they depend upon face-to-face contact and the personal exchange of information (Scott, 1988). It was also argued that new methods of lean supply, such as just-in-time, encourage the clustering of suppliers. However, subsequent studies have raised doubts about whether transactions cost advantages explain agglomeration, and have questioned the strength of local linkages. For example, Angel and Engstrom (1995) found that components in the US personal computer industry are sourced from global networks, and that linkages are predominantly inter-regional and international. It is also difficult to distinguish linkages that influence locational decisions from those that are consequences of such decisions. The general conclusion is that transaction costs provide too static a framework, as the real advantages of flexible districts are dynamic.

Many recent accounts of this dynamism have offered interpretations rooted in institutional sociology with its emphasis on "embeddedness," meaning that economic actions are affected by other actors and by the overall framework of social relations (Martin, this volume). For instance, Cooke and Morgan (1994) argued that the key to successful industrial growth is the presence of networks, which they describe as neither market relationships nor firm hierarchies but as co-operative and reciprocal relations. These co-operative networks are based on relationships of trust between firms which enable information sharing and mutually beneficial contracting (Harrison, 1994a). Networks of producers promote new product development by allowing firms to spread the costs and risks involved, and encourage information exchange and joint problem-solving (Saxenian, 1994). Trustful relations reduce monitoring and contract costs, and may be particularly important to the cheap supply of local credit and to co-operative labor relations. According to some, the adoption of new networks can lead to the revival of older industrial regions (Cooke, 1995).

Some of these arguments exaggerate the co-operative nature of successful clusters, and downplay the importance of rivalry and competition between firms. In Baden-Württemberg, for example, frequently cited as a competitive networked region, Staber (1996) found there is in fact little evidence to suggest that business relations are marked by collaboration and co-operation, and that while some firms experiment with decentralization and collaboration, others are reasserting conventional methods of hierarchical control. While Storper (1997) also highlights the importance of non-market interdependencies, he insists that these cannot be captured by a single formula as they vary significantly between regions. In some cases high trust relations can delay change and represent a form of closure, preventing groups from grasping new opportunities (Scott, 1998). Decentralized networks may also prevent co-ordinated responses to radical technical change (Glasmeier, 1994). Local clusters typically depend on a combination of competition and co-operation, and the key issue is to understand why competition does not take a destructive form and undermine product quality and wage levels (cf. Sheppard, this volume). However, to date there is a lack of comparative systematic work on these issues: much of the analysis has been generalized from single case studies.

Networks are not just created by smaller companies. The largest firms can also decentralize and create all sorts of alliances and networks without losing their power and control (Harrison, 1994b). Changes in corporate form and specialization continue to be one of the key forces shaping local growth, and there are a variety of firm and organizational structures underlying industrial agglomerations (Markusen, 1996). One of the key factors underlying the most successful regions and cities is that multinational firms are strongly embedded in these areas. Large firms are undoubtedly attracted to regions with established specializations and reputations in particular industries. In contrast, direct investment in peripheral regions is typically weakly embedded, with only a small fraction of inputs purchased locally, and the plants and sites created by such investments are inherently vulnerable to closure. As globalization and regional integration continue, and firms concentrate through mergers and alliances, it is likely that larger companies will concentrate their headquarters in world or global cities, while rationalizing employment in less favored regions. Thus regional policies focusing on stimulating endogenous enterprise are unlikely to be a sufficient corrective to the centralization of corporate control. Development policies need to attract higher-quality, more embedded investments, but in many peripheral areas this is a difficult task. At the same time, however, the geographical agglomeration of investment in particular regions and cities does not in itself guarantee the achievement of high rates of long-term growth, as agglomeration is not sufficient to produce a high rate of technological innovation and the generation of new, valuable knowledge.

Innovations and Institutions

The search for a recipe for local economic dynamism, rather than simply static cost advantages, has increasingly been drawn to technological innovation. There is much evidence that some areas are more innovative than others, and most recent explanations highlight the interaction of several factors. Firms rarely innovate in isolation: their performance is determined by their interaction with the local network of actors and institutions. Product innovation tends to be concentrated in places that have a well-developed technological infrastructure, including industrial and university R&D, plus a concentration of related industries and business services (Feldman and Florida, 1994). Such places have been described as innovative milieu,[9] consisting of both physical structures such as configurations of firms, the local labor market, scientific institutions, and the availability of risk capital, as well as non-material factors such as the regional technical culture and know-how, and common representation systems. These may form regional innovation systems in which different types of networks produce different forms of technology transfer and vocational training. The recurrent idea is one of synergy, that is, an elusive combination of factors which needs to be institutionally constructed and guided. Just as technologies follow distinct trajectories shaped by conventions and practices, so it is argued that local milieu also follow distinct trajectories depending on the interaction of such factors as market demand, local and external capital, the adaptability and skill of the labor force, and technological opportunities (Castells and Hall, 1994). However, while painting a rich local picture, some of this work has neglected the importance of central government research policy and the immense influence of military spending.

This emphasis on local infrastructure acknowledges that markets are not free-floating phenomena but are made possible through a framework of institutions and conventions. While markets provide allocative efficiency they are also prone to fail under certain conditions, and collective regional institutions may be necessary (Scott, 1998). For example, firms may have a tendency to under-invest in new technology because it is difficult to prevent knowledge leakage and design imitation, making collective research and design services especially useful. Similarly, market failures are also common in worker training, so that local vocational training and educational institutions, including colleges and universities, may increase growth. Market failures are common also in financial systems, where lenders often find it prohibitively costly to assess and monitor risks, particularly among small firms. Local community and mutual institutions may possess better information about borrowers and may be better able to supply cheap, long-term credit, as has been argued to be the case in the Third Italy and in Germany's Mittelstand (Dei Ottati, 1994). Amin and Thrift (1994) use "institutional thickness" to describe high levels of interaction amongst a wide range of institutions in an area, leading to high levels of contact, co-operation, and information exchange. However, other authors have argued that institutional thickness may itself be a source of lock-in. Old industrial regions in particular may be resistant to diversification and become overly dependent on central government (Grabher, 1993). Thus, while institutional context may be necessary for growth, it may not be sufficient, and the interactions between economic conditions and institutional effects remain poorly understood. We still know little about the feasibility of the geographical transfer of institutional frameworks across regions and cities. Nevertheless, there is an increasing focus on identifying those types of institutional thickness that facilitate collective learning.

Learning and Knowledge Economies

While knowledge has always been an important resource in economies, new information technologies, patterns of flexible production, and rising costs and rapidity of innovation have given learning and knowledge a new significance (Morgan, 1997). All layers of the firm must engage in interactive learning; multi-skilling and networking are critical, and the capability to learn and to rapidly apply this learning to production and sales, are the most important components of firm viability. Again it is argued that interactive learning is facilitated and encouraged by spatial proximity between actors, implying that durable patterns of agglomeration are not the result of cost efficiency but are created by the demand for rapid knowledge transfer. In this view, while information has become abundant, knowledge is a scarce resource. Tacit knowledge, individual skills, organizational routines, and relation-specific skills are difficult to replicate elsewhere as they rely on high levels of trust, and may therefore sustain a region's advantage even in the context of rapid product imitation. Storper (1997) argues that conventions and relations of co-ordination are the means to collective reflexivity, which in turn forms the basis of learning in dynamic regions. These localized and intangible capabilities are not only of relevance to high-technology industries but may also be the basis of competitiveness in low-technology, labor-intensive production, such as the Danish wooden furniture industry (Maskell et al., 1998).

In one sense this learning approach to local economic growth is still in its infancy, nevertheless it has already faced a number of criticisms. In the first place, there is an apparent temptation to represent local cultures too unproblematically and to over-look the politics of representation surrounding any local culture. Instead of examin-ing how cultures are constructed to suit particular groups, and are shaped by political entrepreneurs, some accounts succumb to using culture as a pre-given explanatory factor (Gertler, 1997). Furthermore, the depiction of learning as the most important contemporary economic process could be taken to imply that lack of economic growth can be explained by the relative absence of learning and know-ledge accumulation. There is indeed mounting evidence that poorer neighborhoods suffer from a lack of social capital, and a dearth of social networks connecting them to employed individuals and hence knowledge about the labor market. In disadvan-taged areas, low economic status may be transmitted from generation to generation, and trust and co-operation are often undermined by suspicion, cynicism, and oppor-tunism. Poor economic performance is likely to be reinforced by poor levels of educational attainment. However, these appear to be contributory factors rather than the primary causes of low economic growth, and arguing otherwise risks attributing poverty to dysfunctional behavior and cultures.

If competitiveness means higher returns to current employees only, rather than to the entire local potential workforce, then higher competitiveness can be compatible with employment decline and greater economic inequality. However, excessive inequality may be incompatible with sustained economic growth. Several authors argue that neoliberal economic growth in the south-east of England during the 1980s could not be sustained because of increasing disparities and overheating of the local economy (Allen et al., 1998). It may well be that inequality restrains growth and perhaps intensifies the business cycle. Many recent studies of economic growth in geography have focused on high-productivity, high-cohesion forms of growth, while leaving other less desirable, but widespread, types of growth under-researched. This is illustrated by the debate on service employment and city growth. The knowledge-based, specialized cluster approach is clearly appropriate to some high-status producer service and media industries. Leyshon and Thrift (1997), for example, show that in the City of London the increasing amount of information available has to be interpreted and therefore necessitates more face-to-face contact and localized growth. However, this is only a part of service-based growth. Pollard and Storper (1996) note that service employment growth in the USA is not only occurring in specialized service cities but is also found in a wide range of cities. Furthermore, expansion of producer services in the south-east of England has been dispersed across the region (Coe and Townsend, 1998). To some extent this undoubtedly reflects a decentralization of back office and routine functions away from central city locations, but it also points to other logics of regional cumulative growth, such as new firm formation through spin-offs, and the attractions of more suburban locations.

Perhaps the most important limitation of learning approaches is their tendency to overlook the importance of more conventional price and cost conditions, and exchange and market relations. In many cases social cohesion is as much a result of economic growth as a cause. Moreover, some of these approaches ignore fixed capital and transport infrastructures, which remain crucial for accumulation. Partly as a result, these approaches seem to have an exaggerated sense of the possibilities of local endogenous development and to neglect the constraints imposed by national state and financial regimes. Indeed, regional and urban theorists may be singing a tune that policymakers want to hear (Lovering, 1995). Many of these approaches are being developed in a policy context of heightened territorial competition between authorities for investment, and it undoubtedly suits them to claim that distinctive local capabilities can be the basis for economic growth. However, the learning economy ideas are too imprecise and vague to be interpreted simply as policy derivatives. Indeed, their fuzzy nature often makes their policy implications difficult to pin down (Markusen, 1998).

Conclusion: Models Versus Metaphors

This chapter has argued that while explanations of spatially uneven growth generally have shifted towards softer and knowledge-based factors, such as innovation, training, and learning, economists and geographers have adopted very different approaches and methods in their efforts to explain these factors. While economists continue to use spartan formal models, with the aim of making their assumptions explicit, geographers have used a discursive style in order to represent the complex relationships underlying economic growth. Clark (1998) has recently described this difference as one between economists' reliance on theory-enslaved stylized facts, as against geographers' preference for "close dialogue" with economic actors, allowing a better appreciation of the depth and diversity of local economic circumstances. However, it is also important not to overlook the metaphorical character of recent economic geography.

Key metaphors, such as embeddedness, networks, evolution, and learning economies, have provided the basis of accounts of regional and urban growth. There is nothing necessarily superficial or frivolous about using such metaphors: they can provide new ways of thinking and supply unexpected insights (Barnes, 1996). Metaphors may be useful, in the sense that they can be used to capture and signal processes that are suspected to exist but are not fully understood. However, most metaphors are vague and open-ended and, in my view, their lasting value can only be established over time by systematic empirical research and grounded analyses. Because metaphors are unfinished and open, their implications must be carefully overseen. For instance, evolutionary metaphors appear to suggest that economic conventions and innovations are selected and copied according to how competitive or efficient they are, rather than how they suit the interests of economic decision-makers and powerful social groups. The two are, of course, not always synonymous.

Recent work also seems to use metaphors that smooth out conflicts of interest, implying that the most competitive types of growth are also the most cohesive and equitable, without providing much supporting evidence.

The rush of metaphorical tropes into theories of uneven growth in economic geography has been so rapid that the systematic evaluation and empirical validation of these ideas has not kept pace. While the development of new theories has undoubtedly been enriching and has provided many insights into the causes and conditions of economic growth, there is a widely noted lack of empirical evaluation of their utility. Ironically this may partly be a product of the institutional embeddedness of researchers, who are under increasing pressure to publish quickly rather than to make long commitments to time-consuming empirical analyses. Thus, for very different reasons, both the new location theories and recent economic geography have empirical weaknesses, and some of the key processes behind regional and urban growth continue to remain a mystery.

Endnotes

1. According to Myrdal, spatial flows of capital and labor increase the growth of expanding regions at the expense of lagging regions. Capital and labor are attracted to expanding regions and increase their labor supply and market size, and can also strengthen the backward linkages of firms to their suppliers and forward linkages to firms next along the production chain.
2. The law of diminishing returns states that, beyond a level of input, further increases in the input of a factor of production result in decreases in the additional marginal output of the product per unit of input.
3. A monopoly rent in this context refers to the profits, which a firm can make by selling a new product under monopoly conditions, before the product has been marketed by rival firms.
4. The broadest definition of an externality is where the actions of one actor have unintended consequences for another actor. These may, of course, be both positive and negative. In economics, externalities are often defined as technological externalities whereby one firm's production affects the production process of another firm in the absence of a market transaction between them. Recently, however, it has been argued that pecuniary externalities, which affect prices in market exchanges, may also be important.
5. These occur where the adoption of a technology by several users creates incentives for, and lowers the cost of, its adoption by other users. Where a certain technology becomes dominant within a network of users, entrants to this network will face a strong incentive to adopt the same technology in order to be compatible and to benefit from externalities.
6. A rational choice equilibrium is a stable pattern, which can be derived by modeling the behavior of numerous individuals who are assumed to make rational decisions.
7. Monopolistic competition is a model of market equilibrium based on competition among similar firms producing differentiated products, which are close but not perfect substitutes. It has been widely used as a means of modeling imperfect competition.
8. These ideas were anticipated by Chinitz (1961) who noted that supply factors and social structures and market organization shape local entrepreneurship.
9. Innovative milieu are an example of the embedding of economic activity with the social relations of a particular place.

Bibliography

Allen, J., Massey, D., and Cochrane, A. 1998. *Rethinking the Region*. London: Routledge.

Amin, A. and Thrift, N. 1994. Living in the global. In A. Amin and N. Thrift (eds). *Globalization, Institutions, Regional Development in Europe*. Oxford: Oxford University Press, 1–22.

Angel, D. P. and Engstrom, J. 1995. Manufacturing systems and technological change: The U.S. personal computer industry. *Economic Geography*, 71, 79–102.

Arthur, W. B. 1996. *Increasing Returns and Path Dependence in the Economy*. Michigan: Michigan University Press.

Barnes, T. 1996. *Logics of Dislocation: Models, Metaphors, and Meanings of Economic Space*. New York: Guilford Press.

Barnes, T. 1997. Theories of accumulation and regulation: Bringing life back into economic geography. In R. Lee and J. Wills (eds). *Geographies of Economies*. London: Arnold, 231–47.

Barro, R. J. and Sala-i-Martin, X. 1995. *Economic Growth*. London: McGraw Hill.

Castells, M. and Hall, P. 1994. *Technopoles of the World: The Making of Twenty-first Century Industrial Complexes*. London: Routledge.

Cheshire, P. 1995. A new phase of urban development in Western Europe? The evidence for the 1980s. *Urban Studies*, 32, 1045–63.

Chinitz, B. 1961. Contrasts in agglomeration: New York and Pittsburgh. *American Economic Review Papers and Proceedings*, 51, 279–98.

Clark, G. 1998. Stylized facts and close dialogue: Methodology in economic geography. *Annals of the Association of American Geographers*, 88, 73–87.

Coe, N. M. and Townsend, A. R. 1998. Debunking the myth of localized agglomerations: The development of a regionalized service economy in South-East England. *Transactions of the Institute of British Geographers*, 23, 385–404.

Cooke, P. (ed). 1995. *The Rise of the Rustbelt*. London: UCL Press.

Cooke, P. and Morgan, K. 1994. The network paradigm: New departures in corporate and regional development. *Environment and Planning D: Society and Space*, 11, 543–64.

David, P. 1985. Clio and the economics of QWERTY. *Economic History*, 75, 332–37.

Dei Ottati, G. 1994. Trust, interlinking transactions and credit in the industrial district. *Cambridge Journal of Economics*, 18, 529–46.

Dunford, M. and Perrons, D. 1994. Regional inequality, regimes of accumulation and economic development in contemporary Europe. *Transactions of the Institute of British Geographers*, 19, 163–82.

Dymski, G. 1997. On Krugman's model of economic geography. *Geoforum*, 27, 439–52.

Feldman, M. P. and Florida, R. 1994. The geographic sources of innovation: Technological infrastructure and product innovation in the United States. *Annals of the Association of American Geographers*, 84, 210–29.

Frey, W. H. 1993. The new urban revival in the United States. *Urban Studies*, 30, 741–74.

Gertler, M. 1997. The invention of regional culture. In R. Lee and J. Wills (eds). *Geographies of Economies*. London: Arnold, 47–58.

Glasmeier, A. 1994. Flexible districts, flexible regions? The institutional and cultural limits to districts in an era of globalization and technological paradigm shifts. In A. Amin and N. Thrift (eds). *Globalization, Institutions, Regional Development in Europe*. Oxford: Oxford University Press, 118–46.

Grabher, G. (ed). 1993. *The Embedded Firm: On the Socio-economics of Industrial Networks*. London: Routledge.

Hall, P. 1985. The geography of the fifth Kondratiev. In P. Hall and A. Markusen (eds). *Silicon Landscapes*. Boston: Allen & Unwin, 1–19.

Hanson, G. M. 1996. Economic integration, intraindustry trade, and frontier regions. *European Economic Review*, 40, 941–49.

Harrison, B. 1994a. Industrial districts: Old wine in new bottles? *Regional Studies*, 26, 469–83.

Harrison, B. 1994b. *Lean and Mean: the Changing Landscape of Corporate Power in the Age of Flexibility*. New York: Basic Books.

Harrison, B., Kelley, M. R., and Gant, J. 1996. Innovative firm behavior and local milieu: Exploring the intersection of agglomeration, firm effects, and technological change. *Economic Geography*, 72, 233–58.

Hudson, R. 1997. Regional futures: Industrial restructuring, new high volume production concepts and spatial development strategies in the New Europe. *Regional Studies*, 31, 467–78.

Jacobs, J. 1970. *The Economy of Cities*. London: Jonathan Cape.

Jaffe, A. B., Trajentenberg, M., and Henderson, R. 1993. Geographic localization of knowledge spillovers as evidenced by patent citations. *Quarterly Journal of Economics*, 108, 577–98.

Krugman, P. 1991. *Geography and Trade*. Cambridge, MA: MIT Press.

Krugman, P. and Venables, A. 1990. Integration and the competitiveness of peripheral industry. In C. Bliss and J. Braga de Macedo (eds). *Unity with Diversity in the European Economy*. Cambridge: Cambridge University Press, 56–75.

Leyshon, A. and Thrift, N. 1997. *Money/Space: Geographies of Monetary Transformation*. London: Routledge.

Lovering, J. 1995. Creating discourses rather than jobs: The crisis in the cities and the transition fantasies of intellectuals and policy-makers. In P. Healy, S. Cameron, S. Davoudi, S. Graham, and A. Madani-Pur (eds). *Managing cities: The New Urban Context*. London: John Wiley, 109–26.

Lucas, R. E. 1988. On the mechanics of economic development. *Journal of Monetary Economics*, 22, 3–42.

Markusen, A. 1996. Sticky places in slippery space: A typology of industrial districts. *Economic Geography*, 72, 293–313.

Markusen, A. 1998. Fuzzy concepts, scanty evidence, wimpy policy: The case for rigor and policy relevance in critical regional studies. Center for Urban Policy Research. Working Paper 138. New York: CUPR.

Marshall, A. 1919. *Industry and Trade*. London: Macmillan.

Martin, R. 1999. The new "geographical turn" in economics: Some critical reflections. *Cambridge Journal of Economics*, 23, 65–91.

Martin, R. and Sunley, P. 1996. Paul Krugman's geographical economics and its implications for regional development theory: A critical assessment. *Economic Geography*, 72, 259–92.

Martin, R. and Sunley, P. 1998. Slow convergence? The new endogenous growth theory and regional development. *Economic Geography*, 74, 210–27.

Maskell, P., Eskelinen, H., Hannibalson, I., Malmberg, A., and Vatne, E. 1998. *Competitiveness, Localized Learning and Regional Development: Specialization and Prosperity in Small Open Economies*. London: Routledge.

Massey, D. 1984. *Spatial Divisions of Labor*. London: Macmillan.

Morgan, K. 1997. The learning region: Institutions, innovation and regional renewal. *Regional Studies*, 31, 491–503.

Myrdal, G. 1957. *Economic Theory and Underdeveloped Regions*. London: Gerald Duckworth.

Ottaviano, G. and Puga, D. 1997. Agglomeration in the global economy: A survey of the "New economic geography". Centre for Economic Performance, Discussion Paper 1699. London: CEP.

Piore, M. and Sabel, C. 1984. *The Second Industrial Divide: Possibilities for Prosperity.* New York: Basic Books.

Pollard, J. and Storper, M. 1996. A tale of twelve cities: Metropolitan employment change in dynamic industries in the 1980s. *Economic Geography,* 72, 1–22.

Porter, M. 1990. *The Competitive Advantage of Nations.* London: Macmillan.

Pred, A. R. 1966. *The Spatial Dynamics of US Urban and Industrial Growth, 1800–1914: Interpretive and Theoretical Essays.* Cambridge, MA: MIT Press.

Romer, P. 1986. Increasing returns and long-run growth. *Journal of Political Economy,* 94, 1002–37.

Saxenian, A. 1994. *Regional Advantage: Culture and Competition in Silicon Valley and Route 128.* Cambridge, MA: Harvard University Press.

Schoenberger, E. 1989. New models of regional change. In R. Peet and N. Thrift (eds). *New models in geography: The Political Economy Perspective.* London: Unwin Hyman, 115–41.

Schoenberger, E. 1998. Discourse and practice in human geography. *Progress in Human Geography,* 22, 1–14.

Scott, A. J. 1988. *New Industrial Spaces.* London: Pion.

Scott, A. J. 1998. *Regions and the World Economy: The Coming Shape of Global Production, Competition, and Political Order.* Oxford: Oxford University Press.

Staber, U. 1996. Accounting for variations in the performance of industrial districts: The case of Baden-Württemberg. *International Journal of Urban and Regional Research,* 299–316.

Storper, M. 1997. *The Regional World: Territorial Development in a Global Economy.* New York: Guilford Press.

Chapter 13

Geography and Technological Change

David L. Rigby

As global production becomes increasingly integrated, workers and firms in different regions are forced more directly into competition with one another. How do these firms, and the regions in which they are embedded, compete for the capital and labor required to sustain competitiveness? The dominant strategies of competitive advantage, in one form or another, hinge on technology. More specifically, technological change is the primary determinant of profitability and growth. This much was clear in the 1950s (Abramowitz, 1956; Solow, 1956, 1957) and it remains so today (Grossman and Helpman, 1991; Lucas, 1988; Romer, 1986, 1990).

The role of technical change in fueling economic growth has been the subject of much recent discussion as growth rates have declined in many of the advanced industrialized nations and as certain newly industrializing countries (NICs) are reducing the technological gap (Baumol, 1986; Fagerberg, 1994; Maddison, 1982). Explanations of national differences in innovative performance have turned to focus on industrial organization, on the sectoral and spatial linkages between firms, and on the institutional systems within which they operate (Archibugi and Michie, 1995; Freeman, 1991; Lundvall, 1992; Nelson, 1993; von Hippel, 1988).

Increasingly it is recognized that the motors of national economic performance are sub-national technology districts (Amin and Thrift, 1994; Scott, 1996; Storper, 1992). These innovative regions are characterized by strong ties between regional actors, embedded in institutional structures that reinforce common sets of rules, norms, business cultures, and decision routines (Benko and Dunford, 1991; Brusco, 1982; Grabher, 1993; Granovetter, 1985; Harrison, 1992; Stöhr, 1986; Maillat, 1995; Storper, 1995). With boundaries to the spatial flow of information (Jaffe et al., 1993), region-specific knowledge bases and localized processes of search and learning (David, 1975) are hypothesized to channel technological change along relatively distinct regional trajectories (Aydalot, 1988; Rigby and Essletzbichler, 1997; Rigby and Haydamack, 1998). This gives rise to regional systems of innovation and technological advance.

This chapter examines the geography of technological change. The following section summarizes the importance of technological change, showing how technology shapes broad patterns of growth and income. Attention then shifts to definitions of technology, and the key processes by which technology is altered – namely invention, innovation, imitation, learning, differential firm growth, and turnover. Geographical differences in technology, and the influence of space on the processes of technological change, are analyzed in turn.

Economic Growth and Technological Change

Between 1820 and 1982, the real value of output in the world's largest industrialized economies increased by a factor of seventy. The population of these economies also grew, nevertheless the average worker produced fourteen times more real output at the end of this period than at the beginning (Maddison, 1982). Rates of growth of output and productivity in member nations of the Organization for Economic Cooperation and Development (OECD) generally accelerated over the last 150 years, peaking in the two decades after World War II. From 1950 to 1970, OECD output grew at an annual average rate of about 5 percent. Growth rates decelerated sharply from the late 1960s, popularly since the first oil shock in 1973, but subsequently rebounded after the deep recession of the early 1980s (Maddison, 1987). The growth experience of different groups of countries has varied (Webber and Rigby, 1996). While the advanced, industrialized economies have become richer, and more alike in terms of productivity (Barro, 1991; Baumol, 1986), underdeveloped countries have fallen further behind (Landes, 1998). Newly industrialized countries (NICs) enjoyed unprecedented expansion and productivity growth over the last thirty or so years (Dicken, 1998), though the fragility of this growth has recently been exposed. Some attribute the success of these economies to technological "catch-up" (Barro, 1991; Fagerberg, 1994), though others advance different arguments (Young, 1995).

Since the pioneering work of Solow (1956, 1957), it has become conventional to account for these "stylized facts of growth" using some variant of the following relationship:

$$Y = a \cdot K + (1 - a)L + T \tag{13.1}$$

This equation argues that the rate of growth of output (Y) is equal to the growth rates of capital (K) and labor (L), each weighted by their respective share of net output (a), plus the rate of growth of aggregate productivity (T). This last term, typically represented by time and thus exogenous to the economy, measures the contribution of technological change to output growth. Using this equation, Solow (1957) demonstrated that technological change was the principal determinant of US output growth over the first half of the twentieth century. Over the last forty years numerous economists have augmented Solow's model, notably Denison (1962, 1967), by adding terms measuring changes in the quality of capital and labor inputs, thus reducing the contribution of technology to growth. These challenges notwithstanding, technological change remains the key to long-run economic growth, to rising productivity and income levels.

The Solow model also explains international differences in long-run growth rates (Jones, 1998). In the Solow tradition technology is viewed as a free or public good, and thus growth rate variations between rich and poor countries are linked to national differences in savings and investment rates, in population growth, and (after Mankiw et al., 1992) in rates of accumulation of human capital.

Although technological change was regarded as the primary motor of economic growth it remained exogenous in Solow's model, where it appeared, as Joan Robinson (1953/54) commented, rather "like manna from heaven." In large part this ignorance of the processes driving technological change reflected a theoretical myopia as well as analytical convenience: the neoclassical economic model was simply unsuited to analyze a disequilibrating process like technological change, which is characterized by uncertainty, by asymmetries of information within a heterogenous population of firms, and by markets that are at best imperfect.

After a hiatus of some twenty or so years, interest in growth theory and technology was rekindled in the late 1980s by Romer (1986, 1990), Lucas (1988), and Grossman and Helpman (1991). In their "endogenous growth models," technological change – although largely still a "black box" – was at least conceptualized as an integral part of the economy. Technology is no longer viewed as a public good: firms are able to capture some of the returns from their own research and development efforts. Thus, increasing returns to technology development and adoption emerge as a basis for agglomeration and the maintenance of growth rate disparities. Once more, questions about the meaning of technology and the processes of technological change have moved to the forefront as we try to understand the nature of competition and economic growth.

Technology and Technological Change

In the capitalist economy, production is controlled largely by individual firms. While these firms may adopt a variety of short-term strategies, their fundamental aim is to make a profit. No firm is guaranteed profit, for the market is a chaotic arena where prices for inputs and outputs cannot be determined a priori (Alchian, 1950; Farjoun and Machover, 1983; Nelson and Winter, 1982). Some firms attempt to manage this uncertainty by controlling the market. However, the majority of firms can only control the manner in which they transform inputs into output, seeking to achieve a competitive advantage by increasing the efficiency of their production. For most, efficiency is unknown until they enter the market and are evaluated by their rivals. In this competitive environment firms are compelled to search for new technology, sure only in the knowledge that others are doing the same.

Schmookler defines technology as the "social pool of knowledge of the industrial arts" (1966, p. 1). In this sense, technology represents the set of known ideas or information about the range of products that can be made and the variety of processes that may be employed in their production, including the specific combinations of capital and labor inputs used to produce output, the division of labor (the separation of tasks within and between firms), and the broader, institutional structures within which economic activity is embedded. Technological progress represents the expansion of this pool of information, and technological change occurs when an economic agent uses a new part of this knowledge pool.

The history of technological change is governed by the production of knowledge, by the application of that knowledge, and by how it diffuses throughout the economy. Since Schumpeter (1939), it has become commonplace to distinguish between the processes of knowledge production (invention and learning), of the introduction of that knowledge to the economy (innovation), and the spread of that knowledge through the economy (diffusion). While this conceptual division may be useful for purposes of explication, it is important to remember that these processes are often indistinguishable.

Invention

The production of knowledge and its application have changed considerably over time. Although a division of labor might be considered a prerequisite for the inventive process, advances in science were not closely linked to technology and economic activity much before the nineteenth century (Usher, 1954; though see Musson and Robinson, 1969, for a dissenting opinion). Consistent with this separation of science and economy, invention was initially regarded as a discontinuous process resulting from the inspiration of the occasional genius. Sociologists of invention such as Gilfillan (1935) rejected this view, claiming that invention is more often the result of incremental problem-solving within a relatively familiar framework of ideas and economic relations. Usher (1954) integrated these two approaches, arguing that while invention depends on critical acts of insight, such acts can be encouraged by the creation of appropriate economic and knowledge environments.

"The great invention of the nineteenth century was the invention of the method of invention" (Whitehead, 1925, p. 96). The production of knowledge, or invention, was slowly institutionalized in the research and development (R&D) laboratories of large firms throughout the late nineteenth century. Freeman (1982) traces the emergence of R&D labs in the German chemicals and dyestuffs industries and links the "professionalization" of in-house R&D to the growing costs of technological development. Growth in the system of technology production was also encouraged by consolidation of patent systems throughout North America and Western Europe. Through protection of intellectual property rights and through the system of assignment, by which individual inventors sold the rights to their inventions, the patent system hastened the emergence of a market and trade in technology, leading to a more specialized division of labor in technology production (Lamoreaux and Sokoloff, 1996).

It was not until World War II, and the success of science-based military technology, that a new orthodoxy emerged. This characterized modern technology as applied science (Bush, 1945; Thirtle and Ruttan, 1987). The result was a rapid expansion in private and public support for R&D, and an explosion of economic growth fueled by rising productivity and incomes, by the establishment of new technology-intensive industries, and by the associated growing range of new products. Consistent with this "science-led" picture of technological change, a simple, linear model of the process of invention and technological change became widely accepted (Malecki, 1991). According to this model, basic science, performed in the research labs of universities and government research centers, produced knowledge that was then commercially applied in the research and development labs of modern corporations. From such applications, engineers developed blueprints and prototype commodities that were passed to marketing departments to assess the likelihood of commercial viability. Although this simple model has been roundly criticized for ignoring various feedbacks between the different stages of technology development (Kline and Rosenberg, 1986), and for assuming that science leads industry in the production of ideas for new technologies (Mowery and Rosenberg, 1979), its focus on science–industry links appears increasingly prescient. Scientific activity has become more expensive and therefore more frequently financed by private business, and university–industry linkages have proliferated (Henderson et al., 1998; Malecki, 1991; Saxenian, 1994; Trajtenberg et al., 1997).

This linear model of technology development fails to embrace the variety of means by which individual manufacturing firms and groups of interlinked firms and related institutions generate new technologies. New techniques do not emerge only through the deliberate process of R&D but also are generated by various learning processes, and sometimes are the unintended consequences of problem-solving in production (Arrow, 1962; Lundvall, 1988, 1992). As long ago as 1962, Arrow noted that the experience learned by firms through production was not always appropriated by the firm, but sometimes passed as a public good to society overall. Indeed, these technological spillovers form the basis for the so-called new endogenous growth models of Romer (1986, 1990). Regardless of their origin, whether or not new techniques are adopted in the economy depends upon the process of innovation.

Innovation

Innovation is the application of new ideas to the economy. While innovation depends on invention, the introduction of new technologies is a complex process that also depends on the cost of change and the potential demand for new technology. The rate of introduction of new technology and the characteristics of that technology exert critical controls on economic growth and on the demand for inputs to production such as labor. As growth has slowed and unemployment increased in many parts of the world, there has been intense economic debate about not only the pace of innovation, but also the direction of innovation (i.e. whether new techniques economize on the use of particular inputs to production).

Research on the pace of innovation regards the introduction of a new technology to the economy as a discrete event that can be dated and located. Mensch (1979) and Freeman et al. (1982), following Schumpeter (1939), examine the timing of major innovations. Mensch (1979) argues that major or radical innovations tend to cluster in time, giving rise to periods of intense technological activity akin to Schumpeter's "gales of creative destruction," while other times are characterized by technological stagnation. The periodic clustering of innovations is sometimes used to explain long waves of economic growth (Berry, 1991; van Duijn, 1983). Figure 13.1, taken from Dicken (1998), summarizes these claims.

These arguments have repeatedly been criticized. Freeman et al. (1982) and Kleinknecht (1987) claim that Mensch's (1979) identification of radical or major innovations was *ad hoc*, and dispute his evidence of technological clustering. Kleinknecht also argues that innovation should not be considered a discrete event, because it is merely the prelude to a long series of incremental improvements that may have a far greater economic impact than the introduction of a new technology. Fishlow (1966) has provided support for this claim, documenting the productivity gains associated with a long succession of minor improvements in railroad and

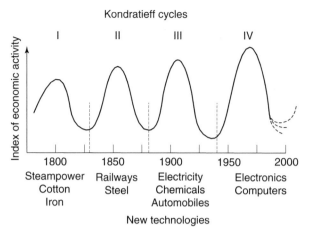

Figure 13.1 Long waves of economic activity
Source: Adapted from Dicken (1998, p. 148)

related technologies, as has Hollander (1965) in studies of Du Pont. In addition to these concerns, proponents of a long-wave, technology-driven model of economic growth do not provide convincing explanations of why technologies cluster, whether or not innovation leads or follows the upswing in economic activity, and why the long cycles of growth have a periodicity of around 50 years.

The impacts of innovation on the nature of technological change, on economic growth, and on the structure of competition are examined further using the concept of the product cycle (Vernon, 1966). In Vernon's (1966) model, as a product ages the factor intensity and skill requirements of production alter. Utterback and Abernathy (1975) explore how the focus of innovation switches from products to new production processes as a commodity matures, as its market expands, and as firms increasingly adopt competitive strategies based upon controlling costs rather than shaping demand. The product-cycle model (figure 13.2) is not without its detractors, however. Taylor (1986) notes that few commodities actually pass through the different stages of the cycle, while Storper (1985) condemns the deterministic nature of the

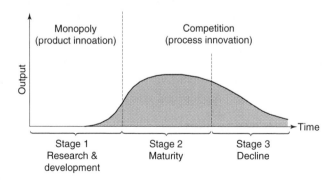

Figure 13.2 The product cycle
Source: Author

whole concept. Debate over the direction of innovation is often traced to Hicks' (1932) claim that firms will tend to adopt techniques that save on relatively more expensive inputs to production. Fellner (1961) disputed this claim, arguing that in competitive markets all inputs are equally costly. Salter (1960) continued the criticism, noting that firms are interested only in reducing costs, and not in the manner of such reduction. Extensive attempts in the late 1960s and early 1970s to link the direction of technological change to the relative prices of inputs in models of induced innovation (see Ahmad, 1966; Kennedy, 1964; and Thirtle and Ruttan, 1987 for a detailed review) were unsuccessful. In large part this is because of a failure to recognize that the processes of searching for and adopting new technologies are expensive. In models where the costs of innovation are considered explicitly, differential returns from reducing inputs with varying costs will encourage firms to economize on more expensive inputs (Binswanger and Ruttan, 1978). A large number of empirical studies of individual industries and economies have examined the induced innovation hypothesis (see Thirtle and Ruttan, 1987). In some of the more well-known, David and van de Klundert (1965) found a labor-saving bias to innovation in the first half of the twentieth century in the US economy, when labor was scarce and thus relatively expensive. Habakkuk (1962) finds consistent results for the nineteenth century, lending general support to the induced innovation hypothesis.

More recently, David (1975) has reformulated the induced innovation model, arguing that the search for new techniques is localized, and that the choice of technique is path dependent. For David, firms accumulate knowledge about technology through experimentation with existing techniques. The local nature of search is conditioned by sharply declining returns to investment in research and development efforts that are relatively dissimilar to existing technology, and by costs of knowledge acquisition that rise steeply beyond the boundaries of existing knowledge bases (Arrow, 1994; Webber et al., 1992). Thus, technological change is increasingly understood as an evolutionary process moving gradually along relatively distinct pathways, subject to interruption by the infrequent development of radically different technological knowledge.

Accordingly, the general history of innovation is perhaps best described as a sequence of radical breaks with past scientific knowledge, and incremental changes along relatively well-defined technological trajectories. Technological trajectories are broadly shaped by a knowledge base that imposes a certain logic of problem-solving, often involving a core technology and an agenda for subsequent improvement. This idea is incorporated in Sahal's (1981) technological guideposts, Dosi's (1982) technological paradigms, Nelson and Winter's (1982) natural trajectories, and Clark's (1985) design hierarchies. Technological trajectories are further shaped by regulatory constraints, common standards, requirements for systems compatibility (railroad gauge and computer operating systems provide simple examples), and other institutional limitations (David, 1985, 1992; Henderson and Clark, 1990; Katz and Shapiro, 1985; Nelson, 1994).

Diffusion

Diffusion, or imitation, is the process by which new technology or knowledge spreads throughout the economy. If the impact of technological change is measured

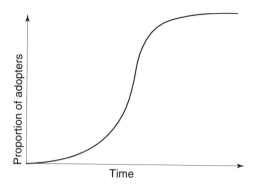

Figure 13.3 The logistic diffusion curve
Source: Author

by aggregate productivity improvements and economic growth, then diffusion is probably the most significant influence on such change (Baldwin and Scott, 1987; Rosenberg, 1982). Perhaps the most well-known studies of diffusion examine determinants of the adoption process. From the pioneering empirical work of Griliches (1957) and Mansfield (1961) came the discovery of the S-shaped, logistic adoption curve, rationalized as a result of the progressive dissemination of information about the technical and economic characteristics of the innovation (see figure 13.3). In this case, new technology is adopted once information about it becomes available, as in Hägerstrand's (1953) study of the adoption of costless techniques in Swedish agriculture. The speed of diffusion in most of these accounts is shown to depend on the anticipated profitability of adoption. This in turn depends upon the distribution of techniques employed within an industry, on the costs of purchasing new technology and learning how to use it, and on the cost of abandoning old techniques (Webber et al., 1992).

The classic, logistic model of epidemic diffusion has two main failings. First, it assumes a constant adoption environment in which new technologies and the characteristics of potential adopters do not change. Second, it focuses on the demand side of the diffusion process, largely ignoring how information about technology is disseminated. "Threshold models," in which changes in economic conditions over time tip the balance in favor of adoption, have been developed in response to the former criticism. For instance, David (1966) accounts for the twenty-year time lag between the development of the mechanical reaper and its widespread adoption in the US Mid-west by a growth in average farm size and increasing labor costs which eventually made mechanization imperative. Olmstead (1975) disagrees, noting that adoption of the mechanical reaper was encouraged by continuous small design improvements. Davies (1979) also discusses the importance of post-innovation improvements in technology that stimulate diffusion, along with differences in the complexity of technology, firm size, and other economic characteristics. Learning by using, which reduces uncertainty about particular technologies, and the emergence of industry standards and growing networks of complementary technologies, have also been viewed as critical in understanding diffusion from a threshold perspective (David, 1992; Nelson, 1994). In addition, Gertler (1993 and 1995) has recently

demonstrated how "cultural" differences between firms retard processes of technology adoption and learning.

Focusing on the demand side of the diffusion process, the logistic model ignores
the different mechanisms by which technological information is transferred (see also
Brown, 1981). First, through the nineteenth century, in the USA at least, trade in
technology became institutionalized as the patent system was consolidated. This
system spurred the transmission of codified knowledge through the practice of
assignment and by public advertisement of new patents (Lamoreaux and Sokoloff,
1996). Second, both codified and tacit knowledge is spread when workers move
between firms (Landes, 1969; Saxenian, 1994). Third, growing specialization and
the division of labor also speed technology diffusion by increasing inter-firm trade.
Product innovations (new products) in capital goods industries become process
innovations (new production processes) for the firms that purchase them (Kuznets,
1930). The purchase of rival firms' goods, and reverse engineering (the disassembly
of commodities and direct imitation of component technologies), allow less efficient
firms to rapidly catch up with "best practice" techniques. Finally, technology can
transfer through firm mergers, acquisitions, and joint ventures which have accelerated in recent years (Dicken, 1998; Malecki, 1991).

Learning

Learning, though often neglected, is another mechanism of technological change.
Learning is simply the acquisition of knowledge. Knowledge about technology and
its use is not generated solely through formal processes of search and adoption, but
frequently originates as a byproduct of other activities. Arrow (1962) notes how
experience accumulated through production results in productivity gains. Rosenberg
(1982) distinguishes a somewhat more general form of learning that occurs through
use of specialized capital equipment. Malecki (1991) identifies a series of other
forms of learning, and Stiglitz (1987) focuses on the importance of learning how
to learn. David (1975) stresses that the localized nature of the learning process
imparts a significant degree of inertia in the process of technology development.
Recent attention given to industrial districts as fonts of knowledge creation (Amin,
this volume) stems from the importance of networks as a means for sharing knowledge, and of the creation of network-based institutions of knowledge governance
that help markets function (Lundvall, 1988; Storper, 1997).

Plant heterogeneity and aggregate technological change

The processes of technological change reviewed above operate at the firm level,
although firms are influenced by broader industrial and geographical influences. At
the aggregate level of the industry or region, technological change is also produced
through selection, or the differential growth of firms, by the entrance of new firms
into the economy, and by the exit of established firms (Nelson and Winter, 1982;
Webber et al., 1992). These aggregate processes of technological change are shaped
by various forces. Selection is governed largely by variations in efficiency, as market
competition rewards more productive firms with increased market share at the
expense of less productive firms. This, in turn, raises aggregate productivity. The

effect of selection on technological change depends upon the degree to which productivity varies between firms, on the size distribution of firms, and on the intensity of competition (Baily et al., 1992; Baldwin, 1998). Firm entry and exit influence technological change if entering and exiting firms have technologies that differ from the average. Entrants often bring new technologies to the market, reshuffling the relative efficiency of existing firms. Evidence is mixed as to whether new firms are typically more or less efficient than average. Exiting firms are typically less efficient than average and so their departure improves aggregate productivity. Entry and exit are influenced by similar factors to selection, although economies of scale, industry concentration, and the costs of advertising and research and development reduce the rate of entry and/or exit (Bain, 1956; Baldwin, 1998; Caves and Porter, 1977).

The Geography of Technology

At any moment in time, a regional economy crudely may be conceived as a collection of economic agents, firms, and workers, embedded within a set of organizational and institutional structures that guide behavior to a greater or lesser extent. As such, regions are also repositories of accumulated knowledge, both codified and tacit. Region-specific knowledge bases consist of familiarity with the production of particular commodities and of specific techniques used in their production. They also include experience with organizational forms, of different ways of separating production through the division of labor, of different ways of managing intra- and inter-firm relations, and experience with institutional structures that regulate the environment within which economic agents operate. Most importantly, these knowledge bases also incorporate behavioral conventions that shape the way in which knowledge is produced or somehow obtained within the region. Because these pools of knowledge differ over space, technology may be differentiated geographically, along with the characteristics and determinants of technological change. This part of the chapter considers geographic differences in technology, and how location affects processes of technological change.

Geographic differences in technology

International differences in technology and the pace of technological change are relatively clear (Dicken, 1998; Webber and Rigby, 1996). Industrial specialization and the overall growth of trade indicate that technological (sectoral) capabilities are nation-specific (Soete, 1987). Cantwell (1991) shows that these capabilities are cumulative, that is, they persist over time. Significant international variations in productivity and in the skill composition of the workforce are reported by the OECD (1996), and Amendola et al. (1992) and Fagerberg (1987) demonstrate the impacts of technology, narrowly conceived, on international competitiveness. Differences in the pace of technological change between countries are usually reported as variations in rates of total factor productivity growth (Denison, 1967; Maddison, 1987; Mankiw et al., 1992). These variations are typically accounted for by rates of R&D spending, by trade in technology, by industry mix and the nature of firm and industry linkages, and by the more general characteristics of national systems of innovation (Archibugi and Michie, 1995; Freeman, 1995; Lundvall, 1992; Porter, 1990).

Industrial specialization and concentration at the sub-national level is also clear (Ellison and Glaeser, 1997). However, the existence of sub-national, regional differences in production techniques remains an open question. A number of studies have documented marked regional variations in labor productivity across a series of industries and at a variety of spatial scales (Casetti, 1984; Hulten and Schwab, 1984; Moomaw, 1983; Rigby, 1992). Others, who view technological change as embodied in capital, and investment as the medium through which new techniques are introduced to the economy, employ differences in the age of capital as a surrogate for geographical differences in technology (Anderson and Rigby, 1989; Rigby, 1995; Varaiya and Wiseman, 1981). Persky and Klein (1975) and Gleed and Rees (1979) provide evidence of regional differences in capital productivity between regions, and Beeson (1987) and Beeson and Husted (1989) reveal differences in total factor productivity across US states, attributing the differences to labor-force characteristics, industry structure, and levels of urbanization. In related work, regional differences in production functions are noted in Lande (1978) and Luger and Evans (1988). Such studies provide useful information about technological variety over space but they remain partial, focused largely on a single input to production, on a few relatively aggregate economic sectors or regions, or on limited time periods. In fact we know surprisingly little about geographical variations in production techniques, and even less about the evolution of technologies over space.

Recent research by Rigby and Essletzbichler (1997) and Rigby and Haydamack (1998) examines regional differences in production techniques within a number of industrial sectors over much of the post-war period. Their analysis suggests that regions tend to occupy broadly similar positions in "technology-space" from one industry to the next. For example, regions with production techniques in one industry that are more capital-intensive than average, or more labor-intensive than average, tend to have production techniques in other industries that exhibit the same characteristics. This is a remarkable finding, indicating that there may be strong geographic tendencies shaping the choice of technique regardless of the manufacturing sector. In related work, Essletzbichler et al. (1998) find significant regional differences in production techniques that persist over time.

Choice of technique at the plant or firm level is closely related to product type, to the characteristics of the market in which a product is traded, and to industrial organization. Industrial organization, another dimension of technology, refers to the social and technical division of labor, to how the processes of production are separated into discrete tasks, and how those tasks are allocated across firms (Williamson, 1975). Since the early work of Coase (1937), that allocation was typically explained by transaction costs and by economies of scale and scope (Scott, 1988). However, by the end of the 1980s it was clear that a simple transactions cost approach was insufficient to account for the varied relationships that bind individual firms and workers to one another and to particular industrial districts. Relations among firms increasingly were seen as governed by various forms of what Storper (1995) calls untraded interdependencies (see also Amin and Thrift, 1994; Camagni, 1991; Camerer and Vepsalainen, 1988; Grabher, 1993). In large part these interdependencies were understood, after Granovetter (1985), as broader sets of social relations that over time coalesce to form regional "cultures," or tacitly understood conventions/institutions that encourage trust, reduce uncertainty, and guide behavior. Thus, the individual firm became less significant as the locus of competitive advantage and technology creation. Case studies of "regional worlds of production" revealed the varied institutional foundations of industrial and regional performance (Saxenian, 1994; Storper, 1993; Tödtling, 1992). The geographical dimensions of industrial organization are explored by Scott and Storper (1987, 1992).

The geographical evolution of technology

Different theoretical visions of the spatial dynamics of production technology exist. Product life-cycle studies suggest that as industries and products mature, technology becomes increasingly standardized and thus more geographically mobile. Models of technology diffusion are also frequently invoked to explain the narrowing of technological variation over space (Brown, 1981; Griliches, 1957; Hägerstrand, 1967), and competition and factor mobility within the neoclassical regional growth model are similarly seen as eroding geographical differences in techniques of production (Borts and Stein, 1964). Technological "catch-up" is commonly thought to underpin convergence in international productivity levels (Fagerberg, 1994).

More recent analysis of technological change in space focuses on the dynamic capabilities of economic agents in different places to generate and sustain a creative milieu that undergirds competitive advantage (Maillat, 1991, 1995; Malmberg, 1996; Marshall, 1920; Myrdal, 1957; Storper, 1997). In this work regional technological change does not take place solely within the boundaries of the firm, but is also generated through interaction with other institutions (Freeman, 1995; Lundvall, 1992; Lundvall and Johnson, 1994; Nelson, 1993). Firms are seen as embedded in overlapping sets of socio-spatial relations including buyer–seller linkages, subcontracting ventures, local business cultures, conventions, and institutions, as well as various types of competitive capital and labor transactions (Granovetter, 1985; Harrison, 1992; Johnson and Lundvall, 1992; Saxenian, 1994; Storper, 1995, 1997; Teece, 1992). Attention has thus shifted toward the shared "technological capital" of industrial districts as the motor of agglomeration, and to the development of a regional variant of the national system of innovation (DeBresson and Amesse, 1991; Freeman, 1991, 1995; Lundvall, 1992; Nelson, 1993).

Accordingly, the evolution of technology is closely tied to the economic and institutional character of particular places. Jaffe et al. (1993) confirm the localized nature of technological progress. The tacit knowledge that often dominates early stages of the innovation process is person- and place-specific, and exhibits strong distance-decay effects (Scott and Storper, 1992). Minimizing the uncertainty of the innovation process demands frequent exchange of information (Teece, 1980, 1986) and this, coupled with the fixed costs of technical choice, encourages spatial, institutional, and technological "lock-in," or inertia (Clark and Wrigley, 1995; Grabher, 1993; Herrigel, 1993; Storper, 1995). The advantages of agglomeration result from a shared knowledge base, enhanced local information exchange and learning (Lundvall and Johnson, 1994; Malmberg and Maskell, 1997; Scott, 1995), multiple sources of innovation (von Hippel, 1988), and the collective sharing of knowledge spillovers (Anselin et al., 1997; Jaffe et al., 1993). As technological and institutional regimes are produced and reproduced in space they are seen as imbued with distinctive geographical and historical characteristics, and this imparts a strong path dependence on future trajectories of regional development (Arthur, 1989; David, 1975, 1985).

These theoretical claims suggest that marked variation exists in the innovative capacity of different regions. Unfortunately, empirical investigation of this claim is difficult as data on the components of technological change are rarely available at the national, let alone the subnational, level. This has prompted the use of a number of proxy measures of regional technological change. Malecki (1979, 1980) explores the geography of R&D spending, considered as an input to the process of technology creation. He reports that R&D activities in the USA are spatially concentrated: putative evidence of the geographic unevenness of technological change. This concentration is explained on the basis of corporate organization, specifically headquarter activities, the existence of large, skilled pools of labor in urban areas, and by industrial agglomeration. Similar findings for the UK are reported by Howells (1984), for Austria by Tödtling (1992), and for industry-specific R&D concentrations by Scott and Angel (1987). Feldman and Florida (1994) provide more recent evidence of the spatial clustering of R&D in research universities and in private industry. In related fashion, Florida and Kenney (1988) and Leinbach and Amrhein (1987), show that the availability of venture capital is limited to relatively few "high tech" regions in the USA.

Measures of technology output in the form of patents are also receiving considerable attention as indicators of technological change. As well as providing detailed information on the nature of new technology, patents indicate the location of the inventor, and whether or not rights to the patent were assigned (transferred) to another individual or organization. They also list citations to existing knowledge in the form of other patents or publications (see Jaffe, 1986). This information has been employed primarily to answer questions regarding the geography of invention by examining the location of inventors, and the existence of knowledge spillovers, although Scherer (1983) and Pavitt (1985) question the reliability of patent data as an index of innovation.

In their examination of the institutionalization of the US patent system, Lamoreaux and Sokoloff (1996) discuss the concentration of patenting in New England and the Mid-Atlantic region through much of the nineteenth century, and its later diffusion as a market for technology developed in the United States. Feldman and Florida (1994) show that US patent activity in 1982 was dominated by the old manufacturing belt, along with California. Jaffe (1989) reports similar inter-state variations in patenting in the early 1980s. O'hUallachain (1998) shows the bias of invention toward large metropolitan areas for much of the 1990s, attributing this to the spatial concentration there of technology-intensive manufacturing industries, well-educated people, and universities and research institutions. Fischer et al. (1994) report significant regional variations in patent activity in Austria.

Motivation for research on knowledge spillovers stems from the new endogenous growth theory (Lucas, 1988; Romer, 1986, 1990; Sunley, this volume). In this literature, spillovers occur when R&D activities undertaken by one firm or industry are used as inputs into the R&D activities of other firms or industries. It is claimed that such externalities bring about continued productivity growth. Griliches (1992) provides an overview, focusing on different methods of measuring spillovers. In aggregate there is considerable evidence of spillovers, though their magnitude varies considerably across studies. Examination of knowledge spillovers embodied in R&D investments weighted by patent data can be traced to the work of Scherer (1982). Analysis of disembodied knowledge spillovers is more frequently associated with Jaffe (1986). Jaffe et al. (1993) examine whether patent citations link geographically proximate inventors. They find that citations are significantly concentrated at a variety of spatial scales, and use this evidence to support their claim of localized knowledge spillovers. Zucker et al. (1998) contest these claims. Anselin et al. (1997) use innovation data for 1982 to examine the geographical boundaries of knowledge spillovers from university research to private sector R&D.

While regional variations in the production of new knowledge are not easily measured, the application of that knowledge in the form of product and process innovations has received attention, limited by the availability of data. In surveys of UK manufacturing establishments, Oakey et al. (1982) distinguish between product and process innovations, arguing that the former provide a reliable indicator of a region's indigenous innovation potential. They show that the incidence of product innovation is considerably higher in core regions than in peripheral regions, and account for this on the basis of R&D costs, the availability of skilled workers, access to information, and manufacturing plant characteristics. These results are echoed by Edwards and Gibbs (1982), Harris (1988), and Tödtling (1992). In the USA, Feldman and Florida (1994) show that innovation is spatially concentrated and attribute this to the usual list of factors for explanation – university and private R&D expenditures, the presence of firms in related industries, and access to producer services.

One problem with the above accounts is that the production of regional competitive advantage through technological change is too narrowly conceived, as dependent upon innovation. We have abandoned the neoclassical model of the representative firm to recognize the heterogeneity of firm characteristics and behaviors, but appear to have too readily adopted a model of the representative region, an innovative territory whose technological dynamism and growth rests upon a rather narrow set of processes and supporting characteristics. Rigby and Essletzbichler (1999) go beyond innovation to examine the relative strength of different sources of aggregate regional technological change: innovation and imitation; changes in plant market share; and plant entry and exit. They show that the geography of aggregate technological change is complex: the absolute and relative sizes of the sources of change vary considerably between regions, and that in many US states innovation and imitation are not the principal determinants of productivity improvements.

Conclusion

At the level of the nation-state, competitiveness is linked to technology and the ability of the elements that define a "system of innovation" to generate and sustain growth (Best, 1990; Nelson, 1993; Porter, 1990). There is increasing recognition, however, that the spatial scale of such systems is local or regional, and that the dynamism of sub-national technology districts is responsible for a considerable proportion of aggregate growth (Amin and Thrift, 1994; Scott, 1996; Storper, 1997). A good deal of academic capital has been invested in searching for the conditions that underpin technological dynamism and the wealth of regions and nations. Not very long ago, the developmental status of a region was linked to the presence or absence of "high-tech" workers, "high-tech" capital, "high-tech" firms, and related research institutions (see the review in Malecki, 1991), and attempts to clone "high tech" spaces such as Silicon Valley and Route 128 in the USA became a widespread foundation of regional policy throughout the world (Castells and Hall, 1994). Walker (1985) and Scott and Storper (1987) outline the weaknesses of these arguments, and Sternberg (1996) suggests that distilling the lessons from successful technology districts produces few generalities.

Current debate on technology and regional growth focuses on social relations. The production and exchange of knowledge that powers technological progress is regarded as a social activity, enhanced by personal interaction, by a common language, and by a common understanding of problems and strategies (Gertler, 1995; Lundvall, 1992; Malmberg and Maskell, 1997). With the "right" infrastructural support, or the "right" mix of tangible elements, the regional problem is now seen as one of generating the appropriate social capital to make those elements operate cohesively. Whether or not "high tech" social capital is the critical ingredient for regional competitive advantage remains to be seen.

Bibliography

Abramowitz, M. 1956. Resource and output trends in the United States since 1870. *Papers and Proceedings of the American Economics Association*, 46, 5–23.

Ahmad, S. 1966. On the theory of induced innovation. *Economic Journal*, 76, 344–57.

Alchian, A. 1950. Uncertainty, evolution and economic theory. *Journal of Political Economy*, 58, 211–22.

Amendola, G., Guerrieri, P., and Padoan, P. C. 1992. International patterns of technological accumulation and trade. *Journal of International and Comparative Economics*, 1, 173–97.

Amin, A. and Thrift, N. (eds). 1994. *Globalization, Institutions and Regional Development in Europe*. Oxford: Oxford University Press.

Anderson, W. P. and Rigby, D. L. 1989. Estimating capital stocks and capital ages in Canada's regions: 1961–1981. *Regional Studies*, 23, 117–26.

Anselin, L., Varga, A., and Acs, Z. 1997. Local geographic spillovers between university research and high technology innovations. *Journal of Urban Economics*, 42, 422–48.

Archibugi, D. and Michie, J. 1995. The globalization of technology: A new taxonomy. *Cambridge Journal of Economics*, 19, 121–40.

Arrow, K. 1962. The economic implications of learning by doing. *Review of Economic Studies*, 29, 22–43.

Arrow, K. 1994. The production and distribution of knowledge. In G. Silverberg and L. Soete (eds). *The Economics of Growth and Technical Change*, Aldershot: Edward Elgar, 9–19.

Arthur, B. 1989. Competing technologies, increasing returns, and lock-in by historical events. *Economic Journal*, 99, 116–31.

Aydalot, P. 1988. Technological trajectories and regional innovation in Europe. In P. Aydalot and D. Keeble (eds). *High Technology Industry and Innovative Environments: The European Experience*. London: Routledge, 22–47.

Baily, M. N., Hulten, C., and Campbell, D. 1992. The distribution of productivity in manufacturing plants. *Brookings Papers on Economic Activity*: Microeconomics, 187–267.

Bain, J. S. 1956. *Barriers to New Competition*. Cambridge, MA: Harvard University Press.

Baldwin, J. R. 1998. *The Dynamics of Industrial Competition*. Cambridge: Cambridge University Press.

Baldwin, W. L. and Scott, J. T. 1987. *Market Structure and Technological Change*. Chur: Harwood.

Barro, R. 1991. Economic growth in a cross section of countries. *Quarterly Journal of Economics*, 106, 407–43.

Barro, R. J. and Sala-I-Martin, X. 1991. Convergence across states and regions. *Brookings Papers on Economic Activity: Microeconomics*, 107–82.

Baumol, W. J. 1986. Productivity growth, convergence, and welfare: What the long-run data show. *American Economic Review*, 76, 1072–85.

Beeson, P. E. 1987. Total factor productivity growth and agglomeration economies in manufacturing, 1959–1973. *Journal of Regional Science*, 27, 183–99.

Beeson, P. E. and Husted, S. 1989. Patterns and determinants of productive efficiency in state manufacturing. *Journal of Regional Science*, 29, 15–28.

Benko, G. and Dunford M. (eds). 1991. *Industrial Change and Regional Economic Development*. London and New York: Belhaven Press.

Berry, B. J. L. 1991. *Long-Wave Rhythms in Economic Development and Political Behavior*. Baltimore: Johns Hopkins University Press.

Best, M. 1990. *The New Competition: Institutions of Industrial Restructuring*. Cambridge, MA: Harvard University Press.

Binswanger, H. P. and Ruttan, V. W. 1978. *Induced Innovation: Technology, Institutions and Development*. Baltimore: Johns Hopkins University Press.

Borts, G. and Stein, J. 1964. *Economic Growth in a Free Market*. New York: Columbia University Press.

Brown, L. A. 1981. *Innovation Diffusion: A New Perspective*. New York: Methuen.

Brusco, S. 1982. The Emilian model: Productive decentralization and social integration. *Cambridge Journal of Economics*, 6, 167–84.

Bush, V. 1945. *Science: The Endless Frontier*. Washington, D.C.: Office of Scientific Research and Development.

Camagni, R. (ed). 1991. *Innovation Networks: Spatial Perspectives*. London: Belhaven Press.

Camerer, C. and Vepsalainen, A. 1988. The economic efficiency of corporate culture. *Strategic Management Journal*, 9, 115–26.

Cantwell, J. 1991. The technological competence theory of international production and its implications. In D. G. McFetridge (ed). *Foreign Investment, Technology, and Economic Growth*. Calgary: University of Calgary Press.

Casetti, E. 1984. Manufacturing productivity and Snowbelt–Sunbelt shifts. *Economic Geography*, 60, 311–24.

Castells, M. and Hall, P. 1994. *Technopoles of the World: The Making of 21st Century Industrial Complexes*. London: Routledge.

Caves, R. E. and Porter, M. E. 1977. From entry barriers to mobility barriers. *Quarterly Journal of Economics*, 91, 241–61.

Clark, K. B. 1985. The interaction of design hierarchies and market concepts in technological evolution. *Research Policy*, 14, 235–51.

Clark, G. L. and Wrigley, N. 1995. Sunk costs: A framework for economic geography. *Transactions of the Institute of British Geographers*, 20, 204–23.

Coase, R. H. 1937. The nature of the firm. *Economica*, 4, 386–405.

David, P. A. 1966. The mechanization of reaping in the Ante-Bellum Midwest. In H. Rodovsky (ed). *Industrialization in Two Systems*. New York: Wiley, 3–39.

David, P. A. 1975. *Technical Choice, Innovation and Economic Growth*. Cambridge: Cambridge University Press.

David, P. A. 1985. Clio and the economics of QWERTY. *American Economic Review, Proceedings*, 75, 332–7.

David, P. A. 1992. Heroes, herds and hysteresis in technological history: Thomas Edison and "The battle of systems" reconsidered. *Industrial and Corporate Change*, 1, 129–80.

David, P. A. and de Klundert, T. van 1965. Biased efficiency growth and capital-labor substitution in the U.S., 1899–1960. *American Economic Review*, 55, 357–94.

Davies, S. 1979. *The Diffusion of Process Innovations*. Cambridge: Cambridge University Press.

DeBresson, C. and Amesse, F. 1991. Networks of innovators: A review and introduction to the issue. *Research Policy*, 20, 363–79.

Denison, E. F. 1962. *The Sources of Economic Growth in the United States and the Alternatives Before Us*. New York: Committee for Economic Development.

Denison, E. F. 1967. *Why Growth Rates Differ*. Washington DC: Brookings Institution.

Dicken, P. 1998. *Global Shift: Industrial Change in a Turbulent World*, third edition. New York: Guilford Press.

Dosi, G. 1982. Technological paradigms and technological trajectories: A suggested interpretation of the determinants and directions of technical change. *Research Policy*, 11, 147–62.

Edwards, A. and Gibbs, D. C. 1982. Regional development, process innovations and the characteristics of the firm. In D. Maillat (ed). *Technology: A Key Factor for Regional Development*. Saint-Saphorin: Georgi, 121–41.

Ellison, G. and Glaeser, E. L. 1997. Geographic concentration in US manufacturing industries: A dartboard approach. *Journal of Political Economy*, 105, 889–927.

Essletzbichler, J. 1999. Regional differences in technology. PhD Dissertation, Department of Geography, University of California, Los Angeles.

Essletzbichler, J., Haydamack, B. W., and Rigby, D. L. 1998. Regional dynamics of technical change in the U.S. structural fabricated metals industry. *Geoforum*, 29, 23–35.

Fagerberg, J. 1987. A technology gap approach to why growth rates differ. *Research Policy*, 16, 87–99.

Fagerberg, J. 1994. Technology and international differences in growth rates. *Journal of Economic Literature*, 32, 1147–75.

Farjoun, E. and Machover, M. 1983. *Laws of Chaos: A Probabilistic Approach to Political Economy*. London: Verso.

Feldman, M. P. and Florida, R. L. 1994. The geographic sources of innovation: technological infrastructure and product innovation in the United States. *Annals of the Association of American Geographers*, 84, 210–29.

Fellner, W. 1961. Two propositions in the theory of induced innovations. *Economic Journal*, 71, 305–8.

Fischer, M. M., Fröhlich, J., and Gassler, H. 1994. An exploration into the determinants of patent activities: Some empirical evidence for Austria. *Regional Studies*, 28, 1–23.

Fishlow, A. 1966. Productivity and technological change in the railroad sector, 1840–1910. In *Output, Employment and Productivity in the US after 1800. Studies in Income and Wealth*, 30. New York: National Bureau of Economic Research, 583–646.

Florida, R. L. and Kenney, M. 1988. Venture capital, high technology, and regional development. *Regional Studies*, 22, 33–48.

Freeman, C. 1982. *The Economics of Industrial Innovation*. Cambridge, MA: MIT Press.

Freeman, C. 1991. Networks of innovators: A synthesis of research issues. *Research Policy*, 20, 499–514.

Freeman, C. 1995. The "National system of innovation" in historical perspective. *Cambridge Journal of Economics*, 19, 5–24.

Freeman, C., Clark, J., and Soete, L. 1982. *Unemployment and Technical Innovation. A Study of Long Waves and Economic Development*. London and New York: Pinter Publishers.

Gertler, M. 1993. Implementing advanced manufacturing technologies in mature industrial regions: toward a social model of technology production. *Regional Studies*, 27, 665–80.

Gertler, M. 1995. Being there: proximity, organization and culture in the development and adoption of advanced manufacturing technologies. *Economic Geography*, 75, 1–26.

Gilfillan, S. 1935. *The Sociology of Invention*. Chicago: Follett.

Gleed, R. H. and Rees, R. D. 1979. The derivation of regional capital stock estimates for U.K. manufacturing industries, 1951–1973. *Journal of the Royal Statistical Society*, 142, 330–46.

Grabher, G. (ed). 1993. *The Embedded Firm*. London: Routledge.

Granovetter, M. 1985. Economic action and social structure: the problem of embeddedness. *American Journal of Sociology*, 91, 481–510.

Griliches, Z. 1957. Hybrid corn: An exploration in the economics of technological change. *Econometrica*, 25, 501–23.

Griliches, Z. 1992. The search for R&D spillovers. *Scandinavian Journal of Economics*, 94, Supplement, 29–47.

Grossman, G. M. and Helpman, E. 1991. *Innovation and Growth in the Global Economy*. Cambridge, MA: MIT Press.

Habakkuk, J. C. 1962. *American and British Technology in the Nineteenth Century*. Cambridge: Cambridge University Press.

Hägerstrand, T. 1967 [1953]. *Innovation Diffusion as a Spatial Process*. Chicago: University of Chicago. English edition 1967.

Harris, R. I. D. 1988. Technological change and regional development in the UK: evidence from the SPRU database on innovations. *Regional Studies*, 22, 361–74.

Harrison, B. 1992. Industrial districts: old wine in new bottles? *Regional Studies*, 26, 469–83.

Henderson, R. and Clark, K. 1990. Architectural innovation: the reconfiguration of existing product technologies and the failure of established firms. *Administrative Science Quarterly*, 35, 9–30.

Henderson, R., Jaffe, A. B., and Trajtenberg, M. 1998. Universities as sources of commercial technology: A detailed analysis of university patenting, 1965–1988. *Review of Economics and Statistics*, LXXXX, 119–27.

Herrigel, G. B. 1993. Power and the redefinition of industrial districts: the case of Baden-Württemberg. In G. Grabher (ed). *The Embedded Firm*. London: Routledge, 227–51.

Hicks, J. R. 1932. *The Theory of Wages*. London: Macmillan.

Hollander, S. 1965. *The Sources of Increased Efficiency: A Study of Du Pont Rayon Plants*. Cambridge, MA: MIT Press.

Howells, J. 1984. The location of research and development: some observations and evidence from Britain. *Regional Studies*, 18, 13–29.

Hulten, C. R. and Schwab, R. M. 1984. Regional productivity growth in US manufacturing: 1951–1978. *American Economic Review*, 74, 152–62.

Jaffe, A. 1986. Technological opportunity and spillovers of R&D: evidence from firms' patents, profits and market value. *American Economic Review*, 76, 984–1001.

Jaffe, A. 1989. Real effects of academic research. *American Economic Review*, 79, 957–70.

Jaffe, A. B., Trajtenberg, M., and Henderson, R. 1993. Geographic localization of knowledge spillovers as evidenced by patent citations. *Quarterly Journal of Economics*, 108, 577–98.

Johnson, B. and Lundvall, B. A. 1992. Closing the institutional gap? *Revue D'Economie Industrielle*, 59, 111–23.

Jones, C. I. 1998. *Introduction to Economic Growth*. New York: Norton.

Katz, M. L. and Shapiro, C. 1985. Network externalities, competition, and compatibility. *American Economic Review*, 75, 424–40.

Kennedy, C. 1964. Induced bias in innovation and the theory of distribution. *Economic Journal*, 74, 541–47.

Kleinknecht, A. 1987. *Innovation Patterns in Crisis and Prosperity: Schumpeter's Long Cycle Reconsidered*. New York: St. Martin's Press.

Kline, S. J. and Rosenberg, N. 1986. An overview of innovation. In R. Landau and N. Rosenberg (eds). *The Positive Sum Strategy. Harnessing Technology for Economic Growth*. Washington, DC: National Academy Press, 275–305.

Kuznets, S. 1930. *Secular Movements in Production and Prices. Their Nature and Their Bearing upon Cyclical Fluctuations*. Boston: Houghton Mifflin.

Lamoreaux, N. R. and Sokoloff, K. L. 1996. Long-term change in the organization of inventive activity. *Proceedings of the National Academy of Sciences*, 93, 12686–92.

Lande, P. S. 1978. The interregional comparison of production functions. *Regional Science and Urban Economics*, 8, 339–53.

Landes, D. S. 1969. *The Unbound Prometheus*. Cambridge: Cambridge University Press.

Landes, D. S. 1998. *The Wealth and Poverty of Nations*. London and New York: Norton.

Leinbach, T. R. and Amrhein, C. 1987. A geography of the venture capital industry in the USA. *Professional Geographer*, 39, 146–58.

Lucas, R. 1988. On the mechanics of economic development. *Journal of Monetary Economics*, 22, 3–42.

Luger, M. I. and Evans, W. N. 1988. Geographic differences in production technology. *Regional Science and Urban Economics*, 18, 399–424.

Lundvall, B. A. 1988. Innovation as an interactive process: from user-producer interaction to the national system of innovation. In G. Dosi et al. (eds). *Technical Change and Economic Theory*, London: Pinter, 349–69.

Lundvall, B. A. (ed). 1992. *National Systems of Innovation*. London: Pinter.

Lundvall, B. A. and Johnson, B. 1994. The learning economy. *Journal of Industry Studies*, 1, 23–42.

Maddison, A. 1982. *Phases of Capitalist Development*. New York: Oxford University Press.

Maddison, A. 1987. Growth and slowdown in advanced capitalist economies. *Journal of Economic Literature*, 25, 649–98.

Maillat, D. 1991. The innovation process and the role of the milieu. In E. M. Bergman, G. Maier, and F. Tödtling (eds). *Regions Reconsidered: Economic Networks, Innovation and Local Development in Industrialised Countries*. London: Cassell, 103–17.

Maillat, D. 1995. Territorial dynamic, innovative milieus and regional policy. *Entrepreneurship and Regional Development*, 7, 157–65.

Malecki, E. 1979. Locational trends in R&D by large US corporations, 1965–1977. *Economic Geography*, 55, 309–23.

Malecki, E. 1980. Dimensions of R&D location in the United States. *Research Policy*, 9, 2–22.

Malecki, E. 1991. *Technology and Economic Development*. Harlow: Longman.

Malmberg, A. 1996. Industrial geography: agglomeration and local milieu. *Progress in Human Geography*, 20, 392–403.

Malmberg, A. and Maskell, P. 1997. Towards an explanation of regional specialization and industry agglomeration. *European Planning Studies*, 5, 25–41.

Mankiw, N. G., Romer, D., and Weil, D. N. 1992. A contribution to the empirics of economic growth. *Quarterly Journal of Economics*, 107, 407–37.

Mansfield, E. 1961. Technical change and the rate of imitation. *Econometrica*, 29, 741–66.

Marshall, A. 1920. *The Principles of Economics*. London: Macmillan.

Mensch, G. 1979. *Stalemate in Technology*. Cambridge, MA: Ballinger.

Moomaw, R. L. 1983. Spatial productivity and city size: a critical survey of cross-sectional analysis. *International Regional Science Review*, 8, 1–22.

Mowery, D. C. and Rosenberg, N. 1979. The influence of market demand upon innovation: a critical review of some recent empirical studies. *Research Policy*, 8, 103–53.

Musson, A. and Robinson, E. 1969. *Science and Technology in the Industrial Revolution*. Manchester: Manchester University Press.

Myrdal, G. 1957. *Economic Theory and Underdeveloped Regions*. New York: Harper.

Nelson, R. R. 1994. The co-evolution of technology, industry structure, and supporting institutions. *Industrial and Corporate Change*, 1, 47–63.

Nelson, R. R. (ed). 1993. *National Innovation Systems*. Oxford: Oxford University Press.

Nelson, R. R. and Winter, S. G. 1982. *An Evolutionary Theory of Economic Change*. Cambridge, MA: Harvard University Press.

Norton, R. D. and Rees, J. 1979. The product cycle and the spatial decentralization of American manufacturing. *Regional Studies*, 13, 141–51.

Oakey, R. P., Thwaites, A. T., and Nash, P. A. 1982. Technological change and regional development: some evidence on regional variations in product and process innovation. *Environment and Planning A*, 14, 1073–86.

OECD 1996. *Technology, Productivity and Job Creation*. OECD: Paris.

O'hUallachain, B. 1998. The metropolitan distribution of patents in the US. Unpublished manuscript, Department of Geography, Arizona State University.

Olmstead, A. 1975. The mechanization of reaping and mowing in American agriculture, 1833–1870. *Journal of Economic History*, 35, 327–52.

Pavitt, K. 1985. Patent statistics as indicators of innovative activities: possibilities and problems. *Scientometrics*, 7, 77–99.

Persky, J. and Klein, W. 1975. Regional capital growth and some of those other things we never talk about. *Papers of the Regional Science Association*, 35, 181–90.

Porter, M. 1990. *The Competitive Advantage of Nations*. New York: Free Press.

Rigby, D. L. 1992. The impact of output and productivity changes on employment. *Growth and Change*, 23, 403–25.

Rigby, D. L. 1995. Investment, capital stocks and the age of capital in US regions. *Growth and Change*, 26, 524–52.

Rigby, D. L. and Haydamack, B. 1998. Regional trajectories of technological change in Canadian manufacturing. *Canadian Geographer*, 42, 2–13.

Rigby, D. L. and Essletzbichler, J. 1997. Evolution, process variety, and regional trajectories of technological change in US manufacturing. *Economic Geography*, 72, 269–84.

Rigby, D. L. and Essletzbichler, J. 1999. Impacts of technical change, selection, and plant entry/exit on regional productivity growth. Unpublished manuscript, Department of Geography, University of California, Los Angeles.

Robinson, J. 1953/54. The production function and the theory of capital, *Review of Economic Studies*, 21, 81–106.

Romer, P. M. 1986. Increasing returns and long-run growth. *Journal of Political Economy*, 94, 1002–37.

Romer, P. M. 1990. Endogenous technological change. *Journal of Political Economy*, 98, S71–S102.

Rosenberg, N. 1982. *Inside the Black Box: Technology and Economics*. Cambridge: Cambridge University Press.

Sahal, D. 1981. *Patterns of Technological Innovation*. Reading, MA: Addison Wesley.

Salter, W. E. G. 1960. *Productivity and Technological Change*. Cambridge: Cambridge University Press.

Saxenian, A. 1994. *Regional Advantage*. Cambridge, MA: Harvard University Press.

Scherer, F. M. 1982. Interindustry technology flows and productivity growth. *Review of Economics and Statistics*, LXIV, 627–34.

Scherer, F. M. 1983. The propensity to patent. *International Journal of Industrial Organization*, 1, 107–28.

Schmookler, J. 1966. *Invention and Economic Growth*. Cambridge, MA: Harvard University Press.

Schumpeter, J. A. 1939. *Business Cycles*. London: McGraw Hill.

Scott, A. J. 1988. *Metropolis: From the Division of Labor to Urban Form*. Berkeley: University of California Press.

Scott, A. J. 1995. The geographic foundations of industrial performance. *Competition and Change*, 1, 51–66.

Scott, A. J. 1996. Regional motors of the global economy. *Futures*, 28, 391–411.

Scott, A. J. and Angel, D. P. 1987. The US semiconductor industry: A locational analysis. *Environment and Planning A*, 19, 875–912.

Scott, A. J. and Storper, M. 1987. High technology industry and regional development: a theoretical critique and reconstruction. *International Social Science Journal*, 112, 215–32.

Scott, A. J. and Storper, M. 1992. Regional development reconsidered. In H. Ernste and V. Meier (eds). *Regional Development and Contemporary Industrial Response*, 3–24.

Soete, L. 1987. The impact of technological innovation on international trade patterns: the evidence reconsidered. *Research Policy*, 16, 101–30.

Solow, R. 1956. A contribution to the theory of economic growth. *Quarterly Journal of Economics*, 66, 65–94.

Solow, R. 1957. Technical change and the aggregate production function. *Review of Economics and Statistics*, 39, 312–20.

Sternberg, R. 1996. Reasons for the genesis of high-tech regions – Theoretical explanation and empirical evidence. *Geoforum*, 27, 205–23.

Stiglitz, J. E. 1987. Learning to learn: Localized learning and technological progress. In P. Dasgupta and P. Stoneman (eds) *Economic Policy and Technological Performance*. Cambridge: Cambridge University Press, 123–52.

Stöhr, W. 1986. Regional innovation complexes. *Papers of the Regional Science Association*, 59, 29–44.

Storper, M. 1985. Oligopoly and the product cycle: essentialism in economic geography. *Economic Geography*, 61, 260–82.

Storper, M. 1992. The limits to globalization: technology districts and international trade. *Economic Geography*, 68, 60–93.

Storper, M. 1993. Regional "worlds" of production: learning and innovation in the technology districts of France, Italy and the USA. *Regional Studies* 27, 433–55.

Storper, M. 1995. The resurgence of regional economies, ten years later: the region as a nexus of untraded interdependencies. *European Urban and Regional Studies*, 2, 191–221.

Storper, M. 1997. *The Regional World: Territorial Development in a Global Economy*. New York: Guilford.

Storper, M. and Walker, R. 1989. *The Capitalist Imperative – Territory, Technology, and Industrial Growth*. New York: Basil Blackwell.

Taylor, M. J. 1986. The product-cycle model: a critique. *Environment and Planning A*, 18, 751–61.

Teece, D. 1980. Economic scope and the scope of the enterprise. *Journal of Economic Behavior and Organization*, 1, 223–47.

Teece, D. 1986. Profiting from technological innovation. *Research Policy*, 15, 286–305.

Teece, D. 1992. Competition, cooperation, and innovation. *Journal of Economic Behavior and Organization*, 13, 1–25.

Thirtle, C. G. and Ruttan, V. W. 1987. *The Role of Demand and Supply in the Generation and Diffusion of Technical Change*. Chur: Harwood.

Tödtling, F. 1992. Technological change at the regional level: the role of location, firm structure, and strategy. *Environment and Planning A*, 24, 1565–84.

Trajtenberg, M., Henderson, R., and Jaffe, A. B. 1997. University versus corporate patents: a window on the basicness of invention. *Economics of Innovation and New Technology*, 5, 19–50.

Usher, A. 1954. *A History of Mechanical Inventions*, second edition. Cambridge, MA: Harvard University Press.

Utterback, J. M. and Abernathy, W. J. 1975. A dynamic model of product and process innovation. *Omega*, 3, 639–56.

Varaiya, P. and Wiseman, M. 1981. Investment and employment in manufacturing in U.S. metropolitan areas 1960–1976. *Regional Science and Urban Economics*, 11, 431–69.

Van Duijn, J. J. 1983. *The Long-Wave in Economic Life*. Winchester: Allen & Unwin.

Vernon, R. 1966. International investment and international trade in the product cycle. *Quarterly Journal of Economics*, 80, 190–207.

Von Hippel, E. 1988. *The Sources of Innovation*. Oxford: Oxford University Press.

Walker, R. 1985. Technological determination and determinism: industrial growth and location. In M. Castells (ed). *High Technology, Space and Society*. Beverly Hills: Sage, 226–64.

Webber, M. J. and Rigby, D. L. 1996. *The Golden Age Illusion: Rethinking Postwar Capitalism*. New York: Guilford Press.

Webber, M. J., Sheppard, E. S., and Rigby, D. L. 1992. Forms of technical change. *Environment and Planning A*, 24, 1679–1709.

Whitehead, A. N. 1925. *Science and the Modern World*. New York: Macmillan.

Williamson. O. E. 1975. *Markets and Hierarchies*. New York: Free Press.

Young, A. 1995. The tyranny of numbers: confronting the statistical realities of the East Asian growth experience. *Quarterly Journal of Economics*, 10, 641–80.

Zucker, L. G., Darby, M. R., and Armstrong, J. 1998. Geographically localized knowledge: spillovers or markets? *Economic Inquiry*, XXXVI, 65–86.

Part III   Resource Worlds

Chapter 14

Resources

Dean M. Hanink

Natural resources are "factors of production." That is, they are employed along with labor and other forms of capital in producing goods and services. The economic geography of natural resources, the places of their production, and the places of their consumption, is fundamentally different from the economic geography of most other production factors. That difference results because unlike labor, unlike machines, and unlike buildings, natural resources actually exist as a form of interaction between the physical environment (nature) and society.

This chapter takes up the economic geography of natural resources using a recurring theme: resources have technological and cultural contexts for their evaluation. That is, there are certain technological and/or social conditions that have to be met before a part of nature is considered valuable in production. The chapter begins with a basic overview of natural resource geographies, focusing on their recognition, classification, and description. The second part of the chapter turns to the analysis of the relative valuation of natural resources using the principle of economic rent, and considers the question of the supply and demand for resources in market economies. The role of technology in effecting economic rent and, therefore, natural resource supply is emphasized. Rent-maximizing behavior on the part of producers is a primary determinant of the locational pattern of resource extraction, which in turn has important implications for natural resource-based economies. Those implications are considered in the third part of the chapter, which ends with a brief discussion of the potential for such economies to achieve sustainability.

Defining and Classifying Natural Resources

In an often-cited statement (e.g. Cairns, 1994, p. 782), Erich Zimmerman wrote "resources are not; they become." The implication is that natural resources are not pre-given, natural factors around which society fashions its production processes, but rather their usefulness is conditioned by cultural, historical, technological (and other knowledge), and geographical circumstances.

For example, almost all of the mineral resources have a technological context to their value. Rees (1989) cites bauxite, for example, which wasn't recognized as a true resource until 1886 when technology was introduced that allowed it to yield alumina (an intermediate resource that is further refined to aluminum). Uranium is another example. It didn't gain significant commercial value as a resource until nuclear fission became controllable. By the late 1950s, uranium ore was highly sought after and was expected to increase in commercial use, and value, as nuclear technological abilities continued to grow. A changing cultural context, however, one in which the environmental and public health costs of using uranium as a resource have become increasingly recognized, has reduced the value of uranium considerably since the late 1970s. Silica, on the other hand, has been recognized as a significant resource in glass production for centuries. Its recognition and value as a resource has increased significantly in about the last 20 years, however, as digital technology has developed and technological change has replaced vacuum tubes with transistors, and transistors with silicon chips.

The same kinds of changing societal contexts affect the recognition of non-mineral resources, too. Wetlands, for example, used to be considered nothing more than sources of disease and land gone to waste. Today, we think of wetlands as valuable protective buffers to flood waters and natural water purification systems. Today, many people are concerned with loss of forests. Beyond their use in goods production, they are a valuable resource that diminishes climate change and provides critical habitat in support of biodiversity. On the other hand, the historian David Landes (1998) writes of the former forests of Northern Europe as a surmountable *barrier* to that region's economic ascendancy.

Once natural resources are recognized as factors of production, they are often described by their supply characteristics. For example, resources are often classified with respect to being exhaustible or renewable. Exhaustible resources are those environmental factors that have a fixed terrestrial or oceanographic supply and therefore will be completely consumed if current rates of use continue. Renewable resources are not in such fixed supply. They may be biotic – forests, for example, or fisheries. Other renewable resources effectively have astrophysical sources, for example winds, tides, and solar energy. The supply of renewable resources is *potentially* limitless over time, precisely because of their renewable quality. While this is true enough for astrophysical resources, it is harder to support in the cases of biotic resources which may, in fact, become exhausted.

The amount, or stock, of a resource is also classified into a series of components (Harris, 1993). The first part is cumulative production, or the amount of the resource that has already been extracted and is, therefore, no longer available. The second part of the stock consists of *reserves*, which include the amount of the resource that is available using current technology and at current prices. The third part of the stock consists of *potential supply*, an effective estimate of future reserves. It is the amount of the resource that will be available if reasonable expectations of technological advance are achieved and/or prices increase. The *recoverable resource* is the combination of reserves and potential supply. It is evident that this classification of resource stock components rests on a technological context and also a context of market preferences as indicated by resource prices.

Because they have both technological and market contexts for their definition, resources don't necessarily decrease over time. Often, as technology and/or prices change, that part of a stock that is potential supply is added to that part that is in reserve at a faster rate than is taken in production. If that occurs, reserves increase rather than shrink. For example, worldwide proven reserves of crude oil increased from about 650 billion (1,000,000,000) barrels in 1975 to about 765 billion barrels in 1985, and then increased to about one trillion (1,000,000,000,000) barrels by 1995 (*Annual Statistical Bulletin 1995*, 1996). That increase in reserves took place concurrently with ever-higher volumes of consumption. It's as if petroleum reserves are increasing the more that they are consumed.

In addition to technological and market contexts, there is also the geographical one. For example, proven oil reserves increased at the global scale from 1975 through 1995, but the record was very uneven at the regional scale (table 14.1). Asia's reserves increased consistently over the period (and will continue to do so as the oil fields of the Caspian Sea region are developed), and so did the reserves of Latin America and the Middle East. In addition, Africa's and Oceania's reserves were greater in 1995 than in 1975. Reserves in Europe and in North America, however, were lower in 1995 than in 1975. Reserves in those regions have declined as a function of actual use. Both those regions exhibit a production history that is common and often predictable (World Resources Institute, 1996, pp. 276–7). Most regional production increases until about one-half of its recoverable resource has been extracted: it peaks at that point, and then begins a decline (see "Natural Resource Economies" below for a description of the related regional economic impacts of the production pattern).

Induced initially by industrial demand and then additional consumer demand, recoverable petroleum resources were developed in Europe and North America early on (by international standards) and drawn down early on as well. Petroleum resources were developed more recently in other regions of the world, so their reserves continue to grow as much by virtue of their economic history as by a fortunate geology. The same geographical pattern of recognition – core regions early, peripheral regions late – is common to a large variety of natural resources (Porter and Sheppard, 1998, pp. 214–18). Even though regional resources decline in supply, it's not necessary that regional consumption follows the same pattern. For example, Asia, Western Europe, and North America are regions with more oil

Table 14.1 World proven crude oil reserves by region, 1975, 1985, and 1995 (thousand million barrels)

Region	1975	1985	1995
Africa	59.1	56.2	75.5
Asia	36.6	36.8	42.0
Eastern Europe	61.9	64.4	59.1
Latin America	36.1	118.6	131.6
Middle East	387.1	431.9	665.1
North America	39.7	33.9	31.7
Oceania	2.0	1.7	2.2
Western Europe	25.6	22.2	21.1

Source: Annual Statistical Bulletin 1995 OPEC (1996)

Table 14.2 World crude oil production, refining, and consumption by region, 1995 (thousand barrels per day)

Region	Production	Refining	Consumption
Africa	6,256.4	2,175.1	1,804.5
Asia	6,132.1	13,284.2	15,220.0
Eastern Europe	7,193.0	5,984.5	5,644.8
Latin America	7,725.0	5,964.1	5,762.8
Middle East	18,856.3	4,826.1	3,225.4
North America	7,936.1	17,560.4	17,910.9
Oceania	640.0	855.4	643.3
Western Europe	5,743.9	14,133.7	12,617.8

Source: Annual Statistical Bulletin 1995 OPEC (1996)

refining than oil production, and more oil consumption than oil refining (table 14.2). Africa, Eastern Europe, Latin America, and most notably the Middle East follow the opposite pattern.

It appears that the location of resource deposits is almost irrelevant to where geographically those resources will be used. That irrelevance is not new, but is the result of the historical importance of the cultural, technological, and market contexts for resource recognition. In the past, industrial cities such as Pittsburgh didn't arise only because of proximate raw material deposits of coal, iron ore, and limestone (for steel production, in Pittsburgh's case). After all, there are many raw material deposits around the world that were known in the nineteenth century but are yet to be "developed" into centers of manufacturing. As noted by Paul Bradley, "some resources 'become' better than others" (Cairns, 1994, p. 782). At the time, the right technology, the right culture, and the right markets, were in the right proximity to recognize the natural resources that certainly facilitated the growth of Pittsburgh. It wasn't the natural resources alone.

Economic Rent and the Exploitation of Natural Resources

One of the earliest and most influential economists to examine resource exploitation was David Ricardo, who wrote in England in the early nineteenth century. Ricardo was most interested in the relationship between crop prices and land prices, but his analysis of their interaction can easily be extended to most natural resources. At the time, a generally held view was that the price of food was high because agricultural land was so expensive. Tenant farmers needed to charge high prices for their products, particularly corn, to cover the high land rents charged by landowners. Ricardo demonstrated that, in fact, the opposite relationship held; the price of land was high because food prices were high. As food prices rose, farmers expanded their production through a process of more intensive cultivation of land already in agriculture and by expanding crop production to land previously left unused for cropping. Ricardo argued that as food prices rose, more and more land could be expected to undergo development for agriculture. In a sense, as prices rose, agricultural land "became" more and more of a natural resource.

Some land "became" a resource better than other land, however, because not all land is suitable for agriculture, and some pieces of agricultural land are better than others. Ricardo observed that qualitative differences in land as a result of differential soil fertility or irrigation requirements had an important bearing on the profits generated from farming. He analyzed variations in land profitability in terms of *economic rent*, a payment over and above what is necessary to stay in business. Rent is defined by the identity:

$$R = Q \cdot (p - c) \tag{14.1}$$

Where R stands for economic rent, Q represents quantity of production (kilograms for example), p is price per unit (kilo) paid in the marketplace, and c is the cost per unit (kilo) of production. As long as a piece of land's economic rent is positive, a farmer can pursue agriculture on it. In contrast, if the economic rent is negative it would imply that costs (Qc) are greater than revenue (Qp) so that farming would be uneconomical.

Ricardo was concerned with what is called *differential rent*, or the difference in rent that is earned by farmers with better agricultural land compared to the rent that is earned by farmers with land of lower quality (Bina, 1989). As food prices rise, farmers of better land enjoy rent increases in greater amounts than would accrue to farmers of poorer land. Because all rents rise with an increase in market prices (or as p increases relative to c in equation 14.1) the market value of land rises as well. The best land has the highest rent and price, and as the quality of the land decreases so do rents and prices.

Differential rents increase as food prices rise. This is because as prices rise additional lower quality land is brought into production as a result of the higher demand for the agricultural good. The consequence is that the difference between the revenue (Qp) generated by the best quality land and the worst quality land increases. That difference, though, is the very basis of differential rent. Hence, as food prices increase so do differential rents.

The shift of the analysis from agricultural land to natural resources in general is easy to make. As described above, reserves of oil, and most other mineral resources for that matter, have been increasing over time. One explanation for such increases is Ricardian; higher prices paid at the market have allowed lower and lower quality portions of stocks to become recoverable resources. Price increases of the type Ricardo observed are often illustrated as a demand shift, which can lead to an explicit expansion of resource production and an implicit expansion of reserves.

Such a shift is shown in figure 14.1, which is a simplified representation of an idealized market for a hypothetical natural resource. Supply of the resource to the market is shown to increase as price rises, indicating not only that more of the resource is produced, but also that it is being produced from more locations. The supply line represents the cost of production (c in the economic rent identity). In a geographical context, its lowest point represents production at the lowest cost location: the place where resource "quality" is the highest. Rising from the lowest point, the supply line indicates that more of the resource is being produced, but at increasing cost as lower and lower resource quality locations are exploited. The limit of production is defined by the price offered for the resource (p in the economic rent identity). Where price and cost are the same, economic rent is equal to zero, so that point defines not only the aggregate quantity of resource production but also the *marginal location* – that place with the lowest quality resource that will be exploited. Where the price is higher than the cost of production, as defined by the supply line, rent is positive. Rent is at a maximum where the gap between the price line above and the supply line below is at its greatest, or at the location associated with the resource's highest quality. Rent decreases as resource quality declines, and the gap between the price line and the supply line narrows and disappears at the marginal location.

The supply line slopes upward in figure 14.1 because more of the resource will be produced at higher prices. Conversely, the demand lines slope downward because higher prices make use of the resource less attractive. The demand lines represent the willingness-to-pay for using the resource. Some people are willing to pay high prices for the resource, but more people would prefer to pay lower prices. Market equilibrium is defined where the demand line crosses the supply line. That intersection defines the price at which the market is "cleared," with the amount of the resource in demand and the amount supplied equalized.

The type of demand shift illustrated in figure 14.1, in which demand rises from level 1 to level 2, could result from population growth (as concerned Ricardo) or even a change in consumer preferences. It indicates a general increase in the willingness-to-pay for the resource that has an impact on both the quantity consumed, at least in the short run, and also the locations of resource exploitation. As the higher market price for the resource is realized, the differential rent (again, price minus cost) increases at the higher-quality resource locations and economic rent is generated in places where it could not have been extracted under the earlier lower price. Ricardo described poorer and poorer quality agricultural land being cultivated as increasing population resulted in increasing food prices. In figure 14.1 the story is similar, with poorer quality resource locations being exploited as the willingness to pay for the resource increases.

Not only can we think of resources and their locations of exploitation "becoming" because of growth in demand, but we can also think of them becoming from the "supply" side, with technological change playing an important role. For example,

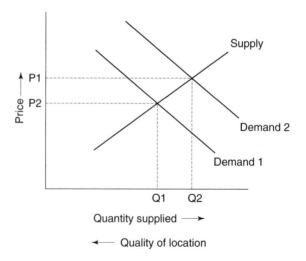

Figure 14.1 An upward demand shift

characteristics such as soil quality or soil moisture can be augmented by capital investment (in fertilizer and in irrigation systems, respectively). Bina (1989, p. 91) quotes Marx as writing, "There is no doubt that as civilization progresses poorer and poorer kinds of land are brought under cultivation. But there is also no doubt that, as a result of the progress of science and industry, these poorest types of land are relatively good in comparison with the former good types." Like Ricardo, Marx had observed that differential rent not only increased because of increasing prices, but that capital investment and technological advance could also cause rents to increase by lowering the costs of production – the supply side of the rent identity in equation (14.1).

It appears that technological change has a profound impact on resource recognition and quality – on resources "becoming" (Dasgupta, 1993). Technological innovations allow resources to be used for particular purposes (silicon, for example). Technological change permits the development of new materials. It makes exploration cheaper and allows extraction to take place at a lower cost. It makes resource use more efficient, permits recycling, and allows lower quality resources to be substituted for higher quality ones.

The effective enhancement of resource quality described by Marx may be illustrated in the impact of a supply shift as diagrammed in figure 14.2. Recalling that the supply line traces production costs, a supply shift to the right, as from Supply 1 to Supply 2 in that figure, represents a general decrease in costs to the resource producing industry. Such a shift would occur, for example, because of a technological breakthrough in resource processing or recovery. Because of the general decrease in costs, locations with resources that were too poor in quality to be exploited under the higher costs represented by Supply 1 can be exploited, with more of the resource being supplied to the market and at a *lower* price.

Differential rent is affected not only by conditions of demand and supply as observed by Ricardo, but also by *location* itself. That contribution to the concept of economic rent was made by J. H. von Thünen, who recognized the importance of transportation cost in the use of agricultural land, and indirectly in the exploitation

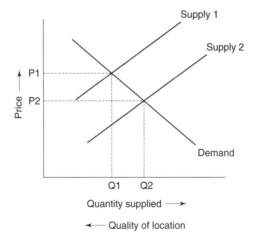

Figure 14.2 A supply shift to the right

of natural resource deposits. The effects of distance on economic rent are illustrated in a *location rent* identity:

$$LR = Q(p - c) - Qtd \qquad (14.2)$$

Where LR stands for location rent, Q, p, and c are as defined above in the economic rent identity, t is a transportation charge per unit of output (cents per kilo, for example), and d is distance from the place of resource production or extraction to the place of resource consumption. Location rent is simply economic rent minus a transport charge. If there is no transport charge, if the resource is consumed at its place of production, then economic rent and location rent are the same thing. In reality, however, resources are often consumed well away from their place of production, so transportation costs may play an important role in whether or not a particular mineral deposit or stand of timber becomes a resource.

Indeed, if transportation charges are ignored, then the sequence of resource exploitation predicted by Ricardo and observed by Marx from higher quality to lower quality land and deposits, does not appear to be consistent. Norgaard (1990), for example, has listed a number of seeming historical contradictions: commercial agricultural expansion by European settlers in what is now the United States and the geographical spread of petroleum production in that country are both listed. In both cases, on-site resource quality seems to have little to do with the timing of exploitation. When transport costs are considered, however, the logic of the Ricardian view is borne out by past experience. Some resources "become better than others" simply because they are closer at hand to their possible use. Many petroleum deposits in Appalachia were less productive than those in East Texas, but the distance of the Texan fields from East Coast markets delayed their exploitation.

Ultimately the evaluation of resource quality, like resource recognition, rests on a number of contexts, or conditions. Where minerals are concerned, characteristics such as purity of deposit, depth of deposit, consumer demand, availability of substitutes, and transport cost are necessary contextual factors that enter the calculus of a resource's economic rent. The evaluation is also influenced by technology because each of the factors just listed is responsive to technological change.

Natural Resource Economies

In the production of goods the United States has high value, ... vast production is made possible by the highly favorable natural conditions prevailing in most of the country. Our forest, mineral, and water resources rank with or above those of any other country... (Colby and Foster, 1940, p. 145).

It's not unusual to see countries or regions spoken of favorably as being "blessed with natural resources." In the past, such a blessing was thought particularly to favor a select group of countries, including the United States as indicated in the preceding quotation. Canada, also, is often considered to have advantage with respect to natural resources. One of its best-known scholars, Harold Innis, focused on sequential natural resource recognition and exploitation in explaining the modernization of the Canadian national economy. Innis' *staple theory of development* concerns the interaction of export markets, physical geography, technological change, and institutional preferences for resources sectors as providing the foundation for the pattern of settlement by Europeans across Canada (Hayter and Barnes, 1990).

Staple production, consisting of direct exploitation and initial processing of natural resources, or staples, began with the Atlantic fisheries in the late fifteenth and early sixteenth centuries, and progressed to the interior with the growth of European demand for fur. Expansion into Canada's interior occurred in response to the development of the timber industries of lumber and pulp. A western progression was driven by the agricultural expansion that followed in the prairies during the late nineteenth and early twentieth centuries. Mineral exploitation and petroleum finds led finally to growth in the western prairies and into the lower Rocky Mountains. The transition of intensive resource exploitation from one type and one region to another type in another region resulted from changes in resource technology, demand in foreign markets, and the increasing accessibility of interior Canada led by railroad expansion.

From one perspective, the staple theory of Canadian development tells a story that is positive in terms of the benefits of natural resource supplies and their exploitation in leading to significant growth in a national economy. Average incomes and standards of living are very high in Canada by international standards. An optimistic view of the source of that country's high income and high living standards is that its exports of natural resource financed the growth and diversification of its national economy in a very beneficial way. From another perspective, and one more in line with Innis' thinking (Barnes, 1994), Canadian development has suffered from a reliance on staple production. According to this more pessimistic view, staples production and its emphasis on resource exploitation for external markets has often left Canada at the mercy of a limited number of often foreign multinational corporations and a few domestic interests that actually retard the economic diversification that would lead to greater economic stability.

If the overall success of the Canadian economy (or that of the USA) is a singularly direct result of its past record of natural resource exploitation and trade, then Canada is in a distinct minority of countries. Most countries of the world that rely on exports of primary products are among the world's poorest and face declining or, at best, uncertain incomes from their foreign markets (Porter and Sheppard, 1998, pp. 382–402). Natural resources usually face *declining terms-of-trade* in international markets, meaning their export value tends to decrease against the value of imported manufactured products and services. Often, even rapid expansion of natural resources exploitation and export fail to compensate for decreasing prices in international markets, resulting in a process of *immiserizing growth* – increasing output with decreasing income (Bhagwati, 1958).

Natural resource prices, however, don't always decline. World oil prices increased dramatically, for example in the mid-1970s, and again in the early 1980s because of organized efforts by an important group of oil exporting countries (OPEC) and a revolution in Iran. Such rapid increases in resource prices can lead to economic booms in exporting countries or regions, at least over the short term. Similar booms are experienced with the development of new natural resource fields because of gains in their economic rent brought about by higher prices, lower costs, or improved accessibility. On the surface, such booms appear positive for the national or regional economies they affect, but in reality they often cause significant problems.

One suite of such problems is called the *Dutch disease* because it was first observed in the Netherlands during the aftermath of the oil price increases noted above (Corden, 1984). One part of the Dutch disease is inflation caused by a rapid increase in spending as a response to the rapid increase in natural resource-based wealth. The spending is often targeted at items that are in short supply or not easily produced in the domestic market, and often at services that increase in price but not in quantity or quality. Another problem is the movement of factors from other economic sectors as they chase the wealth in natural resources. In this case, natural resource wealth leads to concentration and specialization in the economy, and not to the eventual diversification indicated in the optimistic version of Canadian staple theory. Currency appreciation is a third problem. As the value of the exporting country's currency gains due to an increase in resource exports, it becomes more difficult to export the country's other products. Again, economic specialization, rather than diversification, is often encouraged by growth in the natural resource sector.

In addition to the Dutch disease issues, Richard Auty (1995) has also identified a *resource curse* that appears to affect countries that become overly dependent upon their natural resource sectors. The curse is that natural resource wealth hinders economic growth in many countries rather than encourages it, as in the Canadian case. A rich natural resource sector has been used in some countries as the foundation for a government policy of attempted self-sufficiency in economic development. Natural resource export earnings are taken for granted and assumed to be able to provide sufficient foreign exchange and government revenue to offset any retaliation by other countries against a policy that uses high tariffs to protect domestic manufacturing from foreign competition. Such protection often leads to noncompetitive manufacturing that, due to its inefficiency, is hard pressed to make any significant contribution to national income. Unfortunately, resource sector exports are unable to sustain a country's government revenue requirements and foreign exchange requirements for very long, but entrenched interests make it very difficult to alter policy toward manufacturing so that it would contribute more to the national economy. The country's resources that were viewed as a consistent source of wealth prove to be a curse that undermines diversification and growth in other sectors.

The resource curse would revert to a blessing if natural resource prices gained consistently in international markets but the record is more of short-term booms and busts over a long-term price trend of decline. The trend of declining prices is ultimately a function of the trend of declining costs as a result of technological advance. Again, refer to figure 14.2, and its illustration of a supply shift to the right that leads to increased consumption at a lower price. As already represented in table 14.1, however, global trends often mask regional ones in the geography of natural resources. The supply shift describes a general decrease in the cost of production that affects a whole industry, but costs at particular places can rise even though costs in general decline.

Man exhausts fish, forest, grassland, and mineral resources, causing fishing villages, lumbering towns, large ranching centers, and mining camps to decline and disappear (Jones and Darkenwald, 1954, p. 5).

The decline, and even disappearance, of settlements with natural resource-based economies seems almost inevitable. Local ore quality declines as the best is used first, raising costs of exploitation, and even so-called renewable resources such as fisheries and forests are often depleted to a point where their products can no longer support a local economy and population. These results are in fact predictable, when the economic rent identity is placed in the context of a capitalist, or market, economy.

Market economies are marked by profit-seeking behavior on the part of producers, whether they are individuals, firms, or corporations. That behavior often extends to attempts at profit maximization within the bounds of the information that the producers possess. In the context of natural resources, that behavior is effectively manifest in ongoing efforts at increasing, if not maximizing, economic rent. Recall from the rent identities of equations (14.1) and (14.2) above that, for any particular resource deposit, economic rent is increased if the price paid for the resource rises or its cost of production and distribution is lowered.

Producer pricing-power in natural resource markets is rare. De Beers, a South African enterprise, has exercised effective price control over diamonds, but more for gem-quality as opposed to industrial-quality stones. OPEC attempted to control international petroleum prices and was successful for a short time, but its pricing-power dissipated fairly quickly for a number of reasons. Most natural resource reserves are too large in volume and too dispersed to have their supply controlled by a single producer or even a small consortium. Because prices are beyond their control, most producers have to rely on reducing their costs in order to increase the economic rent earned by their natural resource deposits.

...if production in a particular mine or mineral province declines or ends, it rarely happens because the deposits are physically exhausted, but more probably because new discoveries have reduced extraction costs elsewhere or because substitutes have become available (Houthaker, 1990, p. 441).

It appears that investment at a place gives rise to the decline of its own profitability; low wage labor demands higher pay, for example, or markets become saturated as competitive investment is attracted by the success of initial enterprises. As rent declines in an "old" place, new investment is made in new places. Imbalance in regional economies is the result, as relative success is short-lived. The geographical bounds on the swing of capital decrease as transport and communications costs decline and government exercises less and less control over international flows of non-labor factors of production.

The natural resource sector is marked by a technological treadmill with a geographical expression. The *technological treadmill* refers to the constant development of technology required to maintain rent levels as lower and lower quality resources must be exploited in order to maintain production (Roberts, 1992). Technological advance effectively raises resource quality by lowering costs, but the relative decrease of quality in remaining stocks requires that additional technological advance is soon necessary to sustain economic rent. Increases in reserves are largely the result of technological advance that turns potential supply into a current reserve of the resource.

The geographical expression of the technological treadmill involves the constant expansion of natural resource exploration and extraction. As noted earlier in this chapter, mineral and other natural resource stocks were exploited early on in core economies, where demand for their products was large and transport costs were low. High rates of local exploitation raised costs and lowered rents, inducing technological advance in places but also encouraging the spatial expansion of natural resource exploitation from the core to the periphery. The colonization of Africa focused largely on the exploitation of that continent's natural resources by European powers. The relationship still exists in contemporary linkages such as the long-term US support of Mobutu in Zaire (now Democratic Republic of Congo) because of its interest in securing supplies of that country's cobalt and other *strategic minerals* (minerals necessary in production of certain military equipment).

The reach of natural resources from the core to the periphery was the initial phase of real economic globalization. South American tin and copper (for example, see O' hUallachain and Matthews, 1996) and Middle Eastern oil were the basis of the formation of some of the world's earliest and largest multinational corporations. The recent acceleration of economic globalization resulting from technological advance and the government deregulation of most international flows certainly has not escaped the natural resource sector. Its geographical expansion is occurring especially in frontier, environmentally fragile, regions such as the Arctic and near-Arctic of North America and Asian Russia, and Amazonia in South America (Emel and Bridge, 1995). The accelerated globalization of the natural resource sector has dramatically affected a large number of local economies, already subject to the problems of their rising costs (Flora, 1990).

The decline of local natural resource economies is virtually guaranteed by their very establishment. Local costs of extraction are bound to rise and economic rent is bound to fall as the highest quality resources are exploited first. At the local scale, technological advance is more of a holding action than a method of increasing rent, and its effects usually prove temporary. Local economies are affected by a *resource cycle* of initial growth and ultimate decline (Clapp, 1998). The first part of the cycle consists of exploration, discovery, and an initial economic boom as the resource "becomes" and is brought to market. The second stage is one of profitable operation of local reserves, maintained by increasing capital intensity and applications of technological innovation. The third and final stage is depletion in an economic sense, which occurs when the local resource is replaced, as indicated by Houthaker (above), either by other sources or by a substitute material.

Unfortunately, Clapp (1998) has found that renewable resource-based local economies, such as those relying on forestry or commercial fisheries, are as subject to the resource cycle as mining towns. That is not surprising, however, because processes of resource exploitation do not differentiate between renewable and non-renewable resources.

The decline phase of the resource cycle described by Clapp (1998) is consistent with the decline of local resource quality expected under the Ricardian view of rent, and which occurs with continued exploitation. The decline in quality, revealed by increasing costs of extraction, refining, and even distribution, causes economic rent to decline in a place. As rent declines, geographic relocation of resource exploitation either complements or replaces local technological advance. Employment and income declines in the resource sector lead to a loss of support for other parts of the regional economy in a ripple effect, and the formerly booming natural resource region may fall into severe, long-term decline.

W. R. Freudenburg has written of resource-dependent local economies as being "addictive" (1992). Like a person addicted to a narcotic, such communities fail to recognize the hazard of their activities until it is too late. Even once the problem is recognized, addictive local resource economies try to avoid the withdrawal pains of economic change by various methods of retaining natural resource activities. Most of these methods of retention focus on lowering local costs, and therefore raising the local economic rent, of resource exploitation. Tax reductions and other subsidies are commonly offered by local governments, as is loosening of environmental restrictions. Threatened by loss of employment, local labor often becomes more "productive" by forgoing wage increases, and relaxing work rules and safety standards through subcontracting practices.

Conclusion: Resources, Regional Change, and Sustainability

Natural resource economies are unsustainable in both environmental and economic terms because the practice of rent maximization requires that any resource deposit be exploited until it no longer holds any economic reserves. There is, in fact, a so-called *optimal rate of depletion* of a natural resource that is determined by the rate of interest (Barbier, 1989, pp. 62–74). High interest rates warrant more rapid rates of depletion, so that greater income may be earned in non-resource markets, while low rates of interest allow less rapid rates.

In conjunction with seemingly inevitable economic decline, most resource-dependent economies are faced with environmental degradation. Again, rent maximization in competitive markets is most likely to be accomplished by cutting costs of production. Preserving environmental quality, however, by containing acidic runoff from mining operations, for example, or limiting soil erosion in lumbering, is expensive. Often such costs are viewed as unnecessary to actual production, and so are ignored by natural resource enterprises. Typically, government must either impose regulations forcing polluting enterprises to pay environmental mitigation costs ("costly regulations" is a commonly used phrase) or society must pay the financial costs and, too often, the health costs of the environmental degradation.

Sustainability in both economic and environmental senses is achievable, but it requires a change from traditional thinking about resources that is not always easy to make (Reed, 1995). The conventional view of natural resources is that they have value only when employed as factors of production. That view has led to calculations of "optimal" rates of depletion. In addition, that view has led to the implicit recognition of resource depletion as a contribution to national wealth: resource consumption contributes to gross domestic product in national accounts, while conservation does not (Steer and Lutz, 1993). At the local level, resources are viewed in the same way. Jobs and income are generated when natural resources are exploited as factors of production: unemployment and poverty result from their preservation (Power, 1996).

There is, however, a change occurring in the way natural resources are evaluated. Increasingly, recognition of their locally finite nature is resulting in a conscious effort on the part of some producers and many affected communities to limit technological applications to local resource extraction. Technological and capital limitations are being imposed in the Northern Forest of New England, for example, by so-called green producers who have stopped clear-cutting lumber in favor of lower impact selective harvests (Brown, 1998). The same type of limitations are being self-imposed in developing regions, as well, in the interest of providing both economic and environmental security (Litvin, 1998). In some circumstances, mineral exploitation is being forgone in recognition of the net costs of environmental, cultural, and economic despoliation that results from the natural resource cycle (Barre, 1998).

Natural resources are, of course, necessary factors of production and can't be eliminated from use entirely, nor can their contribution to local economies easily be cast aside. Sustainability, however, requires that resource dependency – the development of the resource addiction – be avoided. Such dependency can be avoided through economic diversification that broadens a regional (from local through national scale) economy's base from a narrow focus on resource exploitation to the employment of available resources as inputs to manufacturing and services. The benefits of economic diversification are well known at the national scale, but they exist at regional and local scales as well. By mitigating the boom and bust of the resource cycle, economic diversification can provide a sustainable base of employment and income. By providing economic alternatives, diversification also can enhance environmental sustainability by reducing the exploitation pressure of local resource dependency.

Bibliography

Annual Statistical Bulletin 1995 1996. Vienna: Secretariat, Organization of the Petroleum Exporting Countries.

Auty, R. M. 1995. Industrial policy, sectoral maturation, and postwar economic growth in Brazil: The resource curse thesis. *Economic Geography*, 71, 257–72.

Barbier, E. B. 1989. *Economics, Natural Resource Scarcity and Development*. London: Earthscan Publications.

Barnes, T. J. 1994. Staples theory. In R. J. Johnston, D. Gregory, and D. Smith (eds), *The Dictionary of Human Geography*, third edition. Oxford: Blackwell, 589–91.

Barre, D. 1998. Cold war. *Greenpeace Magazine*, 3, 3 (Fall), 12–14.

Bhagwati, J. 1958. Immiserising growth: A geometrical note. *Review of Economic Studies*, 25, 201–5.

Bina, C. 1989. Some controversies in the development of rent theory: The nature of oil rent. *Capital & Class*, 39, 82–112.

Brown, R. 1998. Small is beautiful. *AMC Outdoors*, September, 20–21.

Cairns, R. D. 1994. On Gray's rule and the stylized facts on non-renewable resources. *Journal of Economic Issues*, 28, 777–98.

Clapp, R. A. 1998. The resource cycle in forestry and fishing. *The Canadian Geographer*, 42, 129–44.

Colby, C. C. and Foster, A. 1940. *Economic Geography*. Boston: Ginn and Co.

Corden, W. M. 1984. Booming sector and Dutch disease economics: Survey and consolidation. *Oxford Economic Papers*, 36, 359–80.

Dasgupta, P. 1993. Natural resources in an age of substitutability. In A. V. Kneese and J. L. Sweeney (eds), *Handbook of Natural Resource and Energy Economics*, Volume III. Amsterdam: Elsevier Science, 1114–30.

Emel, J. and Bridge, G. 1995. The Earth as input: Resources. In R. J. Johnston, P. J. Taylor, and M. J. Watts (eds), *Geographies of Global Change*. Oxford: Blackwell, 318–32.

Flora, B. F. 1990. Presidential address: Rural peoples in a global economy. *Rural Sociology*, 55, 157–77.

Freudenberg, W. R. 1992. Addictive economies: Extractive industries and vulnerable localities in a changing world economy. *Rural Sociology*, 57, 305–32.

Harris, D. P. 1993. Mineral resource stocks and information. In A. V. Kneese and J. L. Sweeney (eds), *Handbook of Natural Resource and Energy Economics*, Volume III. Amsterdam: Elsevier Science, 1011–76.

Hayter, R. and Barnes, T. 1990. Innis' staple theory, exports, and recession: British Columbia, 1981–86. *Economic Geography*, 66, 156–73.

Houthaker, H. S. 1990. Review of *Toward a New Iron Age? Quantitative Modelling of Resource Extraction. Journal of Political Economy*, 98, 440–45.

Jones, C. F. and Darkenwald, G. G. 1954. *Economic Geography*, revised edition. New York: Macmillan.

Landes, D. S. 1998. *The Wealth and Poverty of Nations: Why Some are So Rich and Some are So Poor*. New York: W. W. Norton.

Litvin, D. 1998. Dirt poor: A survey of development and the environment. *The Economist*, March 21, insert, 1–16.

Norgaard, R. B. 1990. Economic indicators of resource scarcity: A critical essay. *Journal of Environmental Economics and Management*, 19, 19–25.

O' hUallachain, B. and Matthews, R. A. 1996. Restructuring of primary industries: Technology, labor and control in the Arizona copper industry. *Economic Geography*, 72, 196–215.

Porter, P. W. and Sheppard, E. S. 1998. *A World of Difference: Society, Nature, Development*. New York: Guilford Press.

Power, T. M. 1996. *Lost Landscapes and Failed Economies*. Washington: Island Press.

Reed, M. G. 1995. Cooperative management of environmental resources: A case study from Northern Ontario, Canada. *Economic Geography*, 71, 132–49.

Rees, J. 1989. Natural resources, economy and society. In D. Gregory and R. Walford (eds), *Horizons in Human Geography*. Totawa, NJ: Barnes & Noble, 364–94.

Roberts, R. 1992. Nature, uneven development and the agricultural landscape. In I. R. Bowler, C. R. Bryant, and M. D. Nellis (eds), *Contemporary Rural Systems in Transition*, Volume 1. Wallingford, UK: C A B International, 119–30.

Steer, A. and Lutz, E. 1993. Measuring environmentally sustainable development. *Finance & Development*, 30, 4, 20–3.

World Resources Institute 1996. *World Resources 1996–97*. New York: Oxford University Press.

Chapter 15

Agriculture

Brian Page

Agriculture was an important component of economic geography through the 1950s. Until that time, research on agriculture and industry within economic geography shared an empirical and descriptive approach strongly influenced by notions of environmental determinism (see Barnes, this volume). But with the rise of model-building and quantitative methods in the 1960s (see Plummer, this volume), the focus of economic geography shifted to studies of industry while studies of farming were shunted into the subfield of agricultural geography. Ironically, this was true even though agriculture provided an early foundation for the new methods of industrial analysis: indeed, two important sources of location theory in the mid-twentieth century – von Thünen's land-use model and Christaller's central place theory – were strongly based upon the study of German farming landscapes.

In the ensuing decades, as agricultural topics became less important within economic geography, research in agricultural geography branched into two separate and distinct areas. The first continued the long-standing tradition of descriptive research by pursuing the study of regional classification and farm characterization. The other branch, meanwhile, began to adapt the techniques of location modeling to a wide range of farm-related research questions. The result was that studies of agriculture within geography were cast either as old fashioned and backward looking, in the case of the former, or derivative of industrial geography, in the case of the latter. In any event, the secondary status of agriculture within economic geography was cemented into place.

Today, the study of agriculture is being brought back into economic geography, though this arguably has less to do with work coming from the tradition of agricultural geography than it does with agriculture-related research conducted in other parts of the discipline, and in other disciplines altogether. In particular, this reincorporation is rooted in political–economic analyses of agriculture which emerged over the past 15 or 20 years and which resonate with new approaches that have been adopted in economic geography more generally (see Swyngedouw, this volume). Such reincorporation, however, is not based on a simple mapping of new industrial theory directly onto the terrain of agriculture. Rather, it is based on a renewed appreciation for the complementarities between farming and manufacturing, and on a recognition that insights gained from the study of agriculture can inform theories of the development of industrial capitalism in some very important ways.

This chapter briefly looks at the shifting position of agriculture within economic geography. Given space constraints, the chapter does not attempt to chart the complete course of this relationship, nor does it attempt to provide a review of research in traditional agricultural geography (but see Grigg, 1995). Instead, the chapter concentrates primarily on the political economy of agriculture in developed capitalist economies, particularly the USA and the UK. That said, the line separating agricultural research in the more developed world and the less developed world has become increasingly blurred. This is because of the interconnection of production and consumption within an increasingly globalized economy, and a convergence in guiding theoretical approaches to the study of farming in industrial and peasant societies.

The chapter begins with a discussion of the evolution of a political economy of agriculture in the social sciences. Next, it turns to a selective review of several current research themes to be found in that literature, paying particular attention to the contributions made by geographers. Finally, it discusses the implications of this new research on agriculture for the wider field of economic geography.

The Development of a Political Economy of Agriculture

The political economy of agriculture first took shape in response to the inability of traditional approaches to provide sufficient explanations of the dramatic changes occurring in agriculture in the more developed world. Because of the dominance of the neoclassical paradigm, research in agricultural economics focused exclusively on issues of farm business efficiency and performance and had little to say about the social consequences of technological change, the rise of large agribusiness firms, the shifting class structure of American farming, or the role of state intervention. Nor could rural sociology comment meaningfully on these trends: the discipline had virtually abandoned the study of agricultural production in favor of descriptive research on rural communities (cf. Friedland, 1982; Newby, 1982). Researchers turned to the classical political economy of Marx – with its characteristic focus on class relations in production as the motor force in the development of industrial capitalism – in order to gain a better understanding of broad, structural change in agriculture.

Marx's writings on the development of capitalism in agriculture comprise a broad comparison of the similarities and differences between farming and manufacturing (Goodman and Redclift, 1985; FitzSimmons, 1986; Mann, 1990). The most cited aspect of Marx's view on the subject, however, was the assertion that wage labor would eventually sweep away all non-capitalist, household-based production in agriculture – a social group that he termed "petty commodity producers" and more commonly known in the US and European context today as family-labor farms. At the turn of the twentieth century, Lenin (1899) extended this aspect of Marx's argument to the Russian context, making the case for the class differentiation of the peasantry into either capitalist farmers or wage laborers. At the same time, Kautsky's (1899) writing on German agriculture made a similar argument, but he also emphasized the factors that had slowed the penetration of capital into agriculture, and raised the possibility that household production was in fact a permanent feature in agriculture – an argument later developed in more detail by Chayanov (1925).

This issue of the whether non-capitalist forms of production are permanent or transitional – the "agrarian question" – was the central concern of those working to re-introduce a political–economic framework into agricultural studies in the 1970s. Some writers adopted the classical Leninist position that the forces of concentration and centralization of capital in agriculture lead to an immutable tendency toward class differentiation in farming, arguing further that post-World War II developments greatly accelerated this tendency and foretold the imminent demise of family labor farms (de Janvry, 1980). Others argued for the continued viability of household producers within the agricultural sector (Mann and Dickinson, 1978). But as the 1980s progressed, key limitations to this literature became apparent. In particular, critics noted that a preoccupation with the agrarian question tilted research toward the study of on-farm social relations and, in corresponding fashion, toward the farm unit as the primary point of analysis. What was missing, they argued, was an appreciation of the ways in which farm production was bound up with wider processes of economic development and capital accumulation (e.g. Buttel, 1982; Goodman and Redclift, 1985; Marsden et al., 1986).

This critique was spurred by the recognition that forces emanating from beyond the farm gate had come to dominate agriculture and food production – a broader perspective that also had roots in classical political economy. Indeed, Kautsky (1899) first observed that capitalist development proceeded not only through the displacement of peasant households by capitalist farm enterprises but also through the articulation of household producers with industry (Watts, 1996). While evident in Kautsky's time, this articulation intensified dramatically in the post-World War II era as farmers became enmeshed in a multiplying web of external linkages tying them to agro-industries (firms supplying farm inputs and marketing or processing farm products) and banks.

One result of this increasing articulation was the rapid expansion of "off-farm" sectors (e.g. farm machinery, food processing) which came to dwarf farming in terms of value-added and employment within the overall food producing system (or "agro-food" system). Another result was a gradual loss of family farm independence. As farmers adopted input-intensive practices, they became dependent upon manufactured products (machinery, seeds, fertilizer, herbicides, etc.), scientific advice and technical assistance, and financial institutions for credit to support continuing operations and business expansion (particularly for purchases of technology). Farmers also became dependent upon downstream agro-industries (commodity shippers, food processors, food retailers) as these firms came to wield increasing influence over farm output levels, farm commodity prices, and farm production standards. In the USA the growth of off-farm sectors was intertwined with federal policy toward agriculture, which: (1) promoted new farm technologies through government-sponsored research and extension; (2) encouraged the deepening of farm–industry linkages through a system of farm price supports which allowed (in fact, compelled) farmers to purchase new technologies and boost productivity; and (3) absorbed and disposed of resulting surpluses in farm products through commodity programs, domestic food programs, and international food aid.

Thus, while the direct displacement of family farms by capitalist farms occurred only to a limited extent, capitalist development in the overall agro-food system nevertheless advanced rapidly as non-farm capital (industry and banks) established an indirect but powerful grip on agricultural production.

This was a seminal insight provided by the political economy approach. It led researchers to shift their analytic focus from the farm to the entire range of activities and relationships surrounding agricultural production. In geography, the political economy approach to agriculture arrived relatively late, but geographers were nevertheless key participants in moving the literature beyond a focus on a narrow agrarian question, and re-positioning it within broader industrial and political contexts (e.g. Marsden et al., 1986; FitzSimmons, 1986; Le Heron, 1988). Through subsequent efforts, geographers have continued to make important contributions to this field of study, one that remains solidly multi-disciplinary in character. For this reason, the following review of current research themes within the political economy of agriculture concentrates not only on the work of geographers, but also on research carried out in other social science disciplines.

Research Themes in the Political Economy of Agriculture

Nature and the industrialization of agriculture

The basic fact that farming is a critical intersection between nature and society has been a central theme in the political economy of agriculture. This was the starting point for Marx, who linked the slow development of capitalism in agriculture (compared to manufacturing) to the physical and biological conditions of farm production. In recent years, scholars have built upon Marx's fragmentary observations to identify the many ways in which natural conditions shape the course of development in agriculture and food production.

Agriculture is a unique branch of industry in that it is constrained by natural processes which act to limit the productivity of labor and restrict capital investment. Here, the role of biology in plant and animal growth is key. There are no industrial substitutes for soil or sunlight, and the biological conversion of energy in plant development and animal gestation cannot easily be accelerated or standardized (Goodman et al., 1987). So, too, on a farm – unlike a factory – it is the biological time necessary for plant and animal growth that dictates the work schedule. As a result, the amount of time in which labor is actively applied to production is just a small fraction of the total time required to produce a farm commodity (Mann and Dickinson, 1978). In addition, the land-based character of farm production poses several constraints to industrialization. In crop farming – again in contrast to a factory – capital cannot be applied to the labor process at a single site where production is expanded or intensified. Instead, increased production requires a spatial extension (and, conversely, decreased production requires a spatial contraction). But because land is a fixed and limited resource, and because land markets are deeply colored by localized social conditions, farmers cannot easily or quickly adjust their investment in land (Marsden et al., 1986).

The natural circumstances of plant and animal growth, however, do not so much stop the industrialization of agriculture as direct it along a distinctive path. Indeed, the historical development of capitalism in agriculture has pivoted around attempts by industrial firms to reduce the importance of nature in all phases of food production. This includes not only constraints associated with the biological properties of agricultural production, but also constraints associated with the biological requirements of food consumption in terms of diet, nutrition, and health. According to a seminal study by Goodman et al. (1987), this process has advanced through two discontinuous but enduring processes, termed *appropriationism* and *substitutionism*.

Appropriationism has occurred as capitalist firms, unable to effect the industrial transformation of agricultural production in its entirety, have instead assimilated discrete aspects of the farm labor process into factory-based industry where they have been rationalized, mechanized, and intensified beyond anything possible on the farm. Here, traditional elements of agricultural production are appropriated by manufacturers, transformed into branches of industry, and then re-incorporated back into agriculture as purchased inputs. Examples include agricultural implements, hybrid seeds, chemical herbicides, and livestock confinement buildings. The piecemeal adoption of technological inputs has allowed the modification of certain biological processes on the farm and has led to a gradual (though only partial) reduction in the importance of nature in production.

Substitutionism has occurred as the outputs from farming have been either replaced or reshaped by industrial work. Here, manufacturers and merchants act to reduce farm products to uniform industrial inputs and, where possible, replace these farm products entirely with fabricated or synthetic non-agricultural inputs to food manufacturing. By substituting industrial for agricultural inputs (for example, the use of high fructose corn syrups instead of sugar), firms avoid problems associated with the variable cost and availability of farm products. Another aspect of this process involves efforts to transform basic foodstuffs into a wide array of processed and packaged food products. In this case, firms have sought to modify the often limiting original form and character of agricultural products through industrial work, thereby increasing the market size and value of food commodities. Examples include the transformation of plain grain into breakfast cereal, or of raw animal flesh into a less perishable product such as luncheon meat.

Capitalist production has thus encroached on agriculture via the expansion of industrial activities surrounding the farm. This process of "agro-industrialization" has moved forward slowly, but over time, natural constraints have been successfully (albeit incrementally) eroded as technological and organizational innovations have been introduced by suppliers of farm inputs and processors and marketers of farm output. Recent advancements in the application of biotechnologies to agriculture and food production highlight the fact that the ongoing "refashioning" of nature comprises the heart of agricultural industrialization. These new technologies represent an enormous and generalized leap forward in the capacity to manipulate natural processes for commercial gain in agriculture, and thus herald a potentially fundamental re-ordering of the social, economic, and spatial organization of the agro-food system (Kloppenburg, 1988; Goodman, 1991).

Agro-food commodity chains

A second theme concerns a finer-grain approach to agriculture carried out through the study of the industrialization process surrounding individual farm commodities.

Commodity-specific analysis of this sort was introduced into the political economy of agriculture through the work of Friedland (1984) on California fruit and vegetable production. This research focused on important differences from crop to crop in terms of the farm labor process, the direction of technological change, the structure of the farm business enterprise, the marketing of farm products, and so on. This type of approach has since been extended to other crops and regions, and has been broadened to focus on entire production systems, or agro-food commodity chains. Agro-food commodity chains are networks of articulated labor and production processes ending with finished food products. In broad outline, such commodity chains consist of five stages: scientific and technical inputs to farming; farm production; processing of farm products; food distribution; and food consumption (Bowler, 1992). The broiler commodity chain, for example, begins with various inputs to chicken production, including manufactured feed and breeding stock, and then progresses through farming and slaughtering/processing before entering a distribution system that delivers chicken meat at retail to consumers. In the big picture, individual commodity chains often intersect and overlap, providing the basic "warp and weft" of the agro-food system if you will.

This approach offers several key insights into the capitalist transformation of agriculture. First, it provides a framework for tracing the multiple paths of industrial development evident within the agro-food system. While in general terms the industrialization of agriculture has progressed through the increased articulation of farming with capitalist industry (as described above), the pace and form of this process varies widely from one commodity to another. Indeed, research in this vein has shown that individual commodity chains (grapes versus beef, for instance) differ markedly with respect to: (a) technological dynamics – that is, the specific types of innovations and strategies pursued by agro-industry as well as the character of technical relations linking stages of the chain to one another; and (b) organizational dynamics – that is, the social character of work and business organization within each stage, as well as the overall pattern of how the various stages are separated or integrated within the division of labor (Page, 1996; Boyd and Watts, 1997).

This variation is rooted in the particular social and institutional histories within (and between) each of these linked stages. Such variation is also strongly associated with the extension of "natural" distinctiveness through the commodity chain. Each commodity differs with respect to the biological qualities of the particular plant or animal in question, the extent of land dependence in production, and the biological properties of the resultant farm product. These differences act to shape the character of industrial transformation not only in farming, but also in input manufacturing, food processing, and food distribution – and thus to channel the commodity chain's overall course of development along a particular path.

Second, the commodity chains approach provides a framework for understanding the connections between producers and consumers of food. Until quite recently, perhaps the key limitation of research on agro-food commodity chains was a concentration on the production end of the chain, particularly the relationships linking farmers to the industries that supply their inputs or process their outputs, and to related questions of the transformation of the farm labor process, the spread of wage labor versus family labor, and the rise of contract farming. Now more attention is being given to the consumption end of commodity chains. Some of this work focuses on corporate concentration in the food retailing sector and the increasingly powerful role that giant retailers play in agro-food commodity chains, particularly with respect to food processing industries (Marsden and Wrigley, 1995). Other work concentrates more on changes in consumer habits and preferences, paying particular attention to the role that diet and food play in shaping and expressing cultural identities (Bell and Valentine, 1997). Here, the shifting position of retailers is linked to a set of changes in consumer practices that have occurred over the past 25 years or so, centering on broad concerns over food quality, personal health, and environmental well-being. These changes in the diet of consumers in the developed world, coupled with occasional consumer activism over specific food issues (e.g. chemical residues, food additives, animal welfare, food labeling, etc.), have reverberated backward through commodity chains. In particular, retailers have responded aggressively to concerns over food quality by promoting the rapid expansion of markets for health-based food products, leading to a shift in the balance of power between retailers and food processing industries in which the retailers are increasingly able to influence the character of food production (Goodman and Watts, 1994).

An example of these processes can be seen today in the struggle over consumer acceptance of bio-engineered food products. Within many commodity chains, the adoption of biotechnologies represents a significant advance in the industrialization process. Yet, traditional food products enhanced by biotechnology (e.g. hormone treated beef or "designer" varieties of corn) are avoided by many consumers because of concerns over quality and healthfulness – they are considered to be "unnatural" and therefore potentially dangerous. Vigorous resistance to such products is visible today in contested discussions over the necessity to label food products containing genetically modified organisms (GMOs). In this case, consumer awareness and resistance threatens to derail, or at least slow down, the transformation of agriculture and food production promised by biotechnology: if consumers demand labeling of GMOs and then refuse to buy these products, farmers will restrict their use of genetically altered seed or livestock, and food processors will restrict their purchase of farm products derived from these sources.

Third, a commodity chains approach provides a framework for grasping the relationships between places and regions. Each agro-food commodity chain exhibits its own technological trajectory, set of organizational forms, and constellation of power relationships. Yet each is also associated with a unique geographic pattern – a spatial division of labor. This is not to imply any sense of immutability or natural determinism, however. Natural conditions (latitude, soils, precipitation, etc.) matter to where agriculture and surrounding activities are located, but such locational constraints can be overcome via the transformation of nature and the creation of new agricultural systems through irrigation, plant and animal breeding, improved production facilities, new transport modes, and so on. Indeed, spatial divisions of labor surrounding agriculture are never static; rather they are constantly being expanded and re-configured in response to competition, market shifts, and persistent social, technical, and political change. At the international scale, an examination of the geography of commodity chains has been used to illuminate the interrelationship of producers in less developed nations with consumers in more developed nations (Friedland, 1994), while at the regional scale, such an examination reveals the mutual dependence of city and countryside (Page and Walker, 1991).

Globalization

A third research theme within the political economy of agriculture is the globalization of the agro-food system. Agriculture and food production have long had a global character, but over the past 20 years this has become more pronounced. Here, research focuses on two related issues: (1) the global character of agricultural regulation; and (2) the increasingly important role played by transnational corporations (TNCs).

With respect to the regulation of the agro-food system, work has concentrated on the internationalization of the US model of agricultural policy after World War II. This policy framework consisted of a host of federal policies and agencies – including agricultural research and extension programs, farm price supports, and subsidized food programs – designed to encourage expanded domestic production and to dispose of resulting surpluses through global food exports. According to Friedmann (1993), these institutions and activities constituted a "food regime" – the broad regulatory apparatus used to establish stable relations between food production and consumption in the post-war era. This food regime became unstable in the 1970s, however, when many other developed nations adopted the US policy framework, leading to overproduction, glutted markets, and the inability to dispose of surpluses. In response to these problems, a new food regime has emerged over the past decade, one characterized by the deregulation of national agricultures in the service of open world trade, coupled with the increasing dominance of transnational firms (TNCs).

The second area of concern, then, is the reworking of global agriculture by TNCs. Here, research concentrates on the involvement of companies such as Coca-Cola, McDonalds, Kellogg, Unilever, and Nestlé in food production and marketing around the world, the standardization of global diets toward processed (or "fabricated") foods produced through global sourcing, and the increasing control of the world's food production in the hands of just a handful of these giant firms (Bonanno et al., 1994; McMichael, 1994).

This research on globalization makes many vital contributions to an understanding of agro-food system dynamics. It directs attention to the essential role of the state in mediating capitalist development in agriculture and food production – a topic that is much less well developed in the agro-food commodity chains literature discussed above. More specifically, it provides great insight into current processes of change through which nation-state regulation is being challenged by a system of supranational regulation. This literature also points to the systematic dominance of the developed world over agriculture in the less developed world, carried out through both the practices of TNCs and the policies of global regulatory institutions. Agriculture in many developing countries has been re-oriented away from food crops for local markets toward either the production of luxury foods bound for the developed world or the production of feed crops for intensive livestock production. As a result, many of these nations are now net importers of staple and processed foods coming from developed nations, and thus find themselves in an increasingly vulnerable position with respect to food prices on the world market.

The globalization literature, however, has also been the subject of criticism. One issue is that the food regimes approach largely ignores the dynamics of industrial change in agriculture. Discussions of successive regulatory regimes leave aside any detailed analysis of industrial change in agriculture: instead, a single technical–organizational model (post-war "productivist" US farming) is superimposed upon the agricultural sector, belying the wide variety of directions taken by industrial agriculture in the developed world, let alone world-wide (cf. Goodman and Watts, 1994; Page, 1996). Another issue is that the food regimes approach privileges the macro-structural scale of analysis when discussing shifts in regulation. Yet, it is probably much too soon to herald the triumph of international capital over the nation-state (e.g. McMichael and Myhre, 1991). Such an interpretation underestimates the lasting importance of nation-states by overestimating the stability of current domestic political alignments and resultant agricultural policy. It also underplays the continuing ability of nation-states to circumvent trade rules, overlooks the importance of regional trading blocs, and ignores the sub-national regulation of agriculture along regional and commodity lines (cf. Goodman and Watts, 1994; Marsden, 1998).

Research on TNCs also has a tendency to over-generalize, leaving the impression that such firms are the sole driving force behind a relentless global restructuring of the agro-food system – able to penetrate every location on earth and every aspect of food production. While TNCs do exercise tremendous power, the still sporadic and selective character of their involvement across various commodity chains and around the world raises many questions concerning their limits and even failures. So, too, the internationalization of agro-industrial firms is too readily collapsed into a single model of globalization thought to follow along the lines of the automobile or electronics industries – that is, one characterized by vertically integrated transnational production systems involving centrally coordinated intra-firm divisions of labor and global sourcing. In fact, few, if any, global agro-industrial firms fit this description: instead, the forms of corporate international production are varied, ranging from the establishment of parallel affiliates in multiple countries to the integration of affiliates and parent firms through limited outsourcing (Watts and Goodman, 1997; Gouveia, 1997).

Important global changes are taking place today with respect to agricultural production, food consumption, and agro-food system regulation. Yet, the term "globalization" remains problematic within the literature because it most often describes processes that are overly broad and encompassing and thus obscures important differences in the experience of agricultural restructuring among different nations and regions. Here, the work of geographers is particularly important as it highlights the articulations between world-scale processes and the character of agricultural transformation in particular places.

Social agency in the agro-food system

The fourth research theme centers on efforts to incorporate the individual and collective actions and interests of people – social agents – into accounts of agro-food system restructuring. This research emerged from a critique of the globalization and industrialization approaches just described. Both traditions, it is argued, favor structural explanations wherein change is carried out by industrial firms, nation-states, or global institutions but overlook (and to varying degrees discount) the role played by human actors across the range of activities linking food production to food consumption.

To date, the main thrust of this research on social agency has been an exploration of the persistent family-based character of farming in the developed world. Contrary to predictions dating back to the posing of the original agrarian question in the nineteenth century, household farm producers have not been swept away by capitalist farms, but continue to persist and compete. Overall, however, household producers have become differentiated into two broad groups: farms that employ industrial production methods, are closely tied to corporate agro-industries, and are integrated into global networks of production and trade; and farms that remain marginal to the agro-food system in terms of on-farm practices and market linkages (Whatmore, 1995).

With respect to the first group of farms – by far the more important from the standpoint of total farm output – the literature treats industrialization not as a top-down imposition of new farm practices by industry, but as a process that is negotiated between family enterprises and corporate agro-industry. Though such negotiation clearly takes place under conditions of uneven power, farmers are nevertheless held to play an active role through household-level decisionmaking – a process in which decisions concerning the adoption of intensive production practices are bound up in a host of other social and cultural issues such as family-life cycles, kinship relations, and gender relations (Whatmore, 1991; van der Ploeg, 1993). Thus, the response to industrial transformation can vary tremendously from one farm to the next given the diversity of household forms, goals, and strategies. For this reason, an understanding of agricultural restructuring cannot easily be read off general tendencies toward industrialization and globalization, but must be based on a careful treatment of the variegated social and technical world of farm households in particular commodity sectors and in particular places.

In terms of the second group – the large number of farm enterprises that remain marginal to the industrialized agro-food system – research again focuses on the great range of "livelihood strategies" employed by a diverse set of households, but also emphasizes the importance of these farms as sites of alternative production strategies. This represents a shift in the treatment of social agency from a focus on the ways in which farmer decisionmaking shapes and conditions (yet, in the end, also advances) the industrialization process, to a focus on the emergence of active resistance to industrial agricultural and food production. In this case, the focus is on the interaction of so-called marginal farmers with new social movements. Over the past 20 years, several important social movements – centering on environmental issues, food contamination issues, animal welfare issues, and so on – have arisen in response to the problems associated with industrialized food production. Though sporadic and mostly uncoordinated, these movements have nevertheless exerted an important influence on consumption habits, thereby encouraging the expansion of traditional production practices and the development of new marketing networks (Buttel, 1997).

Take the case of organic farming. Organic farming accounts for only a small fraction of total farm output in the developed world, but despite this apparently marginal position, organic farmers (particularly in Europe) have been able to influence farm policy decisions beyond their numbers due to their alliance with environmental and consumer groups. From this standpoint, then, organic farming appears not as a fringe activity, but as the leading edge of alternative agro-food relationships rooted in broad social resistance to the industrial model. Yet, it must be noted that because of its increasing importance, organic production has also become the target of agro-industries (at least in the USA) seeking to absorb organic farms into conventional input and marketing channels (Buck et al., 1997).

This issue of the political, or contested, character of change in agriculture has been addressed in a variety of other contexts as well, ranging from explorations of the role of national farmer political organizations in shaping domestic agricultural policy and international farm–industry relationships (Pritchard, 1998) to localized studies of farmers' political struggles over water rights (Roberts, 1996), to the role of place-specific labor–management struggles in shaping trajectories of agricultural industrialization (Wells, 1996). The importance of social agency and resistance is also a central concern of researchers seeking to apply actor–network theory to studies of agro-food system development (Whatmore and Thorne, 1997; Ward and Lowe, 1997). While this sort of approach has received a good deal of attention recently, it has also been criticized for tilting so far toward individual action that it may suppress an appreciation of how TNCs (or other global actors) continue to profoundly affect social equity and human well-being in agro-food systems (Gouveia, 1997; Walker, 1997).

Reincorporating Agriculture into Economic Geography

For years, agriculture occupied a secondary position (at best) within economic geography, a situation that did not change when political economic approaches became prevalent within the subdiscipline. In fact, there remained a deep conceptual schism between agriculture and industry wherein agriculture was comparatively under-theorized as an arena of capitalist development – cast either as a simple backdrop to industrialization processes or as an active brake on wider capitalist growth. Worse, early efforts to re-position agriculture within industrial political economy simply extended theories drawn from the industrial restructuring literature (in this case regulation theory) to cover the farm sector, an effort which, ironically, only obscured the distinctive technological, organizational, and institutional character of agriculture (Kenney et al., 1989).

But agriculture has been retrieved from this conceptual backwater by political economic research that – as discussed above – takes seriously the analytic importance of agriculture's peculiarities, and on that basis has begun to rethink the relationship of agriculture to nature, industry, the state, and the dynamics of regional development. Arguably, the study of agriculture belongs at the very heart of economic geography not only because of the centrality of food production and consumption in modern society, but also because the political economy of agriculture literature engages a range of issues that stand at the center of social science inquiry into the dynamic spatial reconfiguration of contemporary capitalism.

For instance, work on the industrialization of agriculture points to the importance of broadening our vision of social arrangements of production, technical differences, and divisions of labor across all sectors of industry. Recent approaches to industrial restructuring, particularly the regulation and flexible specialization schools, tend to take all industries to be variations on a basic theme drawn from the specific histories and geographies of one or two industrial sectors. The case of agriculture, however, reveals the fallacy of attempting to collapse the diverse experience of industry into a few limited molds. The food producing system exhibits a mix of social relations of production, an incredible breadth of technological development, and a wide range of forms of industrial organization. Such variation – deriving from differences among agro-food commodity chains rooted in the landed basis of production and the natural circumstances of plant and animal growth – suggests a remarkable openness in the evolution of production systems under capitalism and a multiplicity of possible paths of industrialization, and serves to caution against the adoption of overly rigid and reductionist general theories (Goodman and Watts, 1994; Page, 1996).

In addition, debates over globalization within the political economy of agriculture demonstrate the importance of recognizing regional divergence in forms of eco-nomic development despite tendencies toward globalization. Recently, economic geographers have begun to address this issue by paying more attention to the historical and regional contexts of industrial development, concentrating in particu-lar on the ways in which class relations, technical advances, business culture, regulatory structures, etc. emerge from specific local circumstances over time (e.g. Storper, 1997). This focus on the place-specific social context of economic relations and economic institutions resonates with attempts within the agro-food systems literature to make sense of social heterogeneity and the deeply embedded sources of local difference in agriculture, and both mark a re-affirmation of a long-standing research theme in economic geography – namely, that geography matters. After all, interest in locality in political economy was in large measure sparked by Massey's (1984) pioneering analysis of the ways in which contemporary economic restructur-ing is shaped by the accumulated sediments of regional and local history.

Finally, this literature on agriculture points to the importance of building toward a more complete treatment of social agency in studies of economic development. The use of agency-oriented methods to study agriculture reflects a more general quest within economic geography (and the social sciences generally) to move beyond overly structural accounts of economic restructuring. While recent studies of re-gional industrial development offer great insight into the varied social worlds of capitalism, they most often remain abstracted from the realm of human action, motivation, and experience. Yet, as studies of the agro-food system have begun to show, industrial restructuring is always a contested process – one shaped in vital ways by the exercise of power in the micro-politics of everyday life. One strength of this orientation is that it demands a broadening of the concepts of social resistance and struggle to include not only class, but also other dimensions of social difference such as gender, race, and ethnicity. Another is its recognition that struggles over resources, labor, and technological practice that stand at the center of the restructur-ing process are simultaneously struggles over culturally constructed meanings and identities (Hart, 1997). This, in turn, opens up some interesting lines of geographical inquiry (following the cue of cultural geography) concerning how the contest for social power is carried out by encoding the landscapes of everyday life with symbolic meaning. Still, as critiques within the field make clear, great care must be taken so that this focus on agency does not slip into a reification of difference leading to a myopic concern with particularism and local heterogeneity that is blinded to the way in which broader configurations of power within capitalism (for instance, class relations) continue to create systematic imbalances in social equality.

These and other discourses within the political economy of agriculture have the potential to contribute a great deal to the field of economic geography by expanding the scope of the field's inquiry into rural and extractive realms of economic activity, and by providing a valuable new vantage point from which to scrutinize theories of industrial growth and change.

Bibliography

Bell, D. and Valentine, G. 1997. *Consuming Geographies*. London: Routledge.

Bonanno, A., Busch, L., Friedland, W., Gouveia, L., and Mingone, E. 1994. *From Columbus to ConAgra: The Globalization of Agriculture and Food*. Lawrence: University Press of Kansas.

Bowler, I. 1992. The industrialization of agriculture. In I. Bowler (ed), *The Geography of Agriculture in Developed Market Economies*. New York: Longman Scientific and Technical, 7–31.

Boyd, W. and Watts, M. 1997. Agro-industrial just-in-time. In D. Goodman and M. Watts (eds), *Globalising Food*. London: Routledge, 192–225.

Buck, D., Getz, C. and Guthman, J. 1997. From farm to table: The organic vegetable commodity chain of Northern California. *Sociologia Ruralis*, 37, 3–20.

Buttel, F. 1982. The political economy of agriculture in advanced industrial societies: Some observations on theory and method. *Current Perspectives in Social Theory*, 3, 27–55.

Buttel, F. 1997. Some observations on agro-food change and the future of agricultural sustainability movements. In D. Goodman and M. Watts (eds), *Globalising Food*. London: Routledge, 344–65.

Chayanov, J. 1925 [1986]. *The Theory of Peasant Economy*, transl. by D. Thorner, R. Smith and B. Kentlay. Madison: University of Wisconsin Press.

deJanvry, A. 1980. Social differentiation in agriculture and the ideology of neo-populism. In F. Buttel and H. Newby (eds), *The Rural Sociology of the Advanced Societies*. Montclair, NJ: Allanheld, Osmun & Co.

Dupuis, M. 1993. Sub-national state institutions and the organization of agricultural resource use: The case of the dairy industry. *Rural Sociology*, 58, 440–60.

Fine, B. 1994. Toward a political economy of food. *Review of International Political Economy*, 1, 519–45.

FitzSimmons, M. 1986. The new industrial agriculture: The regional integration of specialty crop production. *Economic Geography*, 62, 334–53.

Friedland, W. 1982. The end of rural society and the future of rural sociology. *Rural Sociology*, 47, 589–608.

Friedland, W. 1984. Commodity system analysis. In H. K. Schwarzweller (ed), *Research in Rural Sociology and Development*. Greenwich, CT: JAI Press, 221–35.

Friedland, W. 1994. The new globalization: The case of fresh produce. In A. Bonanno et al. (eds), *From Columbus to ConAgra: The Globalization of Agriculture and Food*. Lawrence: University Press of Kansas, 210–31.

Friedmann, H. 1993. The political economy of food: A global crisis. *New Left Review*, 197, 29–57.

Friedmann, H. and McMichael, P. 1989. Agriculture and the state system. *Sociologia Ruralis*, 29, 93–117.

Goodman, D. 1991. Some recent tendencies in the industrial reorganization of the agri-food system. In W. H. Friedland et al. (eds), *Toward a New Political Economy of Agriculture*. Boulder: Westview Press, 37–64.

Goodman, D. and Redclift, M. 1985. Capitalism, petty commodity production, and the farm enterprise. *Sociologia Ruralis*, 25, 231–47.

Goodman, D. and Watts, M. 1994. Reconfiguring the rural or fording the divide: Capitalist restructuring and the global agro-food system. *Journal of Peasant Studies*, 22, 1–49.

Goodman, D., Sorj, B. and Wilkinson, J. 1987. *From Farming to Biotechnology: A Theory of Agro-Industrial Development*. Oxford: Basil Blackwell.

Gouveia, L. 1997. Reopening totalities: Venezuela's restructuring and the globalization debate. In D. Goodman and M. Watts (eds), *Globalising Food*. London: Routledge, 305–23.

Grigg, D. 1995. *An Introduction to Agricultural Geography*, second edition. London: Routledge.

Hart, G. 1997. Multiple trajectories of rural industrialisation. In D. Goodman and M. Watts (eds), *Globalising Food*. London: Routledge, 56–78.

Kautsky, K. 1988 [1899]. *The Agrarian Question*. London: Zwan.

Kenney, M., Lobao, L., Curry, J., and Goe, R. W. 1989. Midwestern agriculture in U.S. Fordism. *Sociologia Ruralis*, 29, 131–48.

Kloppenburg, J. 1988. *First the Seed*. Cambridge: Cambridge University Press.

Le Heron, R. 1988. *Globalized Agriculture: Political Choice*. Oxford: Pergamon Press.

Lenin, V. I. 1956 [1899]. *The Development of Capitalism in Russia*. Moscow: Foreign Languages Publishing House.

Mann, S. A. 1990. *Agrarian Capitalism in Theory and Practice*. Chapel Hill: University of North Carolina Press.

Mann, S. A. and Dickinson, J. M. 1978. Obstacles to the development of a capitalist agriculture. *The Journal of Peasant Studies*, 5, 466–81.

Marsden, T. 1998. Agriculture beyond the treadmill: Issues for policy, theory and research practice. *Progress in Human Geography*, 22, 265–75.

Marsden, T. and Wrigley, N. 1995. Regulation, retailing, and consumption. *Environment and Planning A*, 27, 1899–1912.

Marsden, T., Munton, R., Whatmore, S., and Little, J. 1986. Toward a political economy of capitalist agriculture: A British perspective. *International Journal of Urban and Regional Research*, 10, 498–521.

Massey, D. 1984. *Spatial Divisions of Labor: Social Structures and the Geography of Production*. New York: Methuen.

McMichael, P. (ed). 1994. *The Global Restructuring of the Agro Food System*. Ithaca: Cornell University Press.

McMichael, P. and Myhre, D. (1991). Global regulation vs. the nation state: Agro-food systems and the new politics of capital. *Capital and Class*, 43, 83–104.

Newby, H. 1982. Rural sociology in these times. *The American Sociologist*, 17, 60–70.

Page, B. 1996. Across the Great Divide: Agriculture and industrial geography. *Economic Geography*, 72, 376–97.

Page, B. and Walker, R. 1991. From settlement to Fordism: The agro-industrial revolution in the American Midwest. *Economic Geography*, 67, 281–315.

Pritchard, W. 1998. The emerging contours of the third food regime. *Economic Geography*, 74, 64–74.

Roberts, R. 1996. Recasting the agrarian question: The reproduction of family farming on the southern High Plains. *Economic Geography*, 72, 398–415.

Storper, M. 1997. *The Regional World*. New York: Guilford Press.

van der Ploeg, J. 1993. Rural sociology and the new agrarian question: A perspective from the Netherlands. *Sociologia Ruralis*, 33, 240–60.

Walker, R. 1997. Fields of dreams or the best game in town? In D. Goodman and M. Watts (eds), *Globalising Food*. London: Routledge, 273–84.

Ward, N. and Lowe, P. 1997. Field-level bureaucrats and the making of new moral discourses in agri-environmental controversies. In D. Goodman and M. Watts (eds), *Globalising Food*. London: Routledge, 256–72.

Watts, M. 1996. Development III: The global agrofood system and late twentieth-century development (or Kautsky redux). *Progress in Human Geography*, 20, 230–45.

Watts, M. and Goodman, D. 1997. Agrarian questions: Global appetite, local metabolism, nature, culture, and industry in *fin-de-siècle* agro-food systems. In D. Goodman and M. Watts (eds), *Globalising Food*. London: Routledge, 1–32.

Wells, M. 1996. *Strawberry Fields: Politics, Class, and Work in California Agriculture*. Ithaca: Cornell University Press.

Whatmore, S. 1991. *Farming Women: Gender, Work, and Family Enterprise*. London: Macmillan Academic and Professional.

Whatmore, S. 1995. From farming to agri-business: The global agro-food system. In R. J. Johnston, P. J. Taylor, and M. J. Watts (eds), *Geographies of Global Change*. Oxford: Blackwell, 36–49.

Whatmore, S. and Thorne, L. 1997. Nourishing networks: Alternative geographies of food. In D. Goodman and M. Watts (eds), *Globalising Food*. London: Routledge, 287–330.

Chapter 16

Political Ecology

Michael Watts

I want to begin by trawling through this week's newspapers, beginning with the occupation by a militant youth wing of the Istekiri people of a number of Chevron oil flow-stations in the Nigerian Niger Delta. Over the last five years, increasingly militant ethnic minorities throughout the oil-producing Delta have aggressively occupied a number of oil installations operated by transnational petroleum companies in the wake of a growing clamor over the control of local petro-revenues by impoverished oil producing communities, and claims for compensation for the ecological destruction associated with 40 years of commercial drilling and pumping. A second story speaks to the question of environmental cancer, and the Blair government White Paper on public health in Britain. It reports on studies that document the extraordinary rise of assorted cancers (of the breast and prostate most notably) over the last 50 years, and the belated public acknowledgment that "pollutants in the environment may cause cancer" (*Guardian Weekly*, July 15, p. 11). The third item marks the release of the new *Human Development Report* (1999) by the United Nations Development Programme (UNDP). Economic globalization, says the UNDP, is creating a dangerous polarization between haves and have nots but little in the way of regulatory structures to counter the risks and threats of globalization (*New York Times*, July 17, p. 4). Central to the UN agenda is the need for a new multilateral environmental agency to regulate the global commons (for example the seas, ozone, and so on). Finally there is a report on the escalating conflicts between, on the one hand, the Brazilian federal ministry of agriculture and coalitions of regional states (led by the Marxist-oriented Rio Grande do Sul), and on the other, local agro-cooperatives over the potential environmental and social consequences of the widespread introduction into Brazil of genetically modified soy by the Monsanto corporation (*Guardian Weekly*, July 15, p. 16).

Environmental issues of this sort are geographical in two senses. First, they are very much the object of study for the field of *political ecology*, which seeks to understand the complex relations between nature and society through a careful analysis of what one might call the forms of access and control over resources and their implications for environmental health and sustainable livelihoods. And second, they display vividly what geographers call the *politics of scale*. These four events encompass a number of political arenas, from *the body* (the rise of breast cancers in the UK) to the locally imagined *community* (ethnic mobilization around corporate irresponsibility and ecological despoliation) to *state and intra-state* struggles (over Monsanto's first harvest of genetically modified soy) to new forms of *global governance* (multilateral regulation for global environmental problems such as climate change).

Struggles over biotechnology or public health may strike you as wholly commonplace and pedestrian, but it is precisely their quotidian character which marks the extent to which "nature" is now so deeply embedded in late twentieth- and early twenty-first century political identities. As it happens, my "green reading" of the popular press comes at a moment when we are shortly to celebrate the thirtieth birthday of a foundational moment in environmental activism, namely the first Earth Day (1970), and subsequently two years later the United Nations Stockholm Conference on the Environment. But has the politics of the environment changed since these defining moments in the late 1960s and early 1970s? One obvious difference is the enhanced knowledge of, and sensitivity to, trans-border and global forms of environmental harm (ozone depletion, climate change), and the extent to which green issues are legislated through inter-state agreements (the Rio Agenda 21 and the Biodiversity Convention of 1992 for example) and multilateral (inter-governmental) organizations. Indeed, one of the striking trends in the last decade has been the "greening" – with limited success it needs to be said – of multilateral institutions like the World Bank (e.g. the Global Environmental Facility), the World Trade Organization, and regional associations such as the European Union and the North American Free Trade Association (NAFTA).

Another difference turns on the restructuring of global capitalism itself, and quite specifically the profound environmental changes associated with the rapid growth and maturity of the newly industrializing countries (NICs) and the collapse of the socialist bloc. The chickens of rapid industrialization in Brazil and Taiwan, and of 50 years of Stalinist hyper-industrialization in the former soviet sphere, came home to roost in the 1990s. And not least, the deepening of the reach of transnational capital, marked incidentally by the rise of a massive corporate and transnational environmental technology industry (Pratt and Montgomery, 1998), has as its counterpoint a proliferation of social movements which typically link economic and ecological justice (the *politics of distribution*) with human rights and cultural identity (the *politics of recognition*). New social movements can be understood as an effort by national and global civil society – social networks and transnational coalitions – to impose some sort of control over transnational corporations and irresponsible or rogue states, most especially the environmental externalities (toxic dumping) and distributional conflicts generated by the export of industry to the Third World via an increasingly deregulated world economy. The road from Stockholm to Rio is littered, then, with new ecological problems and different ecological politics.

In this chapter, I address the ways in which environmental problems have been addressed in the last 30 years, with a particular attention to the field of political ecology. I want to provide a history of the field – it contains a large body of work, possesses its own electronic journal, and as one might expect of a "mature" science contains substantial debates within its ranks – and also an overview of its conceptual toolkit and its theoretical claims. I want to show how, since its formation in the 1970s, political ecology has been challenged – and deepened – both by "internal" theoretical debates and by the "external" environmental and political economic realities it seeks to explain. What is striking about political ecology in the 1990s is the way in which it has, true to its name, grappled with environmental politics, by way of a broader and more sophisticated sense of the *forms* of political contention and a deeper conception of *what* is contended: what I have elsewhere referred to as a "liberation ecology" (Peet and Watts, 1996). Central to the new political ecology is a sensitivity to environmental politics as a process of cultural mobilization, and the ways in which such cultural practices – whether science, or "traditional" knowledge, or discourses, or risk, or property rights – are contested, fought over, and negotiated.

The Intellectual Origins of Political Ecology

What, then, is political ecology? The origin of the couplet – politics and ecology – is instructive in itself since it dates to the 1970s (Watts 1983b) when a variety of commentators – journalist Alexander Cockburn, anthropologist Eric Wolf, and environmental scientist Grahame Beakhurst – coined the term to think about the ways in which questions of access and control over resources (that is to say the toolkit of political economy) were indispensable for understanding both the forms and geography of environmental disturbance and degradation, and the prospects for green and sustainable alternatives. The fact that such writers were concerned to highlight politics and political economy – that is to say a sensitivity to the dynamics of differing forms of, and conflicts over, accumulation, property rights, and disposition of surplus – reflects a concern to distance themselves from other accounts of the environmental crisis which sought to locate the driving forces in technology, or population growth, or culture, or poor land use practice.

Political ecology's originality and ambition lay in its efforts to integrate human and physical approaches to land degradation, through an explicitly theoretical approach to the ecological crisis capable of addressing diverse circumstances (soil erosion in Nepal, water pollution in Delhi) and capable of accommodating both detailed local studies and general principles. As a defining text puts it: "[T]he phrase 'political ecology' combines the concerns of ecology and a broadly defined political economy. Together this encompasses the constantly shifting dialectic between society and land-based resources, and also within classes and groups within society itself" (Blaikie and Brookfield, 1987, p. 17). Less a problem of poor management, inappropriate technology, or overpopulation, environmental problems were *social* in origin and definition. Analytically, the fulcrum of any nature–society study must be the "land manager" whose relationship to nature must be considered in a "historical, political and economic context" (Blaikie and Brookfield, 1987, p. 239). Hence, rapid deforestation in eastern Amazonia, to take one example, needed to be understood in terms of why those who were clearing tropical rainforests did so in the pursuit of economically inefficient and environmentally destructive cattle ranching, and how these social forces – ranchers, peasants, workers, transnational companies – were shaped by larger political–economic forces, not the least of which was the role of the Brazilian state through subsidies, corruption, class alliances, and its backing of the military. In the first generation of political ecology, however, the land managers were almost wholly male, rural, Third World subjects, and curiously unpolitical in their practices and intentions.

What set of ideas and events "produced," as it were, this welding together of ecology and political economy in the first place? To simplify, one can say that efforts to link culture and environment in anthropology and geography arose in part through a combination of Darwinian or evolutionary thinking, the new sciences of ecosystems and cybernetics, the growing political visibility of Third World peasantries (in China and in Vietnam), and the consequences of the Cold War and the atomic bomb. I shall emphasize a post-1945 confluence between three sets of ideas. First, the important connection between cybernetics and systems theory – which derived from the theory of machines and from artificial intelligence developed particularly during World War II – and community ecology. The central figures here were Gregory Bateson and Howard Odum who, while very different in intellectual orientation, provided languages and concepts for thinking about humans in eco- and living systems, the flows of matter, information, and energy that coursed through human practice with respect to the environment, and also the mechanisms – homeostasis, equilibrium, flexibility – by which "adaptive structure" could be maintained in ecosystems.

Second, within anthropology and geography the twin themes of cultural evolution and cultural materialism provided a powerful Darwinian framework for thinking about not only historical change but also patterns of resource use and human adaptation in different environments. In geography this approach was referred to as cultural ecology but it was the Columbia school of ecological anthropology which provided the most sophisticated ideas. Peter Vayda and Roy Rappaport (1967) in the 1960s showed how tribal subsistence people in isolated regions could maintain an "adaptive structure" with respect to their environment. In Rappaport's (1968) terms the natives' "cognized model" of the environment – embodied in various ritual, symbolic, and religious practices – could elicit adaptive behavior understood in terms of the "operational" model of Western ecology. The pig killing rituals of the Tsembaga Maring of highland Papua New Guinea could function as a cultural thermostat, preventing overpopulation by pigs and maintaining some sort of environmental balance with their fragile ecology. Much of this ecological anthropology of the 1960s sought out the "hidden" adaptive functions of culture with respect to the ecosystem, in order to build an abstract model of adaptive structure which existed in all living systems (see Bateson, 1972; Wilden, 1972).

The third lineage is rooted in the social science of the nuclear age and the post-war development of human responses to hazards and disasters. The immediate threat was of course atomic, and the deepening of the Cold War which produced a number of government-funded studies on the perception of, and responses to, environmental threats. Geographers – Gilbert White, Ian Burton, Robert Kates – were very much part of this work in the 1950s and 1960s, focusing on differing sorts of "natural" perturbations – tornadoes, earthquakes, floods – in the United States, and on the perceptions and behaviors of threatened communities and households. Disaster studies centers appeared around the country and sociologists and geographers schooled in survey research, cognitive studies, and behavioralism sought to understand why individuals misperceived or ignored environmental threats, and how communities responded to, say, the threat of tornadoes versus floods or droughts. By the 1970s Clark University, the University of Colorado, and Ohio State University were centers of hazard or disaster research. Much of this work also drew upon organic analogies – adaptation and response – but was also sensitive to cultural perceptions and questions of organizational capacity and flexibility, and access to information. Systems theory was again central to the intellectual architecture of this body of scholarship (Watts, 1983b).

These three approaches – ecosystems/cybernetics, ecological anthropology/cultural ecology, and natural hazards/disaster research – naturally differed in terms of theoretical approach, points of emphasis and method, and geographical sites, but they defined a ground from which political ecology emerged. What triggered the debate within these fields in the 1970s was a debate over the limits of organic analogies, adaptation, and systems/organization theory. In geography and anthropology the challenge came from two related sources. The first was the proliferation of what one might called peasant studies (Shanin, 1970) in which questions of exploitation, social differentiation, and the role of the market among the Third World rural poor were central. Second, and relatedly, was the growth of Marxism within social sciences, and especially in development studies in a variety of guises (world systems theory, dependency, structural Marxism, and so on) during the late 1960s and 1970s (Bryant, 1998). These two tendencies confronted cultural ecology and ecological anthropology by examining not isolated or subsistence communities in harmony with their physical environment, but rather peasant societies marked by the presence of the markets, deep social inequalities, enduring conflict, and forms of cultural disintegration associated with their integration into a modern world system. Here maladaptation rather than adaptation was the order of the day, in which disequilibrium and positive feedback, rather than balance and community maturity, prevailed. While some geographers tried to understand the development of capitalism within peasant communities in ecological terms (Nietschmann, 1972; Grossman, 1984), political economy provided a different set of questions and answers which had more purchase than the evolutionary and Darwinian toolkit. Marginalization, surplus appropriation, relations of production, and exploitation displaced the old lexicon of self-regulation, adaptation, homeostasis, and system response.

The Political Ecological Toolkit

From its very inception, political ecology never represented a coherent theoretical position for the very good reason that the meanings of ecology and political economy, and indeed politics, were often in question. For Watts (1983a) political economy drew upon a Marxian vision of social relations of production as a dialectical arena of possibility and constraint; for Blaikie and Brookfield (1987) a "broadly defined political economy" (p. 17) meant a concern with effects "on people, as well as on their productive activities, of ongoing changes within society at local and global levels" (p. 21). For Martinez-Alier political economy was synonymous with "economic distributional conflicts" (see Guha and Martinez-Alier, 1998, p. 31).

Notwithstanding this diversity of opinion, the work of Blaikie and Brookfield (1987), and their notion of political ecology (PE), can plausibly be taken as an exemplary formulation of the PE perspective. In their view it contains three essential assumptions. The first is interactive, contradictory, and dialectical: society and land-based resources are mutually causal in such a way that poverty can induce, via poor management, environmental degradation which itself deepens poverty (p. 48). Second, Blaikie and Brookfield argue for regional or spatial accounts of degradation which link, through "chains of explanation" (p. 46), local decisionmakers to spatial variations in environmental structure (stability and resilience as traits of particular ecosystems in particular). Locality studies are, thus, subsumed within multi-layered analyses pitched at a variety of regional scales. And third, land management is framed by "external structures" which in the lexicon of PE means the role of the state (p. 17), the core-periphery model (p. 18), and "almost every element in the world economy" (p. 68).

What then was the political ecology conceptual toolkit? The first is a refined concept of *marginality* in which its political, ecological, and economic aspects may be mutually reinforcing: "land degradation is both a result and a cause of social marginalization" (p. 23). Second, *pressure of production on resources* is transmitted through social relations which impose excessive demands on the environment (see Watts 1983a on the "simple reproduction squeeze"). And third, the inadequacy of environmental data of historical depth linked to a chain of explanation analysis compels a *plural approach*. One must, in short, accept "plural perceptions, plural definitions...and plural rationalities" (p. 16). Implicit here is a sense that one person's profit is another person's toxic dump. While it is not explored in depth by the authors, PE opens the possibility for a serious discussion of how nature and environmental problems are represented and the discursive formations which shape policy and practice (Peet and Watts, 1996).

Collectively this body of work has undermined the Malthusian idea that the "pressure of population on resources" causes environmental collapse, and also challenged the idea that distorted or unfree markets, or poor local management by farmers or regulators, generate environmental degradation (see Little and Horowitz, 1987). Rather, relations of poverty and wealth are a major cause of ecological deterioration (Martinez-Alier, 1989). Political ecology has the great merit of focusing on the social relations that shape practice, and in its sympathy with the poor and exploited it addresses the plight of the vulnerable: both their abilities (local knowledges and practices, see Richards, 1985; Zimmerer, 1996), and their constraints (how relations of production make degradation situationally rational). How persons or households are politically and economically vulnerable is a central analytical device in both the work of Blaikie (1985) on why knowledgeable Nepalese peasants mine the soil and in Watts' (1986) account of how herders in West Africa were unable to manage their rangelands. During the 1980s the focus of PE has been largely Third World and "peasant," in which the land manager figures centrally. Curiously the majority of the world's population – those who live in cities and increasingly in the mega-cities, and without direct access to land – have been studiously ignored.

Deepening Political Ecologies

Like any other field of study, political ecology has been an object of debate and contention (see Vayda and Walters, 1999; Escobar, 1999). For purposes of brevity I shall simply take note of four issues. The first is the uneven way in which politics was treated within PE. Analysis focused on how marginalization or production pressure caused soil erosion but less on how peasants struggled and fought over those conditions and how such struggles shaped environmental outcomes and attitudes. An exploration of the ways in which the environment appeared in various political arenas – the household, the workplace, the state – was largely missing. A mapping of the variety of environmental politics and movements – and its relationship to theory – had to await the subsequent work of Guha and Martinez-Alier (1998), and Harvey (1996). Second, the weak specification of political economy – and in PE its vague reference to "exogenous" forces and chains of causation – often produced studies that did not explore such key areas as property rights, the politics of markets, and forms of class power which are central to the materialist basis of environmental problems. In this sense, as Bryant (1998) has noted, perhaps PE was not materialist enough, a deficiency that has subsequently been addressed in the field of "ecological Marxism" (see O'Connor, 1999). Third, while PE and earlier ecological anthropology raised questions of perception and cognition, almost no attention was given to the social constructedness of environment and environmental issues by a panoply of actors (the farmer, the scientist, the regulator, the politician, and so on). One effect of not taking discourses seriously was that it left ecology as an unproblematic category (an arena of "natural laws"). It was here that poststructuralism was to have an impact in the 1990s (Soper, 1995).

As a response to this internal critique, political ecology has moved forward substantially in the last 15 years along a number of key fronts which for convenience I shall discuss under two headings: knowledge, power and practice; and politics, justice and governance.

Knowledge, power and practice

Underlying this new work on knowledge is the recognition that any sophisticated political ecology must contain a phenomenology of nature. That is to say it must take seriously Blaikie's (1985) point that the environmental problem can be "perceived" in a variety of ways. The newer political ecology, however, draws from poststructuralism's concern with knowledge, power, and discourse (see Peet and Watts, 1996). Much of this newer scholarship turns especially on what individuals and groups (and *de facto* communities) know and practice with respect to their local environments (so-called indigenous technical knowledge (ITK) which harkens back to earlier studies of ethnobotany). Perhaps the best political ecological study that addresses the question of peasant experimentation and practice, and the threats which this world confronts, is Zimmerer's book, *Changing Fortunes* (1996) which examines biodiversity and peasant livelihoods in the Peruvian Andes. ITK has been widely explored (and there are a number of international organizations devoted to its generation, propagation, and use) and now widely understood within academic and activist circles (Richards, 1985; Brush, 1996). In problematizing environmental knowledges, political ecology has identified a number of core issues. First, a recognition that environmental knowledge is unevenly distributed *within* local societies; second, that it is not necessarily right or best just because it exists (i.e. it can be often wrong or inappropriate); and third, that traditional or indigenous knowledge may often be of relatively recent invention (which is to say these knowledges are not static or stable but, as Paul Richards (1985) suggests, may be predicated on forms of experimentation). Indeed, it may not be indigenous as such but really is *hybrid* (see Gupta, 1998; Aggrawal, 1999). Indigenous knowledge is of course a tricky idea because most knowledges are not simply local but complex hybrids drawing upon all manner of knowledges – farmers in India may simultaneously employ concepts from Hindu religion and modern Green Revolution technologies. ITK can also take on mystical and ideological forms as in Vandana Shiva's account (1989) of Indian women as "natural" peasant scientists. Insofar as local actors know a great deal about local ecology and that this knowing is typically culturally "institutionalized" and "embedded" in a variety of persons, offices, rituals, and customary practices, the questions become: (i) why has this knowledge been so difficult to legitimate, (ii) under what circumstances can such knowledge/practice be institutionalized without co-optation or subversion, and (iii) how might it be systematized in some way?

Candace Slater's (1994) excellent work on Amazonia reveals another aspect of the knowledge question focusing on how there is a popular imagery of the region (perhaps transnational in appeal), and how this imagery is constructed or made, and how literary, media, and other cultural machinery contributes to what I have elsewhere called a "discursive ecological formation" (Peet and Watts, 1996; see also Guthman, 1997). In her account, the Edenic or naturalized narrative always silences (the Indians have no voice or no voice of their own), and these tropes exclude or distort. Slater ends with the provocation that there is an absence of competing images of Amazonia. True (perhaps?) but under what conditions can competing images *really* compete? In a quite different context, Kuletz's (1997) account of the nuclear damage to the US West also turns on how the landscape is constructed, in part by science, as "worthy" of being subject to nuclear attack (its desirability for the state was its "undesirability" and of course the invisibility of Native American communities). But do these images and constructions of landscape really have the power and effect implicit in these accounts of narratives? Are they "just" images and irrelevant to the hard edges of political economy and environmental destruction?

Another approach to environmental knowledge production targets environmental science and policymaking through the work on epistemic communities, or communities of expertise. Here the knowledge is Western science, and more properly the cosmopolitan scientist-expert-policy maker. Peter Haas (1990) has argued in the context of understanding regional (European Union) and global (multilateral) conventions that the process of consensus building and collective action more generally is *knowledge-based* and *interpretive*. That is to say, international regulatory co-operation is fueled by fundamental scientific uncertainty about the environment which ensures that governments seek out authoritative advisors (experts) who, to the extent that they are part of epistemic communities, are more important to the political solution than the content of the ideas *per se*. Cross-national differences in state behavior are determined by the variation in the penetration and institutionalization of experts (epistemic communities). Biodiversity and stratospheric ozone co-operation are seen in this way as instances of the cognitive and bureaucratic power of scientific experts. This is an argument that has also been made for NAFTA by Benton (1996) who argues that the trade and environmental constituencies brought together around tariff reduction actually created a dialogue – a transnational community of experts – which had not hitherto existed.

The epistemic community idea is not unrelated to new political ecological work that examines particular scientific–policy discourses, "conventional models" as Leach et al. (1997) call them, rooted in particular institutions and practices, which become hegemonic and are then subsequently contested. Some of the most interesting research has examined the politics of colonial and post-colonial conservation. For example, work in Africa has traced debates over soil erosion and land conservation in the 1930s to the complex political struggles among and between the colonial state, white settlers, and the Native Reserves (Mackenzie, 1997). Neumann's excellent book (1999), *Imposing Wilderness*, on the creation of the Arusha National Park in Tanzania and the ideas of landscape and nature which lay behind state appropriations of land from local peasant communities is an especially compelling illustration of how cultural and historical representations of nature intersect with colonial and post-colonial rule. Fairchild and Leach's (1996) reinterpretation of the forest–savanna mosaic is a careful deconstruction of a conventional model in which historical studies coupled with detailed local analysis of agro-ecology confirm what the new "non-equilibrium" ecology posits, namely that climax models of ecological stasis are unhelpful. These static models however do enter into administrative practice (colonial and post-colonial) which reinforces the idea of Guinea's forest cover as "relic" (which Fairchild and Leach see as the basis for driving "repressive policies designed to reform local land use practice") rather than as the outcome of intentional local management practice. Similarly, Swift (1997) has shown how the assumptions about desertification not only rest on remarkably sparce evidence but on questionable models of the dynamics of semi-arid rangelands – their resiliency and stability in other words – which are (i) expressions of linear, cybernetic models of ecological structure and temporalities, and (ii) are attached to neo-Malthusian models of social change. The key here is that conventional wisdom is challenged as an embodied form of knowledge, and the challenge itself reflects a peculiar unity of local knowledge and practice with non-linear models of new ecology. There emerges a concern with pluralism (at the level of truth claims), with democracy (to open up the practices of policymaking to other voices), and complexity/flexibility (of local conditions and historical dynamics).

It remains an open question whether these epistemic communities or conventional models have real power, in contrast to power politics approaches in which inter-state rivalries dominate or indeed, as Raustiala (1997) has shown in examining the differences between Britain and USA at the Biodiversity Convention, whether it is domestic regulatory and political structures and the differing influence of business, not scientists and experts, that matter. The epistemic community is nonetheless especially relevant to the "greening" of multilateral organizations. Kingsbury (1994) shows in his account of the incorporation of environmental issues into the WTO and trade debates, that the process of greening these institutions has only just begun. Robert Wade's (1997) work on the greening of the World Bank and McAfee's (1999) work on what she calls "green developmentalism" show precisely how discourses (like gender and development) are institutionalized in quite specific ways with quite specific institutional powers. Of course much of this discourse turns on how the idea that nature has to be sold to be saved is constructed and legitimated.

Politics, justice and governance: toward an ecological democracy

Political ecology's concern with knowledge, representation, and imagery addresses politics – the politics of knowledge. But politics of another sort, what one might call "ecological democracy," has been addressed by political ecologists explicitly in a number of ways in the 1980s and 1990s (Zerner, 1999). I shall focus on three aspects: gender and resistance, community and governance, and entitlements. Perhaps the most influential studies have been those that have linked questions of cultural studies and everyday resistance with gender. Nancy Peluso's pathbreaking political ecological study of timber and forestry in Indonesia, *Rich Forests, Poor People* (1992a), showed how local communities resisted the incursions of the state, and how the state in turn attempted to "criminalize" local customary rights over access to, and control over, local forest products. Politics, community, and state, were also central to Hecht and Cockburn's (1989) account of Amazonian deforestation in which state subsidies and powerful ranchers and timber companies were key to understanding the dynamics of the frontier violence, and, in turn, relevant to understanding the panoply of social movements (Chico Mendes most visibly) – often with links to left-wing political parties and transnational green NGOs – which resisted loss of local autonomy. The state figures centrally in these accounts: as an instrument through which conservation takes on a coercive or military cast (Peluso, 1992b), and as the means by which land becomes a geostrategic matter (for example, the Brazilian military government accelerating deforestation to "secure" the country's borders).

"Feminist" political ecology (Rocheleau et al., 1996) explores the ways in which environmental concerns are traced through gender roles, knowledges, and practices. Perhaps the most compelling work is drawn from Africa. Mackenzie's book (1997) traces both the erosion – what she calls the "silencing" – of women's environmental knowledge in central Kenya after 1890, and the ways in which women organized and struggled to resist the impact of colonial conservation on their economic liberty, not least through male appropriation of property rights. Richard Schroeder's book, *Shady Practices*, (1999) focuses on the ways in which efforts to create sustainable development projects in drought-prone Gambia – local forest and fruit tree projects – precipitated struggles within the household and often over the obligations and reciprocities of conjugality. Local "traditional" women's work groups become the vehicle for local protest as resistance to male claims over property and access rights spills into a larger public domain.

The community looms large in the new political ecology of the 1990s. But the community turns out to be – along with its lexical affines, namely tradition, custom, and indigenous – a sort of keyword whose meanings (always unstable and contested) are wrapped up in complex ways with the problems it is used to discuss. The community is important because it is typically seen as: a locus of *knowledge*, a site of *regulation* and management, a source of *identity* (a repository of "tradition"), an *institutional nexus* of power, authority, governance, and accountability, an object of *state control*, and a theater of *resistance* and struggle (of social movement, and potentially of *alternate visions of development*). It is often invoked as a unity, as an undifferentiated entity with intrinsic powers, which speaks with a single voice to the state, to transnational NGOs or the World Court. Communities, of course, are nothing of the sort.

One of the problems is that the community expresses quite different sorts of social relations and forms: from a nomadic band to a sedentary village to a confederation of Indians to an entire ethnic group. It is usually assumed to be the natural embodiment of "the local" – configurations of households, lineages, longhouses – which has some territorial control over resources which are historically and culturally constructed in distinctively local ways. A community, then, typically involves a territorialization of history ("this is our land and our resources which can be traced in relation to these founding events"), and a naturalized history ("history becomes the history of my people and not of our relations to others"). Communities fabricate, and refabricate through their unique histories, the claims they take to be naturally and self-evidently their own. This is why communities have to be understood in terms of hegemonies: not everyone participates or benefits equally in the construction and reproduction of communities, or from the claims made in the name of community interest. And this is exactly what is at stake in the current political ecological work on the infamous tree-hugging or Chipko movement in northern India (see Rangan, 1995; and Sinha et al., 1997). Far from the mythic community of tree-hugging, unified, undifferentiated women articulating alternative subaltern knowledges for an alternative development – forest protection and conservation by women in defense of customary rights against timber extraction – we have three or four Chipkos each standing in a quite different relationship to development, modernity, sustainability, the state, and local management. It was a movement with a long history of market involvement, of links to other political organizations in Garawhal, and with aspirations for regional autonomy. Tradition or custom hardly captured what is at stake in the definition of the community.

The community-politics focus has also been central to the work – largely based in the advanced capitalist states – on economic justice, particularly in regard to toxic dumping and hazardous exposure in minority and working class communities. Pulido's book *Environmentalism and Economic Justice* (1996) is an excellent example of how a sensitivity to community struggles over environment and health meets up with larger claims over economic justice and class politics. These sorts of movements were in no simple sense "environmental" since they typically combined human rights, ethnicity/identity, and questions of social justice (Escobar, 1995).

A number of implications stem from the community and justice approaches addressed by political ecologists. First, and most obviously, the forms of community regulations and access to resources are invariably wrapped up in questions of identity. Second, these forms of identity (articulated in the name of custom and tradition) are not stable (their histories are often shallow), and may be put to use (they are interpreted and contested) by particular constituencies with particular interests. Third, images of the community, whether articulated locally or nationally, can be put into service as a way of talking about, debating, and contesting various forms of property (and therefore claims over control and access). Fourth, to the extent that communities can be understood as differing fields of power – communities are internally differentiated in complex political, social, and economic ways – then to that same extent we need to be sensitive to the internal political forms of resource use or conservation (there may be three or four different Chipkos or Love

Canal movements within this purportedly community struggle). Fifth, communities are rarely corporate or isolated which means that the fields of power are typically non-local in some way (ecotourism working through local chiefs, local elites in the pay of the state or local logging companies, and so on). And not least, the community – as an object of social scientific analysis or of practical politics – has to be rendered politically; it needs to be understood in ethnographic terms as consisting of multiple and contradictory constituencies and alliances (Li, 1999, 1996; Moore, 1999). This can be referred to as identifying "stakeholders" – a curiously anemic term – but often what is at stake is something that comes close to class analysis or at the least the identification of wildly different forms of political power and authority.

Kingsbury (1998) has shown beautifully how the contested nature of the community has its counterpoint in international environmental law over the cover term "indigenous" (and one might as well add "tribe" or "ethnicity"). The UN, the ILO and the World Bank have, as he shows, differing approaches to the definition of indigenous peoples. The complexity of legal debate raised around the category is reflected in the vast panoply of national, international, and inter-state institutional mechanisms deployed, and the ongoing debates over the three key criteria of non-dominance, special connections with land/territory, and continuity based on historical priority. These criteria obviously strike to the heart of the community debate which I have just outlined, and carry the additional problems of the normative claims which stem from them (rights of indigenous peoples, rights of individual members of such groups, and the duties and obligations of states). Whatever the current institutional problems of dealing with the claims of non-state groups at the international level (and there are knotty legal problems as Kingsbury (1998) demonstrates) the very fact of the complexity of issues surrounding "the indigenous community" makes for at the very least what Kingsbury calls "a flexible approach to definition," and at worst a litigious nightmare.

Inevitably, in its concern with the community and environmental politics, political ecology has turned to institutions as a necessary starting point to linking socially differentiated communities with biologically differentiated environments. Institutions – understood not simply as the "rules of the game" but as the habituated and regularized "rules-in-use" maintained by human practice and investment performed over time – are typically distinguished from organizations understood as actors or players brought together for a particular purpose. One way to approach institutions and their character is through Amartya Sen's (1980, 1990) theory of entitlement. Leach et al. (1997) suggest that "environmental entitlements" provide a way of linking what Sen calls "capabilities" with institutional design and performance. Entitlements refer to effective command over alternative commodity bundles which derive from a person's endowments (i.e. through direct access to land I can command commodities produced on my land). Environmental entitlements can be seen as the "sets of benefits derived from environmental goods and services over which people have *legitimate effective command* and which are instrumental in achieving well-being" (Leach et al., 1997, p. 9, emphasis added). Environmental entitlements are thus a subset of a larger group of entitlements which collectively provide the means by which basic human needs are met and people experience well-being. All of this sounds very abstract but it highlights the means by which

differentiated social actors gain access to and control over resources through institutionalized practices.

What might all this mean for environmental governance and democracy? Ribot's
work (1998) opens up a number of important avenues for analysis. He examines
state institutional arrangements which shape access to and control of fuelwood in
Senegal. In his view the state deploys law as a form of rural control. Local appointed
authorities backed by the state create fictions in which there is no local representation. Community participation is in fact disabled by forms of state intervention –
and in his view by the continuance of the colonial model of rule through "decentralized despotism." Ribot argues that participation without locally accountable
representation is no participation at all. As he has put it (1998, p. 4) "when local
structures have an iota of representativity no powers are devolved to them, and
when local structures have powers they are not representative but rather centrally
controlled." What passes in Mali or Niger or Senegal as community participation is
circumscribed by the continuing power of chiefs backed by state powers, by the lack
of open and free elections, and by the decentralized despotism of postcolonial
regimes. In the case of institutions that involve state–community linkages, it is
influence and prestige, coupled with authority and money, that fundamentally
frame the forms of governance and hence who participates and who benefits.

Finally, governance has been addressed through the role of NGO and civic action
around green concerns (Princen and Finger, 1994). The work of Peter Evans (1996)
on social capital is especially relevant here because his concern with what he calls
public–private synergies speaks to the ways in which multiple institutions of control
and access associated with the state and with civil society, operating at different
scales and levels, operate synergistically. It poses the question sharply of how public
institutions can be coherent, credible, and have organizational integrity and how the
institutions of civil society can engage in accountable ways with the public sphere. In
the case of environmentalism, however, these public–private synergies cross cut
international boundaries and pose difficult questions for both multilateral regulation
and transaction activism. The work of Keck (1995) on NGO activism and community watershed management in Brazil (and her work on the Acre rubber tappers,
which shows how the movement gained power precisely by presenting their interest
as "worker" interests rather than as ethnic or tribal) and Pezzoli's (1998) book on
community activism in Mexico City pay testimony to both the powers and the limits
of local green activism, and to the difficulty of building new forms of public–private
contract. Baviskar's (1995) account of the Narmada Dam movement reveals the
tensions between sustainable development activists and a community attracted to a
collaboration with the state as "tribal" peoples. Along the way, some constituencies
– the migrant laborers – are left out completely.

Transnational advocacy groups (TNGOs) – and transnational environmental
organizations in particular such as World Wide Fund for Nature or Nature Con-
servancy – also highlight questions of governance and institutional politics (Keck
and Sikkink, 1998; Bryant and Bailey, 1997). Brosius (1997) shows how activists
can be guided by self-centered interests of program building which rest on misleading stereotypes of the community, just as Tsing (1999) documents the ways in which
Meratu community leaders play to a "fantasy" of tribal green wisdom to mobilize
international attention. A number of the large TNGOs have themselves been shaped

by the changing political and market-driven winds in the West producing a sort of in-house corporate environmentalism ("green corporatism") within the larger TNGO community. This itself raises the question of how large TNGOs as major donors: (i) change the domestic politics and structure of the local NGO communities in the South, (ii) how foreign and local NGOs actually build political strategy and alliances, and (iii) how social capital is constructed in North–South inter-NGO collaborations. Bailey's (1998) work on the activities of the WWF in Ecuador highlights the tensions between transnational and local NGO green activism – and that there is a necessary unity of interests between North and South environmentalism.

Political Ecology by Any Other Name: New Frontiers and Questions

I want to conclude with a brief discussion of three emerging fields which join up with the new political ecology. In this border zone there is a complex mixing and hybridity of ideas but it is precisely here in the brackish intellectual water that the rigor of political ecology will be tested, and where new ideas and approaches will be hatched. The first, *reflexive modernization*, associated with the work of Ulrick Beck and Anthony Giddens (Beck et al., 1994), and *ecological modernization* associated with Maarten Hajer (1997) (see also Macnaghten and Urry 1998), draw upon a concern with modernity and green discourses, both of which are distinguished by their efforts to link social theory and the environment. The focus is on the self-reflexive qualities of modernization and on the ways in which the ecological costs and consequences of capitalist modernity are built reflexively into modernity itself; that is to say, it is the environmental consequences of modernity and the scientific understanding of them that constitute a defining quality of modernity itself. Discourses of "risk" or "uncertainty," what Rom Harré (Harré et al., 1999) calls "greenspeak," often constitute the powerful languages in which this self-reflexivity is constituted. These approaches often employ linguistic and discursive analysis rooted in social studies of science, and institutional analyses of regulation. Ecological modernization has been overwhelmingly urban and First World in orientation and has the great merit, like political ecology, of focusing on politics. It draws, however, from a heady mix of social studies of science and discursive institutional analysis (Lash et al., 1996), and has the advantage of examining the corporate sector and firm which have been largely neglected by political ecology (see Mol's research on "refinement of contemporary production technologies" (1996)). If political ecology had as its cornerstones vulnerability, marginalization, and access, ecological modernization has risk, uncertainty, and discourse.

A second body of work focuses on the relations between environment, geopolitics, and violence – the field of "environmental security" (see Dalby, 1996). The central ideas – that the environment is the post-Cold War security issue and that environmental change can cause war and violence – have a long pedigree dating at least to Malthus and Hobbes. In the 1960s, the return of apocalyptic views of food shortage, and of oil scarcity a decade later, brought environmental concerns onto a larger Cold War geo-strategic landscape. There is now a substantial industry around environmental security which arose around a nexus of geo-political conditions: namely, the end of the Cold War, the need of overfunded militaries to legitimize their existence in the face of the clamor for the "Peace Dividend," and the emergence of "new" forms of violence often articulated as identity politics (or the "clash of civilizations") within putatively weak or rogue states which represent "threats" (Islamic terrorism, ethnic cleansing) to peace and security. The most rigorous and systematic effort to theorize environmental security, however, has been provided in Homer-Dixon's book *Environment, Scarcity and Violence* (1999). Here, the debt to Malthus is clear and explicit, and the entire argument rests on a more differentiated notion of "scarcity." The essence of his argument is that environmental scarcity (which means, to him, scarcity of renewable resources) has three forms, namely degradation, increased demand, or unequal distribution. The presence of any of these "can contribute to civil violence" especially through "resource capture" (generally by "elites"), and subsequent "ecological marginalization" of vulnerable or disenfranchised people as a result of resource capture (p. 177). The language of this analysis is replete with ecological systems theory of old – interactive effects, adaptability, thresholds, and so on – and contains a simple model of social friction and conflict, but environmental security does raise important geopolitical questions – of violence and mass conflict – on which political ecology has been remarkably silent.

And finally, there is the field of environment and rights. The attraction here is that it compels political ecology to dig further into philosophies of nature (Soper, 1995), and to link this political philosophy to questions of rights. The emerging geography of animal rights, for example, (Wolch and Emel, 1995; Faber, 1998) suggests other ways in which political ecology can deepen its concern with ecology and a broad-based political economy.

These confluences and inter-mixings suggest, as Peter Taylor (1997) says, that "appearances notwithstanding, we are all doing something like political ecology." Mike Davis's account in *Ecology of Fear* (1997) of the environmental foibles of Los Angeles, or Daniel Weiner's (1999) story of the survival of independent scientist-led citizen's movements for nature protection in the Soviet Union from Stalin to Gorbachev, all confirm, for example, a "family resemblance" to political ecology. Like any family there are complex interdependencies, interactions, conflicts, and negotiations among members. And yet these dialectics of ideas reflect precisely the dialectics of nature and society itself.

Acknowledgment

I am especially grateful to Trevor Barnes for his stalwart efforts to whip this chapter into shape.

Bibliography

Aggrawal, A. 1999. *Greener Pastures*. Durham: Duke University Press.
Anderson, J. 1973. Ecological anthropology and anthropological ecology. In J. Honigmann (ed.), *Handbook of Social and Cultural Anthropology*. Chicago: Rand, 179–239.
Bailey, J. 1998. Green corporatism. Unpublished Manuscript, University of California, Berkeley.
Bateson, G. 1972. *Steps toward an Ecology of Mind*. New York: Ballantine.
Baviskar, A. 1995. *In the Belly of the River*. Delhi: Oxford University Press.
Beck, U. 1992. *The Risk Society*. London: Sage.
Beck, U., Giddens, A. and Lash, S. 1994. *Reflexive Modernization*. London: Polity.

Benton, L. 1996. The greening of free trade? *Environment and Planning A*, 28, 2155–77.

Blaikie, P. 1985. *The Political Economy of Soil Erosion*. London: Longman.

Blaikie, P. and Brookfield, H. 1987. *Land Degradation and Society*. London: Methuen.

Blaikie, P. Cannon, T., Davis, I. and Wisner, B. 1994. *At Risk*. London: Routledge.

Brosius, P. 1997. Prior transcripts, divergent paths. *Comparative Studies in Society and History*, 39, 468–510.

Brush, S. 1996. Whose knowledge, whose genes, whose rights? In S. Brush and D. Stabrisky (eds), *Valuing Local Knowledge*. Washington, D.C.: Island Press, 1–24.

Bryant, R. 1998. Power, knowledge and political ecology in the third world. *Progress in Physical Geography*, 22, 79–94.

Bryant, R. and Bailey, S. 1997. *Third World Political Ecology*. London: Routledge.

Burton, I., Kates, R. and White, G. 1978. *Environment as Hazard*. Oxford: Oxford University Press.

Dalby, S. 1996. The environment as geopolitical threat. *Ecumene*, 3, 472–96.

Davis, M. 1997. *Ecology of Fear*. New York: Vintage.

Escobar, A. 1995. *Encountering Development*. Princeton: Princeton University Press.

Escobar, A. 1999. After Nature. *Current Anthropology*, 40, 30.

Evans, P. 1996. Government action, social capital and development. *World Development*, 24, 1119–32.

Faber, D. (ed). 1998. *The Struggle for Ecological Democracy*. New York: Guilford Press.

Fairchild, J. and Leach, M. 1996. *Misreading the African Landscape*. Cambridge: Cambridge University Press.

Fairchild, J. and Leach, M. 1997. *The Lie of the Land*. London: Curry.

Friedmann, J. 1976. Marxist theory and systems of total reproduction. *Critique of Anthropology*, 7, 3–16.

Grossman, L. 1984. *Peasants, Subsistence Ecology and Development in the Highlands of Papua New Guinea*. Princeton: Princeton University Press.

Grove, R. 1995. *Green Imperialism*. Cambridge: Cambridge University Press.

Guha, R. and Gadgill, M. 1995. *Ecology and Equity*. New Delhi: Oxford University Press.

Guha, R. and Martinez-Alier, J. 1998. *Varieties of Environmentalism*. London: Earthscan.

Gupta, A. 1998. *Postmodern Development*. Durham, NC: Duke University Press.

Guthman, J. 1997. Representing crisis. *Development and Change*, 28, 45–69.

Haas, P. 1990. *Saving the Mediterranean*. New York City: Columbia University Press.

Hajer, M. 1997. *The Politics of Environmental Discourse*. London: Clarendon Press.

Harré, R., Brockmeier, J., and Mulhausler, P. 1999. *Greenspeak*. London: Sage.

Harvey, D. 1996. *Justice, Nature and the Geography of Difference*. Oxford: Blackwell.

Hecht, S. 1985. Environment, development and politics. *World Development*, 13, 663–84.

Hecht, S. and Cockburn, A. 1989. *The Fate of the Forest*. London: Verso.

Homer-Dixon, T. 1999. *Environment, Scarcity and Violence*. Princeton: Princeton University Press.

Keck, M. and Sikkink, K. 1998. *Activists Without Borders*. Ithaca: Cornell University Press.

Keck, M. 1995. Social equity and environmental politics in Brazil. *Comparative Politics*, 27, 409–24.

Keil, R. and Fawcett, L. (eds). 1998. *Political Ecology*. London: Routledge.

Kingsbury, B. 1994. Environment and trade. In A. Boyle (ed.), *Environmental Regulation and Growth*. Oxford: Clarendon Press, 45–78.

Kingsbury, B. 1998. Indigenous peoples in international law, *The American Journal of International Law*, 92, 414–57.

Kingsbury, B. 2000. The international concept of indigenous peoples in Asia, forthcoming in D. Bell and J. Bauer (eds), *Human Rights and Economic Development in East Asia*.

Kuletz, V. 1997. *The Tainted Desert*. London: Routledge.

Lash, S. Szerszyuski, B. and Wynne, B. 1996. *Risk, Environment and Modernity*. London: Sage.

Leach, M. and Mearns, R. 1997. Challenging the received wisdom in Africa. In J. Fairhead and M. Leach (eds), *The Lie of the Land*. London: Curry, 1–33.

Leach, M., Mearns, R., and Scoones, R. 1997. *Environmental Entitlements*. IDS Working Paper, University of Sussex.

Li, T. 1996. Images of Community. *Development and Change*, 27, 501–27.

Li, T. (ed). 1999. *Transforming the Indonesian Uplands*. London: Harwood.

Linkenbach, A. 1994. Ecological movements and the critique of development. *Thesis Eleven*, 39, 63–85.

Little, P. and Horowitz, M. (eds). 1987. *Lands at Risk in the Third World: Local Level Perspectives*. Boulder: Westview.

Mackenzie, F. 1997. *Land, Ecology and Resistance in Kenya*. London: IAI.

Martinez-Alier, J. 1989. Poverty as a cause of environmental degradation. Unpublished Manuscript prepared for The World Bank (Latin American Division), Washington, D.C.

Martinez-Alier, J. and Schulpmann, 1987. *Ecological Economics*. Oxford: Blackwell.

McAfee, K. 1999. Selling nature to save it? *Environment and Planning D: Society and Space*, 7, 155–74.

Macnaughten, P. and Urry, J. 1998. *Contested Natures*. London: Sage.

Mol, A. 1996. *The Refinement of Production*. Utrecht: Van Arkel.

Moore, D. 1999. The crucible of cultural politics. *American Ethnologist*, 26, 3, 1–36.

Neumann, R. 1999. *Imposing Wilderness*. Berkeley: University of California Press.

Nietschmann, B. 1972. *Between Land and Water*. New York: Academic Press.

O'Connor, J. 1999. *Natural Causes*. New York: Guilford.

Peet, R. and Watts, M. 1996. *Liberation Ecologies*. London: Routledge.

Peluso, N. 1992a. *Rich Forests, Poor People*. Berkeley: University of California Press.

Peluso, N. 1992b. Coercing conservation? The politics of state resource control. *Global Environmental Change*, 3, 199–217.

Peluso, N. 1996. Fruit trees and family trees in an anthropogenic forest. *Comparative Studies in Society and History*, 38, 510–48.

Pezzoli, K. 1998. *Human Settlements and Planning for Ecological Sustainability*. Cambridge, MA: MIT Press.

Pratt, L. and Montgomery, W. 1998. Green imperialism. *Socialist Register 1997*. London: Monthly Review Press, 75–95.

Princen, T. and Finger, M. 1994. *Environmental NGOs in World Politics*. London: Routledge.

Pulido, L. 1996. *Environmentalism and Economic Justice*. Tucson: University of Arizona Press.

Rangan, P. 1995. Contesting boundaries. *Antipode*, 27, 343–62.

Rappaport, R. 1968. *Pigs for the Ancestors*. New Haven: Yale University Press.

Raustiala, K. 1997. Domestic institutions and international regulatory cooperation. *World Politics*, 49, 482–509.

Reichel Dolmatoff, G. 1972. *Amazonian Cosmos*. Chicago: Aldine.

Richards, P. 1985. *Indigenous Agricultural Revolution*. London: Hutchinson.

Ribot, J. 1998. Theorizing Access. *Development and Change*, 29, 307–41.

Rocheleau, D., Thomas-Slayter, B., and Wangari, E. 1996. *Feminist Political Ecology*. New York: Routledge.

Schroeder, R. 1999. *Shady Practices*. Berkeley: University of California Press.

Sen, A. 1980. *Poverty and Famines*. Oxford: Clarendon Press.

Sen, A. 1990. Food, economics and entitlements. In J. Dreze and A. Sen (eds), *The Political Economy of Hunger*, Vol. 1. London: Clarendon, 34–50.

Shanin, T. (ed). 1970. *Peasants and Peasant Societies*. London: Penguin.

Shiva, V. 1989. *Staying Alive*. London: Zed Press.

Sinha, S., Gururani, S., and Greenberg, G. 1997. The new traditionalist discourse of Indian environmentalism. *Journal of Peasant Studies*, 24, 65–99.

Slater, C. 1994. *The Dance of the Dolphin*. Chicago: University of Chicago Press.

Soper, E. 1995. *The Problem of Nature*. Oxford: Blackwell.

Swift, J. 1997. Narratives, winners and losers. In J. Fairhead and M. Leach (eds), *The Lie of the Land*. London: Curry, 73–90.

Taylor, P. 1997. Notwithstanding appearances, we are all doing something like political ecology. *Social Epistemology*, 11, 111–27.

Tsing, A. 1999. Becoming a tribal elder and other green development fantasies. In T. Li (ed.), *Transformation of the Indonesian Uplands*. London: Harwood, 159–202.

UNDP 1999. *Human Development Report 1999*. New York: Oxford University Press, for the United Nations Development Programme.

Vayda, P. and Rappaport, R. 1967. Ecology, cultural and non-cultural. In J. Clifton (ed.), *Introduction to Cultural Anthropology*. Boston: Houghton & Mifflin, 477–97.

Vayda, P. and Walters, B. 1999. Against Political Ecology. *Human Ecology*, 27, 1–18.

Wade, R. 1997. Greening the Bank. In R. Kanbur, J. Lewis and R. Webb (eds), *The World Bank*. Washington, DC: Brookings Institution, 611–734.

Watts, M. 1983a. *Silent Violence*. Berkeley: University of California Press.

Watts, M. 1983b. The poverty of theory. In K. Hewitt (ed.), *Interpretations of Calamity*. London: Allen & Unwin, 231–62.

Watts, M. 1986. Drought, environment and food security. In M. Glantz (ed.), *Drought and Hunger in Africa*. Cambridge: Cambridge University Press, 171–212.

Weiner, D. 1999. *A Little Corner of Freedom*. Berkeley: University of California Press.

Wilden, A. 1972. *System and Structure*. London: Tavistock.

Wolch, J. and Emel, J. 1995. *Animal Geographies*. London: Verso.

Zerner, C. 1996. Telling stories about biological diversity. In S. Brush and D. Stabinsky (eds), *Valuing Local Knowledge*. Washington DC: Island Press, 68–101.

Zerner, C. (ed). 1999. *People, Plants and Justice*. New York: Columbia University Press.

Zimmerer, K. 1996. *Changing Fortunes*. Berkeley: University of California Press.

Chapter 17

The Production of Nature

Noel Castree

It may seem strange to include a chapter on the production of nature in a volume about economic geography.[1] After all, according to common-sense understandings of the term, "nature" is the antithesis of society and thus, by definition, incapable of being "produced" by humans within their economic systems (as opposed to, say, being altered or disturbed). Indeed, as if to confirm this, economic geographers have traditionally had relatively little to say about the question of nature. Although geography has long been concerned with human–environment relations, the postwar division of the discipline into human and physical, divided in turn into various thematic specialisms, compartmentalized geographical inquiry. Economic geographers thus pushed to the margins the putatively "non-economic" and, as Martin (1995) notes in a recent survey, organized their research around the twin themes of industrial location dynamics and processes of uneven development, drawing variously upon the theoretical resources of neoclassical, Keynesian, and Marxian economics. Where nature appeared at all, it was usually as part of (an ongoing) minority interest in the spatial and organizational structure of particular resource industries (see, for example, O'hUallachain and Matthews, 1996, on the topic of copper mining). From the perspective of post-war economic geography, then, questions of nature and environment were best left to physical geographers or else to those human geographers who specialized in resource and environmental management. This subdisciplinary blindness to nature did not, however, only reflect the compartmentalization of geographical research. It also reflected the "eco-blindness" of the economists and the economic theories upon which economic geographers drew for inspiration. The neoclassical, Keynesian, and Marxian approaches had little to say about nature and environment, a characteristic also true of less popular approaches such as institutionalist economics. Accordingly, economic geographers treated the economy as a relatively discrete (even closed) system with its own dynamics, the geographical patterning of which was to be the subject of inquiry.

In recent years, though, this has started to change (how else could I write this chapter?). Since the early 1980s a growing number of economic geographers have put nature at the center of their inquiries. A key reason for this shift of focus is the increasingly obvious impact of existing economic systems on the natural world, notably capitalism, which is arguably the dominant mode of producing goods and services worldwide. Some economic geographers see these systems as largely responsible for a plethora of current environmental problems – from local problems like lake eutrophication in UK farmlands to global problems like greenhouse warming – while others consider them central to some of the new and very deep transformations of nature which, until recently, seemed to be the stuff of science fiction. A good example of this is genetically modified (GM) foods, labeled by British critics as "Frankenfoods" during the public controversy over their growth and sale in 1999, and which involve multinational corporations like Monsanto reconstituting nature down to the genetic level in the interests of profitability.

In the face of these epochal eco-transformations, several economic geographers have developed a provocative approach which regards them as instances of the "production of nature." This approach, as we shall see, is associated with Marxian economics and Marxist geography (see Swyngedouw, this volume) and is part of a broader "political-economy" approach to nature and environment. As the couplet of "production" and "nature" suggests, it opens up the typically closed models of economic geography by insisting that economy and ecology are indissolubly intertwined. Accordingly, this approach regards many current ecological problems and transformations less as human tamperings with an external nature than as the planned and unplanned outcomes of a single (if complex) process of the economic production of nature under the dominant global production system, capitalism. This is not to say that nature is only produced within capitalism (for there are non-capitalist modes of producing nature too), but it is to say that, according to the production of nature approach, capitalism is today the most powerful and transformative economic system in relation to nature and environment. Critical, as well as explanatory, the production of nature approach contests the social and ecological consequences of this capitalism. As such, it seeks a new economic order based upon more socially and ecologically just and sustainable principles. To understand it – and its limitations – we begin by considering one of the dominant understandings of nature within geography, one that the production of nature approach opposes.

From Technocentrism to Ecocentrism: Externalizing Nature

The human use of environment and nature preoccupies government, business, and civil society like never before. This recent wave of eco-concern originated in the late 1960s just before the so-called "long post-war boom" enjoyed by Western and several non-Western countries ground to a halt in the face of the global recession of the early 1970s. Writing in 1966, the economist Kenneth Boulding was among the first to warn of the serious environmental costs of unrestricted economic growth. For him, most nations at that time were running a "cowboy economy" rather than a more sensible "spaceship economy" based on an appreciation of the ecological limits to growth. Over 30 years on, as resource depletion continues, as environmental problems multiply, and as developments like GM foods remake nature "all the way down," the burgeoning environmental movement is divided between a "technocentric" wing and an "ecocentric" wing, each with rather different views on nature and its economic usage by humans. (For a detailed consideration of the two wings see O'Riordan, 1989, 1995). Both wings can be found in geography, especially the former which underpins much environmental geography and resource management. And, as will now be shown, both wings share a conception of nature that the production of nature approach seeks to question and criticize.

Technocentrists put humans first, that is, they are anthropocentric. Though they acknowledge that problems exist in the way people use nature and environment, especially in highly industrialized capitalist countries like the USA or UK, they argue that these problems admit technical and administrative solutions. Indeed, some technocentrists have faith that science and technology can even *improve* nature in the interests of human well-being. In other words, technocentrists embrace the existing economic, political, and social order, and propose only to tinker, rather than to dismantle it, in the interests of better environmental management geared to human needs. Broadly speaking, there are three groups of technocentrists. Neo-Malthusians are the most pessimistic, and argue – following the earlier work of the early nineteenth-century economist, demographer, and reverend Thomas Malthus – that drastic reductions in the levels of population and economic growth are needed to avert eco-catastrophe (see, for example, Meadows and Meadows, 1992). At the other end of the technocentric spectrum are optimists, or Cornucopians, who argue that economic growth is an impetus for technical innovation and the timely exploitation of new resources when resource scarcity threatens (see, for example, Simon, 1997). Finally, somewhere in the middle are (in the practical domain) most environmental and resource managers and (in the academic domain) most environmental and resource geographers. In this middle ground one also finds the majority of those advocating the elusive, if appealing, notion of "sustainable development" (see Redclift, 1991).

By contrast, ecocentrists put nature first and argue for a more harmonious human–nature relationship. Ecocentrists worry that we are currently witnessing *The End of Nature* (McKibben, 1989) at the hands of capitalist "growthmania" (as economist Herman Daly, 1973, p. 151 famously called it) and propose to "save" and "preserve" nature. Most ecocentrists lie outside the political mainstream and, though some enjoy degrees of respectability and support (for instance, Greenpeace or the German Green Party), many ecocentric groups (like fox hunt saboteurs, "tree huggers" and the whale-ship ramming Sea Shepherds) are considered more-or-less extreme opponents of existing ways of life. Indeed, many "deep greens" – like the radical organization Earth First! – argue for the wholescale dismantling of today's industrial, technological, capitalist societies in favor of more eco-friendly, small-scale ways of living.

As we shall see, advocates of the production of nature argument try to combine the environmental sensibilities of ecocentrists with a less anthropocentric version of technocentrism. For the moment what is more interesting is the commonality between the seemingly opposed technocentric and ecocentric worldviews. For, from the perspective of the production of nature argument, what the technocentrist "manage/improve nature" and the ecocentrist "save nature" rhetorics share is the questionable assumption of an *external nature*. In both cases "nature" is invoked as a separate realm which acts as source of authority to legitimate existing or even new economic, social, and environmental arrangements. For instance, neo-Malthusians invoke the notion of fixed "natural limits" to growth in order to call for draconian restrictions on births and consumption levels. Likewise, deep greens complain that certain new technological developments – like GM foods – are "unnatural" and should be opposed on ecological and moral grounds. The economic geographer who has most strenuously articulated the production of nature argument – Neil Smith – argues that this putative externalization of nature amounts to nothing less than an "ideology of nature" (Smith, 1984). It is an "ideology" because not only is it incorrect or false but it actively blinds us to the realities of nature within modern capitalism. It is a pervasive ideology, Smith argues, but one that well serves domin-ant capitalist interests. Its ideological content, he suggests, derives from its delib-erate refusal to acknowledge the reality of a nature fully *internalized* within existing socio-economic relations and processes. Of course, there are still parts of the world where a more or less pristine "first nature" remains, but Smith's point is that today this first nature is increasingly subsumed to an economically produced "second nature."

The Production of Nature

At first sight it may seem odd to argue that nature is produced economically. As (Smith, 1984, xii–xiv) concedes, the notion of a "produced" second nature "sounds . . . quixotic and . . . jars our traditional acceptance of what had hitherto seemed self-evident . . . it defies the conventional, even sacrosanct separation of nature and society, and it does so with such abandon and without shame." However, for Smith the production of nature idea only seems odd because the nature–society dualism underpinning technocentric and ecocentric thinking about the environment has such a powerful ideological grip on our imaginations. In dissolving this dualism, the production of nature approach directs our attention not to how modern societies merely "interact with", "interfere with" or "upset" nature and environment, since each of these terms implies an asymmetrical relationship between two ostensibly separate domains. Instead, it seeks to show that nature and society are "inner-related" *from the very start.*

As noted in my introduction, the production of nature approach focuses mainly on the role of modern capitalism in producing nature and, theoretically and intellec-tually speaking, is closely associated with Marxism. Though the approach was first popularized in economic geography by Smith, its origins can be traced back to an influential essay by the geographer David Harvey (1974), entitled "Population, resources and the ideology of science." Writing in the context of the pervasive neo-Malthusianism of the early 1970s, Harvey rejected the commonly accepted argu-ment that "Over-population arises because of the scarcity of resources available for meeting the subsistence needs of the mass of the population" to insist instead that:

. . . there are too many people in the world because the particular ends we have in view (together with the form of social organization which we have) and the materials available in nature, that we have the will and the way to use, are not sufficient to provide us with those things to which we are accustomed (Harvey, 1974, p. 274).

In this way, Harvey sought to draw attention away from the "limits" supposedly dictated by an intransigent, external nature to suggest, instead, that ecological limits were *relative* to the specific socioeconomic systems in place at any one time and place. He thus showed that food scarcity was rarely absolute and that the dubious notion of "over-population" should be replaced with the Marxist notion of "relative surplus population."

Building on this, a set of other Marxist geographers subsequently tried to flesh out the general implications of Harvey's position (Burgess, 1978; Sayer, 1979; Smith and O'Keefe, 1980). This culminated in Smith's (1984) germinal work *Uneven Development*. Not only did Smith seek to offer a theory of the dominant economic system – capitalism – to which Harvey's essay had only gestured, but he was also the first to talk about the capitalist "production of nature." Following Marx, Smith's theory sets out to explain, from a geographical viewpoint, the functioning of capitalism as a specific mode of economic production. Put simply, capitalist production takes the following form:

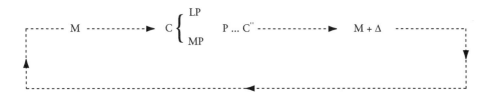

where M (money) is put forward to purchase C (commodities) – namely, MP (means of production: inputs, machines) and LP (labor power) – which are combined in the productive process (P) to produce a new commodity (C′), which is then sold for the original money put forward, plus a profit (Δ). The profit is then reinvested to enable a new round of production. Note that on this reading, capitalist production is processual: it is in motion, not stasis. Note too that "production" refers not merely (and narrowly) to what occurs in the factory, farm, or office, but to the whole, wider *system* of commodity purchase, transformation, distribution, exchange, and sale.

In this system, which Harvey (1985, p. 3) calls "the primary circuit of capital," commodities are produced not simply for their practical – or use – value, but also for their exchange value. After Marx, Smith argued that four cardinal features characterize this system. First, it is inherently growth-orientated: profit, rather than, say, social equity or environmental sustainability, is the primary goal. Second, it is based on competition between capitalists within and between industrial sectors as they fight to sell their products to consumers in regional, national, and world markets. Third, growth and competition set up powerful incentives for individual capitalists to maximize their returns in whatever way possible (e.g. through exploiting new locations or technological innovation). Finally, the origin of profit lies in labor, rather than any other factor of production. The "surplus value" realized at the end of the primary circuit of capital thus originates with laborers, whose work is exploited in the form of surplus labor time over and above that for which they are paid.

Smith innovatively drew out the geographical implications of this capitalist form of production. For him, it embodies opposing tendencies toward the geographical equalization and differentiation of production, or what he calls *uneven development*. Production must always be place-based: it has to occur somewhere. But

Smith's point is that this *differentiated* economic landscape is undermined by an opposed tendency to *equalization* as capitalists eventually look for new locations where they are able to produce more profitably. Uneven development thus emerges as a systematic – rather than incidental – aspect of capitalism.

What has all this to do with nature? At first sight very little. First, as I have said, nature is usually taken to be external to society and so incapable of being produced. Second, nature barely figures in the Marxist theory upon which Smith based his argument. Nonetheless, Smith insisted that Marx's theory is profoundly relevant to nature. Why? Because:

> ... with the progress of capital accumulation ... this material substratum [nature] is more and more the product of social production, and the dominant axes of differentiation are increasingly societal in origin. In short, when th[is] ... immediate appearance of nature is placed in historical context, the development of the material landscape presents itself as a process of the production of nature (Smith, 1984, p. 32).

In other words, notwithstanding Marx's neglect of nature, nature is in fact relevant to his theory from the very start. Smith's point is that once capitalism forges a relation to nature, the latter loses its seeming externality and becomes a socially produced "second nature" which becomes one "moment" within the wider dynamics of capital accumulation. Specifically, the labor process becomes the flashpoint for a socio-natural dialectic in which "nature is mediated through society and society through nature" (ibid., p. 19). The reason is not just because labor is the most *immediate* point of economy–nature interaction. More technically, in Smith's Marxian vision, each specific – or concrete – labor process involving nature (e.g. growing barley in the UK) is measured according to a general – or abstract – labor magnitude which is formed as the average labor time of myriad concrete labors within a given economic sector (e.g. the growth of barley worldwide). In this way, differential labors on different environments in different places become tied to capitalism in general (a local–global dialectic). And in this way, too, specific agricultural and industrial complexes involving the production of nature become subject to the pressures of differentiation and equalization identified by Smith (uneven development).

The "Production" of Nature?

If all this still seems counter-intuitive and confusing, it is perhaps worth saying a little more about the meaning of the term "production" in order to clarify matters further. Production is a portmanteau concept: it carries multiple meanings. It is thus necessary to "unpack" it into its component parts. First, production varies historically and geographically. I mention this again because it is important to distinguish capitalist from *non-capitalist* forms of nature's production. In economic geography, it is the capitalist production of nature that has been the center of attention. Traditional and socialist modes of production produce nature – and resource and environmental problems – very differently, and thus must not be elided with capitalism. Secondly, we can distinguish "weak" from "strong" production. The production of nature argument may seem to imply that nature is infinitely malleable in the hands of capitalism. However, as sectors like agriculture, minerals, and metals show, the material properties of nature remain fundamentally important to the capitalist production system which appropriates them. Thus the weaker version of the production of nature argument simply asserts "that the use of natural substances by people depends upon a whole series of social processes. These include: (1) technologies that alter our capability to use materials, (2) capital investment and disinvestment, (3) markets, (4) transportation, (5) credit institutions, and (6) government...programs" (Roberts and Emel, 1992, p. 260). However, as we approach a new millennium a stronger version of the argument increasingly applies. Today, many nature-based industries have the power to literally reconstitute nature in pursuit of profit. Aside from the example of GM foods already mentioned, one can cite other biotechnological developments such as the use of growth hormones in animal feeds and even, after Dolly the sheep, the cloning of farm animals (see Mannion, 1992). Again, though, this does not deny the material importance of the transformed natural substances. It does, however, point to the remarkable depth of capitalism's production capacities.

Thirdly, it is useful to distinguish capitalist production in the "narrow" and the "extended" sense. I have already referred to this above in passing. Under capitalism, nature is not just produced in the factory or the farm. Rather, what happens in the labor process is directly dependent upon how it is tied to the purchase of inputs, to external markets, to transportation networks, to rival firms, and so on. Production is thus *systemic*. Finally, the production of nature is complex and uneven. At one level this is due to natural differentiation reflecting the myriad different resources and environments on which capitalism draws. However, during the last century this "natural differentiation" has become systematized into patterns of socially produced uneven development by capitalism as different nature-based industries commodify different aspects of nature and seek to sell their products in competitive markets.

As a way of concretizing the rather general, abstract claims of this and the preceding section I now want to present two examples of nature's production under capitalism. Each example highlights different aspects of this production and together both suggest the intellectual and political benefits of thinking about economy–nature relations in a non-dualistic way. Specifically, the first example shows the limits of technocentrist thinking by looking at a common type of environmental problem usually seen as the by-product of economic growth, while the second example shows the limits of ecocentrist thinking by looking at a case (of which there are today more and more) where capitalism materially remakes nature to the point that it is no longer "natural."

Water and Uneven Development in the Southwest USA

The first example is drawn from the work of American economic geographers Roberts and Emel (1992) on irrigated agriculture and groundwater extraction in the post-1945 Southern Plains of Texas and New Mexico. The Plains possess rich soils but are served by unreliable and sparse precipitation. Consequently, in the post-war years the considerable agricultural expansion in the area has depended on irrigation water drawn from the massive Ogallala aquifer which underlies the Plains. Since the early 1970s though, water shortages have resulted from over-use of the aquifer, with water levels falling by as much as 50 meters in some areas. Roberts and Emel show that most water managers have seen these shortages as a "tragedy of the commons" problem. In other words, shortages are seen to stem from the fact that since no one farmer owns the Ogallala water – it is a common or open-access resource – all are free to use it without restriction. Since a given farmer reasons that if s/he does not use the water her/his neighbors will, it is "rational" for him/her to use water without thinking of the overall impacts of this on aquifer levels. The "tragic" result is that in the long term farmers inadvertently harm both themselves and their competitors since water shortages are suffered by almost everyone. In short, what is individually "rational" ultimately proves to be economically and environmentally "irrational."

The "tragedy of the commons" theory is widely used to explain the over-exploitation of natural resources in geography and resource management. The solutions stemming from it entail either privatizing the resources (the assumption being that if someone owns what is presently an unowned resource, like the Ogallala water, they will have a vested interest in conserving it) or getting the state to regulate their use to ensure sustainable exploitation. However, Emel and Roberts argue that there are three problems with these technocentrist explanations and solutions in the Southern Plains case (and the many cases of resource over-use like it worldwide). First, water is seen as a fixed or finite resource which suffers the inadvertent impacts of farmers trying to make a living, as if economy and nature were two realms that exist independently prior to being brought together. Second, water shortages are seen as essentially a property rights problem: it is the lack of property rights in water that is the economic root of the shortages/over-usage. Finally, dealing with water shortages does not entail any fundamental challenge to the structure of agriculture in the Southern Plains but, rather, a slight modification of ownership rules.

Against this, Roberts and Emel seek to put empirical flesh on the theoretical bones of Smith's (1984) argument in *Uneven Development* by seeing Plains water problems as a case of the capitalist production of nature. This may seem odd since water can hardly be "produced" by the farmers in question: it already exists in the Ogallala aquifer. So what do Roberts and Emel mean? To begin, they point out that aquifer water has only become a "resource" in the post-war years when sufficiently powerful mechanized water pumps have been able to tap the Ogallala. Likewise, many of the water shortages from the 1970s onward were not *absolute* but *relative* since they arose from the fact that many farmers could no longer afford to pump deep water because of a sharp rise in fuel prices at the level of the world economy. As Roberts and Emel (1992, p. 60) aptly put it, "The concept of 'natural' resources risks attributing usefulness to this finite substance rather than to a complex interaction between social relations of people and biophysical characteristics." What, then, are the social relations in question? Following Smith, Roberts and Emel argue that looking at property rights is superficial and suggest instead a deeper explanation to do with the form of capitalist production. Given the fact of crops being sold in a competitive market in pursuit of profit, water depletion is for Roberts and Emel best explained by the "cost-price" squeeze that Plains farmers faced from the early 1970s onwards due to competition from cheaper Midwestern farmers, and the sharp increase in fuel prices just mentioned. It was this squeeze, rather than merely a lack of local property rights, which for Roberts and Emel set off a competitive scramble to appropriate Ogallala water and to develop new technological means of doing so. In turn, this competition generated new patterns of uneven agricultural development in which new water-rich areas were farmed until aquifer levels declined, and then still newer areas exploited in a remorseless quest for profitability. For Roberts and Emel, water shortages can only be understood *in relation to* capitalist production, not separate from it. Moreover, because of the environmentally rapacious effects of capitalist production, the ultimate solution to Plains water shortages lies not in allocating property rights but, more fundamentally, in abolishing capitalism altogether (see Castree, 1997, for a similar example of a produced environmental problem).

Producing Food, Privatizing Nature

The second case study is also agriculture-related but is less about an economically produced environmental problem and more about capitalism's increasingly deliberate ability to remake nature "all the way down" in the pursuit of profitability. Where Roberts and Emel examined a case of the "weak" production of nature, in the sense I defined earlier, Jack Kloppenburg's (1998) *First the Seed* investigates an altogether "deeper" production of nature. His book reveals how a set of American agro-foods corporations quite literally "remade" the seeds upon which much US and world agriculture has been based this century. In light of the recent concern over the moral, environmental, and nutritional propriety of developing GM crops and the like, Kloppenburg's analysis stands as a useful counterpoint to ecocentric critiques of these crops which base their arguments on the supposed "unnaturalness" of the foods derived from them. I say this because *First the Seed* shows that GM foods are by no means the first attempts this century to reconstitute nature at the most basic levels in the interests of profitability. For, contrary to popular understandings, many of the foods (such as corn, tomatoes, and soya) that are now being genetically altered through gene-splicing and other new recombinant methods have for a long time been engineered and re-engineered by agro-corporations. In other words, far from being "natural" foods which are only now being "corrupted" by GM technology – as in the ecocentric view – these foods have been "second natural" for decades and part and parcel of twentieth-century capitalism. The point is that the new gene-technologies are only taking a stage further an older, established process of actively producing commercial foods rather than simply "growing" "natural varieties."

First the Seed focuses on the production of hybrid corn in the USA, long a world leader in commercial farming. Grown on a large scale in the USA from the late 1930s onwards, hybrid corn was one of the precursors to the later "Green Revolution" in the developing world and one of the crops responsible for the massive expansion in US food output during the post-war years. However, its impacts extended far beyond food increases to include a set of profound social and environmental transformations which, in turn, stemmed from the commodification of a seemingly innocuous little thing: namely seed. I say "seemingly innocuous" because seed is in fact of central importance to US and world agriculture for obvious reasons. Prior to the 1930s, most seed in the USA was produced by farmers who would collect it from their annual harvest and use it to sow the next year's crop. Where new seeds for new crops were sought after, farmers usually went to the US Government which, through its many agricultural research stations and Land Grant Universities, became the main supplier of alternative – and free – seed types. In the USA, seed was thus a "free good" during the early twentieth century and farmers generally had no need to purchase seed from commercial suppliers. Indeed, prior to the 1930s, this is why there was virtually no commercial seed supply industry in the USA. Until, that is, the discovery of hybrid corn.

Hybrid corn is higher-yielding than non-hybrid corn. This is why within a generation of its invention in the 1930s most US corn farmers had switched to hybrid varieties (by 1965, for example, 95 percent of US corn was hybrid compared to less than 1 percent in the early 1940s, yielding six times more corn per acre than the non-hybrid corn of the 1920s). In its natural, non-hybrid state, corn reproduces by open-pollination, meaning that an ear of corn is fertilized by the pollen of another plant, thus ensuring that a corn population is in a constant state of genetic flux. This flux, prior to the 1930s, constantly confounded efforts to develop and fix superior corn varieties. Hybridization was thus developed as a means of isolating desirable, high-yielding strains of corn. Based on the work of US state-sponsored agricultural researchers, hybridization entailed the controlled cross-breeding of so-called "inbred" corn plants in order to produce seeds that would grow into high-yielding crops. However, while the seeds grew into productive corn plants the seeds those plants produced were, for complex biological reasons, not nearly as productive as the parent seeds.

Though hybrid corn seems to have been an unalloyed good (it was hailed as a "miracle crop" by many), Kloppenburg shows that there was a darker side to the story. Specifically, hybridization became a tool with which new commercial seed companies could produce nature anew in order to privatize seeds and thus create a new market in seed sales where none existed before. First, the fact that hybrid, manufactured varieties produce seed of an inferior quality to their parent seeds became a technical means of dispossessing farmers of their traditional free access to seed for the next year's crop. As Kloppenburg (1988, p. 93) notes, "Hybridisation...uncouples seed as 'seed' from seed as 'grain' and thereby facilitates the transformation of seed from a use-value to an exchange-value. The farmer choosing to use hybrid varieties must purchase a fresh supply of seed each year." Indeed, between 1934 and 1944 a commercial seed supply industry grew up in the USA selling $70 million of hybrid seed-corn. Second, in addition to this way of using "second nature" to extract monies from corn farmers, the new commercial seed companies which grew up in the USA during the 1930s also heavily lobbied the national government in order to obtain patent rights in the invention of new hybrid varieties, beginning with the 1930 Plant Patent Act. As Kloppenburg shows, this was crucial if seed companies were to invest in research into new hybrids and then protect these hybrids from free use by farmers and rival seed companies.

In economic terms the production of hybrid corn was about far more than increased yields, taking a previously on-farm process off-farm, and thereby extracting profit from farmers who had heretofore used their own seed. Environmentally, hybridization also went (and has continued to go) hand-in-hand with a less genetically variable and more monocultural form of agriculture which has left some hybrids susceptible to pests and diseases unless protected by heavy doses of pesticides and herbicides. And, like hybrid seeds, during the post-war years these chemical treatments have been manufactured off-farm by agro-foods multinationals, thus removing a further aspect of farming from farmers' financial control. So it is that

First the Seed shows how the capitalist production of previously "natural" seed varieties became a powerful tool in reconfiguring the economic and ecological make-up of US agriculture (see also Goodman et al., 1987; Goodman and Redclift, 1991).

Intellectual and Political Strengths

The production of nature approach is clearly as rich in its implications as it is provocative in its arguments. Contrary to the technocentric and ecocentric world-views, many environmental problems and transformations of nature under capitalism are seen as produced, not problems and transformations of a separate economy impacting a separate, non-social ecology. This non-dualistic approach possesses some important intellectual and political strengths (see Castree, 1995, p. 19; Smith, 1996, p. 50). First, in disciplinary terms it shows clearly how and why questions of nature and environment ought to be central to economic geography today. Secondly, and less parochially, it reveals the political conservatism or, at best, liberalism of much technocentrism, since technocentrism rarely calls into question the capitalist economic system responsible for so many present day eco-transformations. Third, the production of nature approach also questions the romanticism of much ecocentrism since it shows the fallacy of trying to "save" a "first nature" that is, for the most part, already being reconstituted as a produced "second nature." Fourth, it powerfully historicizes – and thus relativizes – the socioeconomic relation to nature by showing that capitalist production is a phenomenon only of the last one and a half centuries or so and one that is by no means "natural" or inevitable. Fifth, it shows how one production system – capitalism – contains the inner complexity to link local and global produced natures and to generate uneven development between produced natural landscapes. Sixth, it shows the production of nature to have conjoint ecological and social consequences. Specifically, the exploitation of labor and the incidence of uneven development frequently go hand-in-hand with the active production of environmental problems or resource scarcity. Seventh, the production of nature approach shows capitalist production to be a highly political, non-neutral process with questionable socio-ecological outcomes. Finally, the approach opens up the possibility of envisaging a set of post-capitalist natures produced on more socially and ecologically egalitarian and sustainable lines.

Problems, Responses, and Prospects

There are, however, technical, theoretical, ontological, and political problems with the production of nature approach. Technically, technocentrists (like resource managers) and ecocentrists (like Greenpeace and other environmental NGOs) might legitimately complain that, in criticizing capitalism, advocates of the production of nature approach fail to offer any viable economic *solutions* to economically produced environmental/resource problems (Dietz and van der Straaten, 1995) other than the rather unrealistic notion of totally overthrowing capitalism. Theoretically, two problems loom large. First, even sympathetic critics complain that the production of nature approach is *productivist*. That is, it over-emphasizes production at the expense of other processes that simultaneously socialize nature (Braun and Castree,

1998, ch. 1). After all, production is in reality "embedded" in a set of non-economic and non-capitalist relations (Thrift and Olds, 1996). Secondly, far more work needs to be done on translating the abstract logics adumbrated by Smith into meso- and micro-level studies of particular productions of nature in particular places. Aside from the two case studies discussed above, this work is now being pursued in relation to a new "political economy of agriculture" in the developed world, which inserts agro-food production in different locales into wider circuits of global capital (see Brian Page, this volume).

Ontological issues also arise (an ontology is a theory of what exists or what is real). Central here is the suggestion that the production of nature approach is overly *anthropocentric* (and thus, ironically, in this respect similar to the technocentrists it criticizes). The charge has two components: first, that the approach causally prioritizes the capital "side" of the capital–nature dialectic; and second, that it therefore fails to appreciate the ecological and social seriousness of many of the current produced environmental/resource problems found worldwide (such as the currently unknown long-term health and environmental effects of GM foods which are now widely grown in the USA). These criticisms have some validity. After all, the production of nature approach was inspired by Marx's economics and Marx has been shown to be more interested in understanding the human consequences of capitalism than its environmental consequences (Castree, 1995, p. 19). Accordingly, geographers like Smith have looked more at how capitalism produces nature and less at how produced nature affects capitalism. In response, a number of Marxists outside geography have sought, at the theoretical level, to add to Marx's political-economy concepts a set of political-ecology concepts that can help us make sense of the material properties of produced nature (e.g. Altvater, 1993). Additionally, empirical work like Kloppenburg's study shows that not all past Marxian work on the production of nature has ignored nature's physical properties and effects. However, within geography much still needs to be done in this regard (but see Harvey, 1996).

Finally, the production of nature approach is also subject to political problems. One problem is its *Prometheanism* (Soper, 1991) in which, following the general Western Enlightenment view, nature is seen as but an end to human needs or happiness. For ecocentrists the production of nature approach – like Marxism more generally – cannot value nature in its own right (Hayward, 1995, ch. 3). A second problem, more controversially, is that the production of nature approach is *masculinist*. As several feminist critics have pointed out, Marxism's complicity in the Enlightenment question to "master" and control nature – albeit in a non-capitalist form – raises questions about its gender biases and subtexts. As a concept, nature has long been feminized in Western discourses, as a domain to be "conquered," "tamed," and "subdued." By prioritizing paid (predominantly male?) – as opposed to unpaid, domestic (female?) – labor as the force driving nature's production under capitalism, the production of nature approach may inadvertently perpetuate a social and environmental politics blind to women's unequal place in capitalist societies (Plumwood, 1994).

Production of nature advocates concede many of these criticisms (Castree, 1997; Smith, 1998). However – and this is an important point – they would insist that it is necessary and (in principle at least) possible to be *anthropomorphic* without being anthropocentric and Promethean on the one side or, on the other side, ecocentric. Against "greens" who wish to value nature "in its own right" and technocentrists who put humans first, the production of nature approach implies that it is possible to *balance* human and ecological needs by recognizing that all appraisals of nature and what to do with it *are made by humans in the first place* (under capitalism or any other economic system). This anthropomorphic insight avoids the myth of any "return to nature" in itself without necessarily lapsing into the human-centered arrogance (anthropocentrism) of the technocentrics, since the argument is that while it is only possible to value nature in human terms – we simply cannot know what a non-human valuation of nature would look like since we cannot step outside our humanity – this fact does not preclude a more sustainable production of nature in which the environment is respected. And, from a Marxist viewpoint, a more sustainable production of nature will be a post-capitalist one in which unbridled economic growth and the exploitation of working people is a thing of the past.

Criticisms and responses aside, what of the future? Marxian theory has, of late, become less popular in human geography in general and economic geography in particular. Nonetheless, two notable developments promise to further disciplinary interest in the capitalist production of nature and to sustain the relevance of Marxian political ecology into the twenty-first century. The first is the agro-foods and biotechnology industries to which I have referred several times in this chapter. Gene-splicing and other new biotechnologies now promise to take a step further the capitalist production of myriad plant and animal natures as we enter a new millennium. As we have already seen in the UK controversy over GM foods, the social and ecological consequences of this intensified production of agrarian natures is likely to be profound and contested. Secondly, and more worrying perhaps, modern science and industry now collude to remake even the human body through genetic engineering, transplantations, and prosthetics. Is the production of *bodily natures* to become a crucial intellectual and political issue for the twenty-first century? (on this see the recent essays in *Society and Space*, 1998). And, if so, will the corporeal become a serious focus of intellectual and political concern for economic geographers in the years to come?

Endnote

1. In this essay I use the word "nature" unproblematically as, for the most part, a synonym for environment (although towards the end of the essay I extend the term to include the human body). In so doing I am deliberately side-stepping the complex task of defining nature, and refer readers who seek such a definition to Castree (2000).

Bibliography

Altvater, E. 1993. *The Future of the Market*. London: Verso.
Boulding, K. 1966. The economics of the coming spaceship earth. In H. Jarrett (ed), *Environmental Quality in a Growing Economy*. Baltimore: Johns Hopkins University Press, 3–14.
Braun, B. and Castree, N. (eds). 1998. *Remaking Reality*. London: Routledge.
Burgess, R. 1978. The concept of nature in geography and Marxism. *Antipode*, 10, 1–11.
Castree, N. 1995. The nature of produced nature. *Antipode*, 27, 12–47.

Castree, N. 1997. Nature, economy and the cultural politics of theory: The "war against the seals" in the Bering Sea, 1870–1911. *Geoforum*, 28, 1–20.

Castree, N. 2000. Nature. In R. J. Johnston, D. Gregory, G. Pratt and M. Watts (eds), *The Dictionary of Human Geography*, 4th edition. Oxford: Blackwell, 537–40.

Daly, H. 1973. The steady state economy. In H. Daly (ed), *Towards a Steady State Economy*. San Franciso: Freeman Books, 149–74.

Dietz, F. and van der Straaten, J. 1995. Economic theories and the necessary integration of ecological insights. In A. Dobson and P. Lucardie (eds), *The Politics of Nature*. London: Routledge, 118–44.

Goodman, D., Sorj, B. and Wilkinson, J. 1987. *From Farming to Biotechnology*. Oxford: Blackwell.

Goodman, D. and Redclift, M. 1991. *Refashioning Nature*. London: Routledge.

Harvey, D. 1974. Population, resources and the ideology of science. *Economic Geography*, 50, 256–77.

Harvey, D. 1985. *The Urbanisation of Capital*. Oxford: Blackwell.

Harvey, D. 1996. *Justice, Nature and the Geography of Difference*. Oxford: Blackwell.

Hayward, T. 1995. *Ecological Thought*. Cambridge: Polity Press.

Mannion, A. M. 1992. Biotechnology and genetic engineering: New environmental issues. In A. M. Mannion and S. Bowlby (eds), *Environmental Issues in the 1990s*. Chichester: Wiley, 147–60.

Martin, R. 1995. Economic theory and human geography. In D. Gregory, R. Martin, and G. Smith (eds), *Human Geography: Society, Space and Social Science*. London: Macmillan, 21–53.

McKibben, R. 1989. *The End of Nature*. New York: Anchor Books.

Meadows, D. and Meadows, D. 1992. *Beyond the Limits*. London: Earthscan.

O'hUallachain, B. and Matthews, R. 1996. Restructuring of primary industries: Technology, labor and control in the Arizona copper industry. *Economic Geography*, 72, 196–215.

O'Riordan, T. 1989. The challenge for environmentalism. In R. Peet and N. Thrift (eds), *New Models in Geography*, vol. 1. London: Unwin Hyman, 77–104.

O'Riordan, T. 1995. Environmentalism on the move. In I. Douglas, R. Huggett, and M. Robinson (eds), *Companion Encyclopedia of Geography*. London: Routledge, 449–78.

Plumwood, V. 1994. *Feminism and the Mastery of Nature*. London: Routledge.

Redclift, M. 1991. The multiple dimensions of sustainable development. *Geography*, 76, 36–42.

Roberts, R. and Emel, J. 1992. Uneven development and the tragedy of the commons: Competing images for nature–society analysis. *Economic Geography*, 68, 249–71.

Sayer, A. 1979. Epistemology and conceptions of nature and people in geography. *Geoforum*, 10, 19–43.

Simon, J. 1997. *The Ultimate Resource II*. Princeton: Princeton University Press.

Smith, N. 1984. *Uneven Development: Nature, Capital and the Production of Space*. Oxford: Blackwell.

Smith, N. 1996. The production of nature. In G. Robertson, M. Mash, L. Tickner, J. Bird, B. Cuntis, and T. Putnam (eds), *Future Natural*. London: Routledge, 22–34.

Smith, N. 1998. Nature at the millennium: Production and re-enchantment. In B. Braun and N. Castree (eds), *Remaking Reality*. New York: Routledge, 271–85.

Smith, N. and O'Keefe, P. 1980. Geography, Marx and the concept of nature. *Antipode*, 12, 30–39.

Society and Space. 1998. 16, 4.

Soper, K. 1991. Greening Prometheus: Marxism and ecology. In P. Osborne (ed), *Socialism and the Limits of Liberalism*. London: Verso, 271–93.

Thrift, N. and Olds, K. 1996. Refiguring the economic in economic geography. *Progress in Human Geography*, 20, 311–37.

van der Straaten, J. and Gordon, M. 1995. Environmental problems from an economic perspective. In P. Glasgergen and A. Blowers (eds), *Perspectives on Environmental Problems*. London: Arnold, 133–58.

Chapter 18

Single Industry Resource Towns

Roger Hayter

Why have a chapter on single industry resource towns, the only one in this book that refers to a specific kind of place as opposed to a process or economic entity? The very classification of "single industry resource towns" conjures up a stereotypical image with well-known characteristics. First, they boom and they bust, and in this sense the problem with resource-based towns is that they are not resourceful enough. Second, they are located where resources are found, typically in places where other kinds of economic opportunities are limited. These two related features underpin the stereotype of single industry resource towns as remote, specialized outposts comprising populations with limited social and economic options, and vulnerable to the forces of economic destruction whether originating in globally based restructuring or local resource exhaustion, or both.

Peripheral as places, resource towns are also peripheral in mainstream theories of industrialization. Yet industrialization fundamentally depends on resources. For industrialization is as much a search for resources as markets. Moreover, to a significant degree this search is controlled by one specific form of industrial organization, multinational corporations (MNCs), which integrate operations that circumnavigate the globe. For MNCs, strategies of resource exploitation are not simply about "cost minimization." Rather, these strategies are shaped by: internal motives of security, control and stability, and the need to counter the actual or planned behavior of rivals; strategic as well as economic motivations of the major powers which still provide the homes for most MNCs; and the bargains negotiated with host governments seeking to use resources to meet national and regional development goals. If MNCs bring their resources, expertise, and global connections to these bargains, they simultaneously impose dependence on bulk, specialized exports, and non-local control, further contributing to the vulnerability of resource towns.

Global forces, in the form of the international division of labor, resource dynamics, remoteness, and boom and bust, powerfully shape the evolution of single industry resource towns. The link between resource exploitation and social change in resource towns is not straightforward, however (Landis, 1938). Global forces are complex and their interaction with local populations creates varied outcomes.

Indeed, when global markets no longer need a local resource, people may wish to stay in "their" town and resist the implication of bust.

Resource towns, to return to the question posed in the first sentence, deserve to be highlighted within the inquiry of economic geography because they comprise a type of place that exists on the knife edge of the interactions between global and local forces. Indeed, these towns are one of the constituent forces of long-run industrialization. Their evolution was strongly shaped by the imperatives of Fordism in the three decades after World War II, and more recently by the search for more "flexible" forms of production in the present period of restructuring under "post-Fordism." Following a brief historical introduction, this chapter explores resource towns as they reflect the changing imperatives of Fordism and flexibility.

In this chapter, industry-resource towns are defined broadly as towns whose economic base is dominated by the extraction and primary processing of (non-agricultural) natural resources, non-renewable or renewable. Since "single" industry is only the most specialized type of the "industry-resource town" category, these towns are henceforth referred to as industry-resource towns (IRTs). While precise estimates are not available, IRTs are undoubtedly important within the global economy. Among core industrial countries, such as the USA, IRTs remain an important feature of the economic landscape (Krannich and Luloff, 1991). Auty (1990, 1995) notes that there are over 40 developing countries that derive at least 40 percent of their exports from the resource sector, and Canada and Australia are advanced nations whose global economic roles remain strongly defined by resource exports. In Canada, Randall and Ironside (1996, p. 18) note that in the late 1980s the resource sector accounted for 10 percent of Canadian gross national product (GNP) and 40 percent of export values, and that overwhelmingly these exports stem from numerous, specialized towns, often with populations of 20,000 or less. This chapter draws most of its examples from this Canadian context.

Industrialization and Industry-Resource Towns

IRTs represent "pure" export-based forms of industrialization, whether the "exports" are sales to other "core" regions in the same country or to foreign markets. Thus, from their inception, IRTs are unusually exposed and vulnerable to exogenous (global) forces. The viability of IRTs is also intricately affected by "resource cycles," defined by Clapp (1998a) as inevitable long-run patterns of resource exploitation and collapse. While these resource cycles or dynamics are themselves shaped by processes of industrialization, IRTs are also subject to fundamental forces of change that do not affect secondary manufacturing centers. A basic categorization of natural resources is between renewable (flow) and non-renewable (stock) resources. Non-renewable resources are permanently depleted by exploitation, the duration of which is finite, limited by the size of the resource and the rate of exploitation. These limits can be redefined, but not removed, by technological change. Conversely, renewable resources enjoy either permanently or potentially indefinite resource cycles.

In practice, Clapp (1998a) claims resource cycles that end in resource collapse apply to renewable resources, such as forests, as well as to non-renewable resources. In the case of forests, Mather (1990) has noted the existence of "turnarounds" when formerly depleted areas are reforested. Yet, industrial logging inevitably changes the forest resource (Graham and St Martin, 1990). Moreover, second growth forests differ from first growth natural forests both qualitatively and quantitatively, and the transition from the latter to the former typically implies a "fall-down effect" in which timber volumes available to industry are reduced because the new, managed forests comprise smaller trees and are not grown in time to compensate for loss of old growth. Indeed, as higher-quality, more easily accessible resources are exploited, there is a general tendency for resource use to incur higher costs over time, thus continually eroding the competitive advantage of IRTs (Freudenburg, 1992).

Historical perspective

The impacts of industrial and resource dynamics vary among resources, renewable and non-renewable, creating highly divergent patterns of booms and busts that can unfold over centuries. Thus, in the Laurium region, 25 miles south of Athens, Greece, mines employed up to 20,000 slaves leased from the Athenian State to provide silver for coinage for the state, as well as lead and marble (Warren 1973, p. 2). Although discovered earlier, the mines flourished in the fifth century BC, and following the battle of Marathon (490 BC), mine revenues were used to build the navy that beat the Persians at Salamis ten years later. The mines then entered a period of decline, revived briefly around 320 BC until silver prices dropped, and then again in the first century until the slaves revolted in 103 BC. The mines were subsequently closed until the late nineteenth century when French and Greek firms worked the mines for lead, manganese, and cadmium.

The European "age of discovery" extended a ruthless robber mentality to the whole globe. For Spanish conquistadores, the search for gold and silver reflected the belief that precious metals were the basis for wealth and power, defining a crude mercantilist philosophy in which one nation's gain occurred through another's loss. Industry-resource towns, such as the Valenciana (silver) Mine, Guanaguato, Mexico, and the Potasi (gold and silver) mine in Bolivia, both founded in the sixteenth century, were centers of pure exploitation. Slaves provided the labor, with local economies literally ransacked.

The Industrial Revolution marked an enormous surge in the demand for resources and the scale of their exploitation. While gold, silver, and diamonds continued to offer tantalizing promises of instant wealth, massive demands for industrial resources, led by iron ore and coal, provided inputs for further processing as well as a new foundation for economic growth and political power. The "new industrial spaces" of the nineteenth century were resource-based as well as manufacturing-based. Virginia City, Nevada, and the Rhondda, South Wales, illustrate contrasting forms of such industry-resource spaces.

Virginia City epitomized the gold and silver rushes that were vital in opening up the American West, shaping the political economy of the entire country (De Quille, 1974; James, 1998). Indeed, after gold was found in nearby Gold Canyon in the 1850s, Virginia City rapidly emerged as an industrial city of the West. Railways were built, and the town reached a population of almost 20,000 (with adjacent Gold Canyon another 10,000). Within the next 50 years gold and especially silver from the Comstock Lode, mined to depths of over 1000 meters, provided $400 million of wealth that fueled the growth of San Francisco, and kept the Union solvent during the Civil War. President Lincoln expressed his gratitude by granting Nevada statehood within three years of becoming a territory, and without the constitutionally required minimum population. The importance of silver correctly implied to the miners greater permanence than gold mines, a perception reinforced by the California gold rush of 1849. Nevertheless, the bust came within two generations. In 1874 the halt in the use of the silver dollar reduced excavation activity, and in 1875 there was a devastating fire. By 1887 mining effectively stopped. Virginia City had become a ghost town, albeit one destined to become a "living" ghost town.

The Rhondda Valley meanwhile epitomized the "coal rushes" that were vital to British industrialization and hegemony. The Rhondda Valley, a remote "mountainous" area in South Wales, was transformed by the coal mining industry after the 1840s. From just a few sheep farmers in 1830, the population mushroomed to a peak of 160,000 in 1921, distributed among small towns (or villages) such as Penygraig and Tonypandy (Humphreys, 1972). Since then, the Rhondda entered a period of decline in the 1920s from which it has not escaped. The last mines were closed in the late 1960s, and in 1989 the population was 68,000, a level still greater than is justified by local opportunity.

If both were mining landscapes of the same period, Virginia City and the Rhondda are nevertheless worlds apart, culturally and economically, as well as geographically. Virginia City mined "products" prized for their relative scarcity, that contribute to conspicuous consumption, and are a source of wealth and exchange. Prospectors upheld individualism and celebrated capitalism. In the Rhondda, coal was a "commodity" prized for its abundance and low cost – an input and energy source for downstream industrial uses. Socially, the Rhondda communities were working class and deeply socialist, sharing commitments to chapel, family, and union (and, for males at least, the pub). In contrast, the "legend" of Virginia City, christened by a prospector from the State of Virginia during a drinking binge, was of a rowdy, raunchy, and murderous place. The institutions of church, family, and union were not lacking, but their sanctity required strenuous efforts, including not a few hangings organized by a "Vigilance Committee." Even so, to the end, Virginia City retained its reputation for a 24-hour social life.

Cultural differences extended to foodways (Conlin, 1986). In the Rhondda, diets were limited, often insufficient. The miners on the Western frontier, however, received cash windfalls, and their instincts for immediate consumption revealed remarkable tastes for *haute cuisine*, as bacon, beans, and whiskey were quickly replaced by oysters, galantine truffles, and champagne. Conlin (1986, p. 130) claims that "the California gold rush was one of the few booms of its kind in which the hors d'oeuvres were on the scene before the whores." Indeed, the miners' unusual penchant for oysters led to the depletion of the oyster beds in San Francisco Bay, while the subsequent rush to Virginia City helped stimulate Oysterville on the Washington State coast, then considered more important than Seattle (Conlin, 1986, p. 119). Interestingly, refined tastes for food, which contributed to a "bizarre juxtaposition of elegance and rawness" in Western mining towns, did not extend to American logging camps where insistence on plentiful, good food was combined with narrow, traditional "meat and potato tastes" (Conlin, 1986, pp. ix and 129).

In the opening scenes of Charlie Chaplin's classic (silent) film, *Gold Rush* (1925), the image is of an endless line of miners struggling over the Chilkoot Pass on the way to the Klondike gold rush of 1896–7. Likewise, in the Rhondda, miners flooded in from other parts of Wales, England, and Ireland. Virginia City attracted people from Mexico, Europe, China, and many parts of the USA, including miners from the California gold rush which had attracted over 40,000 migrants who had made the journey by sea, either around Cape Horn (especially from Europe) or through the Panama Canal (especially from New York), and by land from Mexico and from easterly directions. Then, and subsequently, resource towns were conceived as migrant towns (Hayter, 1979).

Fordism and Resource Development

In the twentieth century, demands for resources escalated. Between 1918 and 1948, the USA alone consumed more minerals than had the rest of the world in recorded history (Warren, 1973, p. xv). Resource consumption increased even faster during the Fordist long boom from the 1940s to the 1970s. For Western capitalism, Fordism defined a booming, but stable international system anchored by US hegemony and fixed exchange rates, and a general balance between the demand and supply of goods that supported full employment. Fordism also implied a distinct "industrial state" dominated by the related interests of Big Business and Big Government, fully abetted by Big Labor (Galbraith, 1967). Within and among Fordist states, production systems were dominated by horizontally and vertically integrated MNCs exploiting economies of scale to mass produce goods in unionized factories increasingly dispersed around the world. The central thrust of developments in the resource sector during Fordism corresponded closely to this "production template," perhaps even more so than in secondary manufacturing where the model developed. The Fordist model, in turn, had profound implications for the nature of resource towns. First and foremost, resource towns became branch plant towns.

Multinationals and the organization of resource exploitation

As Baldwin (1956) noted, resource exploitation can be organized in two fundamentally different ways: an "entrepreneurial" model and a "plantation" model. These models, which were conceived by Baldwin in different types of agricultural regions, can be readily modified for other resource types. In brief, in the entrepreneurial model, resource exploitation features locally based entrepreneurs (miners, fishers, or loggers), small-scale operations, employees who may become entrepreneurs, strong local linkages, and commitment to local development. In contrast, in the plantation model, resource exploitation is by foreign-owned MNCs, capital intensive, and while labor is well paid (in relation to local standards), job tasks are highly specialized and designed to extract a resource for use in the parent company's operations elsewhere. The business linkages of the plantation model (value-added, services, equipment, profit flows) are primarily international rather than local, and investments in supporting infrastructure, such as transportation facilities and housing, are dedicated to the MNC, rather than publicly available as in the case of the entrepreneurial model.

Baldwin's deductive analysis revealed the superiority of the entrepreneurial model for local development because it offered more hope for local spin-offs and diversification. The plantation model, however, dominated investment throughout the resource sector during Fordism. Thus, in aluminium, coal, copper, diamonds, oil, iron ore, forest products, and other resource activities, MNCs became the dominant form of business organization, and levels of corporate concentration significant. Mines and resource processing mills were created as specialized branch plant operations within integrated corporate strategies, and resource towns were established to house the specialized workforces and their families.

The plantation model took precedence over the entrepreneurial in large measure for economic reasons as resource extraction and transportation exhibited marked economies of scale created by massive fixed costs (in infrastructure as well as plant) and processing efficiencies. In turn, big projects required the expertise, resources, and risk-coping abilities of giant MNCs. At the same time, the calculation of economies of scale is an inexact science, and the plantation model did not dominate solely to enhance efficiency. For MNCs, growth and size were themselves objectives, while governments and global development agencies, such as the World Bank, similarly equated political and social benefits with increasing size. Meanwhile, the "barriers to entry" facing small firms seeking to enter resource sectors became formidable. Moreover, MNCs were better able to locate in remote regions than small entrepreneurial firms. Given government support, notably investments in economic and social infrastructure, and favorable resource rights, MNCs could control branch plants in the remotest of places with all the supporting business services and the necessary coordination from distant head offices.

In general, Fordism witnessed a global dispersal of large-scale exploitation of low as well as high quality resources (Warren, 1973). This dispersal was shaped by spatial variations in cost structures, especially as determined by transportation costs (procurement and distribution) and economies of scale (Hay, 1976). Yet, the dispersal of resource processing activities (and towns) should not be conflated with a simple cost-minimization hypothesis that states that locations that deliver resources to markets at lowest cost are opened first while progressively higher cost sites are opened until revenues fall below costs. In practice, cost (and revenue) structures cannot be assumed as given datum. Readily accessible resources are missed for decades, as was the case for 80 years prior to the discovery of nickel in surface outcrops in Western Australia in a great arc around Kalgoorlie, where gold had been mined since the 1880s (Warren, 1973, p. 22). Indeed, uncertainties about the presence, scale, and quality of resources can feed speculation, including announcements by fraudulent promoters of the discovery of fictitious resources in the hope of obtaining funds from gullible investors.

Moreover, as MNCs emerged to dominate resource sectors during Fordism, powerful institutional forces encouraged the geographic diversification of resource investments while compromising cost-minimizing behavior. Three such forces may be briefly identified. First, the motivations of MNCs to grow while at the same time attempting to spread risks encourage geographic diversification (Vernon, 1971). Further, MNCs often seek to "match" the location strategies of equally large rivals to ensure that they share in newly available resource supplies that may provide significant competitive advantage. MNCs also may not organize material flows to minimize transportation costs but to maintain the viability of activities they own and to manipulate prices so as to reduce taxes (Odell, 1963, p. 112; see Skúlason and Hayter, 1997).

Second, geopolitics and the interests of national development and security have shaped MNC strategies for accessing resources. European colonization in the nineteenth century provides an important precedent in this regard. During Fordism, and the Cold War, the USA became acutely concerned with raw material supply, identified numerous (over 70) resources as "strategic and critical," and developed investment and trade policies to ensure "stable and secure" access to these resources by US-based MNCs (Clark-Jones, 1987; Haglund, 1989; Hayter, 1992). Canada was regarded especially favorably, and Clark-Jones argues that Canada's resource industries were configured principally to serve the USA to create a pattern of "resource continentalism." Simultaneously, Japan, lacking in domestic resources and having lost its empire in 1945, developed "multi-sourcing resource policies," implemented principally by its giant trading companies, which enveloped the globe (Ozawa, 1980; Edgington and Hayter, 1997).

Third, during Fordism, development politics in the developed and developing world highlighted resources as engines of economic growth – as ways to ameliorate balance of payments problems, create jobs, provide taxes, and promote regional development and economic diversification through a variety of multiplier effects in related activities (Auty, 1990, p. 11; Auty, 1995). Such expectations encouraged governments, and key financial institutions, to offer MNCs various incentives, including subsidized economic and social infrastructure, loans, grants and favorable resource utilization terms. New types of resource town were typically part of the development bargain.

Resource towns during Fordism: the Canadian model

Prior to Fordism, company-dominated IRTs provided precedents for Baldwin's plantation model. In these towns, owners and executives were often locally based, the operations were large scale, and the lives of workers and their families were closely dependent on the company. As Landis's (1938) classic study of three iron-mining towns in the USA revealed, the nature of this dependence could differ even within the same region and sector. In many cases, however, the company literally governed the town, controlling housing and services. Moreover, workers' rights were minimal, and worker exploitation and anti-union attitudes were widespread, modified in some towns by company paternalism. Prior to Fordism, company-dominated IRTs were unstable, poor places, and workers had few rights. During Fordism, however, the global proliferation of the plantation model of resource exploitation witnessed the creation of a new type of IRT, purposefully planned as stable, high-income communities, regardless of the degree of isolation. Canada was on the leading edge of this development as numerous planned, new "frontier" and "instant" towns, housing populations from 10,000 to 30,000, were built across the country at great distances from established, "southern" population centers (Robinson, 1962; Bradbury, 1978). Lucas's (1971) generalized model of Canadian resource town evolution reflects back on the Fordist period.

In Lucas's model, "mine (and mill) town Canada" develops through four main stages: namely, construction, recruitment, transition, and maturity. The first three are temporary, socially dynamic, often unstable periods shaped by intense phases of construction activity, temporary housing, high levels of in- and out-migration, and the uncertainties associated with the "start-up" of new, capital-intensive operations in remote locations. For Lucas, these stages were preludes to the more enduring fourth, "final" stage of maturity in which resource towns enjoy prosperity and stability. Indeed, Canadian resource towns during Fordism were quintessentially Fordist.

Thus, Canada's new frontier towns were formally planned as (relatively) big, permanent, "model" settlements, structured as part of broader collaborations between government and MNCs, fully supported by unions. Remoteness and the scale of development demanded such collaboration in the provision of infrastructure. In the case of Schefferville, Quebec, for example, a new 576-kilometer railroad line was required to link the town to (new) port facilities on the St. Lawrence Seaway at Sêpt Isle (Bradbury, 1979). With respect to social infrastructure, governments, supported by MNCs, desired to move away from the company town model of resource towns where companies controlled the housing, many services, and even retail goods. In some regions, such as coastal BC, accommodation in company towns had also been based on ethnic and gender lines. In contrast, the new frontier towns of Fordism were designed according to (then) contemporary urban planning principles as if they were metropolitan suburbs: owner-occupied, large single-family homes were arranged in landscaped, serviced, residentially zoned neighborhoods at suitable distances from the industrial sites and accessible to schools, hospitals, retail outlets, amenities, and other services. By the 1960s, the planned shopping mall was also a feature of resource towns. These towns also received municipal status to provide for local government. For workers and their families, the new resource towns offered high levels of amenity and living conditions, as well as political responsibilities, while removing the stigma associated with company control of social and political life. For MNCs, unwanted responsibilities of supplying housing and services, often a source of social grievance, were removed.

Simultaneously, the comprehensive, physical planning of frontier resource towns during Fordism defined highly structured social environments, all the more transparent by their newness and isolation. Moreover, resource towns were branch plant towns. To a much greater degree than in the past, when decisionmakers and owners were often local or loosely connected with some distant owner, the mines, oil wells, refineries, and sawmills of Fordism were increasingly directly controlled by closely integrated MNCs with head offices located in relatively few, distant metropolitan centers. In such contexts, decisionmaking functions are highly structured and specialized at the bottom of an international hierarchy, and largely limited to responsibilities for operational matters and workforce supervision.

Moreover, resource activities in Canada during Fordism became comprehensively unionized, and resource towns became union towns. Blue-collar jobs were closely structured by the principles of Taylorism, namely job demarcation and seniority. Job demarcation meant narrowly defined jobs and a sharp separation between (specialized) managers, responsible for designing and directing job tasks, and workers engaged primarily in manual tasks. The vagaries of machinery and extracting and converting resources, and the need to solve numerous problems, meant that workers could never be entirely deskilled, but this was the intent of Taylorism. Seniority further structured work relations by linking pay scales, job advancement, and firing and re-hiring patterns according to date of entry into the mine or mill. For managers, Taylorism allowed the development of a stable, specialized, and productive workforce while workers gained dignity and protection in the work place as well as bargaining power which through a succession of collective agreements led to rapidly increasing wages and non-wage benefits, including vacation benefits (linked to seniority) and formal, binding, grievance procedures.

Structured labor relations underpinning workplace stability, in turn, implied community stability. Once hired, firms had little discretion in firing employees. Indeed, some resource firms in remote locations put considerable emphasis on recruiting employees perceived to be stable (Hayter, 1979). Seniority also "locked in" workers to particular lines of progression and accumulating wage and non-wage benefits within mills, and thereby to communities. Workers did quit jobs, but typically "quitters" did not have much seniority, apart from a few experienced workers who occasionally left to join new mills starting up in other communities in an effort to leap-frog the seniority ladder.

As union towns, resource town Canada was thoroughly working class, the labor market was male dominated, and local politics could scarcely ignore union values. Union bargaining also raised the income levels of resource towns. The forest town of Port Alberni in BC, for example, with over 5,000 union workers in a population of 18,000, was consistently listed in the "top ten" Canadian towns on the basis of per capita income during the 1960s and 1970s (Barnes et al., 1999). Not surprisingly, the job expectations of youth within towns such as Port Alberni centered on the dominant resource employers. Indeed, Grade 10 students with virtually no qualifications could expect a high income, and a stable job that would support a comfortable family life. School leavers who considered resource towns as stifling enclaves could always go elsewhere, comfortable in the knowledge that "the mill" could offer well-paid summer employment to help fund education or travel plans.

Resource town Canada during Fordism was not without instability or tension. Labor relations were adversarial, and strikes, lock-outs, and sometimes "poisonous" attitudes split communities from time to time. Resource exploitation occupations were often dangerous; poor health, physical problems, injury, and even death provided a dark under-current to resource town life. Recessionary cycles occurred during Fordism, but lay-offs were typically temporary, the laid-off workers soon hired back, and professional ("white collar") managers were never vulnerable. Smaller resource-based operations were often closed, and older resource regions experienced "structural" unemployment. Yet, in Canada during Fordism, these problems were offset by the creation and expansion of IRTs.

By the early 1970s, Lucas's depiction of Canadian resource towns as stable, prosperous and structured, especially when compared to conditions in the 1930s, reflected the reality of many IRTs. The situation was not to last. Energy crises, stagflation, and the removal of fixed exchange rates heralded a more volatile global economy in which radical technological change, centered on micro-electronics, began to drive restructuring across the industrial spectrum in search of more flexible production structures, work relations, and forms of local development. Indeed, the sources of Fordism's stability – strongly vertically integrated corporations, dedicated production technology, dedicated infrastructure, and structured labor relations – became sources of rigidity and vulnerability. Resource activities and IRTs inevitably became embroiled in this restructuring.

Resource Towns and Flexibility

IRTs during Fordism did not develop exactly the same everywhere. But Fordism celebrates standardization. Thus, Canada's frontier IRTs were collectively shaped by an international division of labor rooted in Taylorism and controlled by MNCs, and shared ideas about model communities. In contrast, the search for flexibility in the contemporary restructuring of IRTs has emphasized themes of vulnerability and differentiation. The experience of Orofino, a timber-dependent community in Northern Idaho, illustrates the implications of the shift from Fordism to post-Fordism in terms of vulnerability (Machlis et al., 1990). Thus, at Orofino, harvest and employment levels have been highly unstable in the 1980s and 1990s, similar to the instabilities recorded in the 1930s, following a period of stability in the 1950s and 1960s. Elsewhere, IRTs have been closed, including model IRTs of Fordism, such as Schefferville, encouraging Bradbury and St Martin (1983) to add a "winding down" stage to the Lucas model of Canadian IRT development (see also Bradbury and Sendbuehler, 1988).

In turn, the expense of creating IRTs, their vulnerability, and the social, political, and corporate costs of downsizing and closure, have stimulated a new model of IRT that reflects the imperatives of flexibility, namely the "fly-in" IRT. As Storey and Shrimpton (1989) document for Canada and Australia, the exploitation of remote resources is increasingly accomplished by the use of temporary workforces that are flown to resource sites, where they live in temporary accommodations for work periods of varying length, separated by trips back to permanent homes elsewhere. Such workforces are typically unionized and well-paid, but flexible; their contracts are closely calibrated with market demands, and their job tasks more broadly defined. Thus the "fly-in" flexible IRT is planned to be temporary and a substantial part of the fixed costs of building permanent settlements (family housing, schools, hospitals, etc.) is saved, thus rendering break-even and profitability thresholds easier to attain. With lower fixed costs, the incentive to maintain production even when prices drop is reduced. The social costs of subsequent closures, resulting from lay-offs, are also dissipated among the communities where workers have their permanent homes.

Admittedly, new instant IRTs, planned and structured along the lines of the Fordist model, have also been built recently. An example is Tumbler Ridge, built in the 1980s in northern BC to house coal miners working mines supplying Japan. Yet, following several profit-deflating price reductions, employment has been reduced substantially and the rumors that the coal mines may close have been confirmed. Tumbler Ridge appears to make the case for the flexible fly-in town.

The causes of resource town closure are complex. Schefferville's iron ore, for example, served iron and steel mills in the USA that were downsizing, and its iron ore faced competition from much lower cost supplies elsewhere, just when its costs of resource extraction were escalating. Throughout the Pacific Northwest, many timber-dependent towns are facing resource depletion and growing environmental concerns for the forests that remain (Freudenburg, 1992, p. 326). Booms and busts of IRTs have also been related to labor bargaining strategies and to geopolitical considerations (Bradbury, 1985; Bradbury and Sendbuehler, 1988). Tumbler Ridge's problems are rooted in Japan's multiple sourcing strategy which includes new mines in the USA and Australia (Gibson, 1990; Parker, 1997). While the mines are new, they are relatively expensive, and as Japanese steel consumers want less coal than once anticipated, overcapacity among affiliated mines has rapidly emerged, and production and price reductions have already been implemented.

In the context of energy politics in Canada, where there are vested interests in oil and nuclear power, the balance of environmental judgment in favor of oil has helped promote developments in Alberta and Newfoundland. On the other hand, the Atomic Energy Company of Canada Limited (AECL) in 1999 decided to close down Pinewa, 100 kilometers to the north of Winnipeg, Manitoba, an "instant" town created in 1963 to support a nuclear reactor and underground research facility that employed as many as 1,400 people, many with PhDs (Flood, 1999). AECL will have completely closed its facility by 2002. Although ostensibly closed as a cost-cutting move, Pinewa was subject to environmental criticism since its inception. Others believe closure of the facility is short-sighted, given that decommissioning work on nuclear reactors is likely to increase and the potential of nuclear power in the world's energy needs. Highlighting the environmental dilemma of nuclear power, the research center, including its above-ground facilities, will be maintained (and heated) to allow the safe break-down of radioactive materials. IRTs are vulnerable to government policy changes regardless of resource depletion concerns.

Many IRTs have seen their dominant industrial base drastically downsize through rationalization and technological changes, often in association with the introduction of more flexible work arrangements (Hayter, 1997; Norcliffe and Bates, 1997; Rose and Villemaire, 1997). In contrast to Fordism, lay-offs have been permanent, and have affected professional as much as "blue-collar" employees (Grass and Hayter, 1989). Inevitably, unemployment has risen, with its attendant indignities, costs, and knock-on effects throughout the community. In the case of Orofino and Wallace, two IRTs in Northern Idaho, Machlis et al. (1990) claim validity for the hypothesis that resource production drives social change in IRTs. The relationships are complicated, however, exemplified by a decrease in marriage rates and increase in arrest rates as production increased, not vice-versa. The reasons for this are unclear.

Meanwhile, retraining possibilities in IRTs are difficult to define in practical terms, and out-migration is not an easy solution (Halseth, 1999a; 1999b). Female labor market participation has risen, but without reaching the income levels of union jobs lost. Attempts to replace Taylorism with flexible work arrangements at existing mills have been contentious, with divisive effects on communities (Hayter, 1997). Unions are less powerful. Migration has accelerated community aging processes. Downsized mills no longer offer jobs to high-school students who now seek to invest more in their education even as job markets are uncertain. The pressures on IRTs to survive, diversify, and create jobs are therefore considerable. These pressures are also contributing towards differentiated growth among IRTs.

In situ differentiation among industry-resource towns

Hayter and Barnes (1997), based on evidence from IRTs in BC's forest economy, which is increasingly moving towards the flexibility of post-Fordism, identify three sources of differentiation. First, the mass production of a relatively few commodities is breaking down into both flexible mass production and flexibly specialized production. In both cases, product differentiation is an important element although economies of scale are less significant in the flexibly specialized networks. Both trends are contributing to local differentiation as flexible mass production takes on different forms in different communities and flexible specialization is highly geographically uneven. Second, in tandem with these production trends, labor agreements increasingly display local variability. Indeed, there are strong pressures to eliminate industry-wide bargaining, the norm since the early 1950s. Third, as forest industries have downsized, forest towns have increasingly attempted to promote economic development through some form of diversification. These efforts feature different approaches and outcomes (Barnes and Hayter, 1994).

Across the USA, many IRTs, as their traditional economic bases have downsized, have attracted in-migrants looking for alternative lifestyles (Blahna, 1990). As Blahna reveals, such newcomers bring new attitudes to IRTs, creating a "culture clash" with long-time residents, and new dimensions to IRTs. In post-Fordism, many new migrants to IRTs are no longer interested in large-scale resource development. In the case of Youbou, BC, a culture clash has been avoided as, in the absence of formal local economic planning, sawmill workers have left to be replaced by in-migrants who work elsewhere, creating an (unplanned) "bedroom" community. Similarly, Pinewa expects to survive following the closure of its radioactive facility (Flood, 1999). As a model community designed for, and limited to, the scientists employed by AECL, Pinewa provides an unusual level of community amenity as well as access to wilderness. Many AECL scientists have chosen early retirement to stay in Pinewa, and now that the town is open to non-AECL residents, others are happy to buy homes in the town.

Some resource communities actively seek diversification. In the case of Chemainus, BC, successful development of tourism, in association with a revitalization of its center and investments in small-scale manufacturing, was led by individual entrepreneurs (Barnes and Hayter, 1992). In the case of Sudbury, Ontario, once the world's biggest nickel mine, serious cutbacks in production and jobs were experienced in the 1970s (Trist, 1979). In response, various community members organized to generate ideas for development and ways of financing, discussions that led to Sudbury's development as a winter sports center. If Chemainus' success was greatly aided by its accessibility to growing metropolitan areas and tourist pools, Sudbury's experience indicates isolation is not an absolute deterrent to diversification.

For Trist, Sudbury's experience illustrates a basic shift in planning philosophy from a centrist, top-down, hierarchical approach, consistent with Fordism, to a bottom-up, community, and networking approach, characteristic of flexibility (Coffey and Polèse, 1985). Typically, bottom-up initiatives, whether originating in the private sector or local public sector, are highly variegated – or "unruly" to use Sjoholt's (1987) label drawn from his Norwegian experience. Collectively, with the differentiated paths of development of existing IRTs, along with the emergence of "fly-in" IRTs, which are planned as temporary settlements, the IRT landscapes of the present period reflect a more varied set of forces than associated with Fordism, especially through enhanced possibilities for local action. Meanwhile, the dominant institutions of Fordist IRTs, the MNCs and unions, have become less influential. In addition, as IRTs diversify, they attract new waves of in-migrants whose lifestyles and values may differ from those of established populations (Blahna, 1990).

Moreover, in many parts of the world, including the USA, the restructuring of IRTs faces new complexities, rooted in the values of environmentalism and aboriginal rights, that were scarcely apparent during Fordism (Buttel, 1992). Expansion of resource frontiers assumed the pre-eminence of industrial values, and environmental considerations were subordinate to the goals of economic growth. It was also felt that aboriginal populations were vestiges of ways of life that would soon disappear altogether. In recent years, however, environmentalism and aboriginalism have emerged as powerful institutions – changing the rules of resource exploitation, preventing some projects, modifying others, and changing the nature and fabric of resource towns (Anderson and Huber, 1988; Hecht and Cockburn, 1989; Barker and Soyez, 1994; Reed, 1995; Hayter and Soyez, 1996; Clapp, 1998b and c). Indeed, environmentalism and aboriginalism are global forces, seeking restrictions on resource exploitation on the one hand, and encouraging enhanced local control and participation (including aboriginal participation) in resource development, on the other hand. In turn, these concerns further encourage the fly-in resource town, and add new tensions in established resource towns, while allowing for locally inspired, differentiated developments in others.

Industry-resource towns: addicted, cursed and trapped?

According to the stereotype, IRTs eventually bust, unable to diversify. For Watkins (1963), resource economies within Canada are "trapped;" for Auty (1990, 1993), resource-based developing countries are "cursed;" and for Freudenburg (1992) IRTs in the USA, and elsewhere, are "addicted" to resource dependence, until they fail. The "addiction," rooted initially in isolation and comparative advantage, is progressively reinforced by institutional structures, attitudes, and resource dynamics. Thus, MNCs operate on the lines of Baldwin's plantation model. Their connections are global and internal rather than local and external. Infrastructure is likewise dedicated to this role. Resource MNCs perform R&D, but rarely in resource towns. In short, the resource operations at the beginning of vertically integrated international supply chains provide little opportunity for local spillovers in the form of forward, backward, and final demand linkages. Moreover, the union culture of IRTs is not conducive to local entrepreneurialism and diversification. Well paid, highly specialized workers, locked into the benefits of seniority, and sharing a collective commitment to adversarial (anti-management) bargaining, lack the incentive, attitude, and training conducive to creating their own businesses. Indeed, in resource towns strong union resistance to non-union, low-wage businesses (outside of small, largely female-employing retail activities) may continue after downsizing. A widespread legacy of Fordism in company and union towns is a weak, non-innovative secondary business sector.

Resource dynamics further militate against plans for diversification (Freudenburg, 1992). Over the long run, resource utilization inherently pushes resource towns into a cost–price squeeze as the cheapest, most accessible resources are exploited first, and as costs rise the emergence of cheaper resources elsewhere adds to capacity and puts pressure on prices. Yet, if the cost–price squeeze is a long-term trend, it is consistently masked by sharp price fluctuations. In booms, prices increase, local expectations are optimistic, and the case for diversification recedes. In busts, prices decline, opportunities for diversification recede, and local attitudes, shaped by past experience, are to "sit tight" and wait for the next boom. The possibility that there may be no "next boom" is unlikely to encourage local investment. Local perceptions of resource size and viability are also often unclear, further confused when mines with abundant resources are closed. Resource towns rarely anticipate closure, and local responses are typically reactive. Indeed, according to Freudenburg (1992, p. 317), there are dangers that extractive regions will become increasingly addicted to resources. He notes that new resource spaces are likely to be in regions increasingly remote from established centers. As capital intensity increases, so do the imbalances in power and scale between extractive enterprise and community, and abilities to diversify may be declining.

In addicted, trapped, and cursed IRTs, possibilities for differentiation are a chimera. From this perspective, contemporary proposals to promote local development based on the "flexible specialization model," which centers on the creation of local agglomerations of innovative networks of small firms (Cooke and Morgan, 1993), are not reconcilable with resource town settings. Thus, O'hUallachain and Matthews (1994) argue, based on trends in copper mining in Arizona, integration and economies of scale continue to underpin highly specialized operations in IRTs, not flexible specialization. Moreover, the quintessential IRT of the new age of flexibility is the fly-in town, a model that fully accepts the implications of addiction.

Yet, IRTs survive. Moreover, their inhabitants see IRTs as desirable places to live as well as work, their advantages often contrasting with the problems associated with metropolitan areas (Gill, 1990). Contemporary restructuring has reinforced, as much as undermined, local commitments to development, and IRTs must now broker environmental and aboriginal, as well as economic, values (Reed, 1995; Hayter and Soyez, 1996). IRTs may be strongly addicted to commodity exports, but the possibility of remedies cannot be ignored (Krannich and Luloff, 1991).

Conclusion

This economic geography of industry-resource towns has revealed the variegated forces underlying their classification. IRTs are special places, vulnerable to industrial dynamics, and to resource dynamics. Situated on the geographic frontiers of capitalism, they are buffeted by the changing strategic needs of core regions and the orchestration of MNCs. Diversification and stability in geographic cores is facilitated by specialization and instability in the periphery, and this contradiction is well illustrated by IRTs. They exist because cores want them; they are often closed down when cores do not. They are vital to the economy but are highly vulnerable. They grow because they are specialized, but their survival depends on diversification.

Many IRTs go bust. In the meantime, they play strategic roles for MNCs and for governments. Their remoteness makes them central to global concerns over environmentalism and aboriginal rights. IRTs are also migrant towns that impart unappreciated social and cultural diversity, even experimentation. Against all odds, and without much attention in geographic theory (but see Auty, 1995), IRTs seek to sustain their roles as "local models" within global economic geography (Barnes, 1987). For economic geography, wishing to integrate economic, political, environmental and cultural processes, the IRT is as good a place to start as any.

Acknowledgments

I gratefully acknowledge the critical eye of Trevor Barnes and the advice of Ivor Winton and Keith Storey.

Bibliography

Amin, A. and Robins, K. 1991. These are not Marshallian times. In R. Camagni (ed.), *Innovation Networks: Spatial Perspectives*. London: Belhaven Press, 105–18.

Anderson, R. S. and Huber, W. 1988. *The Hour of the Fox: Tropical Forests, the World Bank, and Indigenous People in Central India*. Seattle: University of Washington Press.

Auty, R. M. 1990. *Resource-based Industrialization: Sowing the Oil in Eight Developing Countries*. Oxford: Clarendon Press.

Auty, R. M. 1993. *Sustaining Development in Mineral Economies: The Resource Curse Thesis*. London: Routledge.

Auty, R. M. 1995. *Patterns of Development: Resources, Policy and Economic Growth*. London: Edward Arnold.

Baldwin, R. E. 1956. Patterns of development in newly settled regions. *Manchester School of Economics and Social Studies*, 24, 161–79.

Barker, M. and Soyez, D. 1994. Think locally – act globally? The transnationalisation of Canadian resource-use conflicts. *Environment*, 36, 12–20 and 32–6.

Barnes, T. 1987. Homo economicus, physical metaphors, and universal models in economic geography. *The Canadian Geographer*, 32, 347–50.

Barnes, T. and Hayter, R. 1992. "The little town that did": Flexible accumulation and the community response in Chemainus, British Columbia. *Regional Studies*, 26, 647–63.

Barnes, T. and Hayter, R. 1994. Economic restructuring, local development and resource towns: Forest communities in coastal British Columbia. *Canadian Journal of Regional Studies*, 17, 289–310.

Barnes, T. and Hayter, R. (eds). 1997. *Troubles in the Rainforest: British Columbia's Forest Economy in Transition*. Canadian Western Geographical Series No. 33. Victoria: Western Geographical Press.

Barnes, T., Hayter, R. and Hay, E. 1999. "Too young to retire, too bloody old to work": Forest industry restructuring and community response in Port Alberni, British Columbia. *Forestry Chronicle*, 75, 781–7.

Blahna, D. J. 1990. Social bases for resource conflicts in areas of reverse migration. In R. G. Lee, D. R. Field, and W. J. Burch Jr. (eds), *Community and Forestry*. Boulder, CO: Westview Press, 159–78.

Bradbury, J. H. 1978. The instant towns of British Columbia: A settlement response to the metropolitan call on the productive base. In L. J. Evenden (ed.), *Vancouver: Western Metropolis*. Western Geographical Series No. 16, 117–33, Victoria: University of Victoria, Department of Geography.

Bradbury, J. H. 1979. Toward an alternative theory of resource-based town development in Canada. *Economic Geography*, 55, 147–66.

Bradbury, J. H. 1985. International movements and crises in resource-oriented companies: The case of Inco in the nickel sector. *Economic Geography*, 61, 129–43.

Bradbury, J. H. and St Martin, I. 1983. Winding down in a Quebec mining town: A case study of Schefferville. *The Canadian Geographer*, 27, 128–44.

Bradbury, J. H. and Sendbuehler, M. 1988. Restructuring asbestos mining in Western Canada. *The Canadian Geographer*, 32, 296–306.

Buttel, F. H. 1992. Environmentalization: Origins, processes, and implications for rural social change. *Rural Sociology*, 57, 1–27.

Clapp, R. A. 1998a. The resource cycle in forestry and fishing. *The Canadian Geographer*, 42, 129–44.

Clapp, R. A. 1998b. Regions of refuge and the agrarian question: Peasant agriculture and plantation forestry in Chilean Araucanía. *World Development*, 26, 571–89.

Clapp, R. A. 1998c. Waiting for the forest law: Resource-led development and environmental politics in Chile. *Latin American Research Review*, 33, 3–36.

Clark-Jones, M. 1987. *A Staple State: Canadian Industrial Resources in the Cold War.* Toronto: University of Toronto Press.

Coffey, W. and Polèse, M. 1985. Local development: Conceptual bases and policy implications. *Regional Studies*, 19, 85–93.

Conlin, J. R. 1986. *Bacon, Beans and Gallantines: Food and Foodways on the Western Mining Frontier.* Reno and Las Vegas: University of Nevada Press.

Cooke, P. and Morgan, K. 1993. The network paradigm: New departures in corporate and regional development. *Environment and Planning D*, 11, 543–64.

De Quille, D. 1974. *A History of the Comstock Silver Lode and Mines.* New York: Promontory Press (originally published in 1889 by F. Boegle, Virginia City).

Edgington, D. W. and Hayter, R. 1997. International trade, production chains and corporate strategies: Japan's timber trade with British Columbia. *Regional Studies*, 31, 149–64.

Flood, G. 1999. Atomic town stuck in the '60s. *The Vancouver Sun*, March 27, B1–2.

Freudenburg, W. R. 1992. Addictive economies: Extractive industries and vulnerable localities in a changing world economy. *Rural Sociology*, 57, 305–32.

Galbraith, J. K. 1967. *The New Industrial State.* Boston: Houghton Mifflin.

Gibson, K. 1990. Internationalization and the spatial restructuring of black coal production in Australia. In R. Hayter and P. D. Wilde (eds), *Industrial Transformation and Challenge in Australia and Canada.* Ottawa: Carleton University Press, 159–73.

Gill, A. M. 1990. Enhancing social interaction in new resource towns: Planning perspectives. *Tidschrift voor Economische en Sociale Geografie*, 81, 348–63.

Graham, J. and St Martin, K. 1990. Resources and restructuring in the international solid wood products industry. *Geoforum*, 21, 289–302.

Grass, E. and Hayter, R. 1989. Employment change during recession: The experience of forest product manufacturing plants in British Columbia, 1981–1985. *The Canadian Geographer*, 33, 240–52.

Haglund, D. G. (ed). 1989. *The New Geopolitics of Minerals.* Vancouver: University of British Columbia Press.

Halseth, G. 1999a. Resource town employment: Perceptions in small town British Columbia. *Tidschrift voor Economische en Sociale Geographie*, 90, 196–210.

Halseth, G. 1999b. "We came for the work": Situating employment in B.C.'s small resource-based communities. *The Canadian Geographer*, 43, 363–81.

Hay, A. M. 1976. A simple location theory for mining. *Geography*, 71, 65–76.

Hayter, R. 1979. Labour supply and resource based manufacturing in isolated communities: The experience of pulp and paper mills in north central British Columbia. *Geoforum*, 10, 163–77.

Hayter, R. 1992. International trade relations and regional industrial adjustment: The implications of the 1982–86 Canadian–US softwood lumber dispute for British Columbia. *Environment and Planning A*, 24, 153–70.

Hayter, R. 1997. High performance organizations and employment flexibility: A case study of in situ change at the Powell River paper mill, 1980–94. *The Canadian Geographer*, 41, 26–40.

Hayter, R. and Barnes, T. 1990. Innis' staple theory, exports and recession: British Columbia, 1981–86. *Economic Geography*, 66, 156–73.

Hayter, R. and Barnes, T. 1992. Labour market segmentation, flexibility and recession: A British Columbian case study. *Environment and Planning C*, 10, 333–5.

Hayter, R. and Barnes, T. 1997. The restructuring of British Columbia's coastal forest sector: Flexibility perspectives. In T. Barnes and R. Hayter (eds), *Troubles in the Rainforest: British Columbia's Forest Economy in Transition*. Victoria: Western Geographical Series, 181–202.

Hayter, R. and Soyez, D. 1996. Clearcut issues: German environmental pressure and the British Columbia forest sector. *Geographische Zeitschrift*, 84, 143–56.

Hayter, R., Grass, E. and Barnes, T. 1994. Labour flexibility: A tale of two mills. *Tijdschrift voor Economische en Sociale Geografie*, 85, 25–38.

Hecht, S. B. and Cockburn, A. 1989. *The Fate of the Forest: Developers, Destroyers and Defenders of the Amazon*. New York: Verso.

Humphreys, G. 1972. *Industrial Britain: South Wales*. Newton Abbot: David and Charles.

James, R. M. 1998. *The Roar and the Silence: A History of Virginia City and the Comstock Lode*. Reno and Las Vegas: University of Nevada Press.

Krannich, R. S. and Luloff, A. E. 1991. Problems of resource-dependency in U.S. rural communities. *Progress in Rural Policy and Planning*, 1, 5–18.

Landis, P. H. 1938. *Three Iron Mining Towns: A Study in Cultural Change*. Ann Arbor, MI: Edward Bros.

Lucas, R. A. 1971. *Minetown, Milltown, Railtown: Life in Canadian Communities of Single Industry*. Toronto: University of Toronto.

Machlis, G. E., Force, J. E. and Balice, R. G. 1990. Timber, minerals, and social change: An exploratory test of two resource-dependent communities. *Rural Sociology*, 55, 411–24.

Mather, A. S. 1990. *Global Forest Resources*. London: Belhaven Press.

Norcliffe, G. and Bates, J. 1997. Implementing lean production in an old industrial space: Restructuring at Corner Brook, Newfoundland 1884–1994. *The Canadian Geographer*, 41, 41–60.

Odell, P. R. 1963. *An Economic Geography of Oil*. London: Bell.

O'hUallachain, B. and Mathews, R. A. 1994. Economic restructuring in primary industries: Transaction costs and corporate vertical integration in the Arizona copper industry, 1980–1991. *Annals of the Association of American Geographers*, 84, 399–418.

Ozawa, T. 1980. Japan's new resource diplomacy: Government-backed group investment. *Journal of World Trade Law*, 14, 3–13.

Parker, P. 1997. Canada-Japan coal trade: An alternative form of the staple production model. *The Canadian Geographer*, 41, 248–66.

Randall, R. E. and Ironside, R. G. 1996. Communities on the edge: An economic geography of resource-dependent communities in Canada. *The Canadian Geographer*, 40, 17–35.

Reed, M. G. 1995. Co-operative management of environmental resources: A case study from Northern Ontario. *Economic Geography*, 71, 132–49.

Rees, K. G. and Hayter, R. 1996. Flexible specialization, uncertainty and the firm: Enterprise strategies in the wood remanufacturing industry of the Vancouver metropolitan area, British Columbia. *The Canadian Geographer*, 40, 203–19.

Robinson, I. M. 1962. *New Industrial Towns on Canada's Resource Frontier.* Chicago: University of Chicago, Department of Geography.

Rose, D. and Villemaire, M. 1997. Reshuffling paperworkers: Technological change and experiences of reorganization at a Québec newsprint mill. *The Canadian Geographer*, 41, 61–87.

Sjoholt, S. 1987. New trends in promoting regional development in local communities in Norway. In H. Muegge and W. Stohr (eds), *International Economic Restructuring and the Regional Community.* Aldershot: Avebury Press, 277–93.

Skúlason, J. B. and Hayter, R. 1998. Industrial location as a bargain: Iceland and the aluminium multinationals 1962–1994. *Geografiska Annaler*, Series B, 80, 29–48.

Storey, K. and Shrimpton, M. 1989. Long distance labor commuting in the Canadian mining industry. *Centre for Resource Studies*, Working Paper no. 43. Kingston: Queen's University.

Trist, E. 1979. New directions of hope: Recent innovations interconnecting organizational, industrial, community and personal development. *Regional Studies*, 13, 439–51.

Vernon, R. 1971. *Sovereignty at Bay: The Multinational Spread of U.S. Enterprises.* New York: Basic Books.

Warren, K. 1973. *Mineral Resources.* New York: Wiley.

Watkins, M. H. 1963. A staple theory of economic growth. *The Canadian Journal of Economics and Political Science*, 29, 141–48.

Part IV Social Worlds

Chapter 19

Family, Work, and Consumption: Mapping the Borderlands of Economic Geography

Nicky Gregson

This chapter begins with a confession: it's taken me a while to open up the space to produce this piece. This may seem a minor point, but it says much about the nature of academic work in the UK right now and about the intersection of work with other claims: partner, friends, family, pets, leisure pursuits, home decoration, which this chapter, of necessity, engages with too. But it also says something important about the remit of this chapter. In their initial letter, outlining the issues they wanted included here, the editors described their thoughts as a "mishmash." My initial response was to agree. But then I began to mull over why this set of terms – family, work, and consumption – was proving problematic, troublesome. My argument is that this is because they are terms that sit at the boundaries of economic geography. They are markers: terms whose manner of inclusion or exclusion within economic geography, indeed, whose very inclusion/exclusion, have definite, different, even conflicting implications for our understanding/s of what economic geography is about; terms that question the assumptions and the terrain on which and from which economic geography speaks; which say as much about the producing subjects of economic geography as they do about its objects of analysis.

Take "work." Work, the myriad structured and not-so-structured, controlled and not-so-controlled actions, ideas, routines, and interactions that combine to produce goods and services, is central to economic production, however we conceptualize this. As such, work might be expected to figure in at least some economic geographers' analyses. And yet, until recently, it has seldom been visible. Indeed, when we look at many of the key debates that constitute the economic geography landscape over the recent past – for instance, spatial divisions of labor, flexibility, flexible specialization, industrial districts – work is strikingly absent. Lost behind and obscured by "grander" concerns, work is the absent presence; the material effect of "big" processes, yet somehow by comparison insignificant. Recently, this exclusion has been challenged: by feminist geographers, who have stressed the importance of the gendering of workers and occupations/jobs for regional development; by those pursuing a cultural economic analysis, for whom the meaning/s of work, its performance, and the repertoires inscribed within it constitute the primary foci; and by social/cultural geographers, whose interest/s in the body, identities, and subjectivities have led to explorations of disciplining, regulating, and performing various constructions of the working body, as well as to analyses of how work figures in the constitution of identities. So, this economic geography landscape is one where work and workers are not just included, but seen as central to what economic geography is about. That this conflicts with other (earlier) visions of economic geography goes without saying. Writing about "Family, Work and Consumption," then, impels one to navigate the borderlands, the contested "marches" of economic geography. Hence the trouble, the "mish-mash."

So what routes can we follow through this borderland? It would be possible to produce a cartography outlining the economic importance of the family, the changing nature of work, and the growing significance of consumption, but this would make for dull reading. Instead, I prefer a more interwoven mapping that centers the interfusions of home–work and home–work–consumption. But before proceeding, and since one of the critical effects of the literature discussed here has been to question our frequently taken-for-granted assumptions about just what does and does not constitute work, it is necessary to dwell on the term "work."

What Do We Mean by "Work"?

When we in the advanced capitalist economies use the term "work," the accompanying assumptions are that such activities are salaried/waged or financially remunerated and that they occur outside domestic living space. We talk in everyday language about "going (out) to work," readily equating the activities for which we are paid with "work;" the implication being that everything else isn't. Such notions have long been critiqued. They have been shown to pertain only to the advanced capitalist world. Moreover, in assuming as the norm those practices that have been so traditionally only for men, they have been argued to promote a masculine subject position. Correspondingly, "work" is no longer narrowly defined but has been extended to encompass all those activities necessary for social reproduction. So, it includes domestic labor (cooking, cleaning, washing, ironing, etc.) and childcare; and it also includes unwaged labor, such as voluntary work and "family work." And, because it encompasses all these things, the spaces and places of work are not just offices, retail units in malls and high streets or, increasingly rarely, factories, but homes too.

This expansive definition of "work" remains largely unexplored within economic geography, where attention is still primarily confined to work of the paid form (Peck, this volume). But, as we see now, this definition raises critical geographical questions concerning the production, constitution, and relations between the spaces of home and work, as well as key questions about the future shape of paid work. Two examples serve to illustrate these issues. One concerns women home-workers in Southern Europe; the other high-tech (male) scientists in the UK.

Home-working is pervasive in the European South, where it has recently intensified and expanded through the development of diffused, "flexible" production systems and subcontracting networks (Vaiou, 1991). Common in the production of textiles, clothing, leather goods, furniture, electrical appliances, toys, jewelry and

Christmas cards, one of the effects of this "flexibility" has been to incorporate women and ethnic minorities into production. Of greater significance for this chapter though is the way home-working brings the spaces of home and work together: home-working means that the relations of family and of paid work co-incide spatially, which means that for the home-worker they frequently collide. Indeed, Vaiou's work shows that whilst gender relations and familial norms in the European South constitute the conditions for the construction of women as home-workers, providing subcontracting firms with a ready supply of cheap, unorganized, compliant labor, the experience of home-working is that the demands collide with domestic routines. An Athenian home-worker testifies:

I wake up at 6.00 in the morning, prepare food, clean the house and start work as early as I can. I stop for two hours in the afternoon because the neighbors complain about the noise from the [sewing] machine. My food goes down almost unchewed. I do everything like in a car race. I clean the dishes…, I run to the market, then back to the sewing machine till late at night… (Vaiou, 1991, p. 50)

By contrast, a very different study examines men in high-tech, scientific research establishments in Cambridge in the UK (Massey, 1995). Here paid work is talked about in celebratory terms. It is loved; something which these men are not simply interested in or which they find just enjoyable, but which they are all-absorbed by. This has important implications. Indeed, the way these scientists talk suggests the erasure and/or invasion of home space by paid work. Home then is erased by working practices that frequently involve evenings, weekends, and public holidays, and which override holiday entitlements, and it is invaded by studies, home compu-ters, and modems. Moreover, Massey demonstrates how these scientists' minds are often still "at work" even when their bodies are doing "things domestic" – thinking in the bath, whilst playing with children. So, the predominant pattern within these high-tech research environments is one in which one side of the home–work dualism (work) erases the other (home). A minority of men, however, resist this; by keeping work out of home space and by having strict rules about the hours spent at work.

Apart from their intrinsic interest, the above studies are important; for their suggestions about home space, work space, and their interrelations, and for the differences they reveal between different forms of work. In the one instance – the male scientists – paid work invades and erases the space of the home and the domestic labor occurring there. In the other – home-working – paid work and domestic labor coincide in home space and conflict. This difference itself has much to do with the gendering of particular workers. For men, paid work can invade and/or erase the space of home and its attendant work, precisely because of the existence of a gender division of labor and a connotational system which assigns women as those primarily responsible for socially reproductive labor in the home. For women this same division of labor and connotational system means that whilst paid work can invade home space, it can never erase it and its associated work. The testimony of the Athenian home-worker provides ample evidence of this. But so too does that of growing numbers of women working in professional occupations in the advanced capitalist world. Indeed, as several recent studies show, the impossibility of juggling both types of work – without completely re-negotiating the domestic division of labor – means that many professional households are resorting to employing paid domestic workers (almost invariably women) for a range of social reproductive tasks (Gregson and Lowe, 1994; Pratt, 1997). So, when we conceptualize work as including both paid and unpaid labor, we have to acknowledge that men and women have different relations to and "investments" in different types of work, and that these relations and investments in turn impinge on negotiations of the home/work dichotomy, and have considerable implications for the (re)constitution of home space and work space.

Such deliberations also raise questions: about the future of paid work; about the increasing social and economic distance between the work required of the Cambridge scientists and the women they might employ to clean for them for a few hours a week; about the cultural baggage that surrounds high-tech and other professional work – extreme competitive individualism, obsession, present-ism; and about the way this type of work refuses to admit home space within its inner spaces. (Think of how desk photographs of partners and/or children are accepted, often as "hetero-badges," yet how their material presence is marginalized, peripheral, reserved for social occasions and/or the boundary points of the working day, dropping-off or picking-up.) So, should paid work erase and be valued over other forms of work? Should it occur in fragmented, sealed-off workspaces? Is it possible to combine different types of work in ways that acknowledge both their difference and equivalent importance? These are just some of the questions currently being asked and thought about by those seeking a more progressive politics and practice of paid work.

Having demonstrated how work means more than paid work, and examined some of the implications of this understanding, we move on to explore further research that has focused on home–work connectivities.

Home–Work Connectivities

In the previous section the emphasis was on home and work as spaces whose meanings are relationally constructed through gender differences, and as spaces with different potentials for mutual invasion and/or erasure. When we talk about space in this way, whilst there is some connection with physical space (the spaces of actual homes and workplaces), it is important to note that we are thinking of home and work as spaces inscribed with and produced through meaning; spaces that are constituted through discourses – notably of gender, sexuality, family, and work. Nevertheless, much of the recent research on home–work connectivities works from two different takes on spatiality; as physical distance (the absolute space – measurable, quantifiable – between home and work locations, and the friction this exerts), and as contextual place. These two strands comprise the substance to this section.

Home–work and physical distance: containment stories and segmented labor markets

One of the most important recent studies of home–work connectivities is Susan Hanson and Gerry Pratt's *Gender, Work and Space* (1995). Building from earlier research, which emphasized either the work or the home side of the home–work dichotomy, and which stressed respectively either the importance of women's gender roles in constraining female employment and shaping occupational segregation, or the processes of occupational segregation occurring within particular employment sectors, this focuses on the micro scale. It examines how households, communities, and employers in Worcester, Massachusetts, conjoin to produce (and reproduce) particular local labor markets, and, in the process, labor-market segmentation; how tightly constrained ("contained") many Worcester women's labor markets are; how this relates to domestic divisions of labor; and how both local social networks and employers produce restricted employment possibilities for women living in particular areas of Worcester. Along the way we learn how significant physical distance is in shaping women's decisions over what type of employment is possible, and where. Short commutes and the sequential scheduling of partners' employment are seen as desirable and are relatively common practices; transport problems, childcare availability, and domestic divisions of labor serve to reinforce such thinking. So, we see in this study how particular places are characterized by a myriad local labor markets, not one; labor markets that differ for men and women living in the same area of the city.

As some reviewers have indicated, this is a significant intervention in a literature that defines local labor markets in terms of white, male, middle-class experiences, and which conceptualizes their production at a coarser scale. But it is also suggestive regarding context. These stories of containment pertain to white, predominantly but not exclusively working-class women, living primarily in traditional nuclear family households, and in a city which might differ markedly from more important metropolitan areas in terms of spatial extent, political economy, social relations, and local cultures. So, how generalizable is the containment story which Hanson and Pratt tell? Citing more recent work on Columbus (Ohio), Chicago, and Minneapolis-St Paul respectively, they suggest:

Although the cities represented here reach only into the upper midwest and do not yet include examples from the South or West, the overall message . . . is [that] . . . the processes we describe . . . do not seem to be idiosyncratic to that place [Worcester] (Pratt and Hanson, 1996, p. 350).

Containment, then, seems to be not just a Worcester story. Indeed, it isn't just a North American story.

Compare the Worcester study with recent Dutch work for instance. Droogleever Fortuijn (1996) argues that the limits of what is considered possible and the spatial strategies for combining home and work vary between nuclear family households living in different neighborhoods. Professional households living in a high-status inner city area of Amsterdam-South frequently work close to home, travel to work by bicycle, and develop sequencing strategies which enable them to work from home occasionally and to overlap their working hours/days. Comparable households working in the semi-professional occupations and living in a different inner suburban neighborhood of Amsterdam exhibit male full-time working and female part-time working, with one partner always at home with the children at all times. By contrast, the options for a nuclear family household with children living in Almere – a new town 30 km from Amsterdam – are far more restricted: men for the most part commute long distances (by Dutch standards), whilst women confine their job search to the limited possibilities within Almere. This latter situation finds echoes in earlier work on British new towns in Northeast England, where locally based female part-time employment constituted the limits of what was considered possible by women living in nuclear family households, and – again shades of Worcester – constituted precisely the type of labor sought by incoming employers (Lewis, 1984).

Worcester then, as Pratt and Hanson suggest, isn't idiosyncratic, at least so far as its containment stories go. These stories are as familiar within the European context as they are within North America. But this should not be surprising, for there are strong parallels between the households that comprise the primary research base for all these studies; they are almost all white, apparently heterosexual, primarily mid-to-low income, and almost all comprising male/female partnerships with children. They are therefore traditional nuclear family households at a particular life-stage; the very households that would be expected to reveal the containment of women's labor markets. If we seek some different stories, maybe we should be looking elsewhere; at different households whose relation to local labor markets might be more expansive.

The volume of research within economic geography on non-nuclear family households and their relation to labor markets is thin, but there are some studies in social/cultural geography, notably those of lesbian households or that touch on such households, which point to a greater degree of heterogeneity than the containment stories above. Some indication of this is given by a comparison between Adler and Brenner's (1992) lesbian re-make of Castells' ground-breaking study of San Francisco's Castro district and Gill Valentine's (1993) study "(Hetero)sexing space." In the former, the implication is that lesbian households tend to be concentrated in working-class neighborhoods; the inference being that their local labor markets are as contained as those of many heterosexual women living with men and children. But, as Valentine's work conducted in Southeast England suggests, this is an over-simplification. Consider the following interview extract:

Liz works in London and I work in Eastbourne. We spend a couple of nights a week in Eastbourne and a couple of nights a week in London. We did for a while discuss the option of somewhere else in-between, like Dorking or Chertsey. But in-between is middle class suburbia. I would have to think very long and very hard before I could move into an area like the Dorkings of this world. It is so "straight" (middle class, 30s)... (Valentine, 1993, p. 398)

The differences between this home–work relation and that implied by Adler and Brenner are stark. First, and this is where their living arrangement is poles apart from any of the households discussed previously, the members of this partnership do not live in one house. Rather, they oscillate between two. Secondly, and consequently, they spend a number of hours each week traveling. Thirdly, this woman tells us why they haven't chosen to "split the difference." Implicitly, living as a lesbian couple in the suburbia between Eastbourne and London (read: heterosexual nuclear families with children living in small executive "estates") is not an option. That this partnership generates the income to sustain two households and journeying between them, and that they have neither children nor elderly dependents reliant upon local services, is evidently important in enabling such a home–work relation.

Indeed, this relation is well beyond the limits of what would be considered possible, even desirable, by other households. But what is also worth emphasizing is the expansiveness of this narrative. Oscillation, movement, traveling through space; this is what characterizes these women's home–work relation, not the localism of women living in traditional households.

So, when we look at different households, containment stories get disrupted by other narratives. These expansive stories should not be construed as the exclusive preserve of middle-class lesbian partnerships. Media articles in the UK, for example, regularly feature instances of TV presenters/producers, politicians, and so on, who work in, say, London or Birmingham, and live in say Edinburgh or Glasgow. Some of these long-distance commuters are single persons with/without partners; but others are in heterosexual partnerships with children, and more than a few are women. And indeed, when we look a little closer to home, the academic labor force is littered with instances of partnerships that replicate the lesbian partnership above – partnerships that might span East and West Coast USA, or even be transcontinental, but which, simply because of the difficulties of "fixing" two careers in one place, are likely to involve extended home–work relations and frequent long-distance commuting.

One of the effects of the above research has been to demonstrate the emergence of distinctive, unequal geographies of containment and mobility. All of which begs the question, why the emphasis in the home–work literature on containment? For Hanson and Pratt, containment stories reflect their initial insistence on representativeness within the Worcester context. For others, one can only speculate, but the focus on the nuclear family seems to suggest an internalization of this household type as the norm, related to both feminist geographers' preoccupation with the young children lifecycle stage and economic geographers' attention to working-class and/or lower-income occupations. The consequences, however, have been deleterious. What is required is research on home–work interfusions that admits diversity, centers the disruptions other types of household bring to existing understandings of home–work relations, and produces polyvocal narratives to counter the singular containment stories which currently prevail.

Home–work as contextual space

Another way of thinking about home–work relations is to "connect up;" to situate places within the broader context of radical political economy and grander stories about gender relations. Until the mid 1980s, this was one of the main ways of thinking about home–work relations, inspired for the most part by Doreen Massey's *Spatial Divisions of Labour* (1984, 1995). Examples include Linda McDowell and Doreen Massey's (1984) exploration of the intersections between capitalism and patriarchy in four areas in England in the nineteenth and twentieth centuries; Massey's own account of Cornwall; and some of the critical but resonant research conducted by the Lancaster Regionalism Group. More recently, such thinking has fallen out of favor; although traces still remain, witness the vestiges in the Worcester study. Readily identifiable with the grand narratives so berated by the turn to postmodernism, these have become largely forgotten stories.

And yet, as Richard Walker (1996) suggests, there is still space for these stories, provided that they are not the only ones we tell, and that we are sensitive to their partialities. The need to situate places and regions within spatial divisions of labor is as pertinent now as it has always been, precisely because such structures present material limits to what might be considered possible in terms of home–work relations for particular types of households in different places/regions. So, maybe the containment stories of nuclear family households in Worcester have as much to do with Worcester's position within the broader economy of the Northeast USA as with the micro geographies of Worcester; maybe the limits of what is considered possible might not be quite so local for similar households located in different cities? I'm thinking here of identical households to those discussed by Hanson and Pratt but ones located in the Greater London area, where part-time work is less readily available and where long(er) commutes to Central London are as characteristic of women's employment as men's.

Home–work infusions then are one of the means through which certain economic geographers – for the most part feminists – have produced alternative accounts of work. As we have seen, these center the home and household living arrangements and acknowledge the critical effects that these have on work as conventionally understood. Expressed slightly differently, they provide readings situated in the messiness of everyday life and, as such, are positioned in the troublesome borderlands of economic geography. And yet as I write this there is a sense in which some of these debates seem somewhat passé. And that Hanson and Pratt also feel this about their volume is suggested by the insertion of reflexive critical moments within their text. To be sure, I feel a degree of ambivalence about articulating such sentiments. There is, after all, a great deal of importance about such analyses. But academic debates are nothing if not moving feasts, subject to endless "new turns" and re-inflections; responses that say as much about power, performativity, and the discourses that constitute academia as they do about changes in the world which we live in, interpret and represent. And one recent "turn" has introduced a different set of directions to research on family (home) and work to those discussed thus far; the turn to consumption.

Home–Work–Consumption

Unraveling the complexities of the home–work–consumption nexus is an extraordinarily difficult task. Nonetheless, two quotations provide some instructive inroads:

> I often ask beginning geography students to consider where their last meal came from. Tracing back all the items used in the production of that meal reveals a relation of dependence upon a whole world of social labor conducted in many different places under very different social relations and conditions of production... Yet we can in practice consume our meal without the slightest knowledge of the intricate geography of production and the myriad of social relationships embedded in the system that puts it upon our table... (Harvey, 1990, p. 422)

The emphasis here is on relations of dependence; chains and flows that connect places, material commodities, and social labor in a global food system whose end point is a meal. Yet there are subtexts here: consumption is something we apparently do unknowingly; consumption is of secondary importance to relations of production; and food, eating, is about meals, not food per se. Compare this voice, of a well-known geographer writing in the radical political economy tradition, with the following:

The next time you arrive home with a carload of goods from the supermarket, ... before you start packing them away. Subject one or two ... to some lateral thinking. Treat them not simply as mass-market consumables, but as ... cultural artefacts, each with its own ... biography. Have a go at deconstructing your weekly shop. That cheery-looking bunch of bananas for example. Britain doesn't grow bananas – yet – so the odds are they have travelled a long way to grace your fruit bowl. Where did they start their journey? And what about the bunch of grapes? ... The odds are that you will be able to supply remarkably few answers to the questions. The past decade or two have seen the globalisation of the British shopping basket as supermarkets scour the world to satisfy our newly acquired appetite for exotic fruits or out-of-season greens. ... Such global supply lines would have staggered our grandparents. ... but they also mean ... we know less than ever about the way our food is produced ... (Nicholson-Lord, 1997, pp. 58–62).

Although referring to the global food system and the chains of production/distribution that constitute it, there are some important differences between this and the previous quote. Foods are cultural artefacts rather than meals, to be thought of as having biographies that are geographical, about traveling and displacement. And the scene described here is domestic, the routine weekly/monthly food shop being unloaded and stored away at home. Rather than looking back from the meal into global production and distribution chains, this quote connects the latter with knowing (and unknowing) acts of purchase and consumption. That it stems from an article on ethical consumption, then, is probably unsurprising.

Together, these quotes characterize much contemporary geographical writing on home–work–consumption. On the one hand there is the radical political economy tradition, which thinks in terms of chains of global–local dependencies, and of retail spaces as sites/spaces of paid work, but which "brackets-out" the home and the activities of consumption occurring therein. On the other is a culturally oriented version of economic geography, which tends (though not exclusively) to "bracket-out" production considerations, centering instead the geographies, practices, and knowledges of consumption-as-use. For the purposes of this chapter it is the second of these approaches that is more important. It is worth devoting some attention to the first tradition, however, given its import within economic geography.

Much of the research in this tradition is notable for the way it uses the metaphor of the chain to work back from an end point – the purchase of a commodity for direct use – through to the processes and social relations of production. One such example is Ian Cook's work on exotic tropical fruits (1994), which connects Jamaican plantations with the decisions made by UK supermarket executives regarding the visual form of these fruits, whilst similar connections have been made with respect to the production of mangetout (snow peas) in Zimbabwe. Here, regulation size stipulations mean that only a proportion of the crop meets presentational requirements, the remainder becoming cattle feed. Such chains are critical in revealing relations of dependence between First and Third Worlds; in exposing the power of supermarkets in shaping and dictating exchange relations with agricultural commodity producers; and for raising moral questions about a global food system which presents perfect, plastic-packaged mangetout to Western consumers whilst simultaneously consigning huge quantities of less-than-perfect peas to cattle, in a country where levels of basic human food consumption are far from comparable with the West. This is important, but a criticism frequently leveled at this type of work is where it ends. In closing the chain in the supermarket, such research confines consumption to the act of purchase, and denies the role of consumers in shaping the food chain. A plethora of research through the 1990s has argued that the first is a far from sustainable position (Miller, 1987; Jackson, 1993), whilst the second denies the ways in which the globalization of the food system has been hinged to parallel sets of changes in Western consumer food preferences (Goodman and Redclift, 1991; Jackson and Thrift, 1995). So, in "bracketing-out" consumption-as-use and consumers, this type of research re-asserts the primacy of production and retailing over consumption, erasing the home (and the practices occurring therein) from analysis. We are back then with the understandings of work with which we began this chapter; with an economic geography that sees consumption as little more than "surface froth," and with an interpretation of the home–work–consumption nexus that privileges (paid) work over the other two terms.

Another strand of research located loosely within the radical political economy tradition takes consumption more seriously, focusing on retail spaces as sites of paid work and as spaces for the constitution of identities (Leidner, 1993; du Gay, 1996). This research is concerned with the social relations of paid work yet is intrinsically connected with consumption, understood as the act of purchase, and as such goes further into the home–work–consumption nexus than chain-inspired explorations. Yet this still is not far: whilst connections between paid work and consumption are drawn, home space and consumption-as-use/re-use are seemingly off limits – issues, practices, and processes which, by implication, lie beyond the concerns and the subject matter of economic geography.

It is apparent that not all economic geographers share such views, however. Indeed, within current research it is becoming increasingly common not just to extend chains from production and retailing into cycles of use/re-use but to think in terms of disruptive conjunctures of relations and knowledges between consumers and producers. Recent research on twelve North London households epitomizes such tendencies, centering how "our mundane, everyday routines of food shopping, cooking and eating are developed in relation to increasingly internationalised food supply networks" (Cook et al., 1998, p. 163). Moreover, in focusing on consumer knowledges, on what consumers know about the origins of the foods they eat, and about their relations to global food systems, they develop the idea of "structural ambivalence" – that consumers have both a need to know and an impulse to forget the origins of the foods they eat. Of critical importance in enabling the development of this notion is the biographies which foods have within the home. Once bought, individual foods are worked on and incorporated within household relations (and see too Bell and Valentine, 1997). So, food is used to construct the space of the household and the relations and identities that constitute it; practices that construct the home as a relatively autonomous zone. Ironically, though, it is global food systems which are being used to do this; hence the need to forget and "bracket-out" the origins of the foods eaten.

Whilst Cook et al. see the importance of household relations – love, caring, authority, and resistance – and domestic divisions of labor to the biographies of food, it is food as a cultural artefact and culinary cultures which comprise their primary interests. Others, however, have rather more to say about the place of home within the practices of consumption. One such recent study is an interdisciplinary project on shopping, place, and identity, again conducted in North London (Miller et al., 1998). Home and family sit at the heart of this research. Countering the myriad studies that depict the act of purchase, and shopping practices more generally, as hedonistic pleasures, Miller et al. emphasize the embeddedness of shopping within inherently social, and primarily familial, relations. Shopping, consequently, becomes hard work for women, a set of practices that are not just about routine domestic reproduction but which are reproductive of "the family" and family identities. Love, sacrifice, denial, guilt, as well as sheer drudgery then are what this study uncovers about shopping (see also Miller, 1998); a combination of actions and emotions which exemplify the home–work–consumption nexus and the complexities of its intersections.

Thus, when we look at some of the recent geographical work on home–work–consumption, we see how this tends to play up the connectivities between the work entailed in home/household relations and practices of consumption, and to "bracket-out" production considerations as conventionally understood (i.e. as flows of goods and social relations of production). Instead, when production figures at all it is more likely to do so as consumer knowledges. Now part of this is a response to the research shaped by radical political economy. It is an attempt to tell different stories from different starting points, which accord different privileges to the all-too-familiar but still necessary tales of global–local interdependence and social relations of paid work. Nonetheless there are questions worth raising as a critical reflection on this research program. These concern the nature of the commodities which geographical research in this area has honed-in on thus far, and their knock-on effects in terms of thinking about consumption.

One commodity dominates contemporary research on home–work–consumption: food. This is readily understandable, for at one level food is a basic commodity, necessary for human and animal survival and comprising a large part of each household's weekly expenditure. Food has other academic advantages though. In today's context of scares over, for instance, BSE ("mad cow disease"), E-coli, salmonella, and GM (genetically modified) foods, it provides a classic case study for examining producers, retailers, consumers, and the state conjoining in regulating, negotiating, and/or resisting one of social science's current buzz words, risk (Marsden et al., 1998). Moreover, food and the performances accompanying its purchase, preparation, and eating are central to the expression and reconstitution of key social relations; and food is also an admirable commodity to "think through," both metaphorically and in terms of material relations of power and inequality. That it is also inscribed with and simultaneously constitutive of distinctive geographies no doubt helps too! There are, nonetheless, two points I would make about this emphasis on food in the consumption literature.

The first is that food is an unusual commodity in that, once eaten, it is not – at least as regards the norms of human consumption – available for further cycles of use/re-use. Regardless of the length of time it may/may not sit around in our cupboards, refrigerators, and freezers, the length to which it can be put to use is no longer than the time it takes to prepare and eat or, in the case of perishables such as fruit, go moldy in the displayed fruit bowl. This contrasts with many other commodities purchased for household consumption, notably items of home decoration and furnishing, consumer durables, consumables – CDs, books, videos, computer games, toys – or indeed clothing. Most of these are embedded within practices and rituals of ownership/possession by household members. Yet such commodities remain woefully under-researched in geography. This, then, is one of those "hotspots" for future research. For the moment though, the gap is significant, not least in terms of its implications for geographical representations of consumption. Indeed, the emphasis on food tends to foreclose the type of extended analyses of cycles of use/re-use insisted on by most contemporary theoretical representations of consumption (and see too the growing body of empirical work on the second-hand market: Gregson and Crewe, 1997). We should be aware then of the critical effects of focusing on particular types of commodities in geographical writing, and attempt to diversify the range of goods which we "think through."

A second point concerns the lack of reflexivity within much of this writing on food, with embodied subjectivities being a notable absence. This is an important omission with respect to this commodity, as a personal interjection suggests. My relations to food are complex, with hints of anorexia. I go through phases, in turn connected to body image, of minimalist eating, denial, and excessive exercise; I find meals, particularly formal sit-down meals of the course-after-course variety, physically and psychologically challenging; and I am vegetarian, and therefore at certain times/places "a problem" to feed. None of this is unusual; what I describe is part of what years of feminist research on the anorexic body and related eating disorders has documented. But it means that when I come to engage with geographical writing on food I feel a strong degree of revulsion; a desperate need to insist on a space for voices such as mine, and an acknowledgment that this celebratory, almost reverential, style of writing about food be connected to the embodied subjectivities that permit such writing. These are authors whose material bodies and sense of self enable them to engage with food in straightforward ways, through culinary "voyages of discovery" and subjects whose positionality is most definitely middle-class. Such points, and their implications, need to be acknowledged.

Conclusions

Conclusions are part of the performance of academic writing, suggesting it is possible to summarize in a comprehensive, synthetic way, which looks forward and stakes out the terrain of "progress." This is part of the expectation here, I am sure. Student texts, after all, carry such presumptions, both from their audiences and producers/publishers But, and as contemporary geographical writing is beginning to suggest, ending in this way is problematic (Bell et al., 1994; WGSG, 1997). This is because conclusions of this nature revert to privileging one narrative, one voice; a writing tactic that therefore erases as it summarizes; and a style that is inappropriate for expressing geographies of tension, contradiction, and polyvocality. That I want to resist such re-instatements here should be apparent, as should the fact that this is not to duck the issue; what I have attempted to show in this chapter is the multiple nature of contemporary geographical research on the family, work, and consumption, as well as its complex connections.

That there is no conclusion here, then, is important. Instead, I want to end by returning to the trouble with which I began. As I argued there, this chapter represents the borderlands of economic geography; its concerns define its limits, its boundaries, and expose (usually unarticulated) subject positions. And, as with many such challenges, the response on the part of many economic geographers has been either to ignore them (the business-as-usual approach) or to dismiss them as not proper economic geography. Usually this is done by re-inscribing a boundary around paid work and/or retail spaces, thereby excluding the rest as "surface froth," read irrelevant, not economic, cultural even. Thinking about the concerns of this chapter then also requires one to take a position with respect to contemporary debates about "the cultural" and "the economic" (cf. Crang, 1997; Sayer, 1997). But what it has also done, I hope, is to open up the issue of spatiality – how we conceptualize space – for closer scrutiny within economic geography. And it should be apparent that this, a set of issues debated more fully recently in cultural geography, has critical effects on the economic geography we produce. If this chapter has succeeded in making these points, and in creating some trouble, then it will indeed have done its work.

Bibliography

Adler, S. and Brenner, J. 1992. Gender and space: lesbians and gay men in the city. *International Journal of Urban and Regional Research*, 16, 24–34.

Bell, D., Binnie, J., Crean, J., and Valentine, G. 1994. All hyped up and no place to go. *Gender Place and Culture*, 1, 31–48.

Bell, D. and Valentine, G. 1997. *Consuming Geographies: We are Where We Eat*. London: Routledge.

Cook, I. 1994. New fruits and vanity: the role of symbolic production in the global food economy. In L. Busch (ed). *From Columbus to Conagra: the Global Station of Agriculture and Food Order*. Lawrence, KA: University of Kansas Press, 232–48.

Cook, I., Crang, P., and Thorpe, M. 1998. Biographies and geographies: consumer understandings of the origins of foods. *British Food Journal*, 100, 162–7.

Crang, P. 1997. Cultural turns and the (re)constitution of economic geography. In J. Wills and R. Lee (eds). *Geographies of Economies*. London: Arnold, 3–15.

Droogleever Fortuijn, J. 1996. City and suburb: contexts for Dutch women's daily lives. In M. Garcia Ramon and J. Monk (eds). *Women of the European Union*. London: Routledge, 217–28.

du Gay, P. 1996. *Consumption and Identity at Work*. London: Sage.

Goodman, D. and Redclift, M. 1991. *Refashioning Nature: Food, Ecology and Culture*. London: Routledge.

Gregson, N. and Crewe, L. 1997. The bargain, the knowledge and the spectacle: making sense of consumption in the space of the car boot sale. *Environment and Planning D: Society and Space*, 15, 87–112.

Gregson, N. and Lowe, M. 1994. *Servicing the Middle Classes*. London: Routledge.

Hanson, S. and Pratt, G. 1995. *Gender, Work and Space*. London: Routledge.

Harvey, D. 1990. Between time and space: reflections on the geographical imagination. *Annals of the Association of American Geographers*, 80, 418–34.

Jackson, P. 1993. Towards a cultural politics of consumption. In J. Bird, B. Cuntis, T. Putnam, G. Robertson, and L. Tickner (eds). *Mapping the Futures: Local Cultures, Global Change*. London: Routledge, 207–28.

Jackson, P. and Thrift, N. 1995. Geographies of consumption. In D. Miller (ed). *Acknowledging Consumption*. London: Routledge, 204–37.

Leidner, R. 1993. *Fast Food, Fast Talk*. Berkeley, CA: University of California Press.

Lewis, J. 1984. The role of female employment in the industrial restructuring and regional development of the UK. *Antipode*, 16, 47–60.

Marsden, T., Flynn, A., and Harrison, M. 1998. *Consuming Interests: the Social Provision of Food Choice*. London: Taylor and Francis.

Massey, D. 1984, 1995. *Spatial Divisions of Labour*. London: Macmillan.

Massey, D. 1995. Masculinity, dualisms and high technology. *Transactions of the Institute of British Geographers*, 20, 487–99.

McDowell, L. and Massey, D. 1984. A woman's place. In D. Massey and J. Allen (eds). *Geography Matters*. Cambridge: Cambridge University Press, 128–47.

Miller, D. 1987. *Material Culture and Mass Consumption*. Oxford: Blackwell.

Miller, D. 1998. *A Theory of Shopping*. Cambridge: Polity.

Miller, D., Jackson, P., Holbrook, B., Thrift, N., and Rowlands, M. 1998. *Shopping, Place and Identity*. London: Routledge.

Nicholson-Lord, D. 1997. A fair exchange? *Sainsbury's: The Magazine*, 47, 58–62.

Pratt, G. 1997. Stereotypes and ambivalence: the construction of domestic workers in Vancouver, BC. *Gender Place and Culture*, 4, 159–77.

Pratt, G. and Hanson, S. 1996. Response to reviews of *Gender, Work and Space. Antipode*, 28, 349–51.

Sayer, A. 1997. The dialectic of culture and economy. In R. Lee and J. Wills (eds). *Geographies of Economies*. London: Arnold, 16–26.

Vaiou, D. 1991. Homeworking: a woman's job and an ideology. *Synchrona Themata*, 45, 47–53 (in Greek).

Valentine, G. 1993. (Hetero)sexing space: lesbian perceptions and experiences of everyday spaces. *Environment and Planning D: Society and Space*, 11, 395–413.

Walker, R. 1996. *For Better or Worcester: Reflections on Gender, Work and Space. Antipode*, 28, 329–37.

Women Geographers' Specialty Group 1997. *Feminist Geographies: Explorations in Diversity and Difference*. London: Addison Wesley Longman.

Chapter 20

Concepts of Class in Contemporary Economic Geography

David Sadler

The cultural turn in human geography, the re-configuration of the "economic" in economic geography (Thrift and Olds, 1996), and the challenges of postmodern and poststructuralist approaches to social science more generally, posed particularly acute problems for the use of "class" as an explanatory concept in economic geography during the 1990s. This was especially notable given that economic geography had been so fundamental in the evolution of Marxist (that is, class-centered) approaches to human geography during the 1970s and 1980s. This chapter summarizes some of these contributions, and explores some of the reasons for and implications of the limited engagement with class as an explanatory concept within economic geography during recent debates.

The chapter starts with a re-appraisal (or perhaps more accurately a re-statement) of the salient features of class as conceptualized within the classical Weberian and Marxist traditions. It then goes on to examine some of the insights brought by the work within economic geography in the 1970s and 1980s which drew heavily on (and began to contribute to) Marxist class theory. This section exemplifies these contributions and their political implications through a consideration of the class-based campaigns in defense of place, and against closures in a number of major industries, which proliferated in Western Europe and North America during the 1980s. The limits to these campaigns, and the issues that they opened up, are also addressed. Third, I describe recent attempts to re-place class in economic geography, which have sought to de-center class yet retain some of the concept's explanatory value. This section also suggests that such accounts could benefit from paying greater attention to the significance of history and the role of political strategy. The chapter concludes by questioning whether class remains of significant actual or potential relevance to contemporary economic geography, or whether it is destined to remain forever silenced.

Classical Conceptions of Class

In essence, the difference between the classical Weberian and Marxist traditions of class, which have been so influential in social science, comes down to an insistence upon the properties of individuals versus the structural relationships embodied in production. A Weberian perspective emphasizes the role of classes as groups or collections of particular qualities and attributes (such as income or occupation) held in a contingent fashion by individuals. A Marxist viewpoint stresses the way in which the relationship between individuals is structured through the process of producing goods or delivering services. This is not the place in which to enter into a detailed exposition of the relative merits of Weberian and Marxist class theory, however. Rather, because of the centrality of the latter to debates within economic geography over the last two decades, and because I want to suggest that there are still merits to such an analytical framework, the rest of this section focuses on the specific contribution of, and debates around, Marxist conceptions of class in economic geography.

From a Marxist perspective, a key feature of the capitalist system of production is its separation into two classes, capital and labor. Capital is able to appropriate a surplus from the work of labor through ownership of the means of production, whilst labor possesses little more than an ability to perform paid work (see Swynge-douw, this volume). This surplus, or profit, has to be re-invested in further activities if the individual capitalist is not to be overtaken by competitors. In this view of history, there are limits to the long-term stability of the system as a whole, which are set by the contradictory nature of the relationship between the classes. Whilst capital needs labor-power, it also needs to replace it with (more efficient) machinery, creating unemployment. Thus the interests of capital and labor are frequently in conflict. While capital may seem to have the upper hand, its own strategy creates unintended consequences leading to economic crises which in time become increasingly generalized and widespread (Harvey, 1982).

There are of course many variants to this highly simplified Marxist account of class relations, and there have also been many different strands of criticism. At one level the original theory is teleological – it imputes an inevitable trajectory to human existence, even if both practical experience and intellectual debate suggest that there are in fact many different alternative paths. It is functionalist, in that society is held to develop in a certain way because that route is necessary for its existence. Some versions of Marxism are deeply structural, and offer only a limited role for human agency and human consciousness (see, for instance, Althusser, 1969), although others are more sensitive to individual and historical circumstances (see, for instance, Thompson, 1963) – and much of the debate reviewed below relates precisely to this question. Perhaps the key contribution of a Marxist perspective on class is its recognition of the linkages between individualized expressions and experiences of power and inequality, and broader system-wide processes. This was fundamental to the radical movement within human geography (and economic geography in particular) which developed from the late 1960s to the mid-1980s (see Cloke et al., 1991, pp. 28–56).

It was in this period that some of the most productive work in economic geography within a broadly Marxist perspective took place. Such research took class relations as a central starting point in explaining patterns and processes of uneven development within and between cities and regions, and (what were often described at the time as) the continued underdevelopment of the Third World and the legacies of imperialism. In some ways these years could be regarded as the highpoint of

Marxist class-based analyses within economic geography. By the end of the 1980s, however, it was apparent that the tensions created through engagement with social theory – and in particular the challenges posed by postmodernism's disavowal of broader structures – had led to fractious disagreement amongst Marxist economic geographers. This was evident if nothing else in the growing frequency of calls for re-establishment of a collective agenda, as cracks and fissures became increasingly evident in an earlier consensus (see, for instance, Walker, 1989). A body of theory often criticized for its "closed" assumptions – its limited accessibility to alternative ideas – faced a radical challenge in the 1990s, and – I would suggest – proved to be slow to adapt to new times and new intellectual concerns. Thus politically charged concepts of class which had gained a ready audience in the 1980s just as quickly fell from the agenda in the subsequent decade.

Class-centered Approaches to Economic Geography in the 1970s and 1980s: Production, Regional Development, and the Defense of Place

In some ways the decline of politically charged concepts of class was unfortunate, for many useful insights were gained during the period in which Marxist class analysis was commonplace in economic geography. In the course of the 1970s and 1980s, some of the most significant advances within human geography involved engagement with the relationship between systems of production and processes of uneven regional development. I focus on these here (in an admittedly partial fashion) in order to demonstrate some of the contributions made by Marxist-based class analysis to economic geography, and the contributions of geography to class analysis (similar arguments could also be advanced for other strands of research to do with urbanization, development, and imperialism). I argue in this section that Marxist class analysis brought to economic geography a fuller understanding of the implications of class-based contradictions and conflicts within capitalism as a system of production. It also enabled recognition of the social nature of production, involving questions to do with the deployment of labor, its engagement with management, and the range of occupational and technical divisions that might arise within the class of labor in the process of production. In turn, a geographical perspective enhanced Marxist class analysis through a focus on the role of space both in shaping class consciousness and in potentially dividing workers from each other.

The concern with regional inequality was in part a response to the new phase of the global economy ushered in by the recessionary slump of the mid 1970s. It was increasingly recognized that the organization of production was integrally related to questions of location, and that particular local and regional trajectories could only be understood as part of a broader national and international dynamic. At just the same time as the world was becoming economically more inter-connected, so place was ever more clearly of growing significance. Rather than simply conquering space and diffusing development (as earlier formulations would have it), capital was seeking new and more sophisticated means of exploiting and reinforcing the specificities of places, while integrating them into global processes.

One line of enquiry focused on the significance of capitalism as a mode of production, and sought to theorize in the abstract the uneven development of capitalist relations of production. For instance, Harvey (1982) developed a Marxist theory of capitalist crisis with three levels. The first rested upon the fundamental contradiction between capital and labor, and the tendency for capital's strategies of technical change to result in a falling rate of profit (Rigby, this volume). In the second, the financial credit system was a means of (temporarily) resolving such contradictions in the process of production, by guaranteeing the availability of capital for future rather than present use. Ultimately, however – so it was suggested – the financial system could only internalize capitalism's tendency to crisis. The third level introduced a specifically geographical aspect. In this account, the contradictions of capitalism were open to a "spatial fix" (as well as the temporal one through credit) in which geographical expansion into new regions ameliorated crisis tendencies. In this way capitalism was capable of "switching" crises from one region to another, with potentially devastating effects for those people and places left behind. The problem for capitalism as a system of production, however, was that such switches created geographic inertia. Investment in places at one point of time represented fixed capital, which became a barrier to future change. The growth of productive forces therefore increasingly acted as a barrier to rapid geographic restructuring, even though the latter became increasingly necessary. The more the forces of inertia prevailed, the deeper would be the regional impacts at moments of "switching." In this fashion, regional crises would tend to build to global crisis.

Harvey's work was significant for incorporating space into Marxist theorizations of crisis, and in emphasizing the significance of historical–geographical materialism. It was, however, highly abstract and separated from immediate political practice (though see Harvey, 1984). Another strand of enquiry of equal importance to economic geography in this period focused on the connections between the class-based social relations of production and the spatial composition of the economy, epitomized in Massey's (1984) "spatial divisions of labor." Production was seen as an essentially social process, reflecting the class-based hierarchies of decisionmaking and control to be found in large firms. The organization of production within particular factories and offices, in particular places and regions, results from decisions about the training and deployment of labor, and the degree of mechanization. Such choices created specific demands for certain attributes of labor. Within the consumer electronics industry, for example, there was a marked variation between the attributes required for high-level research and development functions at one extreme, and for the routinized tasks of factory-based assembly of mass-produced components on the other. In turn, the supplies of different kinds of labor were temporally and spatially variable, reflecting local labor market characteristics, which stemmed from distinctive regional growth paths.

As a consequence, different kinds of activities from the same firm or economic sector would be located in different places, creating spatial divisions of labor within firms, and often reinforcing differences between places. Attention thus focused on the formation of regional industrial structures through successive rounds or layers of investment, each of which took place on the basis of the legacies of previous layers. This geological analogy opened up several debates, including the primacy attached to economic process, and the extent and nature of the engagement between different rounds of investment (see, for instance, Sadler, 1992). For example, it was clear that production took place on an uneven plane and that decisions about the location of investment would only be taken in the light of a (perhaps partial) knowledge of previous place- and region-specific development paths. As spatially mobile capital sought new locations, management would take into consideration the legacies of prior industries in a region in terms of its workforce's skill characteristics, levels of unionization, and so on. Thus regions with industries in decline might be seen as potential sources of reserves of labor, although precisely which characteristics were attractive to new investors, and which were seen as disincentives to such investment, remained a matter for investigation.

In addition to such applications of Marxist class theory to improve the understanding of the processes behind urban and regional economic restructuring, economic geographers also paid attention to how geography affects the formation and cohesion of economic classes. Two broadly different kinds of contributions can be identified. First, the cohesion of classes (or indeed any social movement) does not just depend on sharing a common set of characteristics but also on being conscious of this. Solidarity involves developing a common consciousness that can overcome difficulties facing collective action, and economic geographers have shown how space can facilitate this. Since the construction of a collective identity depends on communication, the national formation of working-class solidarity – described for example by Thompson (1963) – requires, and should follow, the geographical development of communications systems which are central to the economic geographical landscape (Thrift and Williams, 1987). In addition, the regional agglomeration of large industrial factories and workshops where workers could come together and share their experiences, for example, in the old manufacturing belt of the United States, facilitates a corresponding regional agglomeration of strikes and union activity (Earle, 1992). Finally, on a local level, the traditionally strong nature of working-class solidarity (and the formation of working-class culture) in mining villages reflects their status as highly cohesive and often geographically isolated places based on a single industry and employer.

Second, it was argued that the geographical differentiation of the economy can undermine the cohesion of economic classes. In this view, solidarity between capitalists and workers in the same place, in opposition to workers and capitalists in other places, can arise because of processes of geographically uneven development which allow those residing in one place to prosper at the expense of those living in other places (Harris, 1983; Sheppard and Barnes, 1990; Urry, 1981). Such class-based alliances undermine the cohesion within classes, meaning that paying attention to geography greatly adds to the sophistication of class analysis. The existence of geographical differences in class consciousness (and identity) adds to the social complexity of classes long recognized in Marxist sociology – where in some contemporary accounts, classes are no longer treated as homogeneous categories which result solely from production processes.

Such research brought significant insight to the relationships between class, place, and space as organized through processes of production. In particular, the recognition that production was a social process took economic geography into closer contact with work in industrial sociology on the ways in which factory and office life was constructed through managerial strategies of consensus-building and conflict, involving engagement with the institutions of organized labor (see, for instance, Beynon, 1984). Some of these insights were also deeply political, representing an awareness of the extent to which the economy (and economic geography) is politically constituted. This can be illustrated through a brief consideration of the significance of the class-based campaigns in defense of place that grew in frequency in a number of old industrial regions during the 1980s.

Class-based campaigns in defense of place

During the 1980s, rapid large-scale contraction took place in many of the traditional bases of employment in Western Europe and North America, including coal-mining, iron and steel production, and shipbuilding. The extent of these closures, and the depth of their impact on places that had grown up around these industries, were such as to call into question the future existence of whole communities. Faced with this situation, many proposed closures became the focus of powerful and broad-based campaigns of opposition. These anti-closure movements were built on specific forms of expression of attachment to place – in the sense of settings for human existence – and to class, in the sense of preservation of the opportunity for waged labor (see Hudson and Sadler, 1983, 1986).

Few of the campaigns were successful in preventing closure, partly because of the extent to which national states intervened with both coercive measures (in the form of socially-legitimate force) and consensus-building policies, such as promises of alternative employment creation, superficially attractive terms for withdrawal from the labor market through early retirement, and opportunities for re-training. The campaigns were nonetheless deeply significant, both politically – as key moments in the restructuring of vast swathes of economic activity – and theoretically. In steel towns as far apart as Youngstown in the USA and Longwy in France, and in the coal fields of Britain, for example (see Beynon, 1985; Buss and Redburn, 1983), place and class were starkly revealed as fundamental constituents of economic activity and social life.

Much of the theoretical debate concerned the ways in which those different expressions of identity – attachment to region/community and to class – coexisted (see also Fitzgerald, 1991). In the interpretation of anti-closure campaigns it was necessary to adopt a more differentiated concept of class structure than a simple dichotomy between capital and labor; one which took into account competition *within* classes (although it was notable that practically all anti-closure campaigns were organized in support of the preservation of waged labor, rather than in opposition to the principle). In this way, the possibility of territorially-constituted alliances *between* fractions of locally-bound capital and labor was opened up. This was frequently evident in the ways in which opposition to large-scale closures drew upon support from small businesses and place-specific employers' organizations, although the pattern of coalitions was not always so straightforward. For instance, Herod (1991) described a case in which union leaders supported closure by stifling community-based opposition, whereas the strongly pro-business Governor of West Virginia sought to prevent closure. These and other forms of place-bound alliance can be interpreted in terms of different forms of dependence upon place (see Cox and Mair, 1991), signifying the extent to which the interests of particular class actors are necessarily bound up with those of the place in question at any particular point in time.

Territorial differences within the class of labor were significant to the outcome of many anti-closure campaigns. It was impossible to understand the year-long

1984–85 miners' strike in Britain without taking into account the divide between workers in different parts of the country, for example. That is not to argue, however, that territory and class should be seen as competing bases of social organization. Superficially there might appear to have been a choice between defending place (via specific cross-class alliances) and betraying class (via campaigns through which workers sought to preserve *their* mine or steelworks at the expense of some other mine or steelworks). In practice, I would argue, that is a false opposition. What these campaigns revealed instead was that class interests, organization, and practices are *always* formulated with respect to particular territories and places. Space cannot be added to class as an afterthought; the two are mutually constituted. The task of analysis is therefore to investigate how this happens; the political choices are to do with its preconditions and implications.

Class in Economic Geography in the 1990s

More recently, concepts of class have figured much less centrally on the agenda of economic geography. Research has focused on new and different questions from those of the production-led debates of the 1970s and 1980s, whilst there was also a conscious (and deliberately provocative) challenge to the relevance of Marxist class-based analysis. To take just one prominent example of the latter, Saunders and Williams (1986) complained that a new orthodoxy had emerged in urban geographic research, which over-emphasized the role of class. There was a reluctance, they argued, to accept that class might not be of primary significance in everyday life, or that there might be analytically distinct bases of domination and conflict in society of which class was only one (see also the response by Smith, 1987a).

The turn away from Marxist conceptions of class in economic geography in the 1990s had much to do with developments within human geography and social science in general. Poststructuralist approaches proliferated alongside a concern with identity, postmodernism ushered· in an era where broader structures were ruled to be inconsequential, and feminist critiques brought into question many central tenets of previous accounts – and this is not to mention (only through lack of space) other currents such as postcolonialism. These developments have often loosely been labeled part of the "cultural turn" within human geography, on which there has been much debate already, perhaps even before the turn has begun to near completion (cf. Barnett, 1998).

One of the most direct challenges to class theory came from poststructuralism (Gibson-Graham, this volume). As Barnes (1998, p. 96) put it:

... for poststructuralists, there is no coherent, sovereign individual, there is only a world of differences, of socially constituted identities that are multiple and complex. It is not class politics, but identity politics, and fought out not in the sphere of production but in the sphere of culture where those identities are forged.

As he went on to argue, such an ontological shift not only carried a very different view of the nature of society, but also a contrasting perspective on the role of academic enquiry. For (modernist) Marxists the role of the intellectual is as an ally of the working class, helping to reveal its material interests. For postmodernists, the academy (and not the factory or office as sites of production) becomes the arena of emancipatory politics.

Particularly significant in this regard is the work of Gibson and Graham (cf. Gibson and Graham, 1992; Gibson-Graham, 1996; Graham, 1990). Their post-structuralist Marxism has sought to break down the divide between poststructuralism and Marxism. It has explored three separate but related aspects of political economy in an attempt to re-position Marxism and class theory within geography: anti-essentialism, over-determination, and discursive constructions of the economy.

Essentialism is the intellectual presumption that complex realities are reducible to simple, or essential, realities. A key feature of the poststructuralist critique is that Marxism assigns such an essential property to "the economy," or more specifically to the labor process under capitalism, as something which both underlies the social system as a whole and is "out there" waiting to be uncovered. Such a view has been challenged both for its economic determinism – its attribution of causal primacy to economic process – and for its epistemological reduction of a complex and changing system to knowable essences or properties. (Teleology, introduced earlier in this chapter, is a specific form of essentialism in which the world is seen as governed by a grand design, the essence of historical progress.) It has been argued, however, that it is possible to construct an anti-essentialist Marxism with class as a focus – an entry-point for enquiry – but without attributing the status of universal explanation for everything to class theory (see Graham, 1990). Such an anti-essentialist Marxism involves conscious recognition of the validity of alternative entry-points for enquiry (Gibson and Graham, 1992, p. 114):

We do not wish to contribute to another Marxist knowledge that justifies itself by claiming that class is more fundamental or influential than other aspects of society and that, therefore, a knowledge of class has more explanatory power than other knowledges. Historically, such attempts to marginalise or demote other social processes and perspectives have created irresolvable conflicts and antagonisms between Marxism and other discourses of social transformation.

This acceptance remains of key significance to a more open Marxism within economic geography.

An anti-essentialist and open Marxism is also built on the concept of over-determination: the recognition that each and every social process is uniquely constituted as the effect of all other processes. This carries implications for the way in which the research process is conceptualized, because there are no simple causes that can be identified behind a phenomenon. Each starting point for the process of theoretical development – class, gender, ethnicity, or whatever – will not necessarily produce a *better* understanding (partly because validation criteria are internal to each theoretical framework rather than a function of its relationship to the real world), but a *different* understanding (Graham, 1990, pp. 58–9). In other words, theory is not an activity that clarifies how reality works, but becomes one of the many processes that constitute social life, producing particular knowledges that are necessarily specific and fragmentary.

Class remains central to this project, although precisely how it is conceptualized is itself problematic – and therein lie some of the problems associated with formulating a response to the critique of poststructuralism. In earlier work, Gibson and Graham (1992) argued that class should be understood in relation to social *processes* of exploitation, and that individuals should be seen as capable of participating in multiple class processes. This stress on class as process lay at the heart of the classical Marxist tradition (see above). In subsequent work, however, "re-positioning" class partly involved a view of class as a means of situating individuals into categories, rather than a structural property (cf. Gibson-Graham 1997, p. 90). This later account of class as a concept highlights relationships to property ownership, control over the labor process, exploitation, and organizational capacity, and argues that the problems with this "classification" arise if an individual occupies contradictory positions (for instance as both self-employed and an employee in a part-time job). In this way "class" becomes in part a category into which individuals could be allocated, one among many possible sources of differentiation – thereby eliding into a framework akin to that of classical Weberianism. These distinctions are more than purely semantic, and they inform the chapter's concluding remarks.

The third key feature of this body of work is its concern with the discursive representation of the economy. The extent to which the contemporary economy is "knowable" and therefore amenable to rational decisionmaking processes has been explored elsewhere (Thrift, 1996). For Gibson-Graham (1996, 1997) however, the key discursive problem is political economy's assumption of a single, unified, capitalist totality. They argue that much work in economic geography over the past two decades has involved the construction of a geography of "sameness" and of "class homogeneity," because it emphasizes an economic landscape dominated by capitalism in which non-capitalist class processes are disregarded or downplayed. This discourse has been created through an acceptance of alternative development projects built around the essentialist and masculinist construction of a prosperous (post-Fordist) capitalism as the only route to a non-capitalist alternative. They conclude that only by changing the framework will it become possible to visualize "the end of capitalism (as we knew it)."

These contributions to the debate on Marxist class theory have been significant, sustained, and productive. They have helped formulate a de-centered view of class identity, and an account of the economy that recognizes its multiply-fractured nature. They have enabled connections to be made between class theory and other entry-points into the process of understanding the world. Explicit recognition that class struggle is not the only motor of social change, and that the militaristic image of a unified collectivity of labor perpetually engaged in class warfare with industrial capital is itself of questionable relevance to the contemporary economy, are important advances. Much remains to be done, however.

Few would deny the deceptive power today of discourses proclaiming capitalist hegemony, and recent work has begun to address the discursive construction of the economy and our ways of knowing it. For instance, Schoenberger (1998) explored how the discursive strategies of others affected her own discursive constructions and how these, in turn, affected the material work of academic research. In other words, she engaged with debates about discourse as it affects the relationship between knowing the world and what goes on in the world. The meaning and use of the concept "competitiveness" were explored in two contexts – Nike's dependence on low-wage Asian labor for the manufacture of atheletic shoes, and the living wage campaign in Baltimore. In both cases, the hegemonic discourse of competitiveness opened up a terrain of legitimacy for some kinds of research and closed off other possibilities (such as the extent to which Baltimore's fundamental social problem was one of widespread poverty, or Nike's capacity to make profits even if Asian workers were paid at or close to US wage levels). In so doing, the need to examine academic discourse more closely was made all too apparent, particularly in terms of the way in which an "economic" discourse cemented into place, and drew strength from, the power of the discipline of economics and its influence on policymakers (compare Peck, 1999, who suggests that geographers' attention to local detail restricts the discipline's influence upon policy at a time when economists hold the key – but simple – ideas, or discursive representations, that are congruent with those of business). The implications of this for class theory are significant, for such an "economic" discourse is self-evidently ideological and reinforces the position of particular class interests.

Others have also recently begun to question the desirability of the retreat from Marxist class theory. For instance Castree (1999) seeks to lay the foundations for a renewed political economy, in an essay that reflects on the critiques of Gibson-Graham (1996) and Sayer (1995a), and lays out an "after-modern" Marxist politics of class. This involves recognizing that individuals occupy multiple class positions, and that the fashioning of a class identity is distinct from a structurally-assigned class position. In part the latter is a reaffirmation of Marx's classical distinction between class-in-itself (the objective interests of a class) and class-for-itself (the ability of a class to recognize, and act on, such interests), a distinction central to earlier debates over a body of theory known as rational choice Marxism (see Barnes and Sheppard, 1992).

Perhaps the most fundamental contribution from the debates of the last decade is the recognition that it is possible to retain a Marxist focus on class as one – but not necessarily the only – entry-point for enquiry within economic geography. It is possible to interrogate the ways in which class relations interact and intersect with other aspects of social existence, such as gender, ethnicity, and attachment to place (see for instance Hadjimichalis and Sadler, 1995). To take just one recent example: Gregson et al. (1999) graphically unpacked the diversity of forms of employment in Europe by examining the ways in which "work" (and particularly "atypical" categories of work such as seasonal, part-time, and self-employed) assumed different meanings in different contexts. "Work" therefore can be seen as a series of negotiated and culturally embedded practices, thereby exemplifying the utility of exploring the connections between economic and cultural processes.

At the same time, the continued significance of capitalist class processes, and their implications, should not go unremarked. Even though the economy has become much more differentiated in the later years of the twentieth century, it is presently – and increasingly, on a global scale – capitalist. Yet the class relations of capitalist society are not reproduced automatically. It could be argued that the process of globalization – the shifting flow of capital – is as much a sign of capital's weakness, its inability to subordinate labor, as it is a sign of strength, thereby opening up many different kinds of political opportunity (Holloway, 1995). Whilst it is one thing to recognize such difference, however, it is quite another to treat it in itself as a potential force of transformation. Below, therefore, I focus on the continued potential of class processes, as one amongst many possible sources of social transformation. Two issues are of significance in this context: the place of history, and the role of political strategy.

The place of history

Marxist class analysis is grounded in the principles of historical–geographical materialism. It is based on the premise that an understanding of processes of social transformation and change can only be achieved through analyses of concrete situations of class struggle, which occur in particular times and places, not necessarily of labor's own choosing. Thus a class-based perspective needs to take into account not just the role of geography, nor that of history in isolation, but of the simultaneous co-existence of place and time. This necessitates engagement with debates about the relationship between history and geography. Four different senses in which history has been conceived in human geography were identified by Driver (1988): a series of legacies, an evolutionary motor, a source of agency, and a grounding for theory. He concluded that it was time to bring history back in to the heart of human geography, as an essential part of doing the subject. Such a clarion call has many merits, particularly in the present context.

Informative insights can be gained here from ideas that might at first sight seem tangential, within the field of historical sociology, and in particular in the work of Abrams (1982). He argued that over a period of some 30 years the gap between so-called "empirical" history and "theoretical" sociology (as the two were frequently labeled) had narrowed, as both disciplines increasingly focused on a common project, the problem of *structuring*. Such a claim was exemplified by consideration of Thompson's (1963) view of class as relationship and process rather than object – the way in which the machine of society worked once it was set in motion. As Abrams (1982, p. xiii) put it:

Appreciation of the historicity of class, of class as a relationship enacted in time (with equal stress on all four of those words) is simply not a form of wisdom private to the historian. Nor are the larger insights that time exists in motion and that society is the time-machine working. Sociologists and historians alike need to understand how that maddeningly non-mechanical machine works if the puzzle of human agency is to be resolved.

The core of my argument here is that for "sociologists," we could – and should – just as well read "geographers" into the above. That is to say, understanding the way in which the "machine" of society works is a task that requires appreciation of both its temporal and spatial situatedness. Historical–geographical materialism provides a means of exploring the dynamics of place as an ever-changing construct of class-based relationships. In this time-centered sociology, Abrams duly acknowledged the early work of Giddens (for instance, 1979). Whilst much of Giddens' later work on structuration theory has found its way into human geography, it is unfortunate that some of these earlier insights developed within historical sociology – in particular on the role of history as process – have not yet received as much recognition within human geography in general, and economic geography in particular.

Political strategy

There is little new in the (still-frequent) proclamation that an era of class-based politics has been replaced by a different political order. In a celebrated phrase, Gorz (1982) bid farewell to the working class many years ago (in a book whose title, as has frequently been observed, carried two very different meanings). As Wills and Lee (1997, p. xvi, emphasis added) argued, however, "in the shift away from classical political economy, questions of *political agency* have seemingly been overlooked." In this they were in accord with Sayer (1995b, pp. 79, 82), who commented that a "softer," more pluralist Marxism had emerged following debates with feminism, postmodernism, and new social movement theorists, but that a negative consequence of this was:

> . . . a neglect of basic questions of political economy at the core of Marxism, both in terms of abstract theorising about how capitalism works, and theorising about possible alternative systems. . . . The crisis of Marxism has much to do with the fact that it has become increasingly apparent that even if the problems are structural and present in all versions of the game, it hasn't got a clue as to what would constitute a better game. Postmarxists are torn between criticising the structure of the game and criticising the particular ways individuals and institutions play it.

Thus a problem to be addressed remains that of alternative class-based strategies for, and ways of interpreting, the economy.

Key questions of such political practice were evident in the debate over localities research in the mid to late 1980s, on which much has been written already (see Massey, 1991). The questions remain of significance, however, and for that reason alone it is appropriate to review some of the issues raised then. For instance, Smith (1987b) argued that the problem with such research was that it had become about specified places in and of themselves (partly revisiting earlier debates within geography's disciplinary history concerning its "exceptionalist" attention to the detailed study of particular places and regions, bereft of broader theoretical implications). This criticism had a political significance. There was a prospect that emphasis on the local might lead to "economic microsurgery" (Cochrane, 1987): tinkering with the workings of the economy at a small scale, but neglecting the system as a whole.

In part such criticisms were well-placed, as left-of-center political practice at this time seemed increasingly to resort to the local in the face of the national electoral successes of neo-liberal ideologies. In part, however, they failed to recognize the differences within (what was loosely labeled as) "localities research," for some of this research *did* seek to connect the local with the national and the global, and to explore the class implications of these connections (see, for instance, Beynon et al., 1994). Such a task remains conceptually unfinished, however, and it is addressed further in the concluding comments below.

This section has explored the ways in which the use of Marxist class theory in economic geography was challenged by alternative critiques such as poststructuralism, postmodernism, and feminism during the 1990s. The response to these challenges has been examined through a review of the work of Gibson, Graham, and

Gibson-Graham, focusing on three features: anti-essentialism, over-determination, and discourse. It has been argued that class remains of (under-utilized) significance as an explanatory concept within economic geography, and that two questions warrant further consideration in this regard: the role of history, and the nature of political strategy. The concluding section of this chapter goes on to expand these comments in the light of the earlier review of the achievements within economic geography enabled through the insights brought by Marxist class theory during the 1970s and 1980s.

Concluding Comments

At this point it is appropriate to summarize my own position with respect to the debates reviewed in the previous section. Firstly, the recognition that it is possible to construct an anti-essentialist Marxist theory of class is significant for enabling a more open dialogue between Marxism and other traditions of critical enquiry – some emergent, others longer-established. Secondly, the conceptual framework of over-determination is valuable in so far as it situates class within a broader range of social processes, and encourages exploration of the connections within and between these. The problem remains however of evaluating the merits of different entry-points for academic enquiry, not just in terms of their own internal validity, but also of the extent to which they enable a meaningful story to be told. One view would argue that this is an ineligible question: that in a relativist world, it is not necessary or legitimate to assess the competing claims of different entry-points, precisely because they are not competing. I would hold, however, that there are criteria by which the validity of different entry-points can be established, and that these are in part ethical and in part political.

Thirdly, I have suggested above that it is possible to incorporate a recognition of the significance of discourse without accepting the full implications of Gibson-Graham's (1996) argument in *The End of Capitalism (As We Knew It)*, by reviewing in brief the work of Schoenberger (1998) (and note that her more substantive work on *The Cultural Crisis of the Firm* could also be seen in this light – Schoenberger, 1997). That is to say, I argue that whilst an awareness of the need to challenge hegemonic discourse is a *necessary* condition for transformation, it is not a *sufficient* one, in the sense of automatically enabling such transformation to take place. In part this brings the argument back to the question of political strategy, and to the role of academic enquiry. I would argue that it is necessary to bring class back in – the ongoing class history of real places and of their ever-changing relationship to, and constitution within, broader economic processes (see, for instance, Allen et al., 1998).

These issues can be exemplified with respect to the debate on space and time partly triggered by Harvey's (1989) *The Condition of Postmodernity*. In this, Harvey sought to argue that postmodernism could be seen as a particular condition of historical–geographical materialism that was characterized by space–time compression, in which the speed-up of economic change had overcome spatial barriers, whilst generating an increased fragmentation and differentiation amongst places that had led to militant neo-particularism, an intensified association with the local. Thus Harvey sought to interpret postmodernism as a condition of social existence in which time had transformed space. For Massey, (1992), however, the manner of the emergence of space into the theoretical agenda of social science more generally was problematic, in that the concept had been de-politicized. Drawing on feminist critiques of dualism, she argued that space was *not* something to be defined in opposition to time, but rather should be conceptualized as the simultaneous co-existence of social interrelations at all scales, from the local to the global. This simultaneity is not static, but ceaselessly changing. Such an approach is akin to that envisioned here, enabling class and place to be seen as central elements in an explanatory framework for social transformation.

These insights have partly informed more recent work, which has stressed the way in which scale is a socially-produced and contested process (see, for instance, Swyngedouw, 1997). There is no necessary association between any given process and a particular scale: rather, the way in which that process is scaled is part of the simultaneity of coexistence captured by over-determination. So in interpreting the politics of anti-closure campaigns in the 1980s, for instance, it is necessary to explore the ways in which class (and other) interests were articulated with respect to place at a range of different spatial scales. This interpretation is only possible through a framework that conceptualizes class as process – as dynamic and ongoing – and one grounded in history, not just of any one place but of the changing web that connects different scales.

Thus it is still possible to think about a world in which class is a significant process, albeit in ways very different from earlier formulations. The need for class-based interpretations in economic geography is still starkly evident in the harsh impact on those places left behind by the ebb and flow of capital around the world (see, for example, Dandaneau, 1996). It is productive to explore the connections between class and other processes. Class as an explanatory concept might have been sidelined during the 1990s, but it has by no means been silenced.

Bibliography

Abrams, P. 1982. *Historical Sociology*. Shepton Mallet: Open Books.

Allen, J., Massey, D., and Cochrane, A. 1998. *Rethinking the Region*. London: Routledge.

Althusser, L. 1969. *For Marx*. Harmondsworth: Penguin.

Barnes, T. 1998. Political economy III: confessions of a political economist. *Progress in Human Geography*, 22, 94–104.

Barnes, T. and Sheppard, E. 1992. Is there a place for the rational actor? A geographical critique of the rational choice paradigm. *Economic Geography*, 68, 1–21.

Barnett, C. 1998. The cultural turn: fashion or progress in human geography? *Antipode*, 30, 379–94.

Beynon, H. 1984. *Working for Ford*. Harmondsworth: Penguin.

Beynon, H. (ed). 1985. *Digging Deeper: Issues in the Miners' Strike*. London: Verso.

Beynon, H., Hudson, R., and Sadler, D. 1994. *A Place called Teesside: a Locality in a Global Economy*. Manchester: Manchester University Press.

Buss, T. and Redburn, F. 1983. *Shutdown at Youngstown: Public Policy for Mass Unemployment*. Albany: State University of New York Press.

Castree, N. 1999. Envisioning capitalism: geography and the renewal of Marxian political economy. *Transactions of the Institute of British Geographers*, 24, 137–58.

Cloke, P., Philo, C., and Sadler, D. 1991. *Approaching Human Geography: an Introduction to Contemporary Theoretical Debates*. London: Paul Chapman.

Cochrane, A. 1987. What a difference the place makes: the new structuralism of locality. *Antipode*, 19, 354–63.

Cox, K. and Mair, A. 1991. From localized social structures to localities as agents *Environment and Planning A*, 23, 197–213.

Dandaneau, S. 1996. *A Town Abandoned: Flint, Michigan Confronts Deindustrialization.* Albany, NY: State University of New York Press.

Driver, F. 1988. The historicity of human geography. *Progress in Human Geography*, 12, 497–506.

Earle, C. 1992. The split geographical personality of American labor. In C. Earle (ed). *Geographical Inquiry and Historical Problems.* Palo Alto, CA: Stanford University Press, 346–77.

Fitzgerald, J. 1991. Class as community: the new dynamics of social change. *Environment and Planning D: Society and Space*, 9, 117–28.

Gibson, K. and Graham, J. 1992. Rethinking class in industrial geography: creating a space for an alternative politics of class. *Economic Geography*, 68, 109–27.

Gibson-Graham, J. K. 1996. *The End of Capitalism (As We Knew It): A Feminist Critique of Political Economy.* Oxford: Blackwell.

Gibson-Graham, J. K. 1997. Re-placing class in economic geographies: possibilities for a new class politics. In R. Lee and J. Wills (eds). *Geographies of Economies.* London: Arnold, 87–97.

Giddens, A. 1979. *Central Problems in Social Theory.* London: Macmillan.

Gorz, A. 1982. *Farewell to the Working Class* London: Pluto.

Graham, J. 1990. Theory and essentialism in Marxist geography. *Antipode*, 22, 53–66.

Gregson, N., Simonsen, K., and Vaiou, D. 1999. The meaning of work: some arguments for the importance of culture within formulations of work in Europe. *European Urban and Regional Studies*, 6, 197–214.

Hadjimichalis, C. and Sadler, D. (eds). 1995. *Europe at the Margins: New Mosaics of Inequality.* London: Wiley.

Harris, R. 1983. Space and class: a critique of Urry. *International Journal of Urban and Regional Research*, 7, 115–21.

Harvey, D. 1982. *The Limits to Capital.* Oxford: Blackwell.

Harvey, D. 1984. On the history and present condition of geography: an historical materialist manifesto. *Professional Geographer*, 36, 1–11.

Harvey, D. 1989. *The Condition of Postmodernity: an Enquiry into the Origins of Cultural Change.* Oxford: Blackwell.

Herod, A. 1991. Local political practice in response to a manufacturing plant closure: how geography complicates class analysis. *Antipode*, 23, 385–402.

Holloway, J. 1995. Capital moves. *Capital and Class*, 57, 136–44.

Hudson, R. and Sadler, D. 1983. Region, class, and the politics of steel closures in the European Community. *Environment and Planning D: Society and Space*, 1, 405–28.

Hudson, R. and Sadler, D. 1986. Contesting works closures in Western Europe's old industrial regions: defending place or betraying class? In A. Scott, and M. Storper (eds). *Production, Work, Territory: the Geographical Anatomy of Industrial Capitalism.* London: Allen & Unwin, 172–93.

Massey, D. 1984. *Spatial Divisions of Labor: Social Structures and the Geography of Production.* London: Macmillan.

Massey, D. 1991. The political place of locality studies. *Environment and Planning A*, 23, 267–81.

Massey, D. 1992. Politics and space/time. *New Left Review*, 196, 65–84.

Peck, J. 1999. Grey geography? *Transactions of the Institute of British Geographers*, 24, 131–5.

Sadler, D. 1992. *The Global Region: Production, State Policies and Uneven Development.* Oxford: Pergamon.

Saunders, P. and Williams, P. 1986. The new conservatism: some thoughts on recent and future developments in urban studies. *Environment and Planning D: Society and Space*, 4, 393–9.

Sayer, A. 1995a. Liberalism, Marxism and urban and regional studies. *International Journal of Urban and Regional Research*, 19, 79–95.

Sayer, A. 1995b. *Radical Political Economy: a Critique.* Oxford: Blackwell.

Schoenberger, E. 1997. *The Cultural Crisis of the Firm.* Oxford: Blackwell.

Schoenberger, E. 1998. Discourse and practice in human geography. *Progress in Human Geography*, 22, 1–14.

Sheppard, E. and Barnes, T. 1990. *The Capitalist Space Economy.* London: Unwin Hyman.

Smith, N. 1987a. Rascal concepts, minimalising discourse and the politics of geography. *Environment and Planning D: Society and Space*, 5, 377–83.

Smith, N. 1987b. Dangers of the empirical turn: some comments on the CURS initiative. *Antipode*, 19, 59–68.

Swyngedouw, E. 1997. Excluding the other: the production of scale and scaled politics. In R. Lee and J. Wills (eds). *Geographies of Economies.* London: Arnold, 167–76.

Thompson, E. P. 1963. *The Making of the English Working Class.* Harmondsworth: Penguin.

Thrift, N. 1996. Shut up and dance, or, is the world economy knowable? In P. Daniels and W. F. Lever (eds). *The Global Economy in Transition.* London: Longman, 11–23.

Thrift, N. and Olds, K. 1996. Refiguring the economic in economic geography. *Progress in Human Geography*, 20, 311–37.

Thrift, N. and Williams, P. 1987. *Class and Space: the Making of Urban Society.* London: Routledge.

Urry, J. 1981. Localities, regions and class. *International Journal of Urban and Regional Research*, 5, 455–74.

Walker, R. 1989. What's left to do? *Antipode*, 21, 133–65.

Wills, J. and Lee, R. 1997. Introduction. In R. Lee and J. Wills (eds). *Geographies of Economies.* London: Arnold, xv–xviii.

Chapter 21

Labor Unions and Economic Geography

Andrew Herod

Issues of labor have long been central to understanding the location of economic activities – Weber's (1929 [1909]) locational model, for example, was based, in part, on the geography of labor costs. However, for most of the twentieth century, economic geographers have tended to see labor purely in terms of its costs to capital and how these impacted firms' locational decision-making processes. Only recently has this begun to change as economic geographers have sought to unpack the monolithic category of "labor" to understand how geographic context makes a difference to how workers behave and what this means for understanding how economic landscapes are made (Peck, this volume; Walker and Storper, 1981; Storper and Walker, 1983). Part of this unpacking has involved studying the spatial aspects of labor unionism. In this chapter I outline the principal issues involved in this project. The first section provides a brief overview of several major themes that emerged in the late 1980s and blossomed in the 1990s with regard to studying geographical aspects of workers' and unions' behavior. Its goal is to highlight why understanding the spatial context within which unions operate, together with the impacts unions' activities have on the landscape, is crucial if we are to more fully comprehend not only how the geography of capitalism is currently made but also how it can be made in a more progressive fashion. The second section examines some of the theoretical debates that have shaped the study by geographers of unions and their activities. The third section ponders how some recent developments in the way capitalism is organized geographically may affect unions.

Themes in the Geographic Study of Labor Unions[1]

In the mid-1980s, a number of economic geographers began to study how the geographic context within which labor unions find themselves affects how they operate. Several themes emerged, including how unions were organized geographically; how the state regulates spatially the activities of labor unions; how unions and workers have shaped the evolution of economic landscapes; how new geographical relationships between work and home are affecting the geography of labor organization; how new work arrangements such as the rise of just-in-time (JIT) production and the explosion of service sector employment are forcing unions to adopt different models of organizing; how workers' lives are structured geographically and what this means for unions' abilities to organize; and how workers' "senses of place" shape, and are shaped by, local or regional cultural practices and contexts. Elsewhere (Herod, 1998a) I have suggested that four principal and overlapping foci of research have emerged: the geography of labor union regulation; the relationship between unionism and the economic geography of capitalism; political geographies of union organizing; and how place and local context shape processes of union organizing.

Geographies of labor union regulation

All modes of social regulation have specific, and sometimes unintended, geographic consequences. A concern with the geographic impacts of systems of social regulation, such as labor law, formed the basis for much work within economic geography on labor unions beginning in the mid-1980s. In particular, this research argued that it was important to understand the geographical assumptions built into the laws governing unions if the dynamics of a country's labor unionism were to be understood. Such ideas were taken up in Gordon Clark's work on the National Labor Relations Board (NLRB) in the USA.[2] Principally, Clark sought to show that the geographical assumptions behind US labor regulation affected unions' political and organizational possibilities. He argued, for example, that US labor law has privileged local traditions and social practices, suggesting that while this may allow for collective bargaining and organizing that are sensitive to local conditions, it has also hindered unions in developing national agreements and labor standards (Clark, 1989). As corporations have increasingly become multi-locational they have been able to use variations in labor law and collective bargaining practices in different parts of the USA to extract concessions from labor, a practice that would be considerably harder if national norms prevailed. Similarly, Johnston (1986) examined how the Uniform Commercial Code and the Commerce Clause of the US Constitution encouraged the establishment of a national market for goods, while the local focus of labor law created a spatially heterogeneous set of rules concerning collective bargaining. The result of this has been to give capital great opportunities to engage in "whip-sawing," in which workers are played against each other on the basis of variations in wages, collective bargaining agreements, working conditions, and the like. Clark also showed how new working arrangements such as the growth of "flexible production" and "teamwork" were forcing the NLRB to adapt new models of labor regulation, together with how the Board's varying interpretations of labor law (over time and between places) could shape the evolution of the industrial landscape (Clark, 1986, 1988).

Several other authors have also looked at the ways in which labor law has structured the geography of labor unionism and, in turn, the evolution of economic geographies. Herod (1997a, 1997b) showed how decisions by the NLRB and law courts affected the geography of dual unionism and work location in the East Coast longshoring industry, while Finch and Nagel (1983) illustrated how a change in the regulatory environment concerning teachers in Connecticut led to a greater standardization of contracts across the state. Relatedly, in comparing the strategies typically adopted by unions operating under the National Labor Relations Act with those operating under the Railroad Labor Act (RLA), Baruffalo (1996) has highlighted how different regulatory systems can shape unions' geographical practices. He suggests that whereas the former group of unions often attempts to develop national strikes as a way of pressuring employers, railroad unions frequently attempt to keep strikes local because, under the terms of the RLA, a national shutdown of rail traffic may invite congressional intervention into their industry. Providing a British comparison, Blomley's (1994) analysis of the 1984–85 British coal miners' strike highlighted how the Conservative government used the law to prevent striking miners from traveling between coal fields and so developing solidarity across space. All of these works, in various ways, have illustrated the interconnections between the regulatory environment within which unions must operate and the implications this has for how economic landscapes are made.

Unions and the economic geography of capitalism

A second line of investigation has studied how the geography of labor unionism has shaped the economic geography of capitalism more immediately (as opposed to through the nexus of the regulatory system). An early study by Peet (1983), for example, argued that what he called "the geography of class struggle" had fundamentally determined the way in which the US national economy has developed geographically as high levels of union membership, wages, and strikes (which Peet took to indicate a high level of class struggle) in northeastern states have encouraged the migration of capital to the southern and southwestern states where union membership levels, wages, and strikes have been lower. More recent research questions some of these empirical findings (see Herod, 1997c), but the conceptual argument that labor unions can affect the economic landscape by repelling capital is important for theorizing both how the geography of capitalism is made and labor's role in that process. Other examples include Gordon's (1978) analysis of the historical geography of urbanization in the USA, in which he argued that much of the impetus for the suburbanization of manufacturing in northern industrial regions has been the result of companies' desire to escape unionized urban workforces. Page (1998) made a similar argument with regard to the decentralization of the meatpacking industry from centers of militant union power such as Chicago to smaller towns throughout the US Midwest which had more pliable unions. A similar process appears to have occurred in Britain during the 1980s, as investment flowed from urban to suburban and rural areas in a process designed to replace militant urban unions with less militant suburban/rural ones (Church and Stevens, 1994).

Such analyses illustrate how capital has attempted to negotiate its way in the landscape by avoiding areas of labor militancy – thereby showing how labor unions have indirectly affected the evolution of economic landscapes. A different tack has been taken by other analyses showing how organized labor has more directly shaped the economic geography of capitalism. Parson (1982, 1984), for example, highlighted the role played by unions in the USA in debates over suburbanization and urban renewal, and how they actively shaped these processes as part of a solution to the housing and employment problems faced by many of their members. In his analysis of agriculture in early twentieth-century California, Mitchell (1996) argued that the spatial practices of migratory workers shaped the ways in which the rural landscape was fashioned. In what was perhaps an early hint of the recent "cultural turn" in economic geography, Cooke (1985) analyzed how the cultural practices associated with unionism had been fundamental to the establishment of South Wales as a coherent industrial and political region in the nineteenth century. Turning to more recent events, Hudson and Sadler (1983, 1985, 1986) have argued that the activities of steelworkers, particularly their responses to government efforts in France and Britain to privatize their industry, significantly shaped processes of economic restructuring in these two countries' steel industries. Focusing on the international activities of the US labor movement during the twentieth century, Herod (1997d) has documented how organized labor played an important role in structuring patterns of economic development in Latin America and the Caribbean. To ensure jobs for US workers, many US union officials helped corporations "open up" the region's markets for US products.

A final aspect of this strand of work has examined how the geographical organization of the economic landscape affects the ways in which unions organize. Earle (1992a) suggests, for example, that a geographical examination of strikes in the USA in the 1880s and 1890s highlights the changing characteristics of organized labor. He maintains that as urbanization proceeded, and cities such as Chicago drew in ever larger numbers of migrants from rural and small-town America, together with immigrants from Europe and elsewhere, the bringing together of workers sharing new common experiences facilitated the growth of organized labor's economic and political power. (Thompson (1963) has made similar arguments about how the process of urbanization in Britain during the Industrial Revolution allowed for the "collectivization" of worker protest.) In a quite different situation, Holmes and Rusonik (1991) similarly argued that changes in the economic landscape may dramatically impact unions' structure and capacities to exert political and economic power. For example, the widening of production–cost differentials in the auto industry between Canada and the USA in the late 1970s created a situation in which workers on opposite sides of the border were experiencing very different sets of economic conditions. This, they argued, created such tension between the Canadian and the US segments of the United Auto Workers union that the Canadian workers split from their US colleagues to form a new Canadian Auto Workers union.

The political geography of union organizing

Labor organizing, like any other type of political organizing, is geographically informed. A third avenue of research, then, has been to investigate how space is implicated in union organizing, both in terms of how differences across the economic landscape shape the context within which organizing is occurring and also in terms of the geographical assumptions that are built into different models of organizing. For example, in their discussion of efforts to organize two quite different groups of workers – clerical workers at Yale University and janitors in Los Angeles – Berman (1998) and Savage (1998) have shown how geography informed the tactics and models of organizing adopted by both groups. In the case of Yale this involved dividing the campus up into zones in which different groups of organizers would operate, while in the case of Los Angeles's janitors it involved moving the campaign for a living wage from the private spaces of the workplace (i.e. the individual buildings within which the janitors worked) to the public spaces of the street, thereby bringing public pressure to bear upon the employers.

The geographical mobility of union organizers has also been examined. Indeed, Southall (1988, 1989, 1996) has argued that the mobility of artisans in eighteenth-century Britain laid the basis for the formation of unions in many trades as artisans took ideas about unionism with them to different parts of the country or even abroad. Wills (1998a) has likewise shown how the migration of union organizers between different factories in northern England during the 1960s, 1970s, and 1980s helped both to spread ideas about labor unionism and to develop solidarity between workers in these different plants – a solidarity that became important during a number of industrial disputes. Mitchell (1998) has highlighted the role played by traveling organizers in spatially connecting strikes by agricultural workers in different parts of California in the 1930s. All of these authors have argued, in different ways, that ideas about unionism can be transmitted from place to place through the geographic migration of union activists, helping develop solidarity between workers in different communities. Equally, they have suggested that understanding this process can provide insight into the evolution of cultures of labor unionism in different places in ways that non-spatial approaches cannot. For example, analysis of labor migration, with migrating workers bringing new ideas about unions with them, may help account for sudden transformations in local cultures of unionism in ways that purely historical analyses of such places' internal dynamics do not (this argument is laid out more fully in Herod, 1998b).

A final theme involves understanding how different cultures of work and political life may shape the political geography of union organizing. Earle (1992b), for instance, suggests that through their understanding of the geography of support and non-support for the 1886 general strike in the USA, leaders of the American Federation of Labor (AFL) quickly realized that the US working class was deeply divided along geographical lines. The failure of the general strike in many parts of the country brought the AFL to the view that its own interests were perhaps best served by a narrowly-defined craft unionism focused upon self-identified constituencies in the northeastern industrial heartland, rather than by an industrial unionism that was broader both sectorally and geographically.[3] These leaders determined that a geographically decentralized organizational structure would be a way to minimize the spread of industrial violence that had precipitated widespread state repression of unions after the 1886 strike. The result has been a rather weak federative style of unionism at the national level in the USA, in contrast to the strong centrally run systems in countries such as Germany. (In similar fashion, Charlesworth et al. (1996) suggest that variations in local work cultures and attitudes help explain the geography of industrial protest in Britain between 1750 and 1990.)

Local context and the power of place

Geographers have long regarded the concept of "place" as holding a somewhat vaunted position within the discipline. This is evident in work conducted on labor unionism, which has attempted both to illustrate how union practices vary across the economic landscape but also how the specific geographic context within which organized workers live may shape their practice of labor unionism. For example, Painter (1991) has suggested that public-sector unions' responses to privatization in Britain have been conditioned by local histories and cultures of union activity, economic well-being, and past experiences with privatization. In turn, differences across the landscape in these unions' responses have shaped the process of privatization itself in an on-going manner – greater opposition in some locales has hindered privatization relative to other places where it has been more easily undertaken. Likewise, Wills's (1996) examination of the geography of unionism in the British banking industry highlighted divergent traditions and attitudes towards unionism between Warrington in the north of England and the "new town" of Welwyn Garden City in southern England which are, in turn, affecting patterns of investment and the introduction of new technologies and work practices into the financial industry. Relatedly, Martin et al. (1994a, 1994b) have shown how local context shaped the evolution of industrial relations and union politics in the British engineering industry. As employers have pressured unions to decentralize the geographical scale at which bargaining takes place, nationally uniform contracts are being replaced by a patchwork of locally negotiated ones which may vary considerably from place to place. As a result, capital may be allowed greater opportunities to play plants and localities against one another, exploiting differences in their local conditions and contracts (a situation that also exists in the USA where many national agreements were dismantled in the 1980s (Herod, 1991)).

Theoretical and Methodological Debates

A number of theoretical and methodological debates and discussions have shaped the study of the spatial nature of labor unionism. The most significant conceptual issues have involved debates over the relative importance of "culture" versus "economy" in understanding patterns of unionization (which is itself somewhat reflective of the challenge to traditional economic geography that the "cultural turn" has posed), matters of the geographical scale at which unions carry out their activities, and contrasts in approach between examining the "geography of labor" and "labor geographies." These debates have also raised methodological issues about how to carry out research in this area.

One of the most significant discussions concerning the spatiality of unionism emerged in Britain as a result of efforts to explain the geography of the 1984–85 strike by the National Union of Mineworkers (NUM) in opposition to the government's plans to privatize the coal industry and shut down many unprofitable mines. The debate focused on the relative significance of economic factors and of cultures of unionism in explaining geographical differences in support for the strike and the subsequent secession from the old NUM of miners in the East Midlands. Whereas Rees (1985, 1986) suggested that the pattern of support for the strike had to do largely with the changing geography of investment in the industry – with the National Coal Board shifting production away from peripheral regions such as South Wales and towards lower-cost fields in the East Midlands – Sunley (1986, 1990) and Griffiths and Johnston (1991) maintained that a more significant explanator was the differing cultures of labor politics in regions such as Scotland and South Wales (where miners have a long history of socialist politics) compared to the East Midlands (where miners have tended to be less socialist in outlook).

A slightly different debate about "culture versus economy" in examining the geography of trade unionism in Britain involved an interchange between Massey and a number of her co-workers (Massey, 1994; Painter, 1994; Massey and Miles, 1984; Massey and Painter, 1989) and Martin et al. (1993, 1994c; see also 1996). Much of this interchange involved a methodological debate concerning changing national patterns of union representation. Massey and her colleagues used data on the changing share of unions' membership accounted for by different parts of Britain to suggest that the geography of unionism evident for much of the twentieth century underwent a dramatic transformation in the post-1970 period. They argued that the traditional "heartlands" of unionism such as Scotland, Wales, and Northern England had seen both a relative and an absolute decline in union membership as a result of the geographic reorganization of capital, the failure of unions to organize in regions of new investment, and the relocation of many government functions away from London (which facilitated the growth of public sector unionism in more peripheral regions). Taken together, these developments were seen to have brought about a greater equalization of levels of unionism across the national landscape as many traditional regions of manufacturing unionism were decimated and other regions with little union tradition experienced a growth, particularly in service and public sector unionism. Such developments were understood to suggest that economic restructuring and political reorganization had diminished the influence of local union culture in traditional union heartlands.

In contrast, Martin et al. maintained that the heartlands of British unionism had remained quite resilient during the post-1970s economic restructuring and had not experienced the kind of "smoothing out" suggested by Massey et al. Their opposing view was based on the fact that Martin et al. measured levels of union membership in a different manner, using union *density* (the proportion of workers in different industries and regions who are actually union members) instead of the percentage share of unions' membership accounted for by various regions. Based upon this analysis, they argued that, in fact, economic restructuring had had little effect on relative levels of density for most unions – that is to say, that the heartlands still retained generally higher levels of union density than did the traditionally less-unionized regions. They inferred from this that economic restructuring was less significant in explaining patterns of unionism in Britain towards the end of the twentieth century than were local cultures and traditions of unionism, which seemed to have a certain degree of resilience across time. Though not specifically related to this "culture versus economy" debate, Church and Stevens (1994) suggested that an examination of union density in Britain at the urban and rural – rather than regional – scale of analysis showed patterns different from those described by either Massey et al. or Martin et al.[4]

Such variations in interpretation resulting from analyses at different geographical scales (urban versus regional, for example) highlight the question of geographic scale with regard to examining the spatiality of labor unionism. Certainly, as outlined above, scale has been implicated methodologically in the study of unionism – that is to say, varying the scale of the analysis may lead to differing interpretations of what is being observed, such that an apparent concentration of union membership measured at one scale may appear to be a dispersed pattern at another scale. More interestingly, however, much analysis of the spatial tactics engaged in by workers and unions has also attempted to understand how they have created different geographical scales of organization as an integral part of their political practice, together with how such efforts have been contested by employers, by the state, and by different fractions of workers with different political visions. The ability, for example, to "scale up" from local to national systems of wage bargaining, in which all workers in an industry are paid the same wage, allows unions to eliminate wage competition, thereby limiting employers' abilities to play different groups of workers against each other. In such an example, the ability to construct a new scale of bargaining – national rather than local – represents a significant victory for workers. Such conflicts over scale, though, are not always simply between capital and labor. Rather, they may involve struggles between different groups of workers fighting amongst themselves to organize their activities on a spatial scale that most helps them attain their own goals. For instance, workers in high-unemployment regions may fight against the establishment of national contracts, fearing that nationally set wages may price them out of local labor markets where conditions only support low-wage production. Equally, rival factions within a union might look to achieve different scales of organization. In the case of the British miners' strike, the national NUM leadership attempted to maintain national unity whereas miners in Nottinghamshire broke away to form a regional union (the Union of Democratic Miners) which, they felt, would better reflect their political goals. Social conflicts, then, frequently involve one group of actors trying to confine their opponents to a scale of social operation that limits their abilities to achieve their goals, while making sure their own operations are organized at the appropriate geographical scale to facilitate their own action. How such conflicts are resolved has important implications for the ways in which the economic geography of capitalism is made.

A further issue has concerned the differentiation between what has been termed a "geography of labor" and a "labor geography," a nomenclature intended to distinguish two approaches to examining the spatiality of labor unionism (Herod, 1997e). The term "geography of labor" describes approaches in which workers and their organizations appear in explanations in rather passive terms, as social objects between which capital chooses in making its locational decisions. Such approaches tend, implicitly, to tell the story of how economic landscapes are made from the perspective of capital, which uses such differences to its own advantage. Peet's (1983) analysis of the geography of class struggle in the USA, for example, while highlighting the varying characteristics of labor in different parts of the country, was principally aimed at showing how capital exploited such variations as it reorganized itself in the post-WWII period in the great "Snowbelt–Sunbelt" migration from the old industrial heartland to the new industrial spaces of the South and Southwest. The term "labor geography" was coined to describe approaches that attempt to incorporate a more active sense of workers and their organizations struggling to shape the economic landscape as an integral part of their own social practices. Such approaches seek to tell the story of how economic landscapes are made through the eyes of labor, to show that workers, too, have a vested interest in ensuring that the geography of capitalism is made in some ways and not others. Thus, through their spatial strategies and geographical struggles, workers might attempt to develop a certain "spatial fix" (cf. Harvey, 1982) which they see as central to achieving their social and political goals.

The purpose of this distinction was neither to create a taxonomy of studies, nor to suggest that understanding how the geography of capitalism is made could rely upon only one approach. Rather, it was intended as a corrective to the narrow ways in which economic geographers, informed by both mainstream (i.e. neoclassical) economic theory and Marxist theory, had usually thought about labor. It is important to understand how corporations play different locations against each other based upon their labor characteristics, but it is equally important to understand how workers and their organizations may successfully shape the economic landscape in ways which they prefer – ways that an examination solely of the activities of capital will not reveal. Nevertheless, the two approaches do tend to implicate different methodologies in their explanations, which in many ways reflect Sayer's (1982) distinction between what he called "extensive" and "intensive" research practices. Studies of the "geography of labor" (such as Peet, 1983) tend to be more descriptive of labor, tend to focus upon patterns of labor across the landscape, and tend to lend themselves more easily to quantitative analysis. Those focusing upon "labor geography" (cf. Herod, 1998c) tend toward a more active incorporation of labor, tend to look for causality rather than pattern in explanation, tend more frequently to be case studies of particular situations, and tend to use more qualitative-type research methodologies such as ethnography, personal interviews of key players in particular events, and analysis of archival records which can illuminate causality better than can large-scale statistical analyses (cf. Schoenberger, 1991, on the uses of interviews). Whereas the former tend to produce results that are often generalizable to other situations, the latter tend towards explanations that are not meant to be generalizable in a statistical sense but, rather, are designed to shed light upon broader conceptual issues or economic processes.

Unions in the New Economy

This is not the place for a lengthy discussion of recent trends in the economic geography of capitalism but there are three that appear to augur important consequences for labor unions: the impact of globalization; the growth of the service sector; and the reorganization of work in what has sometimes been termed the transition from "Fordism" to "post-Fordism." I touch on these briefly to show that each has particular implications for how unions organize themselves spatially.

Globalization

The growing integration of the global economy is posing many problems for workers in both the industrialized economies, where they face issues of job loss, and in the developing nations which, though often the recipients of capital investment by transnational corporations, usually have not seen workers' standards of living rise and have tended to remain in a neocolonial economic relationship with Western Europe, the USA and Canada, and Japan. Within such developments, two issues are of particular importance to unions in both the industrialized and the developing worlds. First, clearly, workers in different countries are increasingly working in different branches of the same corporations. Although this is raising many problems for unions, it is also bringing with it new possibilities. Unions representing workers in different countries are increasingly attempting to develop global networks through which they can mount international campaigns against their common employers. There are many problems associated with this – such as widely differing work conditions, levels of remuneration, labor laws, and traditions of unionism – but some unions have achieved a good degree of success in organizing across national boundaries (cf. Moody, 1988, 1997; Hecker and Hallock, 1991; Bendiner, 1987; Frundt, 1987, 1996; Herod, 1995; ICEM, 1998; Johns, 1998; Wills, 1998b; Armbruster-Sandoval, 1999; for an historical account, see van Holthoon and van der Linden, 1988).

Second, globalization and the growing access to telecommunications technology are bringing with them new ways of trying to develop international solidarity between workers, such as corporate campaigns, attempting to get corporations to adopt codes of conduct, consumer activism, and others (Jarley and Maranto, 1990). Although there have been international union organizations since the middle of the nineteenth century, for much of the twentieth century they were split along Cold War lines and were often limited in their capacities to act (for general accounts of international labor activities during the past 150 years, see Windmuller, 1980; Price, 1945; Busch, 1983). The international activities of various nations' labor movements have also typically been the preserve of professional organizers associated with national or international labor union organizations. By their very nature, these have tended to be quite hierarchically organized activities, with unions in one country contacting their national center, which then contacts the international organization, which contacts the national center in the second country, which then contacts the appropriate local union representing workers in the same corporation that employs workers in the first country. The rise of new telecommunications media in general, and the Internet in particular, however, have led some to suggest that in the future rank-and-file workers who are handy with a computer and who can make direct links with their confederates in plants in other countries will increasingly be on the cutting edge of cross-border labor activities – a shift from what Waterman (1993) calls solidarity through paid "agents" of international labor organizations to solidarity through shopfloor "networkers." Such a new model of labor organization brings with it quite different geographical assumptions and ways of thinking about the spatial relationships between workers (such as conceptualizing space in less "hierarchical" terms, viewing connections between workers in terms of their location within cyberspace rather than their actual location in concrete space, and a diminished importance of the national capital cities – where union headquarters are often located – relative to peripheral regions where the workers themselves may be physically located). Indeed, although a relatively new technology, the Internet's interactive nature has already been used successfully in corporate campaigns by several unions (for an account of the international campaign waged by the United Steelworkers of America and the International Federation of Chemical, Energy, Mine and General Workers' Unions against the Bridgestone/Firestone tire company, see Herod, 1998d).

Growth of the service sector

In most industrial economies, employment in services has grown dramatically during the past three decades. This is significant for unions in a number of ways, not least of which is that service sector jobs are often held by women and minorities, groups that labor unions in industrialized countries have either been reluctant to organize or have found difficult to organize. One of the most important implications for unions, however, is that organizing service sector workers may involve different sets of geographical assumptions and spatial strategies than does organizing manufacturing workers.

Savage (1998, p. 231, drawing on Green and Tilly, 1987) has suggested that, at least in the USA, traditional methods of labor organizing have been developed in the context of the manufacturing sector, often relying on models that have certain implicit geographical assumptions about the spatial layout of the workplace. Such models have typically assumed: 1) workers in a workplace "hot shop" have already decided they want to be unionized and contact an organizer; 2) the organizer appeals to "bread and butter" issues such as wages and benefits rather than product quality, worker participation, or broader social justice issues; 3) the organizer appeals to workers on the basis of their identity as workers for a particular employer, not as minorities, women, community members, etc.; 4) the organizer views organizing as a technical matter rather than as a means to engage a broad-based rank-and-file movement; 5) organizing strategies are based on large, centralized workplaces with few entrance gates, regular shift changes, and large, stable workforces (meaning that just a handful of pickets can easily leaflet large numbers of workers about the benefits of unionism); and 6) organizers focus upon winning 51 percent of any union representation vote and then quickly move on to campaigns elsewhere. Such models tend to assume a certain micro-geography to the workplace, such as limited means of egress, and the existence of areas within the plant where workers are more radicalized or have developed solidarity around the defense of certain narrowly defined work skills (perhaps as a result of working at a single task in a particular phase of the production process). They therefore tend to focus attention on organizing at the scale of the workplace itself, a process in which workers are usually disembodied from broader-scale struggles of social justice or from broader notions of geographic community, in which each plant is seen as a separate entity with its own sets of issues, and in which conflicts over organizing and collective bargaining are largely conducted within the relatively private spaces of the shop floor and corporate boardroom.

Organizing in the service sector involves quite different spatial assumptions and micro-geographies of work. Service sector workers often work in much closer physical contact to an employer or supervisor (making it more difficult for organizers to gain access to them). Furthermore, they are often employed in smaller-scale workplaces, are often part-time or temporary workers, and are often employed in places to which non-workers such as customers have access (i.e. in more public spaces than manufacturing workers). In the case of retailing in particular, they also are frequently "multi-tasking" workers who engage in several different tasks during the workday (serving customers, restocking shelves, cleaning up workspaces, doing paperwork, etc.). Savage notes that such differences mean that successful organizing in the service sector must adopt different models of organizing with different spatial assumptions. These include moving the struggle for union rights from the private space of the workplace to the public space of the streets or shopping malls (thereby allowing public and/or consumer pressure to be brought to bear upon employers), and moving beyond site-specific struggles to community-based unionism. Some (e.g. Cobble, 1991) argue that moving the geographical terrain of conflict in this way is particularly effective for workers such as waitresses and janitors who have ties to an occupation but not necessarily to an individual employer and, thus, to a particular workplace.

Post-Fordism, just-in-time production, and teamwork

"Fordist" methods of manufacturing frequently involved firms stockpiling components (the so-called "just-in-case" approach), and dividing the labor process into narrow skill categories as part of the "Taylorization" of work (categories that unions often defend vigorously as a way of maintaining jobs).[5] The rise of what some have termed a "post-Fordist" mode of accumulation in a number of capitalist economies has been marked by the growth of just-in-time (JIT) production (where components arrive at a plant just before they are needed) and "teamwork." The growth of JIT appears to be affecting the geographical relationships between some manufacturers and components suppliers, with suppliers tending to locate closer in both time and space to manufacturers (see Mair et al., 1988, for an example from the US auto industry). Likewise, "teamwork" is affecting the micro-geography of the shop floor as many employers attempt to eliminate job categories and introduce "multi-tasking." This frequently involves redesigning the workplace (e.g. the growth of "modularization" in the auto industry (Weiss, 1999; Juárez Núñez and Babson, 1998)) as well as workers' social and geographical relationships to each other and to the machinery they operate (cf. Parker and Slaughter, 1988, on "teamwork"). The spread of both JIT and teamwork, then, have geographic implications for unions.

Although JIT is often seen as a way for manufacturers to increase profitability by reducing storage costs, and by subcontracting much work previously done in-house to separate suppliers, their reliance upon the timely delivery of components does leave JIT manufacturers vulnerable to labor disputes that disrupt the supply chain. This has become particularly evident in a number of important strikes in the USA in recent years, such as at United Parcel Service in August 1997 (see Coleman and Jennings, 1998) and several strikes by the United Auto Workers against General Motors. There are many aspects of JIT that cannot be covered here, but one that stands out with regard to how unions organize themselves geographically relates to the question of whether firms' increased dependency upon timely delivery of components provides local unions with an opportunity to exert greater power over their employers. Specifically, a local work stoppage may quickly spread throughout the corporate network to become a national, or even international, dispute – as occurred in the 1998 UAW–GM dispute in which virtually all of GM's North American production was brought to a halt within a few days (*Ward's Auto World*, 1998; Herod, 2000). Indeed, this is a strategy the UAW has used effectively during the 1990s (Babson, 1998). Because some local unions representing workers located at strategic points in the production chain are able to cripple a corporation, nationally or even internationally, new geographical relationships may emerge between the local and national union (raising again issues of geographic scale) and between different local unions. The latter may have implications for the geography of union solidarity, as workers in one location may resent being laid off over issues they feel do not affect them and are particular to workers located elsewhere.

The growth of teamwork and the redesign of shopfloor spaces also have implications for unions. Teamwork involves workers switching between many different types of job on the production floor as needed. This may have significant implications for unions in terms of the micro-geographies of the labor process. First, teamwork invariably means that the sharp delimitations between job categories that typified "Fordist" ways of organizing factories, and which often served as the basis for union organization, are eroded as the physical layout of the shop floor is transformed to accommodate the new ways in which teams must now work. This may make it difficult for unions to identify traditional "hot shops" or for workers to develop the kinds of on-the-job solidarity that crystallize out of working long periods together in the same jobs. If workers are switched between different jobs and/or teams, they may never have the extended contact with their colleagues that has usually been necessary for a culture of solidarity to emerge. Second, the introduction of teamwork frequently involves production-line workers taking on supervisory roles, blurring traditional lines between themselves and managers. In some countries such developments have run up against prevailing labor laws which define the appropriate roles of managers and workers more adversarially. In the USA, this has resulted in a growing push to restructure extant labor law, allowing a greater degree of labor–management "cooperation" (Herod, 1997f). It remains to be seen how unions will respond to such developments.

Concluding Comments

In this chapter I have attempted to do two things. First, I have presented an overview of how labor unions have been conceptualized and studied within the field of economic geography. Specifically, I have argued that geographers have increasingly come to view unions as important geographical actors, from which much can be learned about how the economic geography of capitalism is made. In turn, the growing interest in the spatiality of labor unionism has spawned a number of conceptual and methodological debates. Second, I have identified some salient trends in capitalist economies that are affecting how material economic geographies are themselves being made in different ways. Developments such as globalization, the growth of the service sector, and post-Fordist production methods pose new challenges for unions but may also bring with them new possibilities. Unions' responses to such developments will require strategies that are not simply political or economic in nature, but also geographical.

Acknowledgment

Thanks to Eric Sheppard for suggestions on an earlier draft of this chapter.

Endnotes

1. Parts of this section draw upon Herod (1998a). Interested readers should consult this earlier work for a more detailed review.
2. The NLRB was established in the 1930s to settle disputes between unions and employers concerning matters of labor law. For more on the NLRB, see Gross (1974, 1981) and Kammholz and Straus (1987).
3. Craft unionism is when unions organize workers in the same craft (carpenters, bakers, engineers, etc.) into separate unions defined by the specific type of work they do. Industrial unionism is when unions organize all workers in a particular industry (metal-working, autos, etc.) regardless of the actual type of work they do within that industry.
4. These scales were based upon population size as follows: Greater London; conurbations; free-standing towns; large towns; small towns; rural areas.
5. Taylorization refers to the ideas of Fredrick Taylor who, in the early twentieth century, argued that jobs should be broken down into a number of smaller parts so that time and motion studies could determine the most efficient (for the employers) way in which the production line could be organized.

Bibliography

Armbruster-Sandoval, R. 1999. Globalization and cross-border organizing: the Guatemalan Maquiladora industry and the Phillips Van Heusen workers' movement. *Latin American Perspectives*, 26, 2, 108–28.

Babson, S. 1998. Ambiguous mandate: Lean production and labor relations in the United States. In H. Juárez Núñez and S. Babson (eds). *Confronting Change: Auto Labor and Lean Production in North America/Enfrentando el Cambio: Obreros del Automóvil y Producción Esbelta en América del Norte*. Puebla, Mexico: Autonomous University of Puebla, 23–50.

Baruffalo, R. 1996. National handling and U.S. rail consolidation: implications for labor relations. Unpublished manuscript, Department of Geography, University of Kentucky.

Bendiner, B. 1987. *International Labour Affairs: The World Trade Unions and the Multi-national Companies*. Oxford: Clarendon Press.

Berman, L. L. 1998. In your face, in your space: Spatial strategies in organizing clerical workers at Yale. In A. Herod (ed). *Organizing the Landscape: Geographical Perspectives on Labor Unionism*. Minneapolis: University of Minnesota Press, 203–24.

Blomley, N. 1994. *Law, Space and the Geographies of Power*. New York: Guilford Press.

Busch, G. K. 1983. *The Political Role of International Trades Unions*. New York: St. Martin's Press.

Charlesworth, A., Gilbert, D., Randall, A., Southall, H., and Wrigley, C. 1996. *An Atlas of Industrial Protest in Britain 1750–1990*. New York: St. Martin's Press.

Church, A. and Stevens, M. 1994. Unionization and the urban–rural shift in employment. *Transactions of the Institute of British Geographers*, 19, 1, 111–18.

Clark, G. L. 1986. Restructuring the U.S. economy: The NLRB, the Saturn project, and economic justice. *Economic Geography*, 62, 4, 289–306.

Clark, G. L. 1988. A question of integrity: The national labor relations board, collective bargaining and the relocation of work. *Political Geography Quarterly*, 7, 3, 209–27.

Clark, G. L. 1989. *Unions and Communities Under Siege: American Communities and the Crisis of Organized Labor*. New York: Cambridge University Press.

Cobble, S. 1991. Organizing the postindustrial work force: Lessons from the history of waitress unionism. *Industrial and Labor Relations Review*, 44 (April), 419–36.

Coleman, B. J. and Jennings, K. M. 1998. The UPS strike: Lessons for just-in-timers. *Production and Inventory Management Journal*, 39, 4, 63–7.

Cooke, P. 1985. Class practices as regional markers: A contribution to labour geography. In D. Gregory and J. Urry (eds). *Social Relations and Spatial Structures*. New York: St. Martin's Press, 213–41.

Earle, C. 1992a. The split geographical personality of American labor: Labor power and modernization in the gilded age. In C. Earle, *Geographical Inquiry and American Historical Problems*. Stanford, CA: Stanford University Press, 346–77.

Earle, C. 1992b. The last great chance for an American working class: Spatial lessons of the General Strike and the Haymarket Riot of early May 1886. In C. Earle, *Geographical Inquiry and American Historical Problems*. Stanford, CA: Stanford University Press, 378–99.

Finch, M. and Nagel, T. W. 1983. Spatial distribution of bargaining power: Binding arbitration in Connecticut school districts. *Environment and Planning D: Society and Space*, 1, 429–46.

Frundt, H. J. 1987. *Refreshing Pauses: Coca-Cola and Human Rights in Guatemala*. New York: Praeger.

Frundt, H. J. 1996. Trade and cross-border labor organizing strategies in the Americas. *Economic and Industrial Democracy*, 17, 387–417.

Gordon, D. M. 1978. Capitalist development and the history of American cities. In W. K. Tabb and L. Sawers (eds). *Marxism and the Metropolis: New Perspectives in Urban Political Economy*. New York: Oxford University Press, 25–63.

Green, J. and Tilly, C. 1987. Service unionism: Directions for organizing. *Labor Law Journal*. 38 (August), 486–95.

Griffiths, M. J. and Johnston, R. J. 1991. What's in a place? An approach to the concept of place, as illustrated by the British National Union of Mineworkers' strike, 1984–85. *Antipode*, 23, 2, 185–213.

Gross, J. A. 1974. *The Making of the National Labor Relations Board*. Albany, NY: State University of New York Press.

Gross, J. A. 1981. *The Reshaping of the National Labor Relations Board*. Albany, NY: State University of New York Press.

Harvey, D. 1982. *The Limits to Capital*. Oxford: Basil Blackwell.

Hecker, S. and Hallock, M. (eds). 1991. *Labor in a Global Economy: Perspectives from the U.S. and Canada*. Eugene, OR: University of Oregon Labor Education and Research Center.

Herod, A. 1991. The production of scale in United States labour relations. *Area*, 23, 1, 82–8.

Herod, A. 1995. The practice of international labor solidarity and the geography of the global economy. *Economic Geography*, 71, 4, 341–63.

Herod, A. 1997a. Notes on a spatialized labour politics: Scale and the political geography of dual unionism in the US longshore industry. In R. Lee and J. Wills (eds). *Geographies of Economies*. London: Edward Arnold, 186–96.

Herod, A. 1997b. Labor's spatial praxis and the geography of contract bargaining in the US East Coast longshore industry, 1953–89. *Political Geography*, 16, 145–69.

Herod, A. 1997c. Reinterpreting organized labor's experience in the Southeast: 1947 to Present. *Southeastern Geographer*, 37, 2, 214–37.

Herod, A. 1997d. Labor as an agent of globalization and as a global agent. In K. Cox (ed). *Spaces of Globalization: Reasserting the Power of the Local*. New York: Guilford Press, 167–200.

Herod, A. 1997e. From a geography of labor to a labor geography: Labor's spatial fix and the geography of capitalism. *Antipode*, 29, 1, 1–31.

Herod, A. 1997f. Back to the future in labor relations: From the New Deal to Newt's deal. In L. Staeheli, J. Kodras, and C. Flint (eds). *State Devolution in America: Implications for a Diverse Society*. Thousand Oaks, CA: Sage (Urban Affairs Annual Reviews No. 48), 161–80.

Herod, A. 1998a. The spatiality of labor unionism: A review essay. In A. Herod (ed). *Organizing the Landscape: Geographical Perspectives on Labor Unionism*. Minneapolis: University of Minnesota Press, 1–36.

Herod, A. 1998b. Geographic mobility, place, and cultures of labor unionism. In A. Herod (ed). *Organizing the Landscape: Geographical Perspectives on Labor Unionism*. Minneapolis and London: University of Minnesota Press, 123–8.

Herod, A. (ed). 1998c. *Organizing the Landscape: Geographical Perspectives on Labor Unionism*. Minneapolis and London: University of Minnesota Press.

Herod, A. 1998d. Of blocs, flows and networks: The end of the Cold War, cyberspace, and the geo-economics of organized labor at the *fin de millénaire*. In A. Herod, G. Ó Tuathail, and S. Roberts (eds). *An Unruly World? Globalization, Governance and Geography*. New York: Routledge, 162–95.

Herod, A. 2000. Implications of just-in-time production for union strategy: Lessons from the 1998 General Motors – United Auto Workers' Dispute. *Annals of the Association of American Geographers*, 90, 4, 521–47.

Holmes, J. and Rusonik, A. 1991. The break-up of an international labour union: Uneven development in the North American auto industry and the schism in the UAW. *Environment and Planning A*, 23, 9–35.

Hudson, R., and Sadler, D. 1983. Region, class, and the politics of steel closures in the European Community. *Environment and Planning D: Society and Space*, 1, 405–28.

Hudson, R. and Sadler, D. 1985. Communities in crisis: The social and political effects of steel closures in France, West Germany, and the United Kingdom. *Urban Affairs Quarterly*, 21, 1, 171–86.

Hudson, R. and Sadler, D. 1986. Contesting works closures in Western Europe's industrial regions: Defending place or betraying class? In A. Scott and M. Storper (eds). *Production, Work, Territory*. Winchester, MA: Allen & Unwin, 172–93.

ICEM (International Federation of Chemical, Energy, Mine and General Workers' Unions) 1998. Rubber unions go global. *ICEM Update*, No. 65 June 26, 1–2. (Available at http://www.icem.org/update/upd 1998/upd98-65.html).

Jarley, P. and Maranto, C. L. 1990. Union corporate campaigns: An assessment. *Industrial and Labor Relations Review*, 43, 5, 505–24.

Johns, R. 1998. Bridging the gap between class and space: U.S. worker solidarity with Guatemala. *Economic Geography*, 74, 3, 252–71.

Johnston, K. 1986. Judicial adjudication and the spatial structure of production: Two decisions by the National Labor Relations Board. *Environment and Planning A*, 18, 27–39.

Juárez Núñez, H. and Babson, S. (eds). 1998. *Confronting Change: Auto Labor and Lean Production in North America/Enfrentando el Cambio: Obreros del Automóvil y producción Esbelta en América del Norte*. Puebla, Mexico: Autonomous University of Puebla.

Kammholz, T. C. and Straus, S. R. 1987. *Practice and Procedure Before the National Labor Relations Board*. Philadelphia: American Law Institute, American Bar Association Committee on Continuing Professional Education.

Mair, A., Florida, R., and Kenney, M. 1988. The new geography of automobile production: Japanese transplants in North America. *Economic Geography*, 64, 352–73.

Martin, R., Sunley, P., and Wills, J. 1993. The geography of trade union decline: Spatial dispersal or regional resilience? *Transactions of the Institute of British Geographers*, 18, 1, 36–62.

Martin, R., Sunley, P., and Wills, J. 1994a. The decentralization of industrial relations? New institutional spaces and the role of local context in British engineering. *Transactions of the Institute of British Geographers*, 19, 4, 457–81.

Martin, R., Sunley, P., and Wills, J. 1994b. Local industrial politics: Spatial sub-systems in British engineering. *Employee Relations*, 16, 84–99.

Martin, R., Sunley, P., and Wills, J. 1994c. Labouring differences: Method, measurement and purpose in geographical research on trade unions. *Transactions of the Institute of British Geographers*, 19, 1, 102–10.

Martin, R., Sunley, P., and Wills, J. 1996. *Union Retreat and the Regions: The Shrinking Landscape of Organised Labour*. London, Jessica Kingsley.

Massey, D. 1994. The geography of trade unions: Some issues. *Transactions of the Institute of British Geographers*, 19, 1, 95–8.

Massey, D. and Miles, N. 1984. Mapping out the unions. *Marxism Today*, May, 19–22.

Massey, D. and Painter, J. 1989. The changing geography of trade unions. In J. Mohan (ed). *The Political Geography of Contemporary Britain*. Basingstoke, UK: Macmillan, 130–50.

Mitchell, D. 1996. *The Lie of the Land: Migrant Workers and the California Landscape*. Minneapolis: University of Minnesota Press.

Mitchell, D. 1998. The scales of justice: Localist ideology, large-scale production, and agricultural labor's geography of resistance in 1930s California. In A. Herod (ed). *Organizing the Landscape: Geographical Perspectives on Labor Unionism*. Minneapolis: University of Minnesota Press, 159–94.

Moody, K. 1988. *An Injury to All: The Decline of American Unionism*. New York: Verso.

Moody, K. 1997. *Workers in a Lean World: Unions in the International Economy*. London: Verso.

Page, B. 1998. Rival unionism and the geography of the meatpacking industry. In A. Herod (ed). *Organizing the Landscape: Geographical Perspectives on Labor Unionism*. Minneapolis: University of Minnesota Press, 263–96.

Painter, J. 1991. The geography of trade union responses to local government privatization. *Transactions of the Institute of British Geographers*, 16, 2, 214–26.

Painter, J. 1994. Trade union geography: Alternative frameworks for analysis. *Transactions of the Institute of British Geographers*, 19, 1, 99–101.

Parker, M. and Slaughter, J. 1988. *Choosing Sides: Unions and the Team*. Boston: South End Press.

Parson, D. 1982. The development of redevelopment: Public housing and urban renewal in Los Angeles. *International Journal of Urban and Regional Research*, 6, 2, 393–413.

Parson, D. 1984. Organized labor and the housing question: Public housing, suburbanization, and urban renewal. *Environment and Planning D: Society and Space*, 2, 75–86.

Peet, R. 1983. Relations of production and the relocation of United States manufacturing industry since 1960. *Economic Geography*, 59, 112–43.

Price, J. 1945. *The International Labour Movement*. London: Oxford University Press.

Rees, G. 1985. Regional restructuring, class change, and political action: Preliminary comments on the 1984–1985 miners' strike in South Wales. *Environment and Planning D: Society and Space*, 3, 389–406.

Rees, G. 1986. "Coalfield culture" and the 1984–1985 miners' strike: A reply to Sunley. *Environment and Planning D: Society and Space*, 4, 469–76.

Savage, L. 1998. Geographies of organizing: Justice for janitors in Los Angeles. In A. Herod (ed). *Organizing the Landscape: Geographical Perspectives on Labor Unionism*. Minneapolis: University of Minnesota Press, 225–52.

Sayer, R. A. 1982. Explanation in economic geography. *Progress in Human Geography*, 6, 68–88.

Schoenberger, E. 1991. The corporate interview as a research method in economic geography. *Professional Geographer*, 43, 2, 180–9.

Southall, H. 1988. Towards a geography of unionization: The spatial organization and distribution of early British trade unions. *Transactions of the Institute of British Geographers*, 13, 466–83.

Southall, H. 1989. British artisan unions in the New World. *Journal of Historical Geography*, 15, 2, 163–82.

Southall, H. 1996. Agitate! Agitate! Organize! Political travellers and the construction of a national politics, 1839–1880. *Transactions of the Institute of British Geographers*, 21, 177–93.

Storper, M. and Walker, R. 1983. The theory of labour and the theory of location. *International Journal of Urban and Regional Research*, 7, 1, 1–43.

Sunley, P. 1986. Regional restructuring, class change, and political action: A comment. *Environment and Planning D: Society and Space*, 4, 465–8.

Sunley, P. 1990. Striking parallels: A comparison of the geographies of the 1926 and 1984–85 coalmining disputes. *Environment and Planning D: Society and Space*, 8, 35–52.

Thompson, E. P. 1963. *The Making of the English Working Class*. London: Penguin Books.

van Holthoon, F. and van der Linden, M. (eds). 1988. *Internationalism in the Labour Movement 1830–1940*, Volumes 1 and 2. New York and Leiden: E. J. Brill.

Walker, R. and Storper, M. 1981. Capital and industrial location. *Progress in Human Geography*, 5, 473–509.

Ward's Auto World. 1998. Reversal of fortunes: GM's NAO comeback stymied by strikes. July, pages 41–3 and 47.

Waterman, P. 1993. Internationalism is dead! Long live global solidarity? In J. Brecher, J. Brown Childs, and J. Cutler (eds). *Global Visions: Beyond the New World Order*. Boston: South End Press, 257–61.

Weber, A. 1929 [1909]. *Theory of the Location of Industries*. Chicago: University of Chicago Press.

Weiss, L. 1999. Auto makers test labor-cutting strategy in Brazil. *Working Together: Labor Report on the Americas*, 35 (March/April). (Available at http://www.americas.org/sitemap-index.htm)

Wills, J. 1996. Uneven reserves: Geographies of banking trade unionism. *Regional Studies*, 30, 359–72.

Wills, J. 1998a. Space, place, and tradition in working-class organization. In A. Herod (ed). *Organizing the Landscape: Geographical Perspectives on Labor Unionism*. Minneapolis: University of Minnesota Press, 129–58.

Wills, J. 1998b. Taking on the CosmoCorps? Experiments in transnational labor organization. *Economic Geography*, 74, 2, 111–30.

Windmuller, J. P. 1980. *The International Trade Union Movement*. Deventer, The Netherlands: Kluwer.

Chapter 22

State and Governance

Joe Painter

No survey of economic geography would be complete without a consideration of the state. States shape economic landscapes. They seek to regulate the flows of goods, money, energy, people, and information that produce economic geographies. They are major economic actors in their own right, employing large workforces, producing and consuming goods and services, and marshaling large financial and other economic resources in pursuit of state strategies. They play an important role in the production of economic space from state-sponsored local economic development initiatives to efforts to define and promote "national economies," and from the construction of multinational economic territories to dealing with the destabilizing implications of globalization. At the same time, states are, in turn, deeply affected by changing economic geographies. They depend for their resources on the economic prosperity of their territories, and are thus vulnerable to spatially differentiated processes of investment and disinvestment over which they usually have (at best) limited influence. Their ability to govern, and especially their ability to govern democratically, is affected by their legitimacy, which in turn is tied to their fragile attempts to deliver economic success and social welfare. Each of these is affected by economic geography: the spatial unevenness of economic growth and decline produces regions with severe social problems that tax the ingenuity of policymakers, while geographically concentrated growth sectors, such as financial services in the City of London or micro-electronics in California, are held up as models for state policy to promote elsewhere.

To examine the relationships between the state, governance, and economic geography, this chapter is divided into two main sections. The first defines state and governance and considers how they are related to economic processes. The longer second section then applies these ideas to the field of economic geography at a variety of spatial scales.

Understanding the State and Governance

Definitions

Despite its clear importance, the state is difficult to define with precision. Conventional definitions commonly emphasize the following elements. First, the state is a set of public institutions: government ministries, public services, military and police forces, legal systems, and so on. Second, the state is territorial: modern states cannot overlap geographically; they have fixed and clear (though often disputed) borders and, in principle, they exercise power evenly across their territories. Third, the state controls the legitimate means of violence, including the armed forces and imprisonment. This composite definition is an adequate starting point, but there are some problems with it. The first and second parts are especially relevant to economic geography, and thus merit brief elaboration here.

The institutional aspect of the definition seems unremarkable, and yet there can be difficulties in distinguishing clearly between state and non-state institutions. How should we categorize a private utility company, such as a rail network or a water-supply company, that attracts a government subsidy funded from taxation, is subject to state regulation, and has a statutory obligation to provide particular products in particular ways? Or what do we make of a cleaning company that was formed through the privatization of a public service by way of a management buyout, and that generates its entire turnover by selling cleaning services to the same public sector organization of which it was previously a part? In urban economic regeneration, a welter of public–private partnerships have sprung up in many countries to undertake projects aimed at revitalizing rundown neighborhoods, using both public and private resources. Are such partnerships and their projects inside or outside the institutional architecture of the state? It seems that many institutions cannot be neatly labeled as state or non-state. This means that there are difficulties in trying to define the state partly as a particular set of institutions. It also suggests that the state should not be seen as something separate from society, a point I shall return to below.

The territorial aspect of the state is also more complex than appears at first sight. Although territoriality is more important in modern states than earlier ones (Giddens, 1985; Mann, 1988) the territorial integrity of the state is never perfect. States do not, in practice, exercise power entirely evenly across territory, nor does their power cease at their frontiers: state power can be, and often is, extended beyond state borders. Each state claims to be sovereign over its territory (that is, to be the ultimate source of authority). However, true territorial sovereignty is rarely achieved. Moreover, some writers have suggested that it is now being comprehensively undermined in favor of a more complex multi-layered system of political authority in which states are just one among a number of structures with influence over "their" space (Anderson, 1996; Jessop, 1994).

In both institutional and territorial terms, therefore, the state is not as coherent as the conventional definition suggests. According to Corrigan and Sayer the state is not a concrete object (consisting of "institutions plus territory") at all, but is really an *idea*. Quoting from the work of Abrams (1977) they argue that:

"the state" is in an important sense an illusion. Of course, institutions of government are real enough. But "the state" is in large part an ideological construct, a fiction: [...] an ideological artefact attributing unity, structure and independence to the disunited, structureless and dependent workings of the practice of government. (Corrigan and Sayer, 1985, pp. 7–8)

In other words the conventional story of the state – that it is unified, coherent, and territorially sovereign – is a story told by the state itself, and thus not to be taken on trust. Instead the state should be understood as a political project in a continual process of formation, deformation, and reformation. States are complex and heterogeneous and are never fully coherent. Mann (1993) suggests that states are "polymorphous;" they are liable to "crystallize" in a range of different ways, depending on the issue or constituency involved, while Resnick (1990) adopts the metaphor of Proteus, the ever-changing figure from Greek mythology, to convey the perpetual restless transformation of the state.

Finally, states are social arenas. As well as the formal political, legal, and economic structures and processes that make up the architecture of the state, the state consists of myriad informal social processes and networks of social relations. Such networks involve not only elected officials, senior bureaucrats, and social and economic elites, but also the thousands of middle- and low-ranking state employees on whose activities the state depends.

Networks also spread beyond the state and contribute to the blurring of the boundaries of the state noted above. This seems to be an increasing feature of political life, and social scientists have sought new concepts and vocabularies to describe and understand it. One of the most popular is "governance." Governance is a term with several meanings in social science. It is sometimes used loosely to refer to the process of governing in general, whatever form it takes. More specifically it can refer to the involvement of non-state actors (such as firms and voluntary organizations) alongside the state in the process of governing. Within economic geography, however, it is most commonly used to refer to a particular type of relationship between state and non-state organizations. In contrast to hierarchical forms of coordination that depend on top-down decisionmaking, and to market forms of coordination that work through individualized contracts, governance involves coordination through network and partnership. According to Rhodes, governance involves "self-organizing, interorganizational networks" (Rhodes, 1997, p. 53). Rhodes argues that governance is growing in importance relative to conventional government, and identifies four key features of this shift:

1. Interdependence between organizations. Governance is broader than government, covering non-state actors. Changing the boundaries of the state meant the boundaries between public, private and voluntary sectors became more shifting and opaque.
2. Continuing interactions between network members, caused by the need to exchange resources and negotiate shared purposes.
3. Game-like interactions, rooted in trust and regulated by rules of the game, negotiated and agreed by network participants.
4. A significant degree of autonomy from the state. Networks are not accountable to the state; they are self-organizing. Although the state does not occupy a sovereign position, it can indirectly and imperfectly steer networks (Rhodes, 1997, p. 53).

The growth of governance in Rhodes' sense means that both the form and the outcome of the relationships between state activities and economic geography are changing.

State and economy

The social world is not neatly segmented into "economy," "state," and "civil society." That does not mean that society is a unitary whole, with no internal structures or differences. But it does make it impossible to understand "civil society" and "the economy" as spheres of human activity that are separate from the state. State power permeates almost all aspects of social and economic life (and by extension social and economic geography). That is not to say that social and economic life are state-controlled, but they are affected by the state in remarkably deep-seated and enduring ways. Despite the rhetoric of right-wing politicians during the 1980s and 1990s, the state has not been "rolled back" – if that means that more and more social and economic relations are beyond the scope of state power. Indeed, the exercise of state power seems to be becoming more intense with the passage of time, not less (but not necessarily more effective). This can be seen in a simple way in the fact that as time goes on states develop and implement policy on an ever-widening range of issues. New legislation is added to the statute books much more quickly than old legislation is repealed. Each year, new elements of social and economic life become objects of state surveillance, regulation, intervention, and manipulation. Once an issue or a problem is visible to the state, and is taken on as an object of governmental authority, there is no going back.

The widely held (but mistaken) common-sense assumption that the social world is composed of distinct "spheres" (the state and the economy, or governments and markets) is reflected in the separation of political from economic geography. Until recently the state was largely neglected in the writings of mainstream economic geographers. With some honorable exceptions, economic geographers got on with studying the locational decisions of firms, the geographically uneven development of markets, and the spatial flows of goods, finance, and labor, leaving the investigation of the state to the (less numerous) political geographers. Fortunately this lack of attention to the state within economic geography is now being remedied (cf. Knox and Agnew, 1998; Martin and Sunley, 1997; O'Neill, 1997).

The separation of spheres assumption is not just of academic interest. It has important policy implications. This is the assumption that allows government ministers, journalists, and economists to talk of state "intervention" in the economy, as if economic relations are independent of the state except in the specific and noteworthy case when the government "does something" to the economy, for reasons that are typically dismissed by orthodox economists as "meddling," "interference," or "politically motivated."

The geographer Phillip O'Neill explicitly rejects the "separation of spheres" assumption. He challenges

... the politically charged discourse that markets are capable of a separate, private existence beyond the actions of the state's apparatus. Rather, a *qualitative* view of the state is preferred. In this view, the state is seen to play an indispensable role in the creation, governance and conduct of markets, including at the international scale. Consequently, arguments about the *extent* of state intervention are seen as being feeble. Because the state is always involved in the operation of markets, the salient debate should be about the nature, purpose and consequences of the *form* of state action, rather than about questions of magnitude of intervention. (O'Neill, 1997, p. 291)

O'Neill's contribution is particularly important as it emphasizes that the state plays a continuous and major role in the constitution of economies. Thus the conventional focus on "how much" the state should intervene is misplaced. This "quantitative" view of the state should give way to a "qualitative" view that focuses on the nature of the state's role in the economy and, as we shall see in the next section, its geography.

States, Governance, and Economic Geography

The qualitative state and economic geography

O'Neill's emphasis on the qualitative role of the state is very useful for economic geography. Drawing on the work of Polanyi (1957), Cerny (1990), and especially Block (1994), O'Neill sets out four tenets concerning the role of the state:

First, *economy* is necessarily a combination of three events: markets, state action and state regulation. A corollary of this constitution is that there is an infinite number of ways in which an economy can be organized. Second, although economic efficiency is dependent on markets, markets are state-constrained and state-regulated and thereby incapable of operating in a *laissez-faire* environment. Third, neither capital nor the state is capable of achieving its goals simultaneously nor independently. Finally, it should be recognized that any coherence that exists about the idea of *economy* derives essentially from our cultural beliefs, which (in Anglo cultures at least) have led to constructions of economy being overlain with the dichotomy of *planned versus market*, which, in turn, has had the effect of denying the existence of multiple forms of economy. (1997, p. 294)

These tenets have important implications for economic geography. The nature of the combination between markets, state action, and state regulation varies geographically (most notably from state to state). Markets also operate in spatially uneven ways and this geography influences and is influenced by state regulation. Cultural beliefs about the economy and thus about state–economy relations vary significantly from place to place. The importance of the state to the economy, and by implication to economic geography, is summarized by O'Neill in a table (reproduced here as table 22.1) showing the great range of roles that the state undertakes in a modern economy.

Several of the roles listed in table 22.1 are self-evidently geographical, such as land-use planning and regulation, urban and regional development, and the management of territorial boundaries. However, almost all of them have a geography, or are influenced by geography: "economy" and "geography" are mutually constituting. Even such non-spatial state functions as "social wage provision" or "monetary policy" cannot be fully understood without considering their relationship to space and place. Thus the social wage refers to the provision of goods or services by the state to supplement wages obtained in the labor market. It can take many forms, but one of the most important is state subsidization of the housing market, either through the direct provision of public or social housing at below market rents, or

Table 22.1 Roles of the qualitative state in a modern economy

A Maintenance of a regime of property rights
 i. Maintenance of private property rights
 ii. Recognition of institutional property rights
 iii. Basic rules for the ownership and use of productive assets
 iv. Basic rules for the exploitation of natural resources
 v. Rules for the transfer of property rights (between individuals, households, institutions, and generations)

B Management of territorial boundaries
 i. Provision of military force
 ii. Economic protection through manipulation of:
    • money flows
    • goods flows
    • services flows
    • labor flows
    • flows of intangibles
 iii. Quarantine protection

C Legal frameworks to maximize economic co-operation
 i. Establishment of partnerships and corporations
 ii. Protection of intellectual property rights
 iii. The governance of recurring economic relations between:
    • family members
    • employers and workers
    • landlords and tenants
    • buyers and sellers

D Projects to ensure social co-operation
 i. Maintenance of law and order
 ii. Undertake national image-making processes
 iii. Other coercive strategies

E Provision of basic infrastructure
 i. Provision or organization of:
    • transportation and communications systems
    • energy and water supply
    • waste disposal systems
 ii. Assembly and conduct of communications media
 iii. Assembly and dissemination of public information
 iv. Land-use planning and regulation

F Creation and governance of financial markets
 i. Rules for the establishment and operation of financial institutions
 ii. Designation of the means of economic payment
 iii. Rules for the use of credit
 iv. Maintenance of the lender of last resort

G Creation and governance of product markets
 i. Regulation of the market power of firms
 ii. The selection and regulation of natural monopolies
 iii. The promotion and maintenance of strategic industries
 iv. The provision of public goods
 v. The provision of goods unlikely to be supplied fairly

H Production and reproduction of labor
 i. Demographic planning and governance
 ii. Provision of universal education and training
 iii. Governance of workplace conditions
 iv. Governance of returns for work
 v. Social wage provision
 vi. Supply and governance of childcare
 vii. Provision of governance of retirement incomes

I Control of macroeconomic trends
 i. Fiscal policy
 ii. Monetary policy
 iii. External viability

J Other legitimation activities
 i. Elimination of poverty
 ii. Maintenance of public health
 iii. Citizenship rights
 iv. Income and wealth redistribution
 v. Urban and regional development
 vi. Cultural development
 vii. Socialization
 viii. Enhancement of the environment

Source: O'Neill (1997)

through tax relief on the construction and/or purchase of private housing. Since housing markets are subject to enormous inter-regional variations the social wage is in practice geographically differentiated. To take a simple empirical example, there is currently an over-supply of public housing in north-east England and a dramatic under-supply (relative to demand and need) in south-east England. It is thus easier to acquire that element of the social wage in the north-east than in the south-east. Seen another way, public housing tenants in the south-east receive a higher social wage (measured in relation to market prices) than do tenants of similar properties in the north-east.

The same regional geography illustrates the significance of geography to monetary policy. Central banks establish base interest rates that apply across the whole of the relevant territory. However, because local and regional economies typically diverge markedly in terms of sectoral structures, factor prices, and external orientation, a unitary interest rate policy can have spatially divergent impacts. In the UK, for example, manufacturing sectors in old industrial regions such as the north-east were facing recession during the late 1990s while the economy of the south-east, dominated by producer services, was at risk of overheating. This situation led to calls for interest rate cuts on the part of the northern manufacturing interests, who found that the high value of sterling made their exports uncompetitive. High UK interest rates tend to raise the value of sterling relative to other currencies because currency investors buy more pounds to take advantage of the increased interest income available. In a now infamous comment on this situation, the Governor of the Bank of England, Eddie George, declared that unemployment in the north of England was a price that had to be paid to stave off inflation in the south-east. While the interest rate set by the bank is geographically uniform, its effects are not, and neither are the politics that surround it.

Mapping state–economy relations

The examples in the previous section reveal the importance of the state in understanding the geography of economic processes within national borders. There are also major geographical variations between countries arising from differences in the nature of the relations between the state and the market in each. This might seem obvious, but it is worth emphasizing because a number of prominent writers, notably Kenichi Ohmae (1995), have claimed that the nation-state is becoming increasingly irrelevant to economic activity as a result of globalization. By contrast, others such as Hutton (1995) stress the enduring impact of states on the pattern of economic development around the world, and the great importance of variations in state regimes for economic development and prosperity. Hutton focuses on four different systems, summarized in table 22.2. They are American capitalism, Japanese capitalism, the European social market, and British capitalism (Hutton, 1995, pp. 257–84).

Although there are many similarities between the four systems (they all involve wealthy, industrialized, capitalist economies based on production for private profit) Hutton argues that "the similarities disguise vast differences between the social and economic purpose of apparently similar institutions, so that each capitalist structure ends up with very different specific capacities and cultures which are very hard to change" (Hutton, 1995, pp. 257–8).

Table 22.2 Four types of capitalism

Characteristic	American capitalism	Japanese capitalism	European social market	British capitalism
Basic principle				
Dominant factor of production	Capital	Labor	Partnership	Capital
"Public" tradition	Medium	High	High	Low
Centralization	Low	Medium	Medium	High
Reliance on price-mediated markets	High	Low	Medium	High
Supply relations	Arm's-length, price-driven	Close, enduring	Bureaucracy planned	Arm's-length, price-driven
Industrial groups	Partial, defense, etc.	Very high	High	Low
Extent privatized	High	High	Medium	High
Financial system				
Market structure	Anonymous, securitized	Personal, committed	Bureaucracy, committed	Uncommitted, marketized
Banking system	Advanced, marketized, regional	Traditional, regulated, concentrated	Traditional, regulated, regional	Advanced, marketized, centralized
Stock market	V. important	Unimportant	Unimportant	V. important
Required returns	High	Low	Medium	High
Labor market				
Job security	Low	High	High	Low
Labor mobility	High	Low	Medium	Medium
Labour/management	Adversarial	Co-operative	Co-operative	Adversarial
Pay differential	Large	Small	Medium	Large
Turnover	High	Low	Medium	Medium
Skills	Medium	High	High	Poor
Union structure	Sector-based	Firm-based	Industry-wide	Craft
Strength	Low	Low	High	Low
The firm				
Main goal	Profits	Market share, stable jobs	Market share, fulfilment	Profits
Role of top manager	Boss-king, autocratic	Consensus	Consensus	Boss-king, hierarchy
Social overheads	Low	Low	High	Medium, down
Welfare system				
Basic principle	Liberal	Corporatist	Corporatist, social democracy	Mixed
Universal transfers	Low	Medium	High	Medium, down
Means-testing	High	Medium	Low	Medium, up
Degree education is tiered by class	High	Medium	Medium	High
Private welfare	High	Medium	Low	Medium, up
Government policies				
Role of government	Limited, adversarial	Extensive, co-operative	Encompassing	Strong, adversarial
Openness to trade	Quite open	Least open	Quite open	Open
Industrial policy	Little	A lot	A lot	Non-existent
Top income tax	Low	Low	High	Medium

Source: Hutton (1995)

In each of the systems in table 22.2 the state cannot be seen in isolation from other aspects of the model. Like O'Neill, Hutton emphasizes the qualitative relationships between the state and all the other elements. Of the four, the US model appears to give the most limited role to the state. It is a highly flexible and highly market-based system with limited regulation of the labor market and restricted welfare benefits. The corporate sector is highly competitive with an emphasis on short-term profit-ability. At the same time, Hutton points out, the US "has retained important institutional shelters against the full blast of competition and these have become ways of expressing co-operative common purpose" (1995, p. 259). Many of them stem from the New Deal of the 1930s. According to Hutton, both federal and state governments have key roles here with housing, banking, and research and develop-ment sectors all benefiting from state involvement. Even the intensely competitive aspect of the private sector is underwritten by strong anti-monopoly controls put in place by the state.

The second model, that of the European social market, involves a partnership between capital and labor, a financial system that is less oriented to the market than in the USA and more committed to the enterprises in which it invests, a strong and inclusive welfare system, and high degree of formalized power sharing in the political system. This produces, argues Hutton, co-operation, high productivity, and invest-ment. However, because of the high degree of regulation, restructuring is more difficult, though not impossible. This model is apparent in Scandinavia, Germany, and a number of other continental European countries within the European Union.

The third model, that of East Asian capitalism, is typical of Japan and the East Asian "tigers," and increasingly of China. In these systems there is intense competi-tion, but also significant cooperation. There is a close relationship between the key economic institutions such as firms, banks, and unions. Hutton identifies the state as "the architect of these institutional relationships; it seeks to build consensus and then guides firms and the financial system in the direction established by the consensus" (1995, p. 270). Until the Asian economic crisis of the late 1990s, this system saw dramatic rates of economic growth, as geographers Knox and Agnew describe:

Adding government savings to deposits in nationalized banks, the South Korean government controlled two-thirds of South Korea's investment resources during the country's period of most rapid growth in the late 1970s. This power was used to guide investment in chosen directions through differential interest rates and easy credit terms. Korean export expansion, the main method of economic growth, was itself built on an economic base that was stringently protected from foreign imports. Economic growth was orchestrated by activist governments. (Knox and Agnew, 1998, p. 93)

Finally, there is British capitalism. Hutton regards this as the weakest of the four models, and again he is clear that the state plays a key role, although in this case a negative one. As Hutton puts it, "the financial system demands the same high returns from companies, with the same lack of commitment, as in the US; but there is not even the saving grace of statutory regulation and strong regional or state banks to moderate the consequences of such pressure" (1995, p. 281). Although Britain's welfare system is less all-embracing than the social-market model, it does provide a degree of universal coverage not found in the USA. For Hutton, this is again the worst of both worlds, with tax rates higher than the USA but without the levels of social solidarity that compensate for this in the European model. Britain lacks "the tradition of public spirit, common interest and national purpose which variously imbues social market Europe, the US and East Asian capitalism" and the parliamentary system fails to act as a "source of integration and national leadership" (1995, p. 284).

Although his account is not couched in the vocabulary of economic geography, Hutton effectively provides a map of state–economy relations around the industrialized world, and thus a useful approach to understanding the role of the state in the production and transformation of economic geography. The different structures of the four systems involve more than just state institutions and state policies, of course, but, for Hutton, the state is a central determinant of economic performance (1995, p. 20).

With the partial exception of Knox and Agnew, most of the writers mentioned so far derive their accounts of the restructuring of the state explicitly or implicitly from the industrialized Western countries, and particularly from Western Europe and North America. These accounts are, therefore, to a greater or lesser extent, ethnocentric, with relatively little attention paid to the role of the state in Asia, Africa, and Latin America. If our map of state–economy relations is not to be woefully incomplete we must also consider the situation in middle- and low-income countries.

If it is difficult to generalize about patterns of state formation and restructuring within the capitalist core, it is even more problematic to lump together all the so-called "less-developed" countries. Stereotypes abound: inefficiency, corruption, bureaucracy, militarization, and authoritarianism. While examples of each of these can be found, they are hardly unique to the poorer parts of the world, and it would be inaccurate to establish a model of the state in less-developed countries on the basis of them (Bayart, 1993).

Yet some common trends can be observed, related to: the impact of colonialism in many (but not all) cases; the widespread adoption of the discourse of "development" (Escobar, 1995) and associated state strategies; and the sharp and enduring polarization of the global economy, dominated as it is by a small minority of exceptionally and unusually wealthy countries (Amin, 1991). In particular, state legitimacy becomes tied to the ability to promote economic growth, but limited resources and the relative lack of infrastructural power constrains the capacity of the state to deliver. This undermines the legitimacy of the state, which in turn leads to political unrest and, all too often, authoritarian reaction and/or military dictatorship. For example, in the case of sub-Saharan Africa the independent states that emerged from the colonial period appeared to adopt markedly different approaches to economic development "ranging from Marxism-Leninism through African socialism (which emphasized the, supposedly, egalitarian communal values of 'traditional' African society) to an open espousal of free-market capitalism" (Wiseman, 1997, p. 275). In practice, however, Wiseman notes that most African states pursued remarkably similar strategies: "in almost all cases the key feature was the attempt to impose the state as the key player in economic development" (p. 276). In states advocating a capitalist path, this meant that aspiring entrepreneurs had to develop close links with state officials leading to what has been termed "crony capitalism" or clientelism. This limited the development of an independent capitalist class, with the ruling class being closely associated with the machinery of the state and marked by division, instability, and ethnic, religious, and regional differences. As Wiseman writes:

> With control of the state being regarded as vital for individual or group economic prosperity, and frequently for personal survival, the tolerance needed for the existence of legal and legitimate opposition, and of a vibrant and autonomous civil society, which are so fundamental for democracy, has been scarce. African political authoritarianism has had a negative effect on economic development. The model of the authoritarian developmental state, identified in such places as Taiwan, Singapore, Malaysia and South Korea as occupying a key role in promoting economic growth, is not replicated in Africa. (Wiseman, 1997, p. 277)

The 1980s and 1990s have seen significant shifts in state economic strategies in many countries. During the 1980s, international agencies such as the World Bank imposed neoliberal "structural adjustment programs" on many states in exchange for financial assistance. These programs involved cutbacks in government expenditure, reduction in welfare programs, the ending of state subsidies and the opening of domestic markets to international investment. Although strongly defended by their backers as painful but necessary phases of restructuring, their immediate impact was often to produce sharp decreases in living standards for already impoverished people. More recently, and to some extent in response to the excesses of structural adjustment, the same institutions have been sponsoring the shift from government to governance, by promoting the involvement of non-governmental organizations in the delivery of "development" policies.

The rise of the workfare state

Structural adjustment programs in poor countries were one symptom of a worldwide shift towards neoliberal economic strategies that occurred during the 1980s among public policy markets at all spatial scales, from the local to the international. Up to this point in the chapter, the emphasis has been on the ways in which economic processes and outcomes are geographically differentiated both within and between countries, and the importance of the state in accounting for such patterns. The growth of neoliberal policies, by contrast, represented something of a convergence in state economic strategies.

One of the most important targets of 1980s neoliberalism was the welfare state, which had grown up in a variety of different forms in all the advanced industrialized countries during the previous 50 years. The precise nature of state welfare provision varied from country to country but typically included social security payments for those without other sources of income (such as unemployed and retired people), health care, education, and housing. The welfare state was closely linked with economic strategies based on the ideas of the British economist J. M. Keynes. This leads state theorist Bob Jessop to identify the "Keynesian welfare national state (KWNS)" (Jessop, 1997c, p. 516), in which Keynesian demand management and state welfare provision underwrote capitalist accumulation within relatively integrated national economies. According to Jessop (1997a, p. 571; 1997c, p. 516), the Keynesian welfare national state is now giving way to the Schumpeterian workfare (post-national) regime (SWR). In earlier work Jessop (1993) used the term Schumpeterian workfare state: he now favors "regime" to emphasize the increased role of governance mechanisms (see above) to correct market and state failures.

The SWR is "Schumpeterian" after the theorist of economic innovation, Joseph Schumpeter. It emphasizes the re-orientation of policy away from demand management, towards supply-side activities supposed to encourage innovation in products, processes, and markets. These include deregulation, privatization, skills training, and support for entrepreneurial activity. The SWR involves "workfare," not in the narrow sense of 1930s-style food for work programs (though these could be an element in theory), but in the sense that labor-market strategies involve the subordination of social policy to the requirements of international economic competitiveness. Peck (2000) outlines the main components of this shift to workfarism (see table 22.3). In place of the needs-based approach of the welfare state, the workfare regime emphasizes the work ethic. It stresses individual responsibility and requires

Table 22.3 Welfare structures and workfare strategies

	Welfare structures	Workfare strategies
Ideological principles	Entitlement Aid distributed on basis of need	Reciprocity Enforcement of work and work values
Objectives/rationale	Reducing poverty through income transfers Responding to manifest social need	Tackling welfare dependency through promotion of work Maximizing labor market participation
Dominant discourse	Need and entitlement Social work/bureaucratic codes and norms	Work, personal responsibility and self/family-sufficiency Business/employment service codes and norms
Means	Passive income support	Active labor market integration
Labor-regulatory function	Exclusion from wage-labor Socially-sanctioned recipient groups defined on basis of ascribed/categorical characteristics	Inclusion into wage-labor Market-determined treatment groups defined on basis of job readiness
Subject/state	Welfare recipient *On* welfare	Job seeker Moving *through* workfare
Social relations	Determining entitlements of passive subjects	Interventionist case management of active subjects
Hierarchy	Centralized control Limited local autonomy in program design, entitlement and eligibility Management by input controls and sanctions	Centrally-orchestrated devolution Increasing local discretion over program design, entitlement and eligibility Management by output targets and incentives
Delivery	Bureaucratic; line management ethos Process and input orientation Standardized programs	Flexible; local market ethos Output and outcome orientation Variegated programs
Work/work program participation	Limited Voluntary	Extensive Mandatory; ethos of compulsion

Source: Peck (2000)

those seeking work to adopt a flexible approach. The state's role is limited to the provision of skills through education and training, and the organization of programs to encourage, or in some circumstances to coerce, individuals into the labor market.

Jessop identifies three trends in the nature of the state associated with the move from the KWNS to the SWR. First, statehood is being "denationalized" (hence "Schumpeterian workfare *post-national* regime"):

> This structural trend is reflected empirically in the "hollowing out" of the national state apparatus with old and new state capacities being reorganized territorially and functionally on subnational, national, supranational and translocal levels. There is the continuing movement of state power upwards, downwards and sideways as attempts are made by state managers on different territorial scales to enhance their respective operational autonomies and strategic capacities [...] However, countering this trend is the survival of the national state as the principal factor of social cohesion in societies and its associated role in promoting social redistribution. (Jessop, 1997a, pp. 573–4)

Second, the political system is being "destatized." In other words the formulation and implementation of public policy and collective political strategies increasingly involve "governance" in the sense defined earlier, with non-state actors playing an increasing role:

> There is a movement from the central role of official state apparatus in securing state-sponsored economic and social projects and political hegemony towards an emphasis on partnerships between governmental, para-governmental and non-governmental organizations in which the state apparatus is often only first among equals. (Jessop, 1997a, pp. 574–5)

Third, policy regimes are becoming internationalized. State action increasingly takes place in a much wider international context, and policy is influenced by a range of extraterritorial actors and institutions. This can be related to the focus on "international competitiveness" associated with the SWR, and he cites the worldwide shift to neoliberalism (see above) as the most obvious example (Jessop, 1997a, p. 575).

Local state and local governance

The rise of the Schumpeterian workfare regime and the associated shift from government to governance are particularly evident at the local scale. It is certainly at the local level that they have received most attention from economic geographers (see, for example, Goodwin and Painter, 1996; Jessop, 1997b; Peck and Tickell, 1995).

This does not mean, however, that the state is unimportant at the local level. On the contrary, state agencies are frequently central to the mobilization of the partnerships and networks on which governance depends. The term "local state" refers simply to any part or agency of the state that operates over part of the national territory with a degree of autonomy from the central state. It includes, but is not limited to, elected local governments. Local governance refers to collective policy and strategy formation at the local level involving a range of state, para-state, and non-state organizations.

The local state and local governance are important to economic geography because of the role they play in shaping the economic landscape. Local state institutions are significant economic actors in their areas. In some localities the local state is the largest employer, for example, and local authorities are major purchasers as well as facilitating and/or providing a wide range of local goods and services – from leisure facilities to education and from refuse disposal to public parks. In addition to this direct impact, local state institutions frequently pursue economic strategies with varying degrees of autonomy from the center. The constitutional position of local government varies widely from state to state, but the heightened intensity of international economic competition means that local authorities commonly use whatever power may be available to them to influence the economic fortunes of their localities. Typical strategies include: the attraction of inward investment using a variety of incentives; development assistance to local small and medium enterprises; infrastructure provision; land use planning; and training and education.

Each of these strategies can involve the mobilization of a range of actors in the locality, including public sector agencies, private firms, and non-government organizations; in a word, local governance. One important theoretical approach to understanding these relationships is provided by urban regime theory (Lauria, 1997). The concept of urban regime has gained significant prominence in the literature of urban studies and political science, especially in North America (Elkin, 1987; Fainstein and Fainstein, 1983; Lauria, 1994; Orr and Stoker, 1994; Stone, 1989, 1993). Recently it has been gaining popularity in relation to British urban politics (DiGaetano and Klemanski, 1993; Lawless, 1994), while Stoker and Mossberger (1994) have argued that it is possible to use it more widely, including in comparative studies (see also DiGaetano and Klemanski, 1993; Harding, 1994; Stoker, 1995).

Regime theory argues that successful governance almost always depends upon the availability and mobilization of resources and actors beyond those that are formally part of "government." Governing a locality, particularly in the United States where the institutions of elected urban government are relatively weak, relies on the ability to form governing coalitions that bring together the formal agencies of government with interest groups from the wider society. Foremost among these, in the American context at least, are business interests. However, Stone (1989) is at pains to point out that, although the prominence of the business connection is hardly surprising, a regime is not inherently a coalition with business (or with any particular interest come to that).

On the other hand, business interests are central in practice because regime success is evaluated (at least in part) by economic prosperity, and because (in the United States, but not, for example, in Britain) local governments are heavily dependent on local businesses for tax revenues. Outside the United States, other interests may be more central to the coalition. In many European cities, for example, appointed local state officials, technocratic managers, and professionals play a central role. Furthermore, in many cases the most important business interests are not locally embedded to the extent evident in many of the examples discussed in the literature on the USA.

An urban regime may thus be defined as a coalition of interests at the urban scale, including, but not limited to, elected local government officials, which coordinates resources and thus generates "governing capacity." Regimes can pursue a range of different strategies and this has led some writers to note a range of regime types (pro-growth, caretaker, conservationist, etc.). At the risk of some oversimplification, however, we may identify what Harvey (1989) sees as a broad shift from managerial to entrepreneurial strategies of urban governance (see also Hall and Hubbard, 1998). This means a change from running the everyday functions of a locality such as transportation or housing provision (managerialism), to developing active risk-taking strategies to promote the economic fortunes of the locality in an increasingly competitive world (entrepreneurialism).

Harvey argues that there are four options for urban entrepreneurialism. First, "competition within the international division of labor means the creation and exploitation of particular advantages for the production of goods and services" (p. 8), by, for example, seeking to attract inward investment. Second, "the urban region can also seek to improve its competitive position with respect to the spatial division of consumption" (p. 9), by promoting the development of both routine consumption activities and more spectacular festivals. Third, "urban entrepreneurialism has also been strongly coloured by a fierce struggle over the acquisition of key control and command functions in high finance, government, or information gathering and processing" (p. 9). This might, for example, involve attempts to attract government offices undergoing relocation. Finally, "competitive edge with respect to redistributions of surpluses through central governments is still of tremendous importance" (p. 10). Here, Harvey means not only direct government grants, but also lucrative government contracts in, for example, the defense sector. Harvey goes on to suggest that the macroeconomic implications of this kind of inter-urban competition are often actually negative in the short to medium term. Nevertheless he does point to the more progressive possibilities inherent in the generation of renewed collective development strategies at the local scale. Realizing these possibilities depends in part on the construction of appropriately progressive urban regimes.

At present, however, the majority of urban regimes are aimed at promoting local economic growth by prioritizing the interests of businesses above those of other sectors of the community. An example will clarify how this can work in practice. The transition from local government to local governance in the UK has been closely associated with the emergence of new forms of informal power networks and local elites. In addition to the formal partnership organizations that are characteristic of local governance, we are seeing the development of close-knit elite networks, typically dominated by business interests. This is consistent with the idea of the transition from a welfare state to a workfare regime, insofar as the latter encompasses a renewed emphasis on the role of capital and entrepreneurialism within systems of political authority. At the local scale, these business-dominated networks can be thought of as one kind of urban regime. Indeed the original concept of urban regime was developed precisely to describe the relationship between business interests and urban politics (Stone, 1989).

The work of Tickell and Peck on Manchester provides a good example of the transition from local government to local governance and the associated growth in the business-dominance of urban politics (see Peck and Tickell, 1995; Tickell and Peck, 1996). Although Tickell and Peck do not use urban regime terminology, their work does reveal the emergence of the kind of local networks of political actors and interests to which the concept of urban regime refers. In addition to the agencies of elected local government, there are two kinds of formal organizations contributing to the pattern of urban governance in the UK, and evident in the Manchester case. The first are unelected "quangos" (quasi-non-government organizations: public bodies whose members are appointed, usually by central government). They include Urban Development Corporations (of which there were two in the Manchester area, both of which have now come to the end of their ten-year life-spans), training and enterprise councils, regional and district health authorities, and housing action trusts. Secondly,

> ... alongside and deeply networked with this burgeoning system of quangos is a range of other business-led and "partnership" bodies, such as ... the Business Leadership Teams (BLTs) and Business in the Community, which are with the active support of government ministers forcefully presenting the "voice" and "vision" of business as being the legitimate interest in local policy formulation. (Tickell and Peck, 1996, pp. 595–96)

These formal organizations are cross-cut by informal networks of businessmen (and Tickell and Peck emphasize that they mostly *are* men) that often involve multiple memberships of quangos and partnership bodies for specific individuals in the local business community. Politically conservative and male-dominated, these networks represent a significant shift in the power structure of local policymaking from the 1970s and 1980s, when the elected local state in Manchester attempted to implement a range of left-wing policies in the areas of economic development, equal opportunities, and nuclear disarmament. For example, the North West Business Leadership Team, which covers Manchester, consists of thirty self-appointed businessmen and, "has claimed the right to speak for the region, playing a central role in the formulation of the regional economic strategy, the framework for the dispersal of European Union structural funds over the period 1994–6" (Tickell and Peck, 1996, p. 598).

The implication of the shift in Manchester from local government to local governance, and the consolidation of an entrepreneurial and business-dominated urban regime, is that the impact of state and governance at the local level on economic geography is one that is increasingly tied to the interests of men, capital, and private firms, at the expense of women, labor, and the public sector. While the details of the Manchester case are unique, similar trends are apparent in many other UK and US cities (Cox, 1997; Horan, 1997). Indeed the importance of "entrepreneurial cities" is growing worldwide (Hall and Hubbard, 1998).

Conclusion

I have shown in this chapter how economic relations within and between countries are influenced by the state and governance. In contrast to the view that the state's importance is in terminal decline in the contemporary world, I have emphasized that it continues to have an intense involvement in the production and transformation of economic processes and the working out of economic geographies. That being said, and as the work of Jessop showed, governance is becoming less statist, with the growing involvement of a range of non-governmental organizations. Moreover, the emergence of transnational governance regimes, such as the EU, does challenge the centrality of the *nation*-state in the field of economic policymaking and regulation. At the same time, local urban regimes make strenuous efforts to improve the economic status of their localities. However, while the state is losing its taken-for-granted status as the pre-eminent political authority, and a more complex picture is emerging in which the state is one among a number of participants in governance, this hardly heralds its outright demise. States will continue to shape and be shaped by the economic landscape at a variety of spatial scales for many years to come.

Bibliography

Abrams, P. 1977. Notes on the difficulty of studying the state. British Sociological Association Annual Conference.

Amin, S. 1991. The state and development. In D. Held (ed). *Political Theory Today*. Cambridge: Polity, 305–29.

Anderson, J. 1996. The shifting stage of politics: New medieval and postmodern territorialities. *Environment and Planning D: Society and Space*, 14, 133–53.

Bayart, J.-F. 1993. *The State in Africa: the Politics of the Belly*. London and New York, NY: Longman.

Block, F. 1994. The roles of the state in the economy. In N. J. Smelser and R. Swedbert (eds). *The Handbook of Economic Sociology*. Princeton, NJ: Princeton University Press, 691–710.

Cerny, P. G. 1990. *The Changing Architecture of Politics: Structure, Agency and the Future of the State*. London: Sage.

Corrigan, P. and Sayer, D. 1985. *The Great Arch: English State Formation as Cultural Revolution*. Oxford: Blackwell.

Cox, K. 1997. Governance, urban regime analysis and the politics of local economic development. In M. Lauria (ed). *Reconstructing Urban Regime Theory: Regulating Urban Politics in a Global Economy*. Thousand Oaks: Sage, 99–121.

DiGaetano, A. and Klemanski, J. S. 1993. Urban regimes in comparative perspective: the politics of urban development in Britain. *Urban Affairs Quarterly*, 219, 54–83.

Elkin, S. L. 1987. *City and Regime in the American Republic*. Chicago: University of Chicago Press.

Escobar, A. 1995. *Encountering Development: the Making and Unmaking of the Third World*. Princeton, NJ: Princeton University Press.

Fainstein, N. I. and Fainstein, S. S. 1983. Regime strategies, communal resistance, and economic forces. In S. S. Fainstein and N. I. Fainstein (eds). *Restructuring the City*. New York, NY: Longman, 245–82.

Giddens, A. 1985. *The Nation-State and Violence*. Cambridge: Polity.

Goodwin, M. and Painter, J. 1996. Local governance, the crises of Fordism and the changing geographies of regulation. *Transactions of the Institute of British Geographers*, 21, 635–48.

Hall, T. and Hubbard, P. (eds). 1998. *The Entrepreneurial City*. Chichester: Wiley.

Harding, A. 1994. Urban regimes and growth machines: Toward a cross-national research agenda. *Urban Affairs Quarterly*, 29, 356–82.

Harvey, D. 1989. From managerialism to entrepreneurialism: The transformation in urban governance in late capitalism. *Geografiska Annaler*, 71B, 3–17.

Horan, C. 1997. Coalition, market and state: postwar development politics in Boston. In M. Lauria (ed). *Reconstructing Regime Theory. Thousand Oaks: Sage,* 149–70.

Hutton, W. 1995. *The State We're In*. London: Jonathan Cape.

Jessop, B. 1993. Towards a Schumpeterian workfare state? Preliminary remarks on post-Fordist political economy. *Studies in Political Economy*, 40 (Spring), 7–39.

Jessop, B. 1994. Post-Fordism and the state. In A. Amin (ed). *Post-Fordism: A Reader*. Oxford: Blackwell, 251–79.

Jessop, B. 1997a. Capitalism and its future: Remarks on regulation, government and governance. *Review of International Political Economy*, 4, 561–82.

Jessop, B. 1997b. A neo-Gramscian approach to the regulation of urban regimes: Accumulation strategies, hegemonic projects, and governance. In M. Lauria (ed). *Reconstructing Urban Regime Theory*. Thousand Oaks: Sage, 51–73.

Jessop, B. 1997c. Twenty years of the (Parisian) regulation approach: The paradox of success and failure at home and abroad. *New Political Economy*, 2, 503–26.

Knox, P. and Agnew, J. 1998. *The Geography of the World Economy*, third edition. London: Arnold.

Lauria, M. 1994. The transformation of local politics: Manufacturing plant closures and governing coalition fragmentation. *Political Geography*, 13, 515–39.

Lauria, M. 1997. *Reconstructing Urban Regime Theory*. Thousand Oaks: Sage.

Lawless, P. 1994. Partnership in urban regeneration in the UK: the Sheffield central area study. *Urban Studies*, 31, 1303–24.

Mann, M. 1988. *States, War and Capitalism*. Oxford: Blackwell.

Mann, M. 1993. *The Sources of Social Power. Volume II: The Rise of Classes and Nation-states, 1760–1914*. Cambridge: Cambridge University Press.

Martin, R. and Sunley, P. 1997. The post-Keynesian state and the space economy. In R. Lee and J. Wills (eds). *Geographies of Economies*. London: Arnold, 278–89.

Ohmae, K. 1995. *The End of the Nation State: The Rise of Regional Economies*. London: HarperCollins.

O'Neill, P. M. 1997. Bringing the qualitative state into economic geography. In R. Lee and J. Wills (eds). *Geographies of Economies*. London: Arnold, 290–301.

Orr, M. E. and Stoker, G. 1994. Urban regimes and leadership in Detroit. *Urban Affairs Quarterly*, 30, 48–73.

Peck, J. 2000. *Workfare States*. New York, NY: Guilford Press.

Peck, J. and Tickell, A. 1995. Business goes local: Dissecting the business agenda in Manchester. *International Journal of Urban and Regional Research*, 19, 55–78.

Polanyi, K. 1957. *The Great Transformation*. Boston, MA: Beacon Press.

Resnick, P. 1990. *The Masks of Proteus: Canadian Reflections on the State*. Montreal and Kingston: McGill-Queen's University Press.

Rhodes, R. 1997. *Understanding Governance*. Buckingham: Open University Press.

Stoker, G. 1995. Regime theory and urban politics. In D. Judge, G. Stoker, and H. Wolman (eds). *Theories of Urban Politics*. London: Sage, 54–71.

Stoker, G. and Mossberger, K. 1994. Urban regime theory in comparative perspective. *Environment and Planning C: Government and Policy*, 12, 195–212.

Stone, C. 1989. *Regime Politics: Governing Atlanta 1946–1988*. Lawrence, KS: University of Kansas Press.

Stone, C. 1993. Urban regimes and the capacity to govern: A political economy approach. *Journal of Urban Affairs*, 15, 1–28.

Tickell, A. and Peck, J. 1996. The return of the Manchester Men: Men's words and men's deeds in the remaking of the local state. *Transactions of the Institute of British Geographers*, 21, 595–616.

Wiseman, J. A. 1997. The rise and fall and rise (and fall?) of democracy in sub-Saharan Africa. In D. Potter, D. Goldblatt, M. Kiloh, and P. Lewis (eds). *Democratization*. Cambridge: Polity, 272–93.

Chapter 23

Creating the Corporate World: Strategy and Culture, Time and Space

Erica Schoenberger

Let's start by saying that corporations, as they go about their business, create the world in which they – and we – live. This is by way of asserting that they don't just fill up the landscape with their activities because, after all, they have to be *somewhere*. Corporations actively create an entire geography appropriate to their needs and their understandings about how the world works. More strongly, firms need to create this geography not merely as a by-product of their productive and competitive activities, but as part of their ordinary operational and strategic work – as part of the work of surviving and making a profit in a capitalist economy.

We know, of course, that corporations do not have free rein to design any social and economic landscape of their choice. They must deal with the legacy of history and with other individuals, social groups, and institutional agents. One might paraphrase Marx to suggest that firms make their own historical geography, but not in circumstances of their own choosing. But in this chapter, I want to focus on how the world looks from the point of view of this very particular social agent – the corporation – and at what happens when it needs to change its world. I also want to focus on one set of relationships that the firm has with the "outside world" and how it relates to others of its type – the corporations against which it competes. Accordingly, the emphasis here will be on competitive strategy and not so much on production processes, and it will be on life at the top and not so much on how firms construct their relationships with, for example, workers, communities, and governments.

Corporate Culture and Strategy

Where does competitive strategy come from? Usually we think of it as a rational process of decisionmaking that weighs the information available and deduces the appropriate strategy from it (see, for example, Porter, 1980). It is acknowledged that one can't know everything, that not all information is equally good, and that some degree of uncertainty and risk is unavoidable. Similarly, it is recognized that certain kinds of change will trespass against the interests of some individuals who may try to derail a strategy. Consequently, bad decisions are possible but they ought not to be systematic and they ought not to happen *despite* access to very good information. At bottom, it is assumed, corporations act in their own best interests.

Here I want to assert that corporations *do not* always act in their best interests. They frequently develop misguided strategies on the basis of excellent information about what they ought to do. I will describe an example of how this process worked in an individual firm. But I want also to argue that at certain historical and geographical conjunctures large numbers of firms may refuse the implications of the information they have in their hands and, consequently, fail to adapt appropriately to changes in their competitive environment. This will require linking up the explanation of individual corporate strategy formation with a larger story of how and why the environment changed. I will try to make this link through a discussion of how corporate cultures influence strategy-formation.

Corporate culture has lately become a fashionable topic.[1] It is usually viewed as the object of strategy: changing the strategy requires us to change the culture. This is hard because culture is largely a matter of traditional or habitual behaviors and attitudes and these are inherently resistant to change.

I want to approach corporate culture from a different angle – one in which culture is inherently involved in the *production* of strategy. This may allow us to understand why corporations sometimes refuse necessary changes even when they have very good information about what to do.

What, then, is corporate culture, where does it come from, and what does it do? Broadly speaking, corporate culture is an ensemble of material practices, social relations, and ways of thinking. Material practices are how work is actually done – not just the work of people on the production line, but also the work of managers, engineers, accountants and the like. Material practices also include *what* is done – what kinds of things are produced by these different people – and why. The "things" produced include actual products, but also less tangible outputs such as the development of production processes, organizational practices, accounting standards, and strategies about what the firm should do.

Social relations include the ordinary rules of behavior within a community that allow us all to work efficiently with one another. At IBM, for example, the dress code was famously dark suits and white shirts; at Apple the norm ran to T-shirts and jeans. Wearing Apple clothes to IBM or vice versa would mark you instantly as an outsider; someone with whom real communication was impossible because it was patently evident that you did not speak the local language.

More profoundly, social relations underlie how power, rights, and obligations are produced within the corporate community and how they are allocated. In short, where does power or responsibility come from, who has it, and how is it used? In some corporate settings, production workers literally have no power over their own bodies during the workday. They must be physically in their assigned spot at all times, and the individual gestures used in their tasks are choreographed in detail by someone else (Wright, forthcoming). In others, line workers are invited to help design the production process they are engaged in and have the power to stop a process that is, in their judgment, going wrong.

Finally, ways of thinking include not merely ideas and meanings, but also processes of interpretation and the construction of knowledge. How are data and information transformed into something that we know and can act upon? How are values developed about what is important, what kinds of information count, what kinds of activities are productive, who is a member of our community, and who is not? All of these things – work, power, and a whole range of understandings about how the world works and how it *ought* to work – constitute a corporation's culture.

Where does it come from? Everyone in the firm is in some way constructing its culture. But here we will focus on the dominant culture of the firm; which is to say the culture of the dominant – the people who run the firm and develop its strategy. Corporate culture in this sense must necessarily be produced through and be expressive of the material circumstances and understandings of the powerful. It is, accordingly, in important ways, *about* power – about who will have it and what they will use it for. At the top of the corporation, the most important kind of power at stake is the power to exercise one's strategic imagination – to impose one's own view of how the world ought to work. The strategist does this in competition with others of his type both within the firm and with the people running competing firms.[2] This implies that the process of cultural production in the firm is inherently and deeply conflictual. People must, unavoidably, struggle to acquire this power, to use it effectively, and to valorize the results.

Are corporate cultures inherently resistant to change? If this were true, it might explain why corporations have a hard time reacting effectively to major shifts in the competitive environment. But corporations change all the time. They buy and sell divisions, they invest in new equipment, they enter or leave particular markets, they lay off some kinds of people and hire others, and they are constantly reorganizing themselves according to the latest management fads. The problem is not that firms don't change in the face of new challenges. The problem is that they often do not change *appropriately*. This is because the culture can accept certain kinds of change while it is unable to accept others. Cultures are not embodiments of tradition; instead, cultural processes are deeply involved in the selection of which traditions to value and which can be disposed of at need. Corporate cultures *select* some kinds of change and refuse others, although those refused might be exactly what is needed and this fact might be well known within the firm.

The Xerox Corporation and the copier wars

Here's an example.[3] The Xerox Corporation was one of the most successful companies in American history, the first to achieve $1 billion in annual sales within ten years of starting up. It was an innovative company – the first to bring an easy-to-use plain-paper copier to market. It also realized that the real money lay not in selling the extremely expensive machines, but in leasing them cheaply and charging a small sum for each copy made. The company was seemingly untouchable, owning something in excess of 90 percent of the world market for copiers.

Nevertheless, this was a company that worried about its future and actively sought to protect itself against the new and the unexpected. It anticipated, for example, the emergence of competition from IBM and Kodak which began to appear in the 1970s. Like Xerox, Kodak and IBM made large, expensive, technically elegant machines that offered very high performance but cost a fortune and often broke down. From this point of view, they fit perfectly with Xerox's understanding of what the copier market was like and how you made money in it. Xerox did everything it could think of to counter the threat that Kodak and IBM posed. It even overturned its oldest, most sacred corporate taboo to deal with the fact that Kodak had introduced document recirculation. At Xerox, the Prime Directive was "don't damage the original," which meant handling it as little as possible. If Xerox could violate this stricture, it was apparently ready to accept any change to stay on top.

Nevertheless, within ten years, Xerox's market share had slipped to around 15 percent. This was a true catastrophe, but it wasn't because of IBM and Kodak. It was because the market was voting overwhelmingly for the small, slow, inexpensive, reliable machines offered by Japanese competitors. Xerox knew quite a lot about this competition, principally through its subsidiary, Fuji-Xerox, which the company had presciently established in the early 1960s.

Fuji-Xerox felt the force of this competition first. In response, it threw everything into a crash program to develop a machine that would be competitive in the Japanese market. It developed this machine in less than two years, compared with the five to eight year development cycle typical of Xerox. Fuji-Xerox also reorganized the way it made things to produce the new machine with good quality at a low cost. The machine they came up with in 1972, the 2200, was quite successful in the Japanese market.

Rank-Xerox, the company's European subsidiary, was the next to be hit by this new model of competition. In 1977, they responded by buying the machine developed by Fuji-Xerox and selling it in Europe. In the first year, they sold 25,000 – a success that was stunning even by Xerox standards.

Managers from Japan and Europe for years actively urged the parent company to produce the 2200 for the American market. A committee was established to study the problem. This might seem to promise the typical bureaucratic death of a good idea, but after an exhaustive review, the committee recommended unequivocally that the parent adopt the 2200. Top management vetoed the plan. Then the company commissioned McKinsey consultants to re-evaluate the situation. McKinsey enthusiastically endorsed Fuji-Xerox's strategy and its product. Again, top US management rejected the idea.

There were two fundamental reasons for this. One was that small, slow, simple, cheap, low-margin machines had no value or meaning in the dominant Xerox culture. The company had committed itself to large, fast, technically sophisticated, high-margin and expensive products. It understood itself in the world as the bearer of a certain kind of product for a certain kind of market. Accordingly, even if small machines were flooding the market, the only conceivable response was to build a better large machine.

The second reason is that although the dominant culture could recognize IBM and Kodak as potentially serious competition, it could not recognize the Japanese as accredited players in this particular game. This is despite the fact that it was precisely the Japanese who had pushed them to the wall while their own Japanese subsidiary was offering them a nearly guaranteed and essentially free solution. As someone who was on the scene at the time put it, the parent company's attitude was "we taught them everything they know about copiers. How could they have anything to teach us?" (Schoenberger, 1997, p. 200). David Kearns, who is widely credited with eventually turning Xerox around, argues that the company both knew what the problem was and at the same time could not accept this information: "I doubt that it was a case of the information not being available. I think it was purely a matter of denial" (Kearns and Nadler, 1992, p. 122).

Notice, however, that although Xerox had invented xerography and IBM and Kodak had not, people at Xerox didn't assume that IBM and Kodak had nothing to teach them. Similarly, Xerox was never in denial about the competitive threat posed by these two. Its arrogance and its denial were both highly selective. This selectivity meant that information that was literally on the table was, within the frame of Xerox's corporate culture, uninterpretable. It was information that could not be transformed into knowledge and acted upon appropriately. Note that it was Xerox's culture that made the Japanese unreadable, not Japanese "inscrutability."

The effort to turn the company around was a long and arduous one that involved a kind of guerrilla warfare according to Kearns. As part of this struggle, a number of high-level executives left and many positions were filled by people from Rank-Xerox. They brought with them a new sense of who the company was and what it ought to be doing. The conflict, then, produced major shifts in who held power in the company and what they wanted to do with it and, in tandem, deep changes in the company's material practices (the kinds of things it made and how it made and designed them), social relations (decentralization of authority, worker participation), and ways of thinking (about the market, competition, prices, product quality, etc.). In effect, the winners were able to institute a new culture and a new strategy. Note that the strategy changes only in and through cultural change. The old culture did not produce the new strategy.

Second, the process of cultural change within the firm involves struggle and conflict. Peoples' identities and commitments – how they understand themselves in the world and what they think they ought to do in it – are at stake. So, too, are the sources of their social power within the firm and their ability to impose their own sense of who and what the firm ought to be and how it should operate. Little wonder that the struggle is so acute.

Corporate cultures and industrial crisis

Xerox is not the only company about which such a story can be told. The specifics will vary, but it seems plausible to suppose that the intricate relationship between culture and strategy is a normal feature of corporate life. Moreover, each individual corporate story is embedded in a broader industrial culture that is characteristic of a particular time and place. The individuals may be extremely distinctive even within a given industry – Ford is a very different place from GM, for example – but there are shared understandings and practices that have linked a broad swathe of corporations in North America and Western Europe in the twentieth century.

The question is whether there are historical–geographical conjunctures that pose such a deep challenge to the prevailing culture that an entire industrial system may be thrown into crisis. For the answer to be yes, we would need evidence of a competitive challenge that went far beyond the modifications of product type or adjustments in price levels that are normal features of competition in corporate life. It would have to be a challenge that embraced a broad set of practices and understandings related to how markets work, how companies make money in them, who production workers are and what they are for, how things ought to be designed and made, and how different kinds of activity and people within the corporation are to be valued.

I want also to suggest that such a deep competitive challenge necessarily involves a major shift in practices and understandings related to time and space. These may be exceptionally difficult to analyze and respond to effectively because we are unused to thinking about how social practices *create* particular temporal rhythms and spatial formations. But I shall try to show in what follows how the temporal and spatial underpinnings of a particular social order can be rendered quite suddenly obsolete when faced with competition from a new cultural ensemble with a different set of spatio-temporal practices and understandings. Indeed, because time and space appear to be so natural and beyond manipulation – unlike, say, products or production processes – identifying changes in how they operate socially may be the best indicator of the depth of a competitive challenge. In other words, if you can see that time and space are transformed by a new industrial model, you know that it represents a very significant change.

Competition in Time and Space

It might seem obvious that factories came into existence because they were more efficient than other ways of organizing production such as the putting-out system. Why else would anyone invest in a factory, after all? But there is considerable evidence that the transition from putting-out to factory production was not driven primarily by efficiency considerations. The problem with putting-out was not that it was inefficient but that the capitalist had no control over how long people worked. People worked at spinning or weaving for as long as *they* needed to, and then they stopped and did something else. Factories ensured that work went on long enough to provide the surplus that became the capitalist's profit. In other words, the factory made possible a new kind of *time discipline* which was necessary to the whole project of capitalist development and for the profitability of any individual firm.[4]

This new time discipline entailed a new kind of spatial discipline. Work and workers were concentrated in specialized spaces that were separated from the home and increasingly clustered in towns and cities. Workers were assigned specific places in the factory, and their movement around the factory was strictly supervised and constrained.

In short, capitalists must act *on* as well as in time and space as part of their normal business. It is in this sense that we can think of time and space as being socially constructed (rather than natural) and as being subject to the pressures of corporate life.

Over the long run, the tendency in modern capitalist society seems to be that everything speeds up and gets closer together as transportation and communications technologies are improved and cheapened. In the end, as we are constantly informed, we are increasingly harried members of the "global village." Everything has sped up to such a degree that space and distance apparently no longer matter (Harvey, 1989).

In the short run, though, the picture is more complicated. Speeding something up in one arena may actually cause some other process to slow down, or may make transportation a bigger problem than it was before. Someone has to decide whether such a trade-off is worthwhile and contributes to the profitability of the firm. This means that time and space are strategic problems for the firm.

You need look no further than your personal computer to see these trade-offs in action. You probably know that your PC's memory is vastly larger than it would have been ten years ago. This didn't happen because manufacturers crammed more chips into the box, but because they crammed more circuits onto the same chip. To do this, they had to make the circuit line-widths smaller – now substantially below 1 micron. This required billions of dollars in research and development, and billions more in investment in machines and factories capable of operating at this scale. The result is a new bottleneck in transferring all this data into and out of the chip or into and out of the machine. So everyone then must work on increasing bandwidths to accommodate vast quantities of data flowing faster and faster. Also, it turns out that you can make the hardware run faster if you transfer some of the functions into software. But this creates a huge bottleneck in software production: time is compressed in one arena (the functioning of the hardware) but everything slows down in another (writing software).

So there you are, plugged into the Internet – the very essence of a communications technology that is so fast that space and place seem to disappear altogether. You are operating in pure ether, summoning up information and commodities from out of the void.

And yet, if you could look behind the screen, here are things that you would see: warehouses full of stuff with real people driving forklifts around and working the packaging lines, while behind the warehouses are actual factories where people make all that stuff. These are rooted on the ground, and their physical presence and fixity make possible the illusion of utter spacelessness. The corporations necessarily create a real world in tandem with the virtual world.

Even your activities on the Web are mediated through "server farms" where many large computers are crammed into a building processing signals. Server farms are the places that allow us to transcend space, and they are the centralized entities that allow decentralized data processing and information exchange to flourish (Lohr and Markoff, 1999).

But here's the tension. You need to construct physical assets on the ground in order to compete in the ether, which means making large bets about what kind of assets to develop.[5] If you bet wrong, these assets stand to be suddenly vaporized or, to use a more technical term, devalued. From this point of view, e-commerce is not a new business paradigm – it is the normal one speeded up, which means that the risks pile up faster and ramify through the system in complicated ways.[6]

What I've tried to demonstrate is that alterations in time and space are a normal feature of capitalist competition. They may take place so gradually that we're not even aware of them. What happens, though, when a major shift in the spatio-temporal regime takes place in industry?

Time–space Transformations and Industrial Cultures

Something of the sort has arguably happened in the shift out of standardized mass production and into a more flexible production regime. This regime hinges on drastically compressing time in the development of new products and in the firm's ability to change what it is producing on the line. The transition began in the early 1970s and was accompanied by a tremendous crisis in large-scale manufacturing industry in the USA and Western Europe, marked by widespread unemployment and working-class income stagnation, followed more gradually and grudgingly by cutbacks in excess capacity.

The recovery has been highly uneven, with some sectors and firms and regions doing better than others. Though unemployment has declined, notably in the USA, workers ejected from stable, high-wage jobs in manufacturing have not recovered their former levels of job security and rates of income growth. Income distributions have become increasingly polarized, reversing a nearly two-decades-long trajectory of increasing income equality in the leading industrial economies (Mishel and Bernstein, 1998).

What was the nature of this crisis, and why has it been so hard for many firms to recover from it? The crisis was driven in part by the emergence of a new regime of competition and production organized around a very different conception of how markets work and how time and space should be managed.

The "before" picture is a highly stylized outline of the standardized mass production system that originated in the USA and was adopted with modifications in much of Western Europe.[7] Markets were understood to be relatively homogeneous and stable over time. They might be stratified – from Chevy at the low end to Cadillac at the high end – but within these strata and allowing for different colors and other superficial details, everyone bought essentially the same product. Moreover, the product stayed essentially the same for years on end. Fins might come and go annually, but the chassis design was good for ten years and the engine design for fifteen.

Markets were also understood to be stably divided among a small number of known competitors who avoided both cut-throat price competition and constant product innovation. They competed instead on the basis of advertising, brand name loyalty, financing, and distribution networks.

One reason for moderating the pace of product change is that new product development took a very long time and cost a tremendous amount of money. This had to be amortized over a high volume of output – literally millions of units. You needed to sell the thing for years in order to recover all of your costs without pricing the product so high that no one would buy it.[8]

Another reason for moderating the pace of change and for making only a limited variety of products has to do with how mass producers made things. They preferred not to rely on the skills and initiative of workers. Skilled workers could make a variety of things, but they cost more than unskilled ones, and they worked at their own pace and to their own standards of quality and completeness. In the mass production environment, skills were, to the degree possible, transferred to the machines and the pace of work was set by the speed of the moving assembly line. Machines were designed to do one thing extremely well. Dedicated machinery of this sort is quite expensive. You need to make a lot of that one thing in order to recover costs. If you make a huge amount, dedicated machines are extraordinarily efficient. Each thing that comes out is produced quite cheaply. This is the great benefit afforded by scale economies, but it comes at a price. You have to produce at or near capacity in order to keep your costs down, no matter what the market is doing.

In sum, the way to make money in a mass production environment is to make as much as possible of the same thing every year for as many years as possible while convincing your customers that the thing they are buying is entirely new and different from everyone else's. This is a feat that you can only pull off in a reasonably stable competitive environment where everyone plays by the same rules.[9]

Mass production was associated with a characteristic time–space regime. Time was sharply compressed in production. At the inception of the moving assembly line at Ford in 1913, for example, the time it took to assemble a car chassis fell from 12.5 hours to 93 minutes, essentially overnight (Hounshell, 1984, pp. 248–55). On the other hand, as we have seen, time remained stretched out in the realms of product development and in the turnover of fixed capital (all those dedicated machines that made so much so fast). Instead of days and hours, these changed over a period of years – the longer the better.

The temporal stability of the product and its homogeneity within the various price strata in combination underwrote an extraordinary spatial flexibility in the system. In making an endless number of the same thing all the time, it doesn't really matter where you make it. What counts is that product flows continually through the pipeline. In this kind of system, you can make component parts in twelve different countries and assemble them in a thirteenth.

Distance in this system appears to be a solved problem. So long as the cost of moving things around is not prohibitive nothing has to be really near anything else. Instead you can locate different pieces of the production process in wildly divergent locations in order, for example, to gain access to particular kinds of factor inputs (raw materials, labor) at a particular cost. Moreover, since the system is now so incredibly productive, there's a tremendous incentive to spread production over the globe to ensure access to markets that might resist being overwhelmed by a tide of American or European imports and to guard against the emergence of strong local competition.

The system, despite its geographical expansiveness and complexity, is quite robust. Even if a factory is shut down by a strike, there are enough components and finished products piled in warehouses to operate until the strike is over.

Slowing down time in one dimension (product life cycles) allowed the speeding up of time in another dimension (production), and the stretching out of space in yet a third. Mass production implied a very particular spatio-temporal regime. It also entailed a historically specific understanding of what markets were like, how they changed, and how firms competed within them. This went along with pervasive understandings about the fundamental rules of production that guided decisions among competing priorities. In this world, for example, quantity counted more than quality and the ruling maxim was "getting metal out the door." These understandings also guided how different persons and activities were valued. Labor, for example, was looked upon as a disruptive and costly "factor of production" which should be rigorously supervised and disciplined where it couldn't be eliminated. The tremendous supervisory apparatus that this entailed was one of the great cost burdens of the system.

Taken together, this ensemble of practices, social relations, and ways of thinking is an industrial culture. Part of the work that culture does is to help establish the boundaries of the normal and the thinkable. The culture of mass production was apt to find certain kinds of change comprehensible and desirable: speeding up the assembly line, for example, or finding new ways of eliminating labor. But it was likely to find other kinds of change – figuring out ways of rapidly altering the product mix on the assembly line, drastically compressing product life cycles, or asking workers how best to make something – impossible and even meaningless. They couldn't be contemplated, in much the same way that Xerox was unable to contemplate selling the 2200.

Unluckily, the alternative model of production and competition that emerged in the 1970s featured exactly the characteristics that the culture of mass production found abhorrent or unthinkable. This new model, I would stress, emerged out of real, material circumstances that differed sharply from those that supported the rise of mass production. Different versions of it developed in different places, but here I want to concentrate on how it looked in Japan.

It has been easy to suppose that industries developed differently in Japan because Japanese culture is so different from American culture – more harmonious, more group-oriented, more attentive to beauty, and so on. But appealing to the "Japanese-ness" of the Japanese while ignoring, for example, the nature of the Japanese market and how production could be organized within it obscures more than it clarifies. We need to understand the material and historical roots of Japanese industrial culture in order to get at why it looked so different.

For example, the total Japanese market for cars in 1950 was equal to a day and a half's production in the USA. Toyota, under these circumstances, simply had to invent a different way of making cars (Cusumano, 1985; Fruin, 1992). Or consider, with space at a tremendous premium, that Japanese offices couldn't easily accommodate copying machines that were so large they required their own room. Japanese copier makers, accordingly, were under considerable pressure to figure out how to make small copiers that were also inexpensive, and reliable (Schoenberger, 1997).

The Japanese model of "flexible mass production" relies on producing at high volumes in order to reduce costs, but also manages to accomplish two other ends that the American-style system could not: producing a constantly changing mix of products on the same production line, and accommodating the continual, rapid introduction of new or significantly redesigned products. Partly the Japanese model does this by an even more rigorous application of the principles of reducing labor time in production than American firms managed to achieve. At a time, for example, when American producers required upwards of twelve hours to change dies in the huge machines that stamp out large metal parts such as car doors, Toyota could change dies in two to three hours. By 1971, die changing at Toyota took only minutes (Monden, 1981). If it takes twelve hours to change a die, you avoid changing dies like the plague. If it takes three minutes, die changes are a normal part of the job.

How do you coordinate these different product types on a moving assembly line? If you're continually changing what you're making, you need to be able to grab not just any identical part from a huge bin of parts, but *the* part that specifically goes with the particular product that you're making. Here Toyota devised an elegant solution that upended the time-management principles of the US manufacturing system. Toyota's system, known as just-in-time (JIT), also drastically altered the spatial parameters of the production system.

The American system, recall, was based on making as much of the same thing as possible. The production ethos was wholly committed to producing more and more stuff, without reference to the vagaries of the market. In a sense, production was triggered by the imperative to produce. This was possible because you could assume what the market would demand – it didn't have that many choices. Parts and components emerged at a tremendous rate, headed for the assembly line, where they would pile up until they were used. Defective parts would be swept along with the good, and the fact that there might be a problem affecting thousands of units of output would not come to light until they were already assembled into a car or a toaster. The good news, though, was that you never ran out.

The just-in-time system, by contrast, coordinates output closely with the actual profile of market demand. Production is triggered by orders coming in. The underlying ethos here to is make only what you need right now: a need on the assembly line calls for the production of the relevant part. If the immediate task is to put doors on ten light trucks, five with electric windows and five without, you send a message asking for these exact door panels and subassemblies.[10] This message causes them to be built, triggering another set of messages upstream to the people in the stamping plant, the people assembling wiring harnesses, making windows, and so forth. If the next task is doors for sedans, followed by doors for the sports model, you request these accordingly and a whole new round of activity is sparked.

The right doors must arrive just as you need them. They never would, of course, if the people upstream were also thousands of miles away. JIT does not accommodate great distances well. Or, to put this another way, the American system solved the problem of distance in production, and JIT unsolved it.

Suppliers not only need to be reasonably close, they also need to be able to align their own production rhythms with that of the final assembler. They, too, need to be able to change what they do at a moment's notice. In this way, the spatio-temporal characteristics of the model radiate outwards, encompassing everyone – workers and suppliers – who serve it. Notice that what begins as a possibility – if you *can* change dies fast enough, you can change what you make on the line – soon becomes a requirement. To be part of this system, you *must* figure out how to change dies in minutes.

This alters many things. For example, it injects a significant fragility into the system. A strike at a parts plant almost instantly shuts down a whole family of related plants. Someone missing their schedule or producing just a handful of bad parts can stop everyone else in their tracks while the problem is sorted out. Workers and suppliers are now called upon not merely to do their job, but to guarantee that it will be done to specifications and to schedule.

This means that workers have to be self-monitoring and self-motivating. Things happen too fast for the supervisory apparatus of American mass producers to be effective. If a machine went down in the American system, it was possible to wait for someone to be sent to fix it. In JIT, workers must solve their own problems. In a well-functioning JIT plant, an individual worker whose machine is malfunctioning can stop the production line until he or she has fixed the problem. In the US mass production culture, a worker stopping the line was an absolute taboo in the true anthropological sense (Hamper, 1981). Life in this new world is not relaxing. But it is lived quite differently from the old-style mass production line.

Now the underpinnings of an alternative industrial culture can be glimpsed. It is both produced by and expressive of different material practices, social relations, and ways of thinking. Its origins lie in different historical and geographical conditions, and it generates different ways of valuing people and activities. For example, workers in this system are not thought of as unruly, expensive problems that need to be supervised constantly and eliminated where possible, but as self-disciplined, value-adding members of the corporate community who can be relied upon to work *themselves* to the bone.

This point is not reached because, say, Japanese culture is more harmonious and group oriented. Japanese industrial culture hasn't always been harmonious. In the immediate postwar period, it was characterized by violent strikes and tremendous upheaval (Cusumano, 1985). A culture of cooperation on the shop floor is built through a long and difficult historical process, full of struggle and conflict, that works itself out in a very particular fashion.

Faced with this new model of production and competition organized around transformed spatio-temporal practices, many North American and European firms were very slow to respond effectively. They *did* a lot of things. They laid off workers, they closed factories, they re-jigged the division of labor. Mostly, they tried everything in their power to reduce costs. But lower cost was not the most important feature of this new model and does not explain why it was so successful. Greater variety of product, much higher quality, and a more rapid turnover of product generations were the key. Here the older industrial culture had no response.

Arguably, the old culture couldn't comprehend what was really important in the new one. In any case, an alarming proportion of the largest corporations – those that had for generations been dominant in their markets – failed to respond adequately to the new challenge. This failure persisted for years – even decades – as profits evaporated and market shares plummeted. There was no lack of evidence that something needed to be done that was not being done. There was even considerable evidence about what *was* needed. But in case after case, this information was refused although the very existence of the corporation was at stake.

Conclusion

What is striking is that at a particular moment in history so many corporations in North America and Western Europe that had dominated their markets for generations were brought low, with terrible consequences for cities and regions and communities. Decades have passed, and while some companies have recovered, others are still struggling and many have disappeared. To explain the geographical and historical specificity of this crisis, we need to account for two rather different phenomena.

The first has to do with the nature of the challenge. The new model of competition and production must be substantially different from the old – different enough that the old model cannot merely be adjusted but has to be thoroughly overhauled. How does one know when the difference is this big? One way is to count the bodies. The magnitude of the social and economic dislocation involved provides a strong indicator that the difference is big, but it doesn't tell you much about where the difference comes from or why it makes such a difference.

Another way is to analyze the material and social practices and relationships involved in the different models, and how people understand what they are doing within them and why – in short, by analyzing the corporate and industrial culture. This culture necessarily involves the production of a characteristic spatio-temporal regime. As I have suggested, a major historical disjunction in spatio-temporal practices and understandings is itself evidence of the magnitude of a cultural shift and the difficulty of adapting to new kinds of competitive challenge.

In the first instance, the spatio-temporal regime is produced through, and sustains, tremendous investments in the built and social environments. These assets constitute classic sunk costs – they are useful within the industrial regime that produced them but have no value outside of that social order (Clark, 1994). The commitment to change, then, entails enormous cost for the firm and the devaluation of what had formerly been extremely valuable. It would be surprising if this could be accomplished without trauma and great reluctance.

Nevertheless, it is presumably better to incur these costs than to die, and that is what makes the long delay in responding to the new model of competition so intriguing. Many firms very nearly did go to the wall rather than adapt appropriately, even though the nature of the adaptation required was known. The reasons for this refusal to act effectively is the second part of the picture that needs to be filled in.

It is not enough to cite the burden of tradition and habit. As we have seen, firms are not normally prisoners of tradition – they are, rather, normally caught up in tremendous changes all the time. We need to understand why *certain* traditions are selected and sustained against overwhelming evidence of their inadequacy in particular circumstances. I've tried to account for this through the notion of a corporate culture that is actively involved in the production of corporate strategy. This may allow us to understand how certain commitments become so powerful that they can't be challenged, even when they threaten the very existence of the firm and its decisionmakers.

The task here is a complex one. We need to understand how changes in the firm's environment can arise that pose a true life or death challenge for the firm, even though the firm is partly responsible for creating that environment as it competes with other companies and goes about its ordinary business. We need also to understand how, as the firm creates its world through the normal processes of capitalist life, it also creates itself as a configuration of physical and social assets that has real power in some circumstances yet is powerless in others.

This chapter falls short of achieving this task. But it might provide some ways of thinking about how these processes work out and how people – real decisionmakers – understand their situations and act within them. Accordingly, it might also give us some insight into the world we live in and the way we understand who we are and what we are able to do in it.

Acknowledgment

My thanks to Trevor Barnes for his thoughtful commentary and careful editing and to Greg Downey for his comments on an earlier draft. Naturally, I hold them both entirely responsible for the outcome.

Endnotes

1. See, for example, Hampden-Turner, 1990; Kotter and Hesket, 1992; Trice and Beyer, 1993.
2. One ought to be able to say "his or her type" but, as a practical matter, men over-whelmingly hold these positions. Gender-neutral language not only won't change that but gives the misleading impression that the situation might in fact be otherwise.
3. See Schoenberger (1997) for a more detailed account and citations.
4. Good historical sources include Landes, 1970; Lazonick, 1990; Marglin, 1974; and Thompson, 1967.
5. You might think that the bet would be much smaller because you don't have to build retail stores and you don't have to hire salespeople. This turns out not to be so true, and it's one of the reasons why pure Internet companies, despite their incredible stock market valuations, mostly haven't been profitable so far. For one thing, you replace a large number of cheap salespeople with a small number of very expensive programmers. You can never outgrow this need because the web sites need to be upgraded all the time and completely overhauled on a compressed time schedule. For another, it turns out that having real bricks and mortar stores may be a way of making your web site more effective. Even Gateway Computer, a paragon of direct sales, has now started opening demonstration sites in real places (Kaufman, 1999).
6. We can grant that the Internet company doesn't need to own the warehouses and factories. It may need to enter into long-term contractual relationships with providers of space and product in order to function effectively. But somewhere along the food chain, someone has to invest in these physical assets which means putting capital at risk in a situation of complex interdependencies. The Internet firms depend for their success on the suppliers, and the suppliers similarly depend on the Internet firms.
7. Not everyone operated in exactly this way, but nearly everyone was in some way involved in or touched by the system – as small-scale suppliers, for example, or small firms that survived in niche markets ignored by or unsuited to the mass producers. Good sources on how this system worked in general include Dicken, 1998; Hounshell, 1984; Piore and Sabel, 1984; Womack et al., 1990.
8. An indicator of just how long and how expensive the process could be is provided by the development experience for the Ford Mondeo/Contour. It took six years and cost six billion dollars – and this is *after* Ford worked strenuously to reduce the time and money involved in development. If Ford sells only a million of them, each car starts out with $6000 worth of development costs before it is even built.
9. Although basic industries such as steel, chemicals, and energy are characterized by different production techniques than the one sketched here, they also depended on scale economies and high rates of capacity utilization to keep their costs down, and a stable competitive environment to keep their profits up. And since they are closely tied into the final assemblers as suppliers of material inputs in production or use, the rhythms of product change characteristic of those sectors closest to final consumption markets tended to filter back upstream.
10. At Toyota, the messages were sent on small cards known as *kanban*: the system as a whole is often known as the kanban system.

Bibliography

Clark, G. 1994. Strategy and structure: Corporate restructuring and the scope and character-istics of sunk costs. *Environment and Planning A*, 26, 9–32.

Cusumano, M. 1985. *The Japanese Automobile Industry*. Cambridge: Harvard University Press.

Dicken, P. 1998. *Global Shift*. NY: Guilford Press.

Fruin, M. 1992. *The Japanese Enterprise System*. Oxford: Clarendon Press.

Hampden-Turner, C. 1990. *Creating Corporate Culture*. Reading, MA: Addison-Wesley.

Hamper, B. 1981. *Rivethead*. NY: Warner Books.

Harvey, D. 1989. *The Condition of Postmodernity*. Oxford: Blackwell.

Hounshell, D. 1984. *From the American System to Mass Production, 1800–1932*. Baltimore: Johns Hopkins University Press.

Kaufman, L. 1999. Selling backpacks on the web is much harder than it looks. *New York Times*, 24 May, C1.

Kearns, D. and Nadler, D. 1992. *Prophets in the Dark: How Xerox Reinvented Itself and Beat Back the Japanese*. NY: HarperBusiness.

Kotter, J. and Hesket, J. 1992. *Corporate Culture and Performance*. NY: Free Press.

Landes, D. 1970. *The Unbound Prometheus*. London: Cambridge University Press.

Lazonick, W. 1990. *Competitive Advantage on the Shop Floor*. Cambridge: Harvard University Press.

Lohr, S. and Markoff, J. 1999. Computing centers become the keeper of the web's future. *New York Times*, 19 May, A1.

Marglin, S. 1974. What do bosses do? *Review of Radical Political Economy*, 6(2), 60–92.

Mishel, L. and Bernstein, J. 1998. *The State of Working America*. Washington, DC: Economic Policy Institute.

Monden, Y. 1981. What makes the Toyota production system really tick? *Industrial Engineering*, January, 36–46.

Piore, M. and Sabel, C. 1984. *The Second Industrial Divide*. NY: Basic Books.

Porter, M. 1980. *Competitive Strategy*. NY: Free Press.

Schoenberger, E. 1997. *The Cultural Crisis of the Firm*. Oxford: Blackwell.

Thompson, E. P. 1967. Time, work-discipline, and industrial capitalism. *Past and Present*, 38, 56–97.

Trice, H. and Beyer, J. 1993. *The Cultures of Work Organizations*. Englewood Cliffs, NJ: Prentice-Hall.

Womack, J. P., Jones, D. T., and Roos, D. 1990. *The Machine that Changed the World*. NY: Harper Perennial.

Wright, M. (forthcoming). Desire and the prosthetics of supervision. *Cultural Anthropology*.

Chapter 24

Networks of Ethnicity

Katharyne Mitchell

The concept of networks involves relational thinking. What links people together across time and space? How are things and people connected and embedded economically, politically, and culturally? In what ways do goods and information and capital flow and why are they channeled down particular vertices and nodes? The network is a useful way of thinking about cultural links, institutional formations, and general ideas about separation and connectedness; these socio-cultural analyses can then be successfully articulated with theories of the economy (Thrift and Olds, 1996). Thinking in terms of networks forces us to theorize socioeconomic processes as interwined and mutually constitutive.

Networks, furthermore, provide a useful way of thinking about economic relations that don't rely on static, bounded configurations such as the region or the nation, yet also don't ascribe everything to random flows. As Thrift and Olds (1996, p. 333) write: "The network serves as an analytical compromise, in the best sense of the word, between the fixities of bounded region metaphor and the fluidities of the flow metaphor." It is helpful in analyzing the interconnections between things, in tracing links and making translations between objects and people that otherwise are often depicted as pure, separate, and distinct.

Networks of ethnicity are relational social and economic ties based on various commonalities shared by a group of people. These commonalities generally include some combination of traits such as language, culture, religion, and/or home town origin: groups base their sense of social collectivity and cohesion on one or more of these common traits. Any group that identifies itself as sharing a common heritage and belonging together and distinct from other groups can be considered "ethnic;" ethnic *networks* help to extend the group's identity spatially, and are an important facet of social and economic organization, particularly within migrant communities.

Historically, networks of ethnicity have been important in traditional, pre-capitalist economies and in business involving long-distance trade. The economic networks of these trading and other business relationships were established and maintained through a dependence on personal ties such as those of co-ethnicity. Social connections based on ethno-religious commonalities formed the glue that held economic relationships together across space and in times of economic distress. Rather than solely profit-maximizing ventures, ethnic networks operated also with an eye to internal (personal) loyalties and the maintenance of social and economic relationships over time.

Following the nineteenth- and twentieth-century German thinkers Weber, Sombart, and to some degree Marx, many social scientists have argued that the growth of industrial capitalism, urbanization, and the rise of state power has profoundly altered socially embedded economic relationships such as networks of ethnicity. These scholars argue that economic relationships are now based on universal, legal–rational precepts rather than on particularist (personal, nepotistic) ties. Weber (1981 [1922]) and Sombart (1953 [1916]), in particular, depicted a clear transition from business transactions involving communal sentiment and loyalty to those that became completely impersonal; for Weber, this shift was *necessary* for rational bourgeois "modern" capitalism to grow and flourish (see Light and Karageorgis, 1994, p. 647).

While neoclassical economists showed little interest in the impact of social relations on market processes and business organization, those following Weber's lead postulated a profound shift in the interrelationship of social and economic processes from the era of what has been labeled "traditional" capitalism to that of "modern" capitalism. Until relatively recently, the implicit assumption of most social scientists was that ethnic capitalist enterprises were limited to local, small-scale, traditional environments, and mainly found in developing parts of the world. It was thought that advanced or modern capitalism developed from and progressed beyond particularist, fraternalist, and nepotistic forms of capitalism such as those of ethnic networks. If ethnic enterprises continued to exist, it would be only in the context of more primitive economic environments; such environments would be destroyed and replaced by the efficient, regularized, and profit-maximizing form of modern enterprise should the capitalist market ever penetrate there.

In the past two decades, however, there has been renewed scholarly interest in the role of networks of ethnicity in all facets of social, political, and economic life, particularly in the movement and adaptation of migrants. Many scholars now argue that networks of ethnicity have always functioned within modern capitalism and have played a key role in the global economy. Furthermore, these networks are not purely local in operation, as argued earlier, but often stretch successfully across regional and national boundaries. The lack of recognition of the ongoing viability of ethnic networks in preceding years resulted from their invisibility to Western scholars; they were primarily operative in migrant communities and within the informal economy, areas outside the purview of traditional economists.

Now, however, networks of ethnicity are of much greater interest, both in the academy and the popular press (Kotkin, 1992; Seagrave, 1995; Redding, 1990; Weidenbaum and Hughes, 1996). This interest grew initially within sociology and focused on the role and meaning of ethnic economies in immigrant communities. Subsequently it expanded to include the analysis of socio-geographic origins of ethnic economies, the contemporary geographies of ethnic networks, and the role of ethnic networks in the current period of accelerated globalization, flexible accumulation, and transnational migration. This newfound interest in ethnic networks has arisen at least partially as a result of the rapid growth and apparent economic success of the Pacific Rim countries and of transnational movements such as the modern Chinese diaspora. In the following sections I examine the early interest in ethnic economies and their incipient formation, and then move to a discussion of networks of ethnicity in the contemporary period.

Ethnic Economies

The first theorizations of the ethnic economy were promulgated by the sociologists Bonacich and Modell. In an early article (1980), they defined the ethnic economy as one in which the self-employed, the employees, and the employers of the businesses in a particular private economic sector (and controlling a large share of that sector) are coethnics. According to this definition, an ethnic economy is one in which immigrant and ethnic minorities create their own employment opportunities within a specific and separate economic sector, rather than finding employment in the general labor market. In other words, an ethnic economy is an exclusive system of business relations that is created and maintained by a group sharing various cultural commonalities such as language and/or religion (Light and Karageorgis, 1994, p. 648). The constitutive businesses are networked together in various ways depending on the type of industry and on the ethnic group (Zhou, 1998), but an integral component of all ethnic economies is the operation of some form of socio-relational ties among economic agents and firms.

The idea of social relations as an important and powerful glue for the smooth functioning of markets and production is in direct contradiction to neoclassical theories of the firm. In his early work, the neoclassical theorist Williamson (1975), for example, argued that organizational forms arise as a result of superior efficiency in reducing economic transaction costs. In this view, transactions that are frequent or sporadic will be contained within hierarchically organized firms as the most efficient and cost-effective model. (Hierarchically organized firms contain units within the organization that perform functions internally rather than contracting these functions to outsiders.) Patterns within ethnic economies, however, do not necessarily correspond to this type of model for a number of reasons. Frequent or sporadic transactions that may not provide the most cost-effective strategy for an organization in the short term might, in the long term, provide a different set of benefits for the businesses involved. These social or cultural benefits, such as the extension of a network or the maintenance of business relationships over time (and despite reduced profitability in the short term), are difficult to capture in traditional models of "rational" economic behavior. Thus the premise that a particular type of organizational form will arise because of the rationality of business "logic" relies on a definition of rationality that neglects the importance of socio-relational ties in economic transactions (Granovetter, 1985). As a result of this neglect, these types of traditional models often fail to adequately theorize alternative types of organizational forms, such as the flexible subcontracting networks prevalent in many ethnic economies.

For neoclassical economists the logic of business organization follows a purely rational, profit-maximizing conception of economic practices and behavior. Traditional models derived from neoclassical theories emphasize individualism, *laissez faire*, and regulation by market mechanisms such as the interaction of supply and demand. The models also rely on a general economy broadly conceived, and on a Western, legal–rational framework as the conceptual basis for the workings of modern capitalism. They ordinarily do not take into account the types of socio-cultural influences represented by ethnic networks, especially within migrant eco-nomies. However, these types of ties are important for the workings of the ethnic economy (particularly in the realm of subcontracting), and determine the spatial pattern of ethnic economies, as we will see later.

Traditional neoclassical theories of business organization also neglect the large and growing collection of economic activities that are often described as the "in-formal" sector of the economy. This is part of the economy that is not directly regulated by the state, and wherein laborers are not protected or constrained by institutional rules and regulations governing work (Feige, 1990; Castells and Portes, 1989; Portes, 1995). The informal sector is described in both positive and negative terms: negative because as an underground economy it can be the last resort of the poor and low-skilled who are forced to work in unregulated, undignified, and often dangerous positions; positive because it can be an important entrepreneurial site outside the surveillance of centralized agents and an overly regulated and controlled state elite (Portes, 1994; Hart, 1990; De Soto, 1989). In either case, because it exists outside the formal institutions of state regulation and protection, the informal economy must operate on systems of enforcement that are characterized not by state sanction but by force or by trust. Systems of trust, in many societies, are often based on common membership of a group. The normalized sanctions for behavior are understood and shared by each member, and the "business code" must be followed for continued economic connections to be both possible and profitable (Wong, 1988, 1996; Kao, 1996).

Co-ethnicity is one of the most common types of group membership active in informal economies, and in some cities various ethnic networks form the backbone of vast sectors of the informal economy. These co-ethnic networks are generally composed of recent immigrants (Sassen, 1989). Portes (1994, p. 426) gives the empirical example of the informal transportation system of jitneys operating in Miami. These small, uncomfortable but cheap and flexible trucks are owned and driven almost entirely by immigrants, and compete for passengers with the air-conditioned, but inflexible city buses driven by native, unionized drivers. Another example he offers is of a Dominican entrepreneur in the Washington Heights sector of New York, who received capital to launch several businesses from fellow Domin-icans in an informal credit-pooling arrangement. This type of informal arrangement based on ethnic ties manifests the many benefits immigrants and others can derive from retaining some degree of ethnic cohesiveness within cities and even across large geographical areas and borders. Furthermore, as they are not codified in bureau-cratic rules, the types of flexibility that these informal connections can offer are also attractive within various sectors of the formal economy, and there are many cases in the United States of links between the formal and informal economies.

Socio-geographic Origins of Ethnic Networks

How do ethnic economies and ethnic networks develop? One of the key processes involved in the formation of networks of ethnicity is migration. International migration is a socially embedded process that links migrants moving between two societies to those who migrated along the same path at an earlier date. That is, there is a chain migration or path effect that binds together the knowledge, memory, and experiences of migrants traveling between the same two or more points on the globe. The immigration of the earliest cohort from an ethnic group greatly influences the formation of the ethnic community, as well as the ongoing immigrant experience of newcomers to the society (Castles and Miller, 1998). As a result of this constant interchange of information and culture throughout the migratory process, there are numerous economic and socio-cultural advantages to migrants to participate in ethnic networks, and to form and maintain ethnic communities in their destination societies.

Although early theorists of immigrant adaptation postulated a hierarchical and naturalized pattern of social and spatial assimilation following the arrival of immigrants to the city (see, for example, Park et al., 1925), recent critiques of this approach have focused on the internal and external forces that constrain and enable immigrant incorporation. Internal processes include the advantages that may be gained through self-identification as a member of a particular ethnic group – this identification brings with it access to specific resources of capital, labor, and information as described earlier. This type of identification process also limits the individual choice of the immigrant in some ways, but enables him or her to take advantage of larger social networks and the benefits accruing to them at the same time. The ability to access these types of social networks is a kind of good often labeled "social capital" following the writings of the French sociologist, Pierre Bourdieu (1984) and the empirical work of sociologists such as Coleman (1988) and Portes (1994). Portes writes that social capital "refers to the capacity of individuals to command scarce resources by virtue of their membership in networks or broader social structures" (Portes, 1994, p. 14; see also Waldinger, 1995, p. 556). Ethnic networks and economies are thus formed partially on the basis of these self-imposed definitions and identities.

At the same time, powerful external forces relating to state immigration policy and other forms of institutional racism have acted to constrain immigrant choices and operate on the formation of ethnic economies and networks. Discrimination against immigrants and other ethnic minorities affects family ties, work opportunities, residential choices, and ultimately operates on the identity of the group itself. Without question the most severe types of limits that have been imposed on many immigrant groups are state restrictions on the number and type of migrants who may continue to enter the state on a legal basis. In Canada and the United States, for example, periodic exclusions of certain groups, restrictive quotas, and the sporadic denial of entry to refugees and others seeking humanitarian assistance has played a major role in the ways that ethnic communities were formed and maintained over time.

Recent work in geography, such as that of Hiebert (1990, 1993) and Anderson (1988, 1990), explores the relationship between discriminatory laws and covenants and the socio-spatial formation of ethnic communities over time. Anderson's (1988) research on the location and formation of a Chinese ethnic community in Vancouver, British Columbia, for example, demonstrates the formidable impact of state laws that closed off, and then greatly restricted, Chinese immigration between 1885 and

1967, and also municipal health and zoning laws, which served to control both business and residential location. Her work further reveals the key role that ideologies of race and essentialized notions of "Chineseness" played in *maintaining* Chinatown as an ethnic neighborhood throughout the decades. Even following the establishment of official policies of "multiculturalism" in Canada, and the curtailment of formal and informal policies of residential exclusion, the idea of the exotic "Orient" as, for example, a tourist site or dim sum center, aided in the continuation of Chinatown as an ethnic neighborhood perpetually separate from and different than other sectors of the city (see also Lowe, 1996).

Similarly, Hiebert's historical work on Jewish immigrants in Toronto (1990, 1993), shows how labor market discrimination against Jews at the turn of the century prompted collective actions such as the formation of ethnic employment networks. Rates of self-employment and the hiring of co-ethnics in this secondary labor market were far higher than for the general or "primary" labor market as a result of primary sector discrimination. (Because of racial and cultural prejudice, groups who were perceived as non-white at that time were not hired at the same rate or pay scale as "white" laborers and thus were often forced to seek employment outside the general labor market (Gordon, Edwards and Reich, 1982; Peck, 1989; Piore, 1979). Furthermore, *residential* segregation within the Jewish enclave was closely tied to employment opportunities or lack thereof, as many of the jobs in which Jews were employed were necessarily located within walking distance of most of the city's Jewish residents. This pattern of segregation is the result of a number of factors, including the cost of transportation to and from work, the lack of adequate transport services, and/or religious dictates, and it follows the general tendency of immigrant districts to be located near labor-intensive factories (see Scott, 1988; Hiebert, 1993). Thus, both employment and residential segregation were implicated in discriminatory hiring practices in the general labor market. Hiebert writes of this mutually constitutive enclave formation, "The nature of ethnic labor-market segmentation can only be understood when the interaction between ethnic residential settlement and occupational clustering is considered" (1993, p. 255).

The social and economic geographies of segregation in the workplace and at home are also often mutually reinforced by a chain of contacts and institutional sites utilized by most members of ethnic networks. As members of a particular ethnic group are most likely to garner information about the job market from co-ethnics residing near them or involved in the same church or school, their residential segregation immediately impacts the types of employment opportunities they hear about and perceive as open and available to them. As Mark Granovetter (1995) showed in his classic study, *Getting a Job*, social links greatly influence labor markets, and networking is one of the primary processes implicated in labor mobility and the overall access to employment resources. In geography, these social links have been given a spatial dimension. Scholars such as Hanson and Pratt (1995), Peck (1996) and Hiebert (1999) demonstrate how occupational segregation, for example, can be linked to the social *and* spatial relationships between home and work. This networking can be on the basis of gender (as in the work of Hanson and Pratt), but is also often based on networks of ethnicity. Thus ethnic networks of information about the job market are greatly influenced by both social and spatial segregation, which in turn serve to reinforce occupational segregation in a mutually constitutive process.

Some other key constraints influencing the socio-geographic origins of ethnic enclaves pertain to the process of urbanization itself. Urban scholars such as Massey and Denton (1993; 1987), Knox (1994), Harris (1984) and Boal (1987) have shown how discriminatory real estate covenants, discrimination in mortgage lending, red-lining (denial of loans) of poorer (often minority) neighborhoods deemed "blighted" by banks, and the steering of clients to or away from certain sectors of the city by real estate agents have operated to exclude African Americans, Asians, and members of other minority groups from certain desirable neighborhoods. These processes have historical roots, but they also operate frequently in contemporary real estate transactions.

In all of these examples the type of racial formation operative on a particular ethnic group at a particular time is important for an understanding of the manner in which ethnic networks and communities were formed. As Omi and Winant (1986) have shown for the United States, the historical and geographic context of specific laws and ideologies concerning race is crucial for the theorization of employment and residential strategies of minority groups, as well as for an adequate under-standing of how individual and group identities are formed. It is these material and discursive processes operating both externally on the communities in terms of legal restraints and informal sanctions, and within ethnic communities in terms of strategies of resistance and accommodation, that impact the form and mainten-ance of networks of ethnicity over the long term.

Geographies of Ethnic Networks

What are the spatial implications of ethnic networks? Institutional discrimination in areas such as immigration law, zoning, and mortgage lending has clearly influenced the formation of ethnic communities. At the same time, these external forces have operated on, and interacted with, internal processes of ethnic self-identification and community formation. In some cases there have been distinct advantages to working or residing in an ethnic community. As specific types of social and economic networks may be accessible only to those who inhabit the actual physical spaces of the com-munity, proximity to co-ethnics is often considered desirable. Thus the "spaces" of ethnic communities are established and maintained as the product of both external, coercive forces of discrimination, and also of processes internal to the community.

The economic desirability of physical proximity to co-ethnics has been a major component of theories of the "ethnic enclave," which have galvanized economic sociologists and a handful of geographers in the past decade and a half. Ethnic enclaves are ethnic social structures that facilitate (some would claim, propel) business growth through spatial proximity. They are the sites where immigrant groups are spatially concentrated, and where ethnic businesses are organized to serve the co-ethnic population, as well as the general market. According to Portes (1981), one of the earliest theorists of the enclave economy, their primary feature is cooperative behavior among co-ethnic economic actors, especially in the employ-ment of co-ethnics and the help given to start up new businesses. Portes argues that as a result of discrimination in the general economy and labor market, the enclave economy frequently offers equal or better opportunities for immigrant advancement and entrepreneurial achievement than the mainstream economy. This view was challenged in recent years, however, as a number of scholars questioned whether these enclaves help or hinder advancement for immigrants. (For recent empirical work on the ethnic enclave phenomenon see Jiobu, 1988; Portes and Manning, 1986; Portes and Jensen, 1989; Kaplan, 1997; Logan et al., 1994).

Although the position of ethnic enclaves as "positive" sites for immigrant and minority advancement is widely critiqued, mainly from those whose empirical evidence suggests that the enclaves are zones of continuing economic marginality rather than opportunity (Nee and Sanders, 1987), both positive and negative interpretations tend to take the spatial clustering of enclaves as an "essential condition," a static container of activity without need of explanation (Zhou, 1998, p. 229). In the discussions by most immigration sociologists, space is perceived as a stage on which economic activity occurs rather than a constituent of that activity. In other words, there is a lack of understanding of the way that social and spatial relations are intertwined and mutually constitutive, rather than separate spheres. Theorizing the interactivity and mutual constitution of the social and the spatial is important for understanding the formation and maintenance of ethnic enclaves over time, because it reduces the temptation to fix either social or spatial structures in a static, hierarchical position or relationship. Recent work by geographers on the spatial structure of urban ethnic economies was showcased in a special issue of *Urban Geography* edited by David Kaplan (1998).

A second insight about the geography of ethnic enclaves is related to the critique of ethnic communities as homogeneous in character. In a number of recent studies, the idea of ethnic "solidarity," where each member of the community unilaterally advances the fortunes of co-ethnics both economically and socially, is shown to be a facile, romanticized depiction. While ethnic solidarity may occur in some contexts, particularly in the confrontation with forces hostile to the community, there are numerous other examples where class, gender, and generational divides operate equally powerfully. Hiebert (1993) and Pessar (1995), among others, have revealed the often hidden class divisions in ethnic communities, which greatly influence the form of cooperation between co-ethnics, if cooperation occurs at all. Mitchell (1998), Sanders and Nee (1987) and Fong (1994) have demonstrated the significance of generational and other divides. Similarly, Pessar (1987), Hondagneu-Sotelo (1994), Haardy-Fanta (1993), and Jones-Correa (1998, p. 326) build strong cases for conceptualizing the divisions between men and women and engaging a "gendered understanding of immigrant political socialization."

In the case of Latin American immigration to New York City, for example, Jones-Correa (1998) argues that men and women socialize and organize differently, with men tending to favor continuity with the social organizations of the past, while women tend to favor a greater degree of change and adaptation to the new society. These differing patterns of socialization, moreover, are directly implicated in the spatial patterns of the community. For a number of reasons, especially because of their children, women tend to come into contact with more public institutions than men (Jones-Correa, 1998, p. 327). Their greater access to public spaces and institutions means greater mobilization both politically and spatially. Numerous other examples of differences between men and women in the migratory process as well as the process of incorporation into a host society show that the heterogeneity of the ethnic community impacts its geographical formation and transformation.

Another insight of the enclave literature is the over-emphasis on ethnic networks as inherently "minority" networks. Although rarely codified as such, the tenor of enclave scholarship is that these ethnic enclaves are the preserve of the "dominated," who retreat into them as a refuge against discrimination in the labor and housing markets, and thus create an internal resource of capital and labor. It is important to remember, however, that there is a geography of ethnic enclaves utilized by dominant groups. Scholars such as Pierre Bourdieu, for example, examined the uses of social or cultural capital by a French ethnic elite, who derive tremendous advantages from spatial "power" networks formed in exclusionary educational sites (Bourdieu and Passeron, 1998). The enclaves, which are formed, in part, through the exclusion of others (who do not belong to the elite ethnic group), facilitate long-term networks that directly impact access to resources and economic as well as social transactions. These types of elite enclaves and the resultant ethnic networks they promulgated were directly implicated in the formation and maintenance of the colonial empire by the British in the eighteenth and nineteenth centuries. They are also formative in the business success of contemporary elite Chinese and Indian ethnic groups (Ong, 1999; Wong, 1988; Mitchell, 1995).

Geographers have recently begun to offer more sophisticated understandings of the spatial implications of ethnic enclaves by introducing other factors besides "ethnicity" into the spatial equation. For example, in her study of the location strategies of three different kinds of producer service firms in Los Angeles, Zhou (1998, p. 228) analyzed the convergence of two forces – ethnic networks and industrial networks. She found that "While Chinese firms show markedly different spatial patterns from their non-Chinese counterparts, each type of producer service also differs from the others in spatial pattern." She argues therefore that both ethnic networks *and* industrial networks play a significant role in affecting the location strategy of Chinese businesses; thus it is necessary to "go beyond 'ethnic'" as the only factor in understanding the location of Chinese industries. In her research, the New Industrial Space (NIS) thesis concerning the link between industrial networks and territorial agglomeration is also of overriding importance in explaining why these firms locate in particular places.

In Zhou's work it is clear that the rise of high technology, information, and services transformed the types of industry and industrial services that characterize a large share of production and employment opportunities in California. As the types of opportunities available in both the general labor market and in ethnic enclaves shift alongside changes in the global economy, the geography of ethnic networks and ethnic enclaves are also transformed. Any theorization of spatial networks of ethnicity must take these types of broad-based transformations into account. Scale is a key factor in understanding the impact of these changes. As industries evolve and spaces become increasingly compressed, the geographic formation of the ethnic "community" may no longer be limited to a neighborhood based on physical proximity. A sense of community might be connected to a particular neighborhood, but equally there might be strong connections to other neighborhoods in different cities or even different countries. The question then arises, are these transneighborhood, transborder, transnational ethnic communities something new, and if so, how do they operate within the contemporary global economy?

Networks of Ethnicity and Globalization

Long-distance trade between co-ethnic groups has occurred for centuries. The networks of ethnicity that formed to facilitate these types of business transactions across space are manifest in some of the earliest global trading routes such as the Silk Road and the Mediterranean sea links (Abu-Lughod, 1989). Nevertheless, the globalization processes of the contemporary period have ushered in a qualitatively different component to long-distance ethnic networks in the last two decades. With the global extension and spatial fragmentation of production, the ever-increasing deterritorialization of finance capital and credit, and the volume and speed of cross-border movements of people, commodities, and information, contemporary networks of ethnicity are now "transnational" in a new way. This concept requires a brief elucidation of the larger economic context of globalization in the post-Fordist era.

As discussed by numerous scholars in the past few years, the post-war global economy has been characterized by a qualitative transformation in capitalist social formation (Harvey, 1989; Dicken et al., 1997; Giddens, 1990; Amin and Thrift, 1994). Following the oil crisis and severe recession of the early 1970s, new regimes of accumulation were marked by a "startling flexibility with respect to labor processes, labor markets, products, and patterns of consumption" (Harvey, 1987, p. 260). The new flexibility was not limited to the arena of production and consumption, but was aided and abetted by a widespread deregulation of financial systems and markets across the globe. An increasing lack of national control over finance facilitated the rapid movement of capital across international borders, and enabled the simultaneous decentralization and consolidation of economic functions worldwide. Industrial plant closures in former manufacturing centers reflect both the relocation of production systems to new areas, and the overall segmentation of production into a number of geographically separate functions (Bluestone and Harrison, 1982; Piore and Sabel, 1984).

The changing nature of economic activity and social life within this new regime of "flexible accumulation" has been immense. Increasing speed and flexibility in all realms of economic transactions became an overriding concern for many corporations, who faced burgeoning global competition. One of the most important transformative processes vis-à-vis ethnic networks is the overall shift in the production process to faster, more globalized, and more flexible systems that operate on a "just-in-time" rather than a "just-in-case" basis. These new, increasingly flexible production systems, which characterize recent capitalist restructuring, are based largely on subcontracting networks. The smooth functioning of these flexible, high-speed networks relies, at least partially, on social relations characterized by long-term bonds of trust and reciprocity (Saxenian, 1994). Despite the inherent difficulties in extending subcontracting networks across space, a number of long-established ethnic subcontracting networks continue to operate smoothly and efficiently between spatially distant locales. Difficult transactions based on partial information, or business deals that may have proved unprofitable in the short term for one party, can be held together over the long term by the "social glue" characteristic of successful ethnic networks.

In addition to the transformation of production systems, qualitative shifts have occurred in the international financial system that have had major implications for the use of money and credit worldwide. Perhaps the most important change has been the growing elasticity of national borders, as economic and political sovereignty in the monetary realm were undermined by the growing liberalization of international finance (Leyshon and Thrift, 1992, p. 50). With the rapid and extensive restructuring of the finance system, control by the nation-state over money supply, allocation, and value declined, the creation and extension of credit and debt occurred on an unprecedented scale worldwide, and money became increasingly mobile and unconstrained – moving through time and across space with ferocious rapidity.

Through the innovation of new financial instruments, the ability of nation-states to control the production and circulation of money and credit through traditional forms of regulation, including compartmentalization and the restriction of financial institutions to prescribed areas, was greatly reduced. As financial flows became unmoored from national space economies and increasingly global in focus, there was an evolution towards a general "deterritorialisation of credit" (Leyshon and Thrift, 1992, p. 54). In this new fragmented and schizophrenic period of credit creation, the ability of family networks to produce relatively cheap credit for industrial borrowers through internal processes of subsidization was notably efficient (Mitchell, 1995). This internal subsidization occurs when extended family networks are willing to loan money to other network members for lower interest rates and slower turnaround times than the "standard" bank rates. As with the earlier discussion of organizational form, the "rationale" for these loans is not solely economic but based also on factors such as the long-term health of the network as a whole.

With the 1970s global debt crisis and attempts by regulatory authorities to impose measures of financial securitization, numerous international investors became wary of banks as either untrustworthy or costly, and chose to lend directly to borrowers without the "aid" of a banking intermediary (Leyshon and Thrift, 1992, p. 56). The reliance on informal credit channels and direct, unmediated contact between creditor and debtor, made the credit system more personalized and less dependent on institutional norms and sanctions. This type of particularist connection was akin to the type already practiced by a number of ethnic entrepreneurs. Thus when financial flows became less constrained by state regulations during this decade, numerous family-based ethnic networks were well positioned both to compete with and articulate with the increasingly maverick financial institutions.

One prominent example discussed in both academic circles and the popular press was the success of many Chinese "extended" family networks in the global economy. In the practice of seeking informal channels for credit accumulation, Chinese investors and borrowers were far in advance of their Western counterparts because they were able to rely on previously established ascribed and achieved social relations that formed the core of their informal credit networks. This gave them a distinct advantage in a number of business situations in Asia, and, increasingly, worldwide (see Mitchell, 1995; Mitchell and Hammer, 1997; Redding, 1990).

Case studies of Chinese conglomerates lend insight into the complex interlinkages between companies and families. Many of these conglomerates form "interlocking directorates" based on personal trust and ethnic ties that are critical to understanding investment patterns among the largest and most successful companies in East and Southeast Asia (Mitchell and Hammer, 1997, p. 87; Wong, 1996). According to scholars like Backman (1995, p. 161), these linkages lead to "colleagues and not competitors at the top." An examination of some of the leading Chinese conglomerates in Southeast Asia, the Chearavanont, Riady, Kuok, and Liem family businesses, shows the ongoing importance of personal and ethnic connections. For example, Liem Sioe Liong, who emigrated from Fujian to Indonesia in the 1930s, and now controls Indonesia's largest conglomerate, the Salim Group, expanded his business through extensive links with other Hokkien families from China's Fujian province. These ethnic Hokkien links include partnerships with large-scale capitalists and politicians such as Robert Kuok, Khoo Kay Peng, and Mochtar Riady (Rees and Sullivan, 1995, p. 61). Personal, family ties are also common. In the Chearavanont family, for example, the youngest daughter followed the trend among Thai-Chinese family businesses by marrying a member of another powerful family in Thailand's Teochiu Chinese community (Mitchell and Hammer, 1997, p. 91). According to Seagrave (1995), this kind of intermarriage of the scions of Chinese business and finance is common, and is continuing now on a global rather than just a regional scale.

The brief example of the Chinese business networks discussed above is one example of the new ways that transnational ethnic networks are developing in the context of a restructuring global economy. In an effort to capture the capital and remittances of new transnational players such as these, many states have begun to offer the possibilities of dual citizenship and various economic and social incentives to its citizens living and working overseas (Basch et al., 1994; Smith and Guarnizo, 1997). Governance itself is thus clearly affected by ethnic networks that cross increasingly porous state boundaries. Transnationalism has led to new kinds of migrant experiences, new meanings of citizenship and state–society relationships, and new types of ethnic networks for migrants and businesspeople working in more than one national locale.

Conclusion

Networks of ethnicity have always been an important part of economic systems worldwide. Early research emphasized the role of ethnic networks in local systems and immigrant enclaves, but it became increasingly apparent that ethnic networks function effectively within "modern" and global capitalist economies as well. At the same time, scholarly interest in networks of all kinds has grown markedly in the past few years, and what was initially a field dominated by migration studies branched out to include large sectors of the social sciences as a whole.

Although they were always a key component in long-distance trading relationships and immigrant economies, ethnic networks are now important facilitators of an ever-increasing transnational movement of people, information, and capital. With the changes brought about by global restructuring, understanding the role of ethnic networks has become increasingly important in analyses of global commodity chains and transnational migration, and in theorizations of how capitalism operates internationally. In contrast with earlier assumptions of ethnic economies as static and fundamentally local in their operations, current research indicates ethnic networks to be both dynamic and international in scope. The "particularist" ties that Weber and others postulated would keep ethnic economies from developing in conjunction with modern capitalism have, in contrast, been shown to be both resilient and adaptable to global restructuring and the modern international economy.

Bibliography

Abu-Lughod, J. 1989. *Before European Hegemony: The World System A.D. 1250–1350*. Oxford: Oxford University Press.

Amin, A. and Thrift, N. 1994. Living in the global. In Amin, A. and Thrift, N. (eds). *Globalization, Institutions and Regional Development in Europe*. Oxford: Oxford University Press, 1–22.

Anderson, K. 1988. Cultural hegemony and the race-definition process in Chinatown. *Environment and Planning D: Society and Space*, 6, 127–49.

Anderson, K. 1990. *Vancouver's Chinatown: Racial Discourse in Canada*. Montreal: McGill-Queen's University Press.

Backman, M. 1995. *Overseas Chinese Business Networks in Asia*. Canberra: AGPS Press.

Basch, L., Glick Schiller, N., and Blanc, C. 1994. *Nations Unbound: Transnational Projects, Postcolonial Predicaments, and Deterritorialized Nation-States*. New York: Gordon and Breach Publishers.

Bluestone, B. and Harrison, B. 1982. *The Deindustrialization of America*. New York: Basic Books.

Boal, F. W. 1987. Segregation. In Pacione, M. (ed). *Progress in Social Geography*. London: Croom Helm, 90–128.

Bonacich, E. and Modell, J. 1980. *The Economic Basis of Ethnic Solidarity*. Berkeley: University of California Press.

Bourdieu, P. 1984. *Distinction: A Social Critique of the Judgement of Taste*. Translated by Richard Nice. Cambridge, MA: Harvard University Press.

Bourdieu, P. and Passeron, J. C. 1998 [1977]. *Reproduction in Education, Society and Culture*. Translation by Richard Nice. London: Sage Publications.

Castells, M. and Portes, A. 1989. World underneath: The origins, dynamics, and effects of the informal economy. In Portes, A. Castells, M. and Benton, L. A. (eds). *The Informal Economy: Studies in Advanced and Less Developed Countries*. Baltimore: The Johns Hopkins University Press, 11–37.

Castles, S. and Miller, M. 1998. *The Age of Migration*. New York: Guilford Press.

Coleman, J. 1988. Social capital in the creation of human capital. *American Journal of Sociology*, 94, 95–120.

De Soto, H. 1989. *The Other Path*. Translated by June Abbott. New York: Harper and Row.

Dicken, P., Peck, J. and Tickell, A. 1997. Unpacking the global. In Lee, R and Wills, J. (eds). *Geographies of Economies*. London: Arnold, 158–66.

Eccles, R. 1981. The quasifirm in the construction industry. *Journal of Economic Behavior and Organization*, 2, 335–57.

Feige, E. 1990. How big is the irregular economy? *Challenge*, 22, 5–13.

Fong, T. 1994. *The First Suburban Chinatown: The Remaking of Monterey Park, California*. Philadelphia: Temple University Press.

Giddens, A. 1990. *The Consequences of Modernity*. Stanford: Stanford University Press.

Gordon, D., Edwards, R. and Reich, M. 1982. *Segmented Work, Divided Workers: The Historical Transformation of Labor in the United States*. Cambridge: Cambridge University Press.

Granovetter, M. 1985. Economic action and social structure: The problem of embeddedness. *American Journal of Sociology*, 91, 481–510.

Granovetter, M. 1995 [1974]. *Getting a Job: A Study of Contacts and Careers*. Chicago: University of Chicago Press.

Haardy-Fanta, C. 1993. *Latina Politics, Latino Politics: Gender, Culture, and Political Participation in Boston*. Philadelphia: Temple University Press.

Hanson, S. and Pratt, G. 1995. *Gender, Work, and Space*. New York: Routledge.

Harris, R. 1984. Residential segregation and class formation in the capitalist city: A review and directions for research. *Progress in Human Geography*, 8, 26–49.

Hart, K. 1990. The idea of the economy: Six modern dissenters. In Friedland, R and Robertson, A. F. (eds). *Beyond the Marketplace: Rethinking Economy and Society*. New York: Aldine de Gruyter, 137–60.

Harvey, D. 1987. Flexible accumulation through urbanization: Reflections on "post-modernism" in the American city. *Antipode*, 19, 3, 260–86.

Harvey, D. 1989. *The Condition of Postmodernity*. Oxford: Blackwell.

Hiebert, D. 1990. Discontinuity and the emergence of flexible production: Garment production in Toronto, 1901–1930. *Economic Geography*, 66, 229–53.

Hiebert, D. 1993. Jewish immigrants and the garment industry of Toronto, 1901–1931: A study of ethnic and class relations. *Annals of the Association of American Geographers*, 83, 243–71.

Hiebert, D. 1999. Local geographies of labor market segmentation: Montreal, Toronto and Vancouver, 1991. *Economic Geography*, 75, 339–69.

Hondagneu-Sotelo, P. 1994. Regulating the unregulated? Domestic workers' social networks. *Social Problems*, 41, 50–64.

Jiobu, R. M. 1988. Ethnic hegemony and the Japanese of California. *American Sociological Review*, 38, 583–94.

Jones-Correa, M. 1998. Different paths: Gender, immigration and political participation. *International Migration Review*, 32, 326–49.

Kao, C.S. 1996. "Personal trust" in the large businesses in Taiwan: A traditional foundation for contemporary economic activities. In Hamilton, G. (ed). *Asian Business Networks*. Berlin: Walter de Gruyter, 61–70.

Kaplan, D. 1997. The creation of an ethnic economy: Indochinese business expansion in Saint Paul. *Economic Geography*, 73, 214–20.

Kaplan, D. 1998. The spatial structure of urban ethnic economies. *Urban Geography*, 19, 489–501.

Knox, P. 1994. *Urbanization: An Introduction to Urban Geography*. New Jersey: Prentice Hall.

Kotkin, J. 1992. *Tribes: How Race, Religion and Identity Determine Success in the New Global Economy*. New York: Random House.

Leyshon, A. and Thrift, N. 1992. Liberalization and consolidation: The single European market and the remaking of European financial capital. *Environment and Planning A*, 24, 49–81.

Light, I. and Karageorgis, S. 1994. The ethnic economy. In Smelser, N. J. and Swedburg, R. (eds). *The Handbook of Economic Sociology*. Princeton: Princeton University Press, 467–71.

Logan, J., Alba, R. and McNulty, T. 1994. Ethnic economies in metropolitan regions: Miami and beyond. *Social Forces*, 72, 691–724.

Lowe, L. 1996. *Immigrant Acts*. Durham: Duke University Press.

Massey, D. and Denton, N. 1987. Trends in the residential segregation of blacks, Hispanics, and Asians, 1970–1980. *American Sociological Review*, 52, 802–25.

Massey, D. and Denton, N. 1993. *American Apartheid: Segregation and the Making of the Underclass*. Cambridge, MA: Harvard University Press.

Mitchell, K. 1995. Flexible circulation in the Pacific Rim: Capitalisms in cultural context. *Economic Geography*, 71, 364–82.

Mitchell, K. 1998. Reworking democracy: Contemporary immigration and community polit-ics in Vancouver's Chinatown. *Political Geography*, 17, 729–50.

Mitchell, K. and Hammer, B. 1997. Ethnic Chinese networks: A new model? *Hong Kong Bank of Canada Papers on Asia*, 3, 73–103.

Nee, V. and Sanders, J. M. 1987. On testing the enclave-economy hypothesis. *American Sociological Review*, 52, 771–3.

Omi, M. and Winant, H. 1986. *Racial Formation in the United States*. New York: Routledge.

Ong, A. 1999. *Flexible Citizenship: The Cultural Logic of Transnationality*. Durham: Duke University Press.

Park, R., Burgess, E., and McKenzie, R. 1925. *The City*. Chicago: University of Chicago Press.

Peck, J. 1989. Reconceptualizing the local labor market: Space, segmentation and the state. *Progress in Human Geography*, 13, 42–61.

Peck, J. 1996. *Work-place: the Social Regulation of Labor Markets*. New York: Guilford Press.

Pessar, P. 1987. The Dominicans: women in the household and the garment industry. In Foner, N. (ed). *New Immigrants in New York*. New York: Columbia University Press, 103–30.

Pessar, P. 1995. The elusive enclave: Ethnicity, class, and nationality among Latino entrepre-neurs in Greater Washington, D.C. *Human Organization*, 54, 383–92.

Piore, M. 1979. *Birds of Passage: Migrant Labor in Industrial Societies*. Cambridge: Cam-bridge University Press.

Piore, M. and Sabel, C. 1984. *The Second Industrial Divide*. New York: Basic Books.

Portes, A. 1981. Modes of incorporation and theories of labor immigration. In Kritz, M., Keely, C. and Thomas, S. (eds). *Global Trends in Migration*. New York: Center for Migration Studies, 279–97.

Portes, A. 1994. The informal economy. In Smelser, N. J. and Swedburg, R. (eds). *The Handbook of Economic Sociology*. Princeton: Princeton University Press, 426–49.

Portes, A. 1995. Economic sociology and the sociology of immigration: A conceptual over-view. In Portes, A. (ed). *The Economic Sociology of Immigration*. New York: Russell Sage Foundation, 1–41.

Portes, A. and Jensen, L. 1989. The enclave and the entrants: Patterns of ethnic enterprise in Miami before and after Mariel. *American Sociological Review*, 54, 929–49.

Portes, A. and Manning, R. 1986. The immigrant enclave: theory and empirical examples. In Olzak, S. and Nagel, J. (eds). *Comparative Ethnic Relations*. New York: Academic Press, 47–68.

Redding, S. G. 1990. *The Spirit of Chinese Capitalism*. New York: Walter de Gruyter.

Rees, J. and Sullivan, M. 1995. Work hard, make money. *Far Eastern Economic Review*, August 31, 61–2.

Sanders, J. and Nee, V. L. 1987. Limits of ethnic solidarity in the enclave economy. *American Sociological Review*, 52, 745–73.

Sassen, S. 1989. New York City's informal economy. In Portes, A., Castells, M. and Benton, L. A. (eds). *The Informal Economy: Studies in Advanced and Less Developed Countries*. Baltimore: The Johns Hopkins University Press, 57–72.

Saxenian, A. 1994. *Regional Advantage: Culture and Computing in Silicon Valley and Route 128*. Cambridge: Harvard University Press.

Scott, A. 1988. *Metropolis: From the Division of Labor to Urban Form*. Berkeley: University of California Press.

Seagrave, S. 1995. *Lords of the Rim: The Invisible Empire of the Overseas Chinese*. New York: G. P. Putman's Sons.

Smith, M. P. and Guarnizo, L. 1997. The locations of transnationalism. In Smith, M. P. and Guarnizo, L. (eds). *Transnationalism From Below*. New Brunswick: Transactions Publish-ers, 3–34.

Sombart, W. 1953 [1916]. Medieval and modern commercial enterprise. In Lane, F. C. and Riersma, J. C. (eds). *Enterprise and Secular Change*. Homewood, IL: R. D. Irwin, 25–40.

Thrift, N. and Olds, K. 1996. Refiguring the economic in economic geography. *Progress in Human Geography*, 20, 3, 311–37.

Waldinger, R. 1995. The "other side" of embeddedness: A case-study of the interplay of economy and ethnicity. *Ethnic and Racial Studies*, 18, 555–80.

Weber, M. 1981 [1922]. *General Economic History*. Translated by Frank Knight. New Brunswick, NJ: Transaction.

Weidenbaum, M. and Hughes, S. 1996. *The Bamboo Network: How Expatriate Chinese Entrepreneurs are Creating a New Economic Superpower in Asia*. New York: Free Press.

Williamson, O. 1975. *Markets and Hierarchies*. New York: Free Press.

Wong, S. L. 1988. *Emigrant Entrepreneurs: Shanghai Industrialists in Hong Kong*. Hong Kong: Oxford University Press.

Wong, S. L. 1996. Chinese entrepreneurs and business trust. In Hamilton, G. (ed). *Asian Business Networks*. Berlin: Walter de Gruyter, 13–26.

Zhou, Y. 1998. Beyond ethnic enclaves: Location strategies of Chinese producer service firms in Los Angeles. *Economic Geography*, 74, 228–51.

Part V   Spaces of Circulation

Chapter 25

The Economic Geography of Global Trade

Richard Grant

Trade is one of humankind's oldest activities. The earliest long-distance trade can be documented in the third millennium BC, when ships and caravans carried pottery, cloth, tools, precious metals, and other commodities that were traded by the Greeks, Egyptians, Mesopotamians, Romans, and Chinese with other peoples of the world.

Patterns of trade and specialization are most often understood in the framework of comparative advantage, which itself is based on neoclassical assumptions of free trade and perfect markets. A country has a comparative advantage in the production of a good when the opportunity cost of that good is lower than the trading partner's cost. When trade is possible, each country specializes in the export of a good in which they perform relatively better than others, according to differences between the countries in technology, labor productivity, and factor endowments (capital, land, labor, and natural resources).

The theory of comparative advantage argues that differences between places should be exploited to the maximum through specialization and trade to achieve economic development. Patterns of trade and specialization are linked in comparative advantage at all geographical scales, and in turn connect places together economically. Layers of trade geographies are thus created at a variety of spatial scales; global, world regional, national, and sub-national.

Although the theory of comparative advantage has provided the framework for understanding trade patterns and specialization, it has its limitations. It typically incorporates a geographic component by including considerations of transport costs as well as of a distance decay principle (whereby the intensity of trade increases with propinquity). Yet the low priority given to the geographical component in comparative advantage theory means that in practice it is an aspatial approach, emphasizing static notions of location and distance. In order to explain contemporary trade patterns, economic geographers attempt also to analyze the connections among government–business relationships, trade policy, and changes in the organization of production.

Most trade researchers accept that free trade and perfect market assumptions, the bases of comparative advantage, are inadequate for understanding contemporary global trade (Yarbrough and Yarbrough, 1997). Although there are many practical reasons for trade theorists and governments to continue their adherence to free trade theory, practical reasons do not validate the theory. It is widely accepted that in practice the system of global trade bears little resemblance to the presumed neoclassical ideal. Even free trade proponents (e.g. Bergsten, 1998, p. 3) state that as much as 40 percent of global trade is not free trade.

In the 1990s, we find that the literature on the geography of global trade is little more than an amalgam of specialist studies of portions of the trade system (e.g. country, commodity, and regional studies), and few researchers synthesize or integrate these findings. There are several challenges in accounting for trade patterns and specialization that conventional trade theory cannot explain. These can be grouped into three broad categories: sectoral composition and types of trade globally; trade barriers; and the roles of governments and firms and their interrelationships. These challenges are detailed in the sections of this chapter: here I summarize major points within each. After reviewing each of the challenges I assess the role of the state in trade at the millennium, and highlight the challenges and opportunities for economic geographers interested in undertaking trade research.

The first set of challenges involves the search for explanations of the sectoral composition of trade and the shifts in types of trade that predominate in the global system. Different types of trade and individual countries' particularities require different explanations. For instance, trade in manufactured goods responds to influences different from those in primary or service goods. Furthermore, the spatial organization of global trade is no longer solely a horizontal country–country trade of finished commodities (whereby countries fully produce and trade final goods). Instead, vertical specialization (whereby countries specialize and trade in one stage of the production process) is spatially (re)organizing trade, and accounting for 20–25 percent of total trade (Hummels et al., 1998).

The second set of challenges for trade theory is to account for the various barriers to trade. At the establishment of the General Agreement on Tariffs and Trade (GATT; the organization that sets rules for trade between nations) in 1947, free trade primarily referred to exports and imports of goods unrestricted by barriers at the border (tariff barriers), a conception of trade that no longer suffices. Trade barriers are expanding to include many that are more difficult to measure and quantify, such as non-tariff barriers (NTBs), competition policies, and economic corruption. The salience of these impediments to trade supports the argument that free trade assumptions need to be greatly modified.

The third set of challenges for trade theory is to account for the roles of government and firms, and their interrelationships. There are some researchers who now attempt to go beyond the simple economic-geographical relationships and basic decisionmaking processes assumed in comparative advantage based on free markets. Governments actively intervene in the trade policy arena, shaping comparative advantage and modifying the "inherited" factors of production. Patterns of comparative advantage are also products of the world economic structure (the hierarchical core, periphery, and semi-periphery structure), which are, in effect, human creations reflecting economic and trade policy emphases. Trade across firms (intra-industry trade) and trade within firms (intra-corporate trade) is estimated to be between 30 and 50 percent of the total trade of OECD (Organization for Economic

Cooperation and Development) states (Grant et al., 1993), and is determined less by market forces than by the decisions of multinational companies in conjunction with government policies.

It is clear that comparative advantage can no longer be understood with static notions or with simple profiles of national or industrial patterns. The globalization of production has resulted in the emergence of distinct trade orientations – regional, intracorporate, intraindustry, high-technology, and electronic commerce (e-commerce) – that are shaping new geographies. These trade orientations now intersect to create a complex global trade mosaic. The contemporary macroeconomic geography of trade involves a kaleidoscope of individual dynamic geographies of commodities. Thus trade theory, and in particular the geography of trade, needs to take on the three sets of challenges presented here.

The Changing Sectoral Composition of Global Trade

Both qualitatively and quantitatively, trade has undergone fundamental changes over time. Economic value and the dominant types of trade accounting for most of the value-added in cross-border trade have proceeded through distinctive eras over time (table 25.1), with technology as a constant driving force. In the first phase, primary commodities (i.e. raw materials, minerals, and food products) dominated. In the second phase, beginning with the Industrial Revolution, manufacturing goods accounted for the largest share of international trade and was the most lucrative sector. Trade in primary commodities during this phase still took place but the technological upgrading of products made manufacturing trade of standardized products more attractive. In the third phase, services became global commodities accounting for a growing share of cross-border trade.

Twin technological revolutions in information and transportation have fueled the growth and upgrading of commodities in trade. Transport costs have continued to

Table 25.1 Trade eras

	1450–present	1840–present	1970–present	2000+
Economic units	Primary commodities	Manufacturing goods	Services	Experiences
Economy	Agrarian	Industrial	Service	Experience
Spatial organization	Local	Regional/ international	International/global	Global
Economic function	Extract	Make	Deliver	Stage
Nature of offering	Fungible	Tangible	Intangible	Memorable
Key attribute	Natural	Standardized	Customized	Personal
Method of supply	Stored in bulk	Inventoried after production	Delivered on demand	Revealed over a duration
Seller	Trader	Manufacturer	Provider	Stager
Buyer	Market	User	Client	Guest
Factors of demand	Characteristics	Features	Benefits	Sensations

Source: Based on Pine and Gilmore (1998)

fall throughout the twentieth century. Advances in communications technology (telephone, facsimile, electronic mail, and video conferencing) have facilitated the coordination and monitoring of production in diverse locations. Improvements in trans-Atlantic cable capacity and the corresponding increasing capabilities of global communications have also facilitated more interactions among firms in different countries.

The contemporary sectoral composition of trade, profiled in table 25.2, shows strong growth for capital goods, especially machinery and transportation and commercial services. By contrast, trade in primary commodities continues to fall, and its relative share of trade has been declining, especially over the last 30 years, because of the cyclical decline of commodity prices relative to manufacture and services trade. Manufacturing trade still predominates, but this market is becoming more differentiated due to the customized delivery of products.

The strong growth registered for services is a recent phenomenon. Services accounted for 25 percent of global exports by 1996. Service trade is qualitatively different from manufacturing trade in that services are infinitely expansible and potentially weightless; many people can use them at the same time, and once the goods are produced, they can be replicated at a low cost. For some companies (such

Table 25.2 Composition of world trade, 1965–96

GATT/WTO breakdown[1] (shares of total world trade)	1970	1980	1990	1996
Merchandise				
Agriculture	16.5	12.5	10.0	11.4
Mining	12.0	22.0	11.5	11.2
Manufactures	50.0	45.5	57.0	73.3
(Not specified)	2.5	3.0	2.5	3.0
Capital goods	29.5	26.5	37.0	39.0
Services[2]	19.0	17.0	19.0	24.6

World Bank breakdown (shares of total world merchandise imports)	1965[3]	1979[4]	1990[5]	1995
Food	18.0	12.0	9.0	12.2
Fuels	10.0	20.0	11.0	9.0
Other primary commodities	17.0	9.0	8.0	7.7
Manufactures	55.0	58.0	73.0	72.0
Machinery, transport	23.0	25.0	34.0	30.9

[1] GATT, 1992: table 1; 1990; table 8; 1989: 9; WTO, 1997: table 11.2, p. 9.
[2] Services include shipping and other passenger, port, and transportation services; travel goods and services other than passenger services acquired by persons staying for a year or less in an economy where they are not resident; and other private services (communications, advertising, brokerage, management, professional, and technical services). Services do not include investment income on unrequited transfers, whether official or private (e.g. migrants' transfers, workers' remittances). See GATT, 1989, box 2.
[3] World Bank, 1992: table 15.
[4] Total-import weighted average of averages for industrial market economies (weight = 0.758), middle-income economies (weight = 0.209), and low-income economies (weight = 0.034) (World Bank, 1982, table 10, pp. 128–9).
[5] Total-import weighted average of averages for industrial market economies (weight = 0.735) and developing economies (weight = 0.265) (World Bank, 1987, table 12).
Sources: WTO (1998), World Bank (1997, table 14, p. 188)

as IBM), more profits are generated from providing services than from producing goods in the traditional sense.

Some analysts have suggested that the highest rewards in twenty-first-century trade, and the next competitive battleground beyond services trade, will lie in staging and trading experiences (Pine and Gilmore, 1998). They predict that commodity trade will be further upgraded and that the highest rewards from trade will involve the selling of lifelike interactive or virtual reality experiences globally (table 25.1). Presently, cross-border experience trade is negligible, and its expansion will be contingent on the development of new technologies to facilitate and deliver its transfer. It is anticipated that the key attribute will be personalization, rather than customized services or standardization. It is also expected that experience trade will spread beyond theaters and theme parks to many aspects of global business. Like other commodities that result from R&D, experiences derive from an iterative process of exploration, scripting, and staging – capabilities that aspiring experience merchants will need to master (Pine and Gilmore, 1998, p. 99).

The recent globalization of production also has altered the types of commodities that are traded. For instance, a significant proportion of trade is now constituted as component parts rather than finished goods. Vertical specialization takes place when countries acquire expertise in particular stages of the production process: a country may import a good from another country to use for the production of its own good and then export its good to the next country. The sequence only ends when the final good reaches its destination. Vertical trade involves, for example, the skill-intensive design and manufacturing of a microchip in one country and its labor-intensive assembly onto a memory board in another, whereas horizontal trade entails completing all stages of computer manufacturing in a single country.

Vertical trade has not been widely studied, but available data suggest that smaller economies are most involved. For instance, it accounted for 35 percent of total trade in the Netherlands and more than 25 percent of total trade in Ireland, Denmark, South Korea, Spain, Malaysia, and the Philippines (Hummels et al., 1998, p. 88). Large economies (such as the USA, Germany, and Japan) are least involved in vertical trade because they generally find it easier to achieve scale economies and retain every production stage of a good. Vertical specialization also varies widely across industries: those with the highest levels of vertical trade are motor vehicles, shipbuilding, aircraft, chemicals, and nonferrous metals. Vertical trade is a growing sector of global trade but still accounts for less overall growth than horizontal trade, suggesting diversification in trade patterns rather than the replacement of one by the other (table 25.3).

This discussion of the composition of global trade is based on limited data, and three cautionary notes should be added. First, there is much more trade information available for OECD economies. Second, global production of foreign firms is now estimated to be more than $3 trillion, greater than the value of global trade. This is important because a US firm operating in South Korea and exporting components to its US affiliate is counted as a South Korean export and not a US export. Trade statistics seldom distinguish trade in intermediate goods and finished goods, so we can only estimate how commodity chains affect the structure of global trade. Third, global service companies have been shown to serve the markets that they locate in to a greater extent than manufacturing firms, which are more likely to

Table 25.3   Growth in trade (%)

Country[1]	Vertical trade	Horizontal trade
Australia	13	87
Canada	44	56
Denmark	27	73
France	28	72
Germany	19	81
Japan	3	97
Netherlands	47	53
United Kingdom	30	70
United States	12	88

[1] Most data series run from early 1970 to early 1990.
Source: Adapted from Hummels et al. (1998, p. 92)

export (Fieleke, 1995), because for a number of services (e.g. banking and consulting) proximity of suppliers to consumers is required.

Barriers to Global Trade

Traditionally, trade analysis has dealt with the overall "openness" of countries, especially how tariff barriers limit the free flow of trade and why some sectors are more open than others. More recent perspectives in global trade have amplified these concerns in addition to focusing on newer avenues of government intervention in trade, especially NTBs, competition policies, and corruption practices, all of which are a function more of unique national political economic histories than any global convergence in trade practices (see subsections). Our concept of trade barriers needs to be expanded to accommodate several counter-liberalizing trends in the global economy.

Trade openness is often measured by the ratio of goods and services exports to GDP, revealing a strong global trend toward openness for major trading economies (except Japan; table 25.4). The small export-oriented economies of Singapore, Hong Kong, Malaysia, and Ireland stand out from the rest in their openness. In addition, there has been a strong opening up of fairly closed economies, such as China and South Korea, and trade liberalization policies, adopted as part of structural adjustment policies (SAPs), are also opening up developing economies like Ghana and India. These trends indicate that insulation is declining: global trade is becoming an ever more important influence on domestic prosperity.

Trade liberalization has been consistently promoted under the GATT, and since 1995 the GATT has been updated under the umbrella of the World Trade Organization (WTO). Average tariff rates on industrial producers in 1947 were 40 percent: by 1997 they had been lowered to 6.3 percent (WTO, 1998). Average tariffs will fall to 3.8 percent by 2003, once the commitments made under the Uruguay Round are fully implemented (Daly and Kuwahara, 1998, p. 207). Reductions in tariff barriers have fostered growth in all types of trade, especially vertical trade since it involves multiple border crossings.

Importantly, barriers to trade are not limited to international border measures. Governments can also target specific states for retaliatory purposes (political or

Table 25.4 Openness ratios, 1965, 1980, and 1997 (ratio of exports of goods and nonfactor services to GDP, multiplied by 100)

By income level and country	1965	1980	1997
Low-income economies	8	13	19
Middle-income economies	17	22	26
High-income economies	12	19	20
Argentina	8	5	9
Canada	19	28	38
China	4	6	20
Ghana	17	8	25
Hong Kong	71	90	132
India	4	7	12
Ireland	35	48	75
Japan	11	14	9
Korea, Rep. of	9	34	38
Malaysia	42	58	90
Singapore	123	215	187
United States	5	10	11

Source: World Bank (1992, table 9, pp. 234–5; 1998, table 13, pp. 214–15)

economic). Two specific means by which this occurs are unilateral retaliatory policies and trade sanctions. For example, Super 301 is a unilateral trade policy that is targeted specifically and politically toward recalcitrant states named by the US Government as hindering American exports. Economic sanctions range from aid cutbacks and trade bans to trade embargoes. The US Government, for example, targeted 26 countries for sanctions (including China, Cuba, India, Iraq, and Nigeria) in 1997, indirectly affecting half of the world's population (Hufbauer, 1998, p. 1).

Nontariff barriers (NTBs)

The proliferation of NTBs since the 1970s represents a counter-liberalizing trend from which few countries have been exempt and few sectors spared. NTBs are discretionary barriers applied to specific commodities, often in the form of quantitative restrictions, packaging and labeling requirements, health standards, etc. In 1986, 50 percent of all products traded were subject to some type of NTB, with the food category especially highly protected (Grant et al., 1993, p. 21). The completion of the Uruguay Round of the GATT resulted in a marked decline in NTBs, partly due to the elimination of "voluntary" export restraints, the phasing out of Multifiber Agreement (MFA) quotas on textiles, and "tariffication" of agricultural NTBs (the conversion of EU agricultural NTBs into tariff rates).

NTBs vary widely among countries (table 25.5). According to the measures currently employed, anti-dumping and countervailing duties are the predominant types of NTBs in the USA. In Japan non-automatic licensing arrangements are most common, and price controls and variable charges are most widely used in the European Union (EU). NTBs also vary with specific commodities. For instance, high NTBs are employed on live animals and textiles in Japan, on vegetable products and prepared food in the EU, on leather, textile, footwear, and base metals in the USA, and footwear and machinery in Canada (Daly and Kuwahara, 1998, p. 228).

Table 25.5 Pervasiveness of different types of NTBs (%)

NTB categories	USA			EU			Japan			Canada		
	1989	1993	UR[1]	1989	1993	UR	1989	1993	UR	1989	1993	UR
All NTBs	24.1	23.7	8.6	27.3	27.7	8.2	14.14	13.4	11.9	12.5	12.2	4.6
Core NTBs[2]	24.1	23.6	8.5	25.9	25.2	4.3	13.7	12.6	11.1	10.3	9.5	1.6
A. Quant. restrictions	18.3	18.7	0.6	20.4	20.1	1.7	12.9	11.7	10.5	7.6	7.4	0.1
Export restraints[3]	17.3	13.5	0.0	15.9	15.0	0.0	0.3	0.1	0.0	5.8	6.0	0.0
Nonautomatic licensing[4]	0.0	0.0	0.0	5.1	5.6	1.5	10.1	10.1	9.8	2.7	0.2	0.0
Other QRs	6.1	5.8	0.6	0.2	0.2	0.2	2.8	1.6	0.6	0.8	1.2	0.0
B. Price controls	15.4	9.6	7.6	13.2	10.6	3.2	0.8	0.9	0.7	2.8	2.1	1.6
Variable charges[5]	0.1	0.0	0.1	9.2	9.5	2.4	0.8	0.9	0.6	0.0	0.0	0.0
AD/CVS[6]	15.3	9.5	7.6	1.1	1.1	0.9	0.0	0.0	0.0	2.8	2.1	1.6
Other price controls	0.1	0.1	0.1	3.1	0.1	0.1	0.0	0.0	0.0	0.0	0.0	0.0

[1] Levels to be achieved on the basis of the Uruguay Round agreement.
[2] Core NTBs consist of two broad measures: quantitative restrictions and price control measures.
[3] Export restraints are arrangements between importers and exporters, whereby the latter agree to limit exports to avoid the imposition of mandatory restrictions by importing countries.
[4] Nonautomatic licensing is the practice of requiring, as a condition for importation, a license which is not granted automatically and which may be used on a discretionary basis or specific criteria.
[5] Variable charges bring the market prices of imported products close to those of corresponding domestic products.
[6] Anti-dumping and countervailing actions involve investigations by anti-dumping authorities, usually in response to domestic producers' complaints, the purposes of which are to determine whether dumping or subsidies exist and, if so, whether the practice causes "material injury" to domestic producers. When the complaint is upheld, a duty is levied on imports from the country named.
Source: Daly and Kuwahara (1998, pp. 226–7).

NTBs will require careful monitoring in the future to ensure that measures phased out or prohibited under WTO agreements do not surface in a different form. Given the fungibility of NTBs, more data are needed on health, safety, and environmental regulations and standards that impede trade.

Competition policies

Barriers that operate behind national borders and fall into the domain of competition policy have grown more significant. In the broadest sense, competition policy determines the institutional mix of competition and cooperation that gives rise to the market system. Surprisingly, it is not competition that is the objective of competition policy, but efficiency (market functioning) and fairness (market entry), sought after in unique blends worldwide. For instance, the USA de-emphasizes fairness while the UK and France give it more prominence (Graham and Richardson, 1997, p. 7). In some instances domestic firms are sheltered, exert monopoly power, and earn above-normal profits because foreign firms are hindered from competing in the market. In areas like airlines, telecommunications, financial services, insurance, and intellectual property, international differences in regulatory regimes and rules create highly varied competition structures.

A formal codified competition policy is not found for all countries, but all have informal competition conventions. As markets have globalized, differing competition policies and conventions have come into contact and have led to conflict. For instance, the US Government has prosecuted cases against the governments of China and Japan to the WTO, for impeding US companies' access to their markets (e.g. Kodak's case against Fuji of Japan).

The debate over competition policies in global trade is still at the conceptual stage. No study has succeeded in mapping out their vast heterogeneity, and criteria for assessing how competition polices affect trade have not yet been developed. As a result cross-national data are limited to high-profile competition disputes.

Corruption

Corruption is a third barrier impeding the free movement of goods and services. Former US Commerce Secretary Mickey Kantor called corruption "a virus threatening the health of the international trading system" (quoted in Elliott, 1997, p. 153). It is now viewed as a global problem, and researchers are moving beyond categorizing regimes as "corrupt" or "free" to frame corruption in the context of marketization.

Corruption is difficult to measure because it can surface in two ways: by government inaction and by government action. For instance, the Ghanaian Government in the 1980s was criticized for not doing enough to eliminate corruption, whereas the Kenyan Government in the 1990s has been an active co-conspirator in manipulating the trade environment for personal and political gains (Grant, 2000). To indicate levels of corruption, Transparency International (TI) conducts surveys on business people's experiences in doing business in a particular country to solicit opinions about the degree to which improper practices (such as illegal payments) are necessary to facilitate business transactions. Results show that wealthier countries, especially those in Western Europe, are perceived to be less corrupt, whereas poorer

Table 25.6 A partial ranking of corruption around the world,[1]
1998 (descending order)

Least corrupt	Most corrupt
Denmark	Cameroon
Finland	Paraguay
Sweden	Honduras
New Zealand	Tanzania
Iceland	Nigeria
Canada	Indonesia
Singapore	Colombia
The Netherlands	Venezuela
Norway	Ecuador

[1] Transparency International surveyed 85 countries. Numerous countries
were excluded from the survey because of insufficient reliable data.
Source: Transparency International (1998)

countries, especially those in Sub-Saharan Africa and Latin America, are perceived
to be most corrupt (table 25.6).

The net impact of corruption on trade has not been determined. Depending on the
circumstances, corruption may either increase (by adding another restriction) or
decrease impediments to trade (by allowing exporters to bypass tariff barriers),
affecting the volume as well as the composition of trade. Barriers to trade are raised
if corruption is out of control, too costly, or primarily in the form of distortion.
Trade is reduced, for example, when a customs official retaliates for nonpayment of
a bribe and allows a perishable commodity shipment like bananas to rot in storage.
On balance, corruption might increase trade at the margins (Elliott, 1997).

Anecdotal evidence suggests that tax evasion is the most common motive of
corruption. Other determinants include political culture, government type (non-
democratic governments can be more tolerant of corruption), postcolonial syn-
dromes (where staying in power is more important than governing) and inadequate
wages for customs officials.

A WTO binding agreement is needed to eliminate all forms of corruption. Difficulties arise, however, because corruption varies by means (smuggling, rent-seeking
behavior, tax evasion, and illegal payments), country, commodity and even by
individual agents (customs officials, bureaucrats, and exporters). Yet the net impact
of corruption on global trade patterns is significant. Overall, the incorporation of
national economies into global trade is neither routinized, nor as open or free as
trade proponents would have us believe.

The Role of Governments

Nation-states

Besides diverting trade via barriers, governments in the 1990s became active in
creating trade over and beyond what is typically considered in comparative advantage. Governments influence the geography and organization of production by
joining in regional trade arrangements and by emphasizing strategic trade policies.
Economic globalization has facilitated more trade and foreign direct investment, and

in the process has redefined the role of the state as an effective manager of the national economy.

Governmental intervention (and ultimate influence on comparative advantage) ranges from specific export promotion and import prohibition to broadly drawn infrastructural policies, R&D, and targeting of particular industries for special treatment (e.g. the aircraft industries in the USA and the EU). In Asian newly industrialized countries (NICs), for example, the role of policymakers is essential for explaining the inconsistencies between their domestic factor endowments and the relatively sophisticated behavior of their firms in global markets. Each NIC has found a powerful engine to drive its economy: specifically, immigrant entrepreneurs in Hong Kong, multinational corporations (MNCs) in Singapore, small labor-intensive firms in Taiwan, and giant subsidized conglomerates in Korea. By contrast, Japan gives the highest priority to R&D and small and medium-sized enterprises; France focuses more on export promotion; Germany emphasizes regional development and sectoral aid; and the USA spends half of its industrial support on investment incentives (Malkin, 1993).

The interrelationship between governments and global trade patterns moves us away from such unrealistic assumptions as "fair" and "free" trade, to highlight policy environments and imperfect markets. Firms, especially MNCs, operate within the frameworks of differential state regulatory (and deregulatory) policies in trade and industry as well as in foreign investment, and of the global regulatory policies adopted by the WTO. MNCs take advantage of national differences in regulatory regimes to pursue global competitive advantage. A symbiotic relationship is evident, as the changing nature of the regulatory surface is, in part, a government response to the strategies of MNCs.

Regional emphases in trade policy

A special scale of government intervention is the trend toward regionalism. Already the WTO has been notified of 180 regional trading arrangements (WTO, 1998). All but 3 of the 132 WTO members (Japan, Hong Kong, and South Korea) belong to at least one regional bloc. The WTO is supposed to vet regional blocs for conformity with multilateral rules, but regional groupings consider it in their best interest to keep bloc criteria and procedures vague.

Any inability of the WTO to solve new issues in global trade may encourage countries to look for regional solutions. The general objective of regional trading blocs is to increase efficiency and gain for members by creating more trade. They are erected by policies that liberalize and harmonize trade among members and in some cases by the creation of a supranational organization that regulates and mediates the bloc. The most well-developed regional trading bloc is the European Union, with its fifteen members trading more within the bloc than outside of it. Part of the impetus toward regional integration is a reaction to the positive news emerging from the EU experiment. Other trading blocs are in earlier phases of development (e.g. NAFTA), and the internal ties are not as tight. The majority of regional economic groupings fall into the categories of free trade areas (whereby tariffs are liberalized) or customs unions (tariffs are liberalized, and member countries agree to have a common external tariff with nonmembers).

Only the EU comes close to full economic union and harmonization of economic policies.

Geographical proximity, if not contiguity, appears to be one of the most important factors motivating bloc formation. Doing business at a distance involves three types of costs. Transport costs and time elapsed in transporting (important with perishable commodities and with just-in-time production) are clearly related to distance. The other cost correlated with distance lies with human and physical capital investments (knowledge about a partner's language, culture, markets, and business practices) among geographically and culturally proximate states. Some of these investments may be in public goods (e.g. the information generated by a regional trade promotion agency), and others may be in regional linkages (e.g. investments in a port). Petri (1992) concludes that the full impact of human and physical capital investments may be highly cumulative: regional linkages may increase exponentially as each investment stimulates subsequent investments and additional trade, increasing the likely profitability of further investments.

A similar level of economic development (facilitating intra-industry trade), compatible trade polices, and the presence of a leading trading state may also be important determinants of bloc formation. Other motivational factors toward regionalism appear different in various regions. For instance, political and historical factors seem to be the most important driving forces in the EU, foreign direct investment factors appear most significant in East Asia, and fears of both exclusion from globalization and contagion from global financial crises drive bloc formation in the developing world.

Regional trading blocs are discriminatory by nature and, as such, go against the general principle of nondiscrimination established by the GATT. Most regional blocs have a strongly defensive character: they represent an attempt to gain size advantage in trade by creating large markets for their producers and protecting them, at least in part, from outside competition. Consequently, the largest regional trading blocs have considerable influence on global trade. Regional trading blocs affect the geography of trade by trade diversion (member countries import goods from each other rather than from nonmember countries, as done previously) and by increasing the level of intra-regional trade. Frankel (1997, p. 113) estimates that regional trade blocs boost intra-regional trade by at least half (e.g. the EU) and by fivefold at the maximum (e.g. ASEAN). In the process they reduce trade with the rest of the world.

Because members of each group trade more with each other than they do with nonmember countries, trade is becoming more geographically concentrated in many parts of the world. Concentration in trade has steadily increased among EU members, and the Latin American Mercosur and the Andean Pact groupings have the highest trade concentration rates of any regional blocs in the global economy. Asian concentrations are high but have been declining in recent years, showing the fears of a Yen bloc forming in Asia to be unfounded. The lowest levels of intra-regional trade occur in African and Caribbean regional trade groupings (Frankel, 1997).

City governments

At the sub-national level, metropolitan governments have become active in implementing policies to promote their position in the global economy (Noponen et al.,

1997). Local governments typically link investment and trade policies in an attempt to shape the trade profile of cities. They also play important roles in enticing MNCs (through trade missions, incentive packages, export-processing zones, infrastructural provisions, etc.) and in fact target particular types of MNCs to locate within the city. In some instances, cities' promotion policies complement national and regional trade policies (e.g. Miami's promotion as a command city for the Caribbean and Latin America); but in many cases they do not.

Metropolitan governments have come to realize that cities have different stakes in national trade policy as a result of their unique position within the national and global economies. National export promotion policies disproportionately aid those cities relatively well positioned with high-technology complexes, major research universities, or an oligopolistic position in global trade (e.g. Boston, Anaheim-Santa Ana) (Noponen et al., 1997, pp. 83–5). The economic revitalization of Dublin, Ireland, acclaimed as "the best business location in Europe" (Grant and White, 1998), can only be explained by a synergy of national and local government policies that have promoted technology and new industries (business software in particular). By contrast, for many older manufacturing cities (e.g. Cincinnati, OH, and Syracuse, NY) and for goods with a high-input content (e.g. wood, cement) the pursuit of domestic markets is more rewarding than a focus abroad, because the friction of distance and costs of doing business overseas constitute significant transaction costs (Wolf, 1997).

The role of city policies and sub-national regional governments (provinces/states) in promoting trade is virgin territory for research. Why some cities are more active and how city size, location, historical conditions, and technology interact to shape cities' geography of trade needs to be explored. The city level of analysis in trade suffers from obstacles in obtaining local trade data, but many city governments, ports, and airports are now beginning to collect these.

The Role of Firms

Firms influence the economic geography of trade through both intra-industry and intra-corporate trade. They have complex interconnections, often operate under conditions of imperfect competition, and shape trade patterns. In addition to these complexities, there are serious data problems in the study of firm activities in global trade. Not only is much of the pertinent information concealed in companies' annual reports, but data are usually limited to OECD states.

Intra-industry trade

Intra-industry trade occurs among firms in the same industry. Wide variations are evident in the scale and scope of firm operations, and in the organization, technology, and geography of commodity chains. Unfortunately, because of a large bias toward OECD firms in the literature, we know nothing about firms in the periphery of the global economy or about the spatial organization of small and medium-sized firms.

Intra-industry trade occurs primarily in intermediate goods rather than in finished consumer goods. A distinguishing feature of intra-industry trade is that the buyers are often other firms rather than individuals; thus, items must meet firms' needs of quality, durability, and serviceability rather than individual consumer taste or fashion. Another characteristic of intra-industry trade is its allegedly low adjustment costs in the face of trade liberalization. It is well documented, for example, that the EU registered early liberalization successes because of intra-industry trade. Few industries disappeared and many firms rationalized production by reducing the number of their product varieties and by lengthening the production runs of those varieties that were retained (Grant et al., 1993).

The strength of intra-industry trade varies by industry, country, and with adjustment processes like regional integration. Although the evidence of across-industries-across-countries trade is fragmentary at best, Grant, Papadakis, and Richardson (1993) find higher levels of intra-industry trade among high-technology firms and developed economies, such as France and the USA, and lower levels for Indonesia, the Philippines, and Thailand. Ray (1991) finds differences in intra-industry trade between the USA and its trading partners. For example, intra-industry trade between the USA and Japan primarily involves intermediate goods produced in small plants using labor-intensive techniques, whereas US trade with Mexico involves similar goods but produced at large plants. Intra-industry trade within the EU mostly pertains to labor-intensive industries, intermediate goods produced in small plants, and science-based industries (Brulhart, 1998). It has been growing fastest in the peripheral areas of the EU (e.g. Ireland and Spain), suggesting economic convergence and regional integration of Europe's poorest states with its wealthiest regions in the center (Brulhart, 1998, p. 347).

Important determinants of intra-industry trade are scale economies (share in marketing, planning and support cost for each industry) and geographical concentration. For high-technology industries, intra-industry trade is heavily localized (Brulhart, 1998). While the behavior of firms best explains intra-industry trade, governments' local content legislation, affecting what is defined as a national product, also impacts intra-industry trade levels. In addition, patterns of national specialization in intra-industry trade have emerged, especially in the EU since the 1980s (e.g. office data products in Ireland, leather goods in Italy; Brülhart, 1998, p. 336). More research on the determinants of intra-industry trade is necessary. Indeed Krugman (1994) reports that little progress has been made with empirical work on intra-industry trade in general.

Intra-corporate trade

It is estimated that an increasing proportion of global trade is intra-corporate trade (within MNCs) rather then interstate. MNCs and multiproduct firms, which have expanded in both number and scope of activities since the 1960s, have encouraged the development of intra-corporate trade as a means to overcome geographical, financial, and technological limitations. MNC behavior varies cross-nationally. For example, Urata (1993) notes that affiliates of US firms have a higher propensity to export than do Japanese affiliates. Significantly, in this category of global trade, intra-corporate transactions are less controlled by conventional, "arm's length" price/cost determinants of regular trade; transactions between affiliates are least likely to be market determined; and they are sensitive to international decisions of MNCs. Intra-corporate trade also responds differently to changes in economic conditions than does trade between unrelated parties (Zeile, 1997, p. 23).

MNCs "value" multinationality for a number of reasons, including tax avoidance (transfer pricing, tax havens), lower cost inputs, and managerial objectives that may differ from share-price maximization, organizational efficiency (operating offices in each of the world regions as opposed to a single global office), avoidance of NTBs, and consumer preference for domestic over foreign varieties of a good (Lundbäck and Torstensson, 1998). For example, Guinness Ghana is an MNC that produces specific Guinness products, such as "Guinness Malta," for the West African market. In addition, Morck and Yeung (1991) emphasize the role of intangible assets – marketing abilities, superior production skills, managerial skills, and consumer goodwill – in encouraging firms to go multinational. Firms increase their value by internalizing markets for these assets, the value of which is enhanced in direct proportion to the scale of the firms' markets. Because these intangibles are based predominately on proprietary information, they cannot be efficiently exchanged at arm's length with other firms. Firms, therefore, augment their value by expanding overseas, if the gain from applying these intangibles in a foreign market outweighs the additional costs of operating a subsidiary overseas (Morck and Yeung, 1991). MNCs are able to transfer and disseminate technology, to measure and compare costs, and to detect vagaries of how best to compete in price, quality and service across countries (Grant et al., 1993). The nature of the market may also be a factor. US intra-corporate trade is mainly connected with manufacturing, while foreign intra-corporate trade in the US market is mainly connected with marketing and distribution activities (Zeile, 1997).

MNCs have complex interconnections with governments, some adversarial and others increasingly nonadversarial. Over the last two decades, government–MNC relationships have evolved from conflict toward a "new partnership" (Dunning, 1993). Governments are now less concerned with the impact of MNCs on balance of payments and trade deficits, and more focused on the extent that MNC activity helps upgrade domestic resources, capabilities, and access to global markets. Governments interpret MNCs through the lens of global competition and politics. For example, European MNCs are actively participating in the EU institutional policy-making process. The European Commission and European MNCs' representatives now strive to build a consensus for their proposals, especially in high technology (Peterson and Sharp, 1998). In the developing world, since the introduction of SAPs, foreign companies, including those from the South, are also forging new linkages with governments and profoundly altering development goals (Wells, 1998).

The Nexus of Governments and Firms at the Turn of the Century

High-technology trade

High-technology industries are organizations in which knowledge is a prime source of competitive advantage, there is above-average spending on R&D, and high proportions of scientists and engineers in employment are the norm. Good examples are the aircraft, electronics, aeronautical, and pharmaceutical industries. The rise of new high-technology industries is a persistent feature in OECD states. Since the 1980s, it has grown faster than manufacturing trade as a whole and, in 1992, accounted for 31 percent of US exports, 17 percent of EU exports, and 27 percent of Japanese exports (Peterson and Sharp, 1998, p. 123).

Empirical evidence confirms that success in high technology bestows national benefits on productivity and high-wage job creation. As a consequence, high-technology industries are major determinants of national competitiveness, and trade in high-technology is not generally "free" in the traditional sense because trade outcomes are heavily manipulated by policy instruments such as industrial targeting. A country's competitive position in high-technology industries is less a function of its national factor endowments than of the strategic interactions between its firms and governments and the innovation networks of other countries. Between 1985 and 1997, firms across the EU collaborated on 668 projects which received 16.7 billion Euros in funding (Peterson and Sharp, 1998, p. 91). The EU's technology collaboration policies have had important political effects: they have created a technology community by routinely encouraging European firms to work with firms beyond their national borders. Airbus is a good example of a government-backed consortium of cross-national firms from France, UK, Germany, the Netherlands, and Spain.

Governments help determine high-technology trade by giving research and industrial subsidies to high-technology industries, targeting, procuring, patent protecting, setting rules that govern market access, and unilaterally defending high-technology interests with trade policies designed to manage global competition. At times, governments also emphasize geographical concentrations of production in distinct subnational regions of their host countries or "technology districts," such as Silicon Valley's command in semi-conductors and Dublin's command in European business software services (the "silicon bog"). Research in the EU has shown that high-technology trade is localized within intra-regional trade patterns, with a movement toward geographical concentration under way since the 1990s (Brulhart, 1998, p. 346).

In high-technology industries, firms influence comparative advantage by emphasizing a combination of factors, such as innovative capabilities, intellectual property rights, and product-based technological learning (PBTL), rather than by concentrating on the introduction of an innovation or imitation, the factors considered crucial in conventional trade theory. The aim in PBTL is to maintain advantages in the share of trade by preserving technological leads – in other words, by creating the world's "best practice" (Storper, 1992). By contrast, trade share advantages are shorter lived for imitators because the catch-up process is already in motion. Firms not only consider the supply side of comparative advantage (based on cost advantage and price competitiveness) but also focus on the demand-driven nature of competitiveness (e.g. price, quality, service, channels of distribution, consumer tastes) in high-technology trade.

E-commerce

E-commerce is the latest medium for trade, although presently its market is small and US-centered, accounting for $26 billion of global sales in 1997 (OECD, 1998). However, it is expected to expand by a factor of ten by the year 2000 and to grow exponentially thereafter (OECD, 1998). E-commerce has penetrated sectors unevenly. The media have hyped consumer goods, especially online merchants selling books (e.g. amazon.com), wine, and computers as well as travel and financial services. However, the OECD (1998, p. 3) notes that the largest e-commerce market involves businesses supplying products to other businesses, where the transactions of just a few firms exceed all estimates of the business- to-consumer market.

E-commerce offers enormous spatial implications for the entire trade system, primarily because it shrinks the economic distances between producers and consumers. It is particularly revolutionary in three respects: level of participation, speed of change, and availability of information. First, participation in global commercial transactions is expanded beyond large corporations and banks to certain individuals (often in particular places) at the click of a mouse. Consumers in different countries can load "hypercards" with electronic cash, and engage in anonymous card-to-card transfers (Korbin, 1997, p. 66). In addition, new intermediaries are required (e.g. network access providers, electronic payment systems, and services for authentication and certification of transactions) that are footloose and far less labor-intensive than traditional intermediaries. Secondly, the speed of technological change within industries that facilitate e-commerce is rapid. For example, at IBM the technological advancement is so rapid that a "web-year" is equivalent to three months of conventional time. Third, e-space makes information available worldwide and facilitates the expansion of trade by making it infinitely easier for buyers and sellers to find each other, although the complexity of the web can make searching a very involved and unscientific process, with the same search yielding different results. E-commerce could result in trade approximating the ideal of perfect competition: low transaction costs, low barriers to entry, improved access to information for consumers, decentralization and no hierarchy, and no conditions for authoritarian or monopoly control. At the same time it offers the potential to improve quality and prices. Cyber-trade may also increase vertical specialization by allowing companies to disperse different phases of the industrial life cycle among different countries depending on local conditions.

E-trade will be regulated very differently from conventional country–country trade. It allows for its users to circumvent governments, which complicates identification of where sales taxes should be paid. States are active in shaping comparative advantage in e-commerce. Already the US and EU World Wide Web consortiums are leading by cooperating on e-space R&D (Sassen, 1997). In Ireland, for example, Telecom Eireann (the government-owned telecommunications company) has provided $23 million worth of Internet infrastructure (multimedia computers and high-speed connections) to businesses and homes in the "Information Age Town" of Ennis (population 17,000) (*The Irish Times*, 1997, p. 1). A growing consensus among international organizations, governments, businesses, and citizens supports a tariff-free zone for goods and services delivered electronically (e.g. software data), but maintains that physical goods ordered electronically and delivered through conventional means should be subject to generally applicable duties.

The main determinant of e-commerce is, obviously, technology, which goes beyond the narrow operations of computers to include the full range of technologies that allow e-space to function. The barriers to e-trade are related more to level of economic development than to national or international regulations. The geography of e-commerce is dependent on the number and location of hosts. Presently there are 16.1 million hosts globally, with the USA and Scandinavia standing out as the most

Internet-connected, averaging 40 hosts per 1,000 population (*The Economist*, 1997, p. 98). Yet 29 countries (e.g. Libya, Laos, Congo, North Korea) have no Internet connection (Matrix Information and Directory Services, 1998, p. 3).

Thus, despite the rhetoric of more evenly distributed power in e-space, the reality is far more likely to reflect a geography of centrality focused on information cities rather than on dispersed locations. Such cities are developing comparative advantages through heavy concentrations of infrastructure, information technology specialists, and buildings (Sassen, 1997). The sharpening of inequalities in the distribution of e-space infrastructure will result in developing countries becoming further marginalized in trade. Countries like Ghana have attempted to build up their transportation infrastructure for 40 years to compete better in conventional trade. If the history of infrastruture building is anything to go by, developing countries are "web-decades" from widespread participation in e-commerce. Foreign capital will be necessary for them to develop the information technology infrastructure, which may well be ruled and shaped by investors' goals. This is, of course, reminiscent of railroad development during colonialism, which facilitated imperial trade rather than the internal economic organization of societies.

Conclusions

Traditional notions of trade have been undermined by several recent trends in the trade, policy, and intellectual environments. In the trade environment, capital, intermediate, service, and high-technology goods now constitute a large and growing share of global trade. Firms, as opposed to traditional consumers, play a greater role in explaining trade relationships, and often firms forge strategic alliances or contracts (intra-industry or intra-corporate) that ensure that their actions are less influenced by market transactions. Combined, these different types of trade flows in the global economy produce labyrinthine trade geographies.

In the global policy environment, the WTO has been only moderately successful in keeping pace with the growth, changing structure, and complexity of trade. The challenge of the first 50 years of the GATT was to manage a world divided: the more difficult challenge for the next 50 years is to manage deepening integration. Freer markets have meant more rules, and the number of regulations that govern trade has mushroomed. This is leading to the universalization of many laws that govern cross-border commerce and to the harmonization of legal codes and international standards (e.g. accounting practices) that allow national economies to function. However, this convergence of national economic regimes has also meant a lower tolerance for system divergence, and more trade friction. It is clear that the WTO architecture needs to be strengthened to fit the economic reality of the twenty-first century.

In the intellectual environment, most researchers acknowledge that conventional trade theory is inadequate to explain contemporary global trade patterns. As the nature of trade has evolved and as trade globalization has proceeded, the playing field for states has not leveled out. Places that harness technology are much more likely to be command centers in trade. National and municipal governments, policymakers, and businesses have taken on important roles in creating and then maintaining the global competitiveness of places. Some more than others, however, have the power to arbitrarily determine comparative advantage and to ensure that the right comparative advantages are created for the right industries (Porter and Sheppard, 1998). Geography still matters for governments and firms, even in the increasingly flexible and globally interdependent market.

Research on trade needs to be refocused in three areas. First, geographers can contribute to trade research in a number of important ways: by undertaking detailed case studies of trade regions, by elucidating the relationships among the different scales (global, regional, state, subregional, city, and local) of the global trade economy, and by examining in detail the nexus of government–firm relations, the processes of economic globalization, and the relationships between trade liberalization and globalization in the developing world.

Second, more trade evidence should be gathered across the global economy. Researchers have been particularly biased in their selection of case studies, concentrating on OECD economies because of data availability. Many more case studies need to be undertaken on developing countries. Typically macro-studies of regions like Africa have been used to make generalizations about individual country experiences in global trade (Grant and Agnew, 1996), but this is an example of the individualist fallacy in which the character of the whole is drawn from the one or other of its parts.

Third, trade researchers need to become data gatherers. For too long they have used secondary national data to analyze trends from their comfortable offices in the North. Trade data need to be collected at a variety of geographic scales, such as industry (e.g. intra-industry), the region (intra-regional), the city, and the web. Moreover, the categories for which commodity trade data have been assembled reflect the earlier commodity trade system, where resource and manufacturing trade dominated. These categories should be reformulated to more accurately reflect global, regional, and local realities. Data need to be gathered at the micro level to examine how globalization is affecting individual traders. Poor official data quality and availability should not preclude research on the developing world. Trade researchers have an obligation to play their part and collect primary local trade data in order to understand better how trade affects the majority of the world's population.

Bibliography

Bergsten, C. F. 1998. Fifty years of the GATT/WTO: lessons from the past and strategies for the future. *Institute for International Economics Working Paper 98–3*. Washington D.C.: Institute for International Economics.

Brulhart, M. 1998. Trading places: industrial specialization in the European Union. *Journal of Common Market Studies*, 36, 319–46.

Daly M. and Kuwahara, H. 1998. The impact of the Uruguay Round on tariff and non-tariff barriers to trade in the "Quad." *The World Economy*, 21, 207–34.

Dunning, J. H. 1993. Towards a new partnership? *International Economic Insights*, July/August, 23–5.

The Economist. 1997. Top internet countries. February 15, p. 98.

Elliott, K. A. 1997. *Corruption and the Global Economy.* Washington D.C.: Institute for International Economics.

Fieleke, N. S. 1995. The soaring trade in "nontradeables." *New England Economic Review*, 28, 25–36.

Frankel, J. A. 1997. *Regional Trading Blocs in the World Economic System*. Washington D.C.: Institute for International Economics.

General Agreement on Tariffs and Trade. Various years. *Annual Report*. Geneva: GATT.

Graham E. M. and Richardson, J. D. 1997. *Competition Policies for the Global Economy*. Washington D.C.: Institute for International Economics.

Grant, R. 2000. Economic globalization, politics and trade policy in Ghana and Kenya. In N. Kliot and D. Newman (eds). *Geopolitics and Globalization: The Changing World Political Map*. London: Frank Cass House, forthcoming.

Grant, R. and Agnew, J. 1996. The representation of Africa in world trade, 1960–92. *Annals of the Association of American Geographers*, 86, 729–44.

Grant, R. and White, M. 1998. Celtic tiger or paper tiger? The Republic of Ireland in the global economy. Paper presented at the Annual Meeting of the Association of American Geographers, Boston, Mass., March 25–29.

Grant, R., Papadakis, M. and Richardson, J. D. 1993. Global trade flows: old structures, new issues, empirical evidence. In C. F. Bergsten and M. Noland (eds). *Pacific Dynamism and the International System*. Washington D.C.: Institute for International Economics, 17–63.

Hufbauer, G. C. 1998. Sanctions-happy USA. *International Economics Policy Briefs*. Washington D.C.: Institute for International Economics.

Hummels, D., Rapoport, D. and Yi, K-M. 1998. Vertical specialization and the changing nature of trade. *Federal Reserve Bank of New York* (Economic Policy Review), June, 79–99.

The Irish Times. 1997. Ennis wins 15 million pounds prize as Information Age Town. September 25, p. 1.

Korbin, S. J. 1997. Electronic cash and the end of national markets. *Foreign Policy*, 107 (Summer), 65–77.

Krugman, P. 1994. Empirical evidence on the new trade theories: the current state of play. In *New Trade Theories*. CEPR Conference report. London: CEPR.

Lundbäck, E. J. and Torstensson, J. 1998. Demand, comparative advantage and economic geography of international trade: evidence from the OECD. *Weltwirtschaftliches Archiv* [Review of World Economies], 134, 230–49.

Malkin, D. 1993. Industrial policy in OECD countries. *International Economic Insights*, March/April, 22–23.

Matrix Information and Directory Services. 1998. Internet State, January 1998. http://www.mids.org/mmg/501/pub/ed.html

Morck R. and Yeung, B. 1991. Why investors value multinationality? *Journal of Business*, 64, 165–87.

Noponen, H., Markusen, A., and Driessen, K. 1997. Trade and American cities: who has comparative advantage? *Economic Development Quarterly*, 11, 76–87.

OECD. 1998. Dismantling the barriers to global electronic commerce. http://www.oecd.org/search97cgi/s9

Peterson J. and Sharp, M. 1998. *Technology Policy in the European Union*. New York: St. Martin's Press.

Petri, P. 1992. The East Asian trading bloc: an analytical history. Paper presented at the NBER Conference on the United States and Japan in Pacific Asia, Del Mar, Calif., April 2–5.

Pine, J. and Gilmore, J. H. 1998. Welcome to the experience economy. *Harvard Business Review*, July–August, 97–105.

Porter, P. W. and Sheppard, E. S. 1998. *A World of Difference. Society, Nature and Development*. New York: Guilford Press.

Ray, E. J. 1991. US protection and intra-industry trade: the message to developing countries. *Economic Development and Cultural Change*, 40, 169–88.

Sassen, S. 1997. Electronic space and power. *Journal of Urban Technology*, 4, 1–17.

Storper, M. 1992. The limits to globalization: technology districts and international trade. *Economic Geography*, 68, 60–93.

Transparency International. 1998. The Transparency International Corruption Perceptions Index 1998. http://www.gwdg.de/~uwvw/icr_serv.html

Urata, S. 1993. Foreign direct investment and economic development in Pacific Asia. In C. F. Bergsten and M. Noland (eds). *Pacific Dynamism and the International System*. Washington, D.C.: Institute for International Economics, 273–97.

Wells, L. T. 1998. Multinationals and the developing countries. *Journal of International Business*, 29, 101–14.

Wolf, H. C. 1997. Patterns of intra- and inter-state trade. *National Bureau of Economic Research Working Paper 5939*. Cambridge, MA.

World Bank (various years). *World Development Report*. New York: Oxford University Press.

World Trade Organization 1997. *Annual Report 1996*. Vols. 1 and 2. Geneva: WTO.

World Trade Organization 1998. *Annual Report 1997*. Vols. 1 and 2. Geneva: WTO.

Yarbrough, B. V. and Yarbrough, R. M. 1997. *The World Economy. Trade and Finance* (4th edition). New York: Dryden Press.

Zeile, W. J. 1997. U.S. intrafirm trade in goods. *Survey of Current Business*, 77 (February), 23–78.

Chapter 26

Money and Finance

Andrew Leyshon

"Money is what you'd get on beautifully without if only other people weren't so crazy about it," Margaret Case Harriman (quoted in Dunkling and Room, 1990, p. 156).

"The value of a dollar is social, as it is created by society," Ralph Waldo Emerson (quoted in Furnham and Argyle, 1998, p. 1).

Until relatively recently, economic geography seemed quite content to conform to the first part of Margaret Case Harriman's observation, as the subdiscipline evolved in virtual isolation from the world of money and finance. Reasons can be mobilized to explain why economic geographers have tended not to pay much attention to money. Thus Martin (1999) suggests that economic geography's long-term neglect of money and finance was in part a product of an unfortunate turn of historical fate, in that August Lösch, one of the founding fathers of the subdiscipline, died before he could complete a follow-up to his landmark *The Economics of Location*. Martin argues that this book would have focused directly upon money and location, and, if completed, would have changed the post-war history of economic geography. But it wasn't, and the growing importance of a neoclassical economic perspective within economic geography and its sister discipline of regional science from the 1950s onwards militated against a serious consideration of the geography of money and finance. Neoclassical economics assumes that there is "free and costless movement of capital and labor and perfect and ubiquitous information flows between regions," which has the effect of "assum[ing] away any regional role for money" (Martin, 1999, p. 3).

While this explanation is perfectly plausible, economic geography's neglect of the subject still remains something of an historical disciplinary puzzle, especially given that the nature of money means that it has always been an unusually appropriate subject for economic geographers to study. Even the most cursory examination of the history of money reveals it to be a social innovation which owes its success to its exceptional ability to solve problems of economic co-ordination over space and time (Leyshon and Thrift, 1997). Money has been at the center of economic exchange within human societies for an extraordinarily long period of time (Davies, 1994), and it has so effectively crowded-out non-monetary systems of exchange – such as barter – that where non-monetary systems exist they become curiosities for comment and debate. A measure of money's success as a social institution is that most people hardly give its existence a second thought (beyond the fairly universal feeling that more of it would be helpful). It is usually only at times of monetary and economic crisis that the relative fragility of the social institution of money becomes apparent, along with a recognition of how much its survival relies upon the persistence of social norms and expectations.

Recent advances in social theory have drawn attention to the ways in which social practices and material objects combine together in networks of relations (Whatmore, 1999). From its earliest incarnations, money has been constituted through social relations and naturally occurring materials, which in combination then act to extend social relations and money itself over geographical space. Over time, the nature of these materials has changed, and in general the tendency has been for the material form of money to become less important as the social norms and expectations surrounding money became more important. Therefore, over time the "value" inherent in money has become less dependent upon the physical form of the money, such as gold or silver coins, and more dependent upon some institutional authority, such as banks (as in checks[1] drawn on banks) and the state (as in bank notes underwritten by a central bank). Through all these social–material transformations, money has remained a deeply geographical phenomenon (Leyshon and Thrift, 1997).

Historians and anthropologists of money have long been aware of this, but only since the 1980s has there been a rapid growth of research within economic geography. This chapter does not seek to provide a comprehensive overview of all the work undertaken in this area (see Leyshon, 1995, 1997, 1998; Martin, 1999). Rather, what I seek to do is focus upon four areas of research, which represent the interception of two broader themes within geographical research on money and finance (table 26.1).

The first of these themes is theoretical in nature. For much of its brief history, the geography of money and finance has sought theoretical inspiration from a broadly political economy perspective. Following Barnes (1996), it is possible to describe such an approach as "progressive," for it is motivated by a belief in social progress, which in turn is linked to a quest for understanding, through the application of rationality and reason (cf. Bassett, 1999; Lowe and Short, 1990). Such rationality and reason is used to uncover a hidden order which is seen to govern social action in the world, and in turn may be acted upon to elicit social progress. Progressive approaches such as political economy have the confidence to suggest interventions within social life to achieve progress, because they assume there exists a universal

Table 26.1 Geographies of money and finance

	Progressive geographies of money	Social Constructivist geographies of money
Place	Place and the urban dynamics of finance	Making up financial places
Space	Money and economic transformation	Creating alternative financial space

truth that holds for all times and all places, "something that lies outside the changing external context and therefore could provide the permanence necessary to judge claims to knowledge" (Barnes, 1996, pp. 7–8).

More recently, however, geographies of money and finance have been written from a perspective that takes up a counter position to "progressive" accounts of social change, rejecting the notion of progress and the value of rationality. This broad theoretical position, which Barnes describes as "anti-Enlightenment," but which I refer to here as a movement of *social constructivism*, rejects smooth, linear accounts of social change, favoring instead disjunctive, ruptured, and non-linear accounts. Such approaches dismiss attempts to uncover "the truth" in social science. Truth is seen not as transcendental and foundational, but as contingent and contextual, varying from time to time and from place to place. Rather than seeking a rational, foundational order in the world, that may be uncovered through better, more powerful theories and concepts, social constructivists argue that such orders are in fact constructed through our attempts to understand the world. It is more productive, therefore, to think of order as a verb rather than a noun, something that is made or constructed, albeit temporarily.

The second theme revolves around concerns about the relationship between money, space, and place. Considerable effort has been expended in investigating the ways in which money and finance act upon particular *places*, including studies of phenomena such as financial centers, on the one hand, and places abandoned by mainstream financial services industries – inner cities or peripheral housing estates – on the other hand. This work tends to have practical and concrete concerns, with a particular emphasis on the processes and practices that unfold within particular places. It is complemented by research that has considered money more as an abstraction, and the ways in which money may be seen both to create and to distort understandings and perceptions of *space*.

The remainder of this chapter is divided into three main parts. The first considers the growth of a geographical literature on money and finance from the 1970s onwards, for the most part work informed by progressive social theory – particularly a broad political economy. The second looks at the rise of a set of alternative geographies of money and finance, seeking to extend existing concerns with the places and spaces of money, but in ways informed by social constructivism. The third constitutes some concluding remarks.

Towards a Geography of Money and Finance

Geographical research on money and finance only began in earnest following a broader disciplinary shift towards political economy and an interest in Marxism. New areas of research emerged through asking different kinds of questions about economic geography. Nevertheless, despite the central importance of money within Marxist theory, it took another ten years or more before interest had broadened sufficiently to be able to identify a distinctive subfield of economic geography concerned with the relationship between money, space, and place.

In what follows I examine research on geographies of money and finance undertaken beneath the umbrella of political economy. I do this in two stages. First, I deal with research focusing on the relationship between money and place, paying particular attention to work on the urban dynamics of money and finance. I then go on to look at work that has focused more on the relationship between money and space, and in particular on the role of money and finance within broader processes of capitalist transformation.

Place and the urban dynamics of finance

The publication in 1973 of David Harvey's *Social Justice and the City* is a major staging point in the trajectory of human geography, as the discipline followed its author along the journey from quantification, model-building, and "objective" social science towards a more politically aware and "radical" academic agenda. The book was an outcome of Harvey's attempts to reconcile existing geographical theory, to which he had been a major contributor, with the material evidence of socio-spatial inequality of the US inner city.[2] Harvey gave a theoretical voice to a growing concern for "relevance" among many human geographers who were dismissive of the normative, "ivory tower" nature of work undertaken within the tradition of quantification (Peet, 1998).

Harvey sought to develop a means by which the creation of socio-spatial inequity could not only be understood, but actually overturned in the interests of social justice. He argued for more progressive and interventionist research that would, literally, change the world. Interestingly, in making such an argument we re-encounter the figure of August Lösch, identified earlier by Martin (1999) as a kind of spectral antecedent to geographical research on money and finance. Lösch is approvingly quoted by Harvey (1973, p. 148) for claiming that his "real duty [is] not to explain our sorry reality, but to improve it." Harvey's goal was similarly ambitious, boldly claiming that one of the objectives of his research was no less than "to eliminate ghettos" (1973, p. 137).

Social Justice and the City encouraged geographers to begin exploring geographies of money and finance in new ways. Marxist theory provided a conceptual key that promised to unlock the causes of long-run material processes creating uneven development within the city. In this view, the search for profit is the overriding objective within the capitalist city, leading Harvey and others to explore both the theoretical insights that could be gleaned from urban rent theory, and the ways in which competition for space has a financial outcome. A related interest was the key role played by financial gatekeepers in directing capital flows within the urban infrastructure. To illustrate the ways in which thinking along these lines seemed to provide explanations to apparent anomalies within the urban property market consider the following passage, in which Harvey uses the logic of capital to explain the persistence of large quantities of low-quality rented housing stock within inner city Baltimore:

There are some curious features about ghetto housing. One paradox is that the areas of greatest overcrowding are also the areas with the largest numbers of vacant houses. There are about 5,000 vacant structures in Baltimore – a good many of which are in reasonable condition – and they are located in areas of greatest overcrowding. Other cities are experiencing something similar...[T]he rents...landlords charge are very high in relation to the quality of accommodations, while properties, if they do change hands, do so at negligible prices. The banks, naturally, have good rational business reasons for not financing mortgages in inner city areas. There is greater uncertainty in the inner city and the land is, in any case, frequently regarded as "ripe" for redevelopment. The fact that failure to finance mortgages makes it even riper is undoubtedly understood by the banking institutions, since there are good profits to be reaped by redevelopment under commercial uses (Harvey, 1973, p. 140).

In other words, the anomaly of a supply of empty but unavailable properties located cheek-by-jowl with high-density residential accommodation is "explained" through "deeper" causes – the circulation of capital and a generalized search for profit. Individuals and families wishing to buy these properties are denied the opportunity because they live in what are perceived to be "high risk" areas. Such behavior may be "rational" for financial institutions but at a systemic level the outcomes are less benign. It creates "the paradox of capital withdrawing from areas of greatest need to provide for the demands of relatively affluent suburban communities" (Harvey, 1973, p. 112).

Inspired by such observations, and by Harvey's encouragement to look for deeper, structural reasons for divisions within urban space, geographers began to investigate the city anew. In doing so, they turned their attention to the role played by financial and related institutions in channeling capital to and from particular places. Much of this work analyzed the housing market, and in particular how banks and property agents discriminated against the least economically powerful members of society, through such practices as credit rationing, redlining, and "blockbusting" (Boddy, 1976a, 1976b, 1980, 1981; Harvey and Chaterjee, 1974; Williams, 1976). Credit rationing and redlining reinforced the class and ethnic segregation of the city by denying credit to individuals and families within certain parts of the city, and at times prevented class and ethnic mixing by directing certain groups away from areas to retain the "character" of the neighborhood. Blockbusting encouraged ethnic mixing in the anticipation that the movement of ethic minorities into particularly inactive local housing markets would generate turnover as middle-class families left in a flight to new suburban areas.

After this initial flurry of activity, however, economic geography's interest in the distributional effects of money and finance within the urban environment began to wane. It was not until the 1990s that attention turned once more to the role of money and finance in shaping the economic and social geography of the capitalist city. This revival again stemmed from concerns about the role of the financial system in deepening "spatial apartheid" in the US city of the 1980s and 1990s (Davis, 1992). Initially, the impetus came from beyond geography, as researchers in African-American studies, urban economics, and policy studies began to document the problems of financial discrimination within inner city areas (e.g. Bates, 1991, 1993; Caskey, 1994; Grown and Bates, 1992; Squires, 1992). Such studies pointed to a change in the institutional fabric of inner cities, which had continued to deteriorate in the 20 years or so since Harvey had begun writing about the inner city "ghetto." The US inner city had once been a place "in which all of the institutions of the dominant society were reproduced in parallel, and which individuals excluded from the dominant society were active in the parallel set of institutions" (Lash and Urry, 1994, p. 156). But by the 1980s and 1990s, a form of institutional collapse meant that even these parallel institutions began to disappear.

In seeking to explain this process, geographical research developed the concepts of "financial exclusion" (Leyshon and Thrift, 1997) and "financial dynamics" (Dymski and Veitch, 1996) to describe the process by which poor and disadvantaged groups are directly and indirectly denied access to mainstream retail financial services. Financial exclusion plays an active part in the production of urban (and rural) poverty, because those who experience the most difficulty in gaining access to financial services tend to be those experiencing multiple forms of social deprivation.

Geographical research on financial exclusion echoes the 1970s concerns with social justice by focusing upon the equity effects of the financial services industry. This recent work is distinctive, however, because it is more concerned with the impact of money and finance on everyday life, not just on the housing market. Access to mainstream financial services within contemporary capitalist societies is crucially important because many economic exchanges are mediated through banks and other financial institutions in direct transfers between accounts. Without a bank account, individuals and households may have to pay more for certain services to cover the extra cost and risk incurred for handling cash. Excluded individuals also find it difficult to obtain affordable credit, and may be forced to resort to more expensive credit facilities provided by "predatory" financial services firms, such as money-lenders (Dymski and Veitch, 1996; Leyshon and Thrift, 1996, 1997). Moreover, an inability to obtain affordable insurance means that households cannot shield themselves from risk, and are forced to bear the full financial consequences if they become victims of accidents and environmental hazards (such as floods or storms).

Considerable effort has been made to track the effects of financial exclusion, the most visible signs of which have been the extensive branch closure programs of retail financial services firms. Closures in Britain and the USA were spatially uneven, with branches closing fastest in areas of social and economic deprivation, particularly in inner cities with large ethnic minority populations. The programs were undertaken to cut costs and refocus business on the more profitable parts of the customer base in an increasingly competitive market. In the USA, this meant that entire communities were abandoned by the financial services industry, so that large parts of inner cities and many rural areas lost much of their financial infrastructure (Christopherson, 1993). In Britain, where one fifth of all bank and building society branches closed during the 1990s, the fastest rate of closure also occurred in socially deprived inner city areas. Problems of access were exacerbated for the poor, the elderly, and those who have physical disabilities, whose mobility over space is constrained, because the thinning out of branch networks increased the average "journey to bank." These excluded populations are spatially concentrated in poor places, and belong to the social groups who are also discriminated against by such developments as the rise of telephone and Internet banking. In many socially deprived areas, large sections of the population are too poor even to afford a telephone (Graham and Marvin, 1996), let alone a personal computer.

Both the pioneering studies of Harvey and others in the 1970s, and recent research into financial exclusion in the 1990s, identify the logic of capital competition and the search for profit as being central to processes of financial exclusion (in inner cities) as well as financial inclusion (in the suburbs). Moreover, both types of research share a concern for social justice, and for the negative distributional effects of a financial system that would appear to discriminate against individuals and places on the grounds of class, in some cases combined and amplified by racial difference.

More recent work in this area has continued to draw attention to the ways in which the financial services industry discriminates between individuals and social groups, but in a way that goes beyond the narrow economism of some earlier accounts. It draws attention to the ways in which culture is mediated within the consumption of financial services, and in how the behavior of relatively affluent financial consumers helps steer the financial industry in particular directions (see, for example, Leyshon et al., 1998; Leyshon and Thrift, 1999).

Money and economic transformation of space

A second strand of progressive research on the geography of money and finance is concerned more with the broader transformation of *space*. This research is very much "big picture" geography, and here too the work of David Harvey is important, but from a later period. In *The Limits to Capital* (1982), Harvey provides a superb analysis of the dynamics of the financial system under capitalism, outlining the centrality of money and finance within the circulation and accumulation of capital. The role of money and finance is even more central to his analysis in *The Condition of Postmodernity* (1989). In a wide-ranging review of economic and social change over some 25 years, Harvey homes in upon an increasingly powerful and auto-nomous financial system which has brought about a broad transformation in the capitalist system as whole. He argues that the imperatives of the financial system have brought about a world characterized by *time–space compression*; that is, a world that has shrunk in terms of time–space. The world shrinks because capitalist producers are locked into a perpetual race to exploit new markets and to reduce what Marx described as the turnover time of capital; that is, the amount of time taken to convert investments into profitable returns.

Harvey argued further that this race for "fast money" was not only bringing about an acceleration in the pace of economic life, but had in effect also brought about a major transformation in the nature of capitalism as a whole (Harvey, 1989). In so doing he made contact with a wider body of research within economic-orientated social science, known as regulation theory, which for over a decade had been attempting to explain why capitalism appeared to be increasingly beset by crisis, instability, and volatility from the late 1960s on (see Martin, this volume). Accord-ing to Harvey, the relative stability of the "Fordist–Keynesian" era had been dis-located by the rise of an era he described as "flexible accumulation," brought into being by an acceleration of time–space compression "that has had a disorienting and disruptive impact upon political-economic practices, [and] the balance of class power, as well as upon cultural and social life" (Harvey, 1989, p. 284).

Although volatility and ephemerality are seen to be endemic to this new era of capitalist accumulation, the monetary and financial system is held up not only as a good exemplar of the character of the age, but also as one of the main causes of it. In doing so, Harvey goes beyond mainstream regulation theory to connect with a broader international political economy literature that had begun to come to similar conclusions (Corbridge and Agnew, 1991; Clarke, 1988; Cox, 1987; Overbeek, 1993; van der Pijl, 1984, 1989; Thrift and Leyshon, 1988). This work examined the way in which the international monetary and financial system constitutes a kind of "regulatory space," which holds together broader processes of accumulation. In doing so, this research drew attention to the changing scale of financial activity, which had become increasingly globalized. This was enabled by advances in both information and communications technologies, allowing money to be transferred around the globe almost instantaneously and at relatively little cost, and later by increasingly liberal governmental attitudes towards financial institutions, which became identified as a source of revenue and well-paid employment.

In addition, this work began to point to the importance of the Bretton Woods system, created by the USA and the UK towards the end of World War II to bring monetary stability to a world recovering from conflict. Supra-national organizations, such as the International Monetary Fund (IMF) and the World Bank, were established to oversee the international monetary system, and to allow for accommodation between an open and liberal system of trade on the one hand, and relatively autonomous national economic policies on the other. These organizations were supposed to be so large and powerful that they would effectively "sterilize" the influence of private financial interest, which many held responsible for the collapse of the world economy during the 1920s and 1930s and the political and economic instability that led eventually to global conflict (Cox, 1987; Helleiner, 1993; Leyshon and Tickell, 1994).

From the early 1960s, the Bretton Woods system began to unravel as private financial capital flows began to grow and overtake those of the international institutions set up to manage and direct the international monetary system. The system of fixed exchange rates upon which it was based finally collapsed in the early 1970s, and the ability of national governments to resist the power of financial interests went into a steep decline. This, more than anything else, brought about the end of the post-war period of generalized economic affluence and stability. Gradually, through foreign exchange pressure, governments were forced to adjust their monetary policies to become anti-inflationary (Yergin and Stanislaw, 1998), largely because low inflation best defended financial assets and wealth over the long term. Relatively inflationary economies were beset by foreign exchange crises as international investors began to move their money elsewhere.

More recent research within this tradition has explored the transformation of financial space in more detail, going beyond the tendency to overgeneralization which can characterize such accounts. Geographers have examined the growth of "offshore" financial spaces with permissive regulatory environments, created as havens from the restrictions that national governments placed on financial activity (e.g. Hudson, 1998a, 1998b, 1999; Roberts, 1994, 1995). Work also has been undertaken on the implications of the rise of a "money manager capitalism," which has turned investment institutions into influential agents in urban and regional development through their control over large volumes of mobile capital (Clark, 1997; Martin and Minns, 1995).

Social Constructivist Geographies of Money

This section deals with a set of alternative approaches to the relationship between money, space, and place that fall under the umbrella of social constructivist approaches. Again, the section is divided into two parts: a discussion of money and *place*, specifically through geographical work on financial centers; and a consideration of money and *space*, exemplified by "local" currency systems.

Making up financial places

The increase in financial activity in the 1980s, particularly within North America, Europe, and South-east Asia, resulted in rapid employment growth in financial centers. Geographers began to investigate the urban and regional consequences of international and regional financial centers (Leyshon et al., 1989; Leyshon and Thrift, 1997). Initial research pursued a fairly orthodox political economy line, drawing on prevailing research on the rise of "producer services" (Daniels, 1993), and on the role played by "world cities" in co-ordinating the global economy within what Castells termed the "space of flows" (Castells, 1989).

The limitations of a political economy interpretation of the workings of financial centers were soon exposed, however. It became apparent that financial centers could not be understood merely through an application of abstract economic theory; rather, financial centers were active, dynamic, and purposeful places, within which matters of culture, social networks, and the embodied nature of the workers themselves effectively "make up" such places. One of the reasons for going beyond the political economy approach was that the logic of "the space of flows" and of time–space compression suggested an inexorable effacement of place within the international financial system (see, for example, O'Brien, 1991). However, although the time–space horizons of the international financial system have shrunk remarkably, financial centers show remarkable persistence and remain important concentrations of financial activity.

How to explain this paradox? One way out of this puzzle is to apply a more social and cultural interpretation of the financial center than had been used hitherto, and to move away from an abstract political economy approach to see centers as active and dynamic ensembles of networks; of bodies, of machines, of concepts, of interpretations, of information. This is the approach developed by Nigel Thrift (1994, 1996), who argues that it is not particularly helpful to see the financial center as a kind of rigidly ordered, machine-like entity, autonomously marshaling the movement of money and capital worldwide. The problem with such accounts, particularly favored by progressive critics of the workings of the financial system, is that they give the impression that the financial center conforms to a rationality that can be understood and is external to the financial center itself. Thrift argues for a flatter and more contextual understanding of social action, whereby notions of "rationality" and "order" are not externally determined, but are assertions that help actors to cope with the chronic uncertainties and contestability of social life. In so doing, he draws directly upon social constructivist accounts to develop what he describes as a kind of "contingent foundationalism" which he brings to bear in his reformulation of the "logics" of the financial center.

This reformulation focuses upon the centrality of monetary information. Information is central to the workings of financial systems, so that the role of the financial center is to generate, capture, interpret, and represent the vast amount of information about economic, social, political, and cultural change that flows through it daily. The financial center thus acts as a collective means of coping with these torrents of information that continuously flow around the global economy. In this context financial centers may be seen as concentrations of expertise, produced through a "complex division of labor embodied in the skills of the workforce, in machines, texts, and so on" (Thrift, 1994, p. 375). This collectivity of expertise is capable not only of handling large volumes of information but, more importantly, is able to make sense of the monetary world, by providing interpretations and narratives that explain what all the information might actually mean.

Therefore, financial centers can be said to be made up of place-specific networks of actors, technologies, and texts, which are constantly being revised and updated, both to discern and to impose understanding and meaning on the financial world (see also Dodd, 1994). From this perspective financial centers persist because successful monetary networks within financial centers are bound to particular places for relatively long periods of time. To go beyond the network is to step outside the loop of information, knowledge, and interpretation; it is to take a step into the unknown. In a financial world that has speeded up, due to the effects of time–space compression, such a move could be extremely costly, if not financially ruinous.

The limitations of the political economy approach to money and place in general, and to an understanding of financial centers in particular, have been underlined by Linda McDowell, who has also drawn attention to the embodied nature of much financial practice through her research into gender segregation in the City of London (McDowell, 1994, 1997a, 1997b; McDowell and Court, 1994a, 1994b, 1994c). This research draws attention to the role of the body within financial centers, and how bodily deportment and presentation acts as an important "regulator" of what is and what is not construed as "appropriate" financial knowledge (see also Leyshon and Thrift, 1997). Financial services jobs are particularly dependent upon the successful representation of an appropriate body, if such jobs are to be done successfully. As McDowell (1994, p. 440) observes, success here depends upon "radiating an air of confidence, of selling one's self-image as part of the produce." In the end it is difficult to tell when the self of the seller ends and the service or product sold begins.

If one accepts that financial understanding is made up of a network of actors and resources, then an understanding of financial centers requires attention to the types of people and bodies that are seen as particularly appropriate carriers and interpreters of the information and knowledge that flow through such centers. Within financial centers in the past, these bodies were not only male, but produced through a privileged class background, and effusive of a particular kind of paternalistic masculinity. The nature of the changes undergone in the City over the past two decades or so, however, has seen this traditional masculinity challenged, both by a new kind of aggressive masculinity, which is less reliant upon traditional channels of recruitment (see Jones, 1998), and by the entry in recent years of the first generation of senior women. As McDowell illustrates, these women face all kinds of performative challenges to assert their authority in what has traditionally been a homosocial place, wherein women were normally only admitted into relatively low-status positions such as secretaries and typists. But in carving out careers for themselves in this often hostile environment, women are bringing new sensibilities and attitudes to the interpretative mix of the financial center.[3]

Making financial space

In moving the geography of money and finance beyond the political economy approach to encompass a broadly social constructivist approach, the research described above has drawn upon the concept of monetary networks to interrogate the relationship between money and place. This concept also has utility in exploring the relationship between money and space, particularly in the variant of the monetary network developed by Nigel Dodd (1994). Dodd's purpose is to attempt to provide a flexible but robust theory of money that is "sufficiently abstract to enable . . . cases to be compared without generalizing to such an extent that important variations cannot in fact be accounted for or explained" (Dodd, 1994, p. xxvii). In fact, Dodd attempts to meld the best elements of anthropological accounts of money, emphasizing the material and symbolic properties of money, with progressive social theory accounts, which focus upon the general and abstract qualities of money.

In practice, however, Dodd remains closer to social theory accounts than to more substantive anthropological accounts of money. He insists that all monetary systems contain five essential abstract qualities: (i) a system of *accountancy*, which enables "money" in the network to function as a medium of exchange, a store of value, and measure of account – the three concrete requirements of all money forms; (ii) a system of *regulation*, to protect and defend these functions of money; (iii) *reflexivity*, by which past experience of the network enables participants to develop expectations of the future which, in combination with a system of regulation, enable the network to develop as a means of deferred payments and thus to project forward in time; (iv) the existence of *sociality*, so that it is possible for information about exchange and value to circulate between actors within the network; and (v) *spatiality*, which means that monetary networks will have specific types of territoriality (Dodd, 1994). According to Dodd, this specific combination of abstract properties has enabled money to work in a wide range of forms, in different places, and at different times. Indeed, he seems to suggest that variations in the relative strength of the essential abstract qualities explains why some monetary networks manage to survive longer than others.

Although to date no empirical work has been undertaken to directly test the veracity of this theoretical framework, it is nevertheless a provocative and suggestive formulation, which allows for proliferative and overlapping understandings of monetary systems. Thus, there may be many competing monetary networks in existence at any one time, each with distinctive systems of accountancy, regulation, reflexivity, sociability, and accountancy.

To illustrate the diversity of monetary networks we can return to the point at which we began this analysis of different theoretical treatments of money, space, and place, with analyses of geographies of financial exclusion. It is possible to imagine the mainstream financial system as constituting its own monetary network. In terms of the processes described above, the properties of this network have clearly undergone some modifications in recent years. Its sub-systems of regulation and sociability have been reworked, both because of changes in statutory regulation, and because of the switch to a form of governance that has seen institutions become dependent on new forms of information which travel through information techno-logies, computer software, and databases (Leyshon et al., 1998; Leyshon and Thrift, 1999). One of the most interesting changes, however, has been to aspects of the spatiality of the mainstream monetary network. In theory, the territoriality of retail financial services firms extends right up to the geographical borders of the economy within which they are licensed to operate, but in practice the spatiality of such firms is discontinuous and partial. Indeed, a distinguishing factor of the mainstream financial system in recent years is the way in which it has made an effort to withdraw from specific spaces, in order to reconcentrate in others. This withdrawal occurs not just through such well-publicized episodes of physical infrastructure withdrawal as bank branch closure. The use of remote means of information gathering and service provision provides firms with a greater possibility to serve customers over space. However, these technologies are being used to concentrate only upon those custo-mers that have the right market potential. Therefore, this network has changed the nature of its spatiality; it has reduced its dependence upon direct physical access to consumers, and withdrawn entirely from some places. But at the same time, through information and communications technology, the network has increased its scope and flexibility, and in so doing has become stronger.

The transformation of this particularly dominant monetary network has occurred at the same time as a set of alternative monetary networks has emerged, whose growth is related in part to the withdrawal of the mainstream monetary network from particular places. Although still limited in size and scale, they present a potentially radical alternative to mainstream monetary networks. Phenomena such as credit unions, local exchange and trading systems, and time dollar schemes are all examples of monetary networks that contain the same abstract qualities as other networks, although the precise nature of these properties differs in a number of ways (Bowring, 1998; Lee, 1996; North, 1999; Purdue et al., 1997; Thorne, 1996; Williams, 1998). We can compare different monetary networks as a means of analyzing different monetized social relations that constitute the extremes of such networks over space. By way of illustration, and as a comparison to mainstream monetary networks, consider the example of local exchange and trading systems (LETS), local currency systems that first emerged in Canada in the early 1980s as a local response to the economic and social problems caused by the withdrawal of money following economic recession. Members of such systems offer goods and services for sale which can be paid for in a local currency unit, which is then either credited or debited from the LETS accounts of the two parties concerned.

LETS have spread worldwide: by 1999 there were over 1,300 LETS operating in 15 countries around the world (Williams, 1999). They are distinctive monetary networks along all five of the main abstract properties. First, the system of *accoun-tancy* is one that revolves around a special, local, credit money, which may be commensurate with, but is not usually exchangeable for, the mainstream currency.[4] Such local money serves as a measure of value and a medium of exchange, but it does not function very well as a store of value. No interest can be earned on credit accumulated, and the system is an accumulation of credits with no external value other than when they are activated through expenditure; that is, they have to be in action to be worth anything.[5] It is the absence of interest on credit which many supporters of LETS claim to be part of the radical potential of such systems. In mainstream monetary networks debt incurs financial penalties, but within LETS debt incurs only a requirement of reciprocity, and a willingness to reduce debt through doing work for others, which is encouraged. In that sense, these networks are celebrated for their potential to foster a sense of community through interaction and mutual benefit.

Secondly, these networks have a fairly informal system of *regulation*, one that mostly relies upon goodwill and the collective moral suasion of LETS as a whole. The major moral hazard is seen to occur with the problem of the "free-rider;" that is, a major debtor who makes little or no attempt to reduce their debt through undertaking work for other members of the system, and/or who leaves the system or the area before doing so. Given the implicit aim to foster community through reciprocity, however, the general consensus within LETS is that a more punitive form of regulation would be counterproductive.

Thirdly, the issue of *reflexivity* relates here to the problems of creating trust and of generating expectations of the long-term survival of such novel systems into the future. Without sound expectations that credits accumulated may be expended in the future, it will be difficult for actors to justify spending much of their time earning LETS currency.

Fourthly, the *sociality* of such systems is partly institutionalized, as all LETS are part of larger, national organizations, and have a local office that helps to encourage contact between members by advertizing the goods and services offered and sought by LETS members. This helps disseminate information about exchange through the network. Since social outcomes, such as the fostering of community, are considered one of the prime objectives of LETS, the value created by such networks may be better measured in terms of social capital (the activities that members of a community invest their time in, and which enhance its cohesiveness and livability) than financial capital.[6]

Finally, such networks have quite constrained *spatial* boundaries. Although LETS exist worldwide, each operates as an inherently local institution, signified by the way LETS draw upon specific local features or sayings to name the unit of currency used. LETS are able to create an alternative financial space, which resounds to a quite different timbre and resonance than do mainstream monetary networks, but these networks simply do not extend very far. Their limited spatial reach acts as a constraint upon their effectiveness, for they are limited to the resources that exist within their area. For these networks to be properly effective as alternatives to mainstream monetary networks, they have to develop the same spatial reach, and same ability to draw from a heterogeneous array of actors and resources. Unless they do so, the fortunes of such alternative monetary spaces will be hostages to the places in which they are formed. In some cases, this may not be a problem, as the nurturing of social capital may bring sufficient benefits to a local community to overcome the inherent limits of a relatively closed system. But in other cases, the outcome may be less propitious.

To emphasize this point, let us return to David Harvey, where we began in many ways. Harvey has observed that promoting local self-help in an environment of uneven development contains within it a chronic weakness. Such initiatives "will simply result in the poor controlling their own poverty while the rich grow more affluent from the fruits of their riches. The redistributional implications are clearly regressive" (Harvey, 1973, p. 93). Even in the midst of more socially constructivist approaches to the geography of money and finance, therefore, echoes of the progressive concern for social justice resound.

Conclusions

I have provided a brief overview of a body of research on geographies of money and finance since the early 1970s. During this time, work has developed along a theoretical trajectory from progressive accounts of social change concerned with identifying an underlying order and rationality to such changes, towards a more social constructivist position which sees order as more contingent and contextual. With this shift, the geography of money and finance has moved beyond narrow economic and class-based political issues, to include a greater attention to culture, gender, ethnicity, and corporeality.

Despite a late start, geographers have thus covered much ground on this subject. We have now reached the "end of the beginning" as far as the geography of money and finance is concerned (Leyshon, 1995). Economic geography is taking money seriously, and in doing so demonstrating the importance of space and place to the world of finance. It is also providing new insights into how we should think about scale in economic geography, and the relationship between the global and the local in monetary and financial matters. The financial system is more global than ever, as money travels ever faster and more effortlessly in the form of investments and, crucially, divestments. In recent years there has been a crop of serious financial crises, covering Asia, Russia, and South America, as global money has first discovered, and then abandoned, places as sites of investment, leaving chaos in the wake of withdrawal. But although they are seen as global issues, with the investment decisions made by the managers of large financial institutions, the funds are actually managed on behalf of millions of individual contributors, who make regular pension or other investment payments. Therefore, the "global" financial system is also irrevocably "local," in that its power comes from millions of individual and household decisions to seek out a high return on investments. It is this that gives fund managers the discretion to move money in and out of investments as appropriate. It also means that anyone who has a pension or other kinds of investments is implicated in the actions of the global financial system to some extent. We are implicated in other ways too, which are even more transparent than the actions of fund managers in global financial centers managing our collective investments. As financial consumers become more financially literate and move their money to more specialized, higher paying accounts, they participate in the "creaming off" of retail financial from mass market accounts which are less able to subsidize less affluent customers. The act of seeking a better deal for ourselves can have implications for distant others. Therefore, there is no escape from the geography of money and finance.

Acknowledgment

I am grateful to Eric Sheppard for his extremely helpful comments on an earlier draft of this chapter.

Endnotes

1. "Cheques" in the UK.
2. Actually inner city Baltimore. Harvey moved there in the late 1960s from Bristol in England to take up a job at Johns Hopkins University. According to Peet (1998), Harvey's observations of life in Baltimore led him to reject the carefully crafted, but political neutral, philosophy of science which he had developed in his earlier book, *Explanation in Geography* (Harvey, 1969).
3. It is interesting that one of the areas within which women have been particularly successful is fund management, and particularly ethical fund management, where most of the leading funds are run by women managers.
4. That is, the "legal tender" of the economy e.g. US dollars, Canadian dollars, French francs, pounds, etc.
5. Compare this to conventional "savings." If this is kept on deposit at a bank, for example, it is pooled and lent out to borrowers. The money is effectively "working" elsewhere in the economy to generate the additional value that will be paid back in interest on the savings.
6. Social capital refers to the ways in which members of a place invest their time and effort in community-related activities in order to enhance its cohesion and livability.

Bibliography

Agnew, J. and Corbridge, S. 1995. *Mastering Space: Hegemony, Territory and International Political Economy*. London: Routledge.

Barnes, T. J. 1996. *Logics of Dislocation: Models, Metaphors, and Meanings of Economic Space*. London: Guilford Press.

Bassett, K. 1999. Is there progress in human geography? The problem of progress in the light of recent work in the philosophy and sociology of science. *Progress in Human Geography*, 23, 27–47.

Bates, T. 1991. Commercial bank financing of white- and black-owned small business start-ups. *Quarterly Review of Economics and Business*, 31, 64–80.

Bates, T. 1993. *Banking on Black Business: The Potential of Emerging Firms for Revitalizing Urban Economies*. Washington D.C.: Joint Center for Political and Economic Studies.

Boddy, M. 1976a. The structure of mortgage finance. *Transactions of the Institute of British Geographers*, NS 1, 58–71.

Boddy, M. 1976b. Building societies and owner occupation. In M. Edwards et al. (eds). *Housing and Class in Britain*. London: CSE.

Boddy, M. 1980. *The Building Societies*. London: Macmillan.

Boddy, M. 1981. The property sector in late capitalism: the case of Britain. In M. Dear and A. Scott (eds). *Urbanization and Urban Planning in Capitalist Society*. London: Methuen.

Bowring, F. 1998. LETS: an eco-socialist initiative. *New Left Review*, 23, 91–111.

Caskey, J. P. 1994. Bank representation in low-income and minority urban communities. *Urban Affairs Quarterly*, 29, 617–38.

Castells, M. 1989. *The Informational City: Information Technology, Economic Restructuring, and the Urban-Regional Process*. Oxford: Blackwell.

Christopherson, S. 1993. Market rules and territorial outcomes: the case of the United States. *International Journal of Urban and Regional Research*, 17, 274–88.

Clark, G. L. 1997. Pension funds and urban investment: four models of financial intermediation. *Environment and Planning A*, 30, 997–1015

Clarke, S. 1988. *Keynesianism, Monetarism and the Crisis of the State*. Aldershot: Edward Elgar.

Corbridge, S. and Agnew, J. 1991. The US trade and budget deficits in global perspective: an essay in geopolitical economy. *Society and Space*, 9, 71–90.

Cox, R. W. 1987. *Production, Power, and World Order: Social Forces in the Making of History*. New York: Columbia University Press.

Daniels, P. W. 1993. *Service Industries in the World Economy*. Oxford: Blackwell.

Davies, G. 1994. *A History of Money: From Ancient Times to the Present Day*. Cardiff: University of Wales Press.

Davis, M. 1992. Who killed LA? A political autopsy. *New Left Review*, 197, 3–28.

Dodd, N. 1994. *The Sociology of Money*. Cambridge: Polity.

Dunkling, L. and Room, A. 1990. *The Guinness Book of Money*. London: Guinness.

Dymski, G. and Veitch, J. 1996. Financial transformation and the metropolis: booms, busts, and banking in Los Angeles. *Environment and Planning A*, 28, 1233–60.

Furnham, A. and Argyle, M. 1998. *The Psychology of Money*. London: Routledge.

Graham, S. and Marvin, S. 1996. *Telecommunications and the City: Electronic Spaces, Urban Places*. London: Routledge.

Grown, C. and Bates, T. 1992. Commercial bank lending practices and the development of black-owned construction companies. *Journal of Urban Affairs*, 14, 25–41.

Harvey, D. 1969. *Explanation in Geography*. London: Edward Arnold.

Harvey, D. 1973. *Social Justice and the City*. London: Edward Arnold.

Harvey, D. 1982. *The Limits to Capital*. Oxford: Blackwell.

Harvey, D. 1989. *The Condition of Postmodernity: An Enquiry into the Origins of Cultural Change*. Oxford: Blackwell.

Harvey, D. and Chaterjee, L. 1974. Absolute rent and the restructuring of space by governmental and financial institutions. *Antipode*, 6, 22–36.

Helleiner, E. 1993. When finance was the servant: international capital movements in the Bretton Woods order. In P. Cerny (ed). *Finance and World Politics: Markets, Regimes and States in the Post-Hegemonic Era*. Aldershot: Edward Elgar, 20–48.

Hudson, A. 1998a. Placing trust, trusting place: on the social construction of offshore financial centers. *Political Geography*, 17, 915–37.

Hudson, A. 1998b. Reshaping the regulatory landscape: border skirmishes and the Bahamas and Cayman offshore financial centers. *Review of International Political Economy*, 5, 534–64.

Hudson, A. 1999. Offshores onshore: new regulatory spaces and real historical places in the landscapes of global money. In R. Martin (ed). *Money and the Space Economy*. Chichester: Wiley, 139–54.

Jones, A. 1998. (Re)producing gender cultures: theorizing gender in investment banking recruitment. *Geoforum*, 29, 451–74.

Lash, S. and Urry, J. 1994. *Economies of Signs and Space*. London: Sage.

Lee, R. 1996. Moral money? LETS and the social construction of local economic geographies in Southeast England. *Environment and Planning A*, 28, 1377–94.

Leyshon, A. 1995. Geographies of money and finance I. *Progress in Human Geography*, 19, 531–43.

Leyshon, A. 1997. Geographies of money and finance II. *Progress in Human Geography*, 21, 278–89

Leyshon, A. 1998. Geographies of money and finance III. *Progress in Human Geography*, 22, 433–46.

Leyshon, A. and Thrift, N. 1996. Financial exclusion and the shifting boundaries of the financial system. *Environment and Planning A*, 28, 1150–56.

Leyshon, A. and Thrift, N. 1997. *Money/Space: Geographies of Monetary Transformation*. London: Routledge.

Leyshon, A. and Thrift, N. 1999. Lists come alive: electronic systems of knowledge and the rise of credit-scoring in retail banking. *Economy and Society*, 28, 434–66.

Leyshon, A. and Tickell, A. 1994. Money order?: The discursive construction of Bretton Woods and the making and breaking of regulatory space. *Environment and Planning A*, 26, 1861–90.

Leyshon, A., Thrift, N., and Pratt, J. 1998. Reading financial services: texts, consumers and financial literacy. *Environment and Planning D: Society and Space*, 16, 29–55.

Leyshon, A., Thrift, N., and Tommey, C. 1989. The rise of the British provincial financial center. *Progress in Planning*, 31, 151–229.

Lowe, M. and Short, J. 1990. Progressive human geography. *Progress in Human Geography*, 14, 1–11.

Martin, R. 1999. The new economic geography of money. In R. Martin (ed). *Money and the Space Economy*. Chichester: Wiley, 3–27.

Martin, R. and Minns, R. 1995. Undermining the financial basis of regions: the spatial structure and implications of the UK pension system. *Regional Studies*, 29, 125–44.

McDowell, L. 1994. Social justice, organizational culture and workplace democracy: cultural imperialism in the City of London. *Urban Geography*, 15, 661–80.

McDowell, L. 1997a. *Capital Culture*. Oxford: Blackwell.

McDowell, L. 1997b. A tale of two cities? embedded organizations and embodied workers in the City of London. In R. Lee and J. Wills (eds). *Geographies of Economies*. London: Arnold, 118–29.

McDowell, L. and Court, G. 1994a. Missing subjects: gender, power, and sexuality in merchant banking. *Economic Geography*, 70, 229–51.

McDowell, L. and Court, G. 1994b. Gender divisions of labor in the post-Fordist economy: the maintenance of occupational sex segregation in the financial services sector. *Environment and Planning A*, 26, 1397–1418.

McDowell, L. and Court, G. 1994c. Performing work: bodily representations in merchant banks. *Environment and Planning D: Society and Space*, 12, 727–50.

North, P. 1999. Exploring the politics of social movements through "sociological intervention": a case study of local exchange trading schemes. *The Sociological Review*, 46, 564–82.

O'Brien, R. 1991. *Global Financial Integration: The End of Geography*. London: Pinter.

Overbeek, H. 1993. *Restructuring Hegemony in Global Political Economy: The Rise of Transnational Neo-Liberalism in the 1980s*. London: Routledge.

Peet, R. 1998. *Modern Geographical Thought*. Oxford: Blackwell.

Purdue, D., Dürrschmidt, J., Dowers, P., and O'Doherty, R. 1997. DIY culture and extended milieux: LETS, veggie boxes and festivals. *The Sociological Review*, 45, 645–67.

Roberts, S. 1994. Fictitious capital, fictitious spaces: the geography of offshore financial flows. In S. Corbridge, N. Thrift and R. Martin (eds). *Money, Power and Space*. Oxford: Blackwell, 91–115.

Roberts, S. 1995. Small place, big money: The Cayman Islands and the international financial system. *Economic Geography*, 71, 237–56.

Squires, G. D. (ed.) 1992. *From Redlining to Reinvestment: Community Responses to Urban Disinvestment*. Philadelphia: Temple University Press.

Thorne, L. 1996. Local exchange trading systems in the United Kingdom. *Environment and Planning A*, 28, 1361–76.

Thrift, N. 1994. On the social and cultural determinants of international financial centers: the case of the City of London. In S. Corbridge, N. Thrift and R. Martin (eds). *Money, Power and Space*. Oxford: Blackwell, 327–55.

Thrift, N. 1996. *Spatial Formations*. London: Sage.

Thrift, N. and Leyshon, A. 1988. "The gambling propensity": banks, developing country debt exposures and the new international financial system. *Geoforum*, 19, 55–69.

van der Pijl, K. 1984. *The Making of an Atlantic Ruling Class*. London: Verso.

van der Pijl, K. 1989. Ruling classes, hegemony, and the state system. *International Journal of Political Economy*, 19, 7–35.

Whatmore, S. 1999. Hybrid geographies: rethinking the "human" in human geography. In D. Massey, J. Allen and P. Sarre (eds). *Human Geography Today*. Cambridge: Polity, 22–39.

Williams, C. C. 1998. Helping people to help themselves: the potential of LETS. Political Economy Research Centre, Policy Paper Number 14, University of Sheffield.

Williams, C. C. 1999. The potential of LETs in tackling social exclusion amongst young people. In S. Garcia (ed). *Inclusion Through Participation: Policies to Promote Social Inclusion*. Brussels: European Commission, DG12.

Williams, P. 1976. The role of financial institutions and estate agents in the private housing market: a general introduction. Centre for Urban and Regional Studies, Working Paper No. 9, University of Birmingham.

Yergin, D. and Stanislaw, J. 1998. *The Commanding Heights: The Battle Between Government and the Market Place that is Remaking the Modern World*. New York: Simon & Schuster.

Chapter 27

The Political Economy of International Labor Migration

Helga Leitner

While migration has been part of human history from the beginning, both the sheer numbers of people on the move in different parts of the world and their real or alleged impact have induced scholars to characterize the contemporary period as "the age of migration" (Castles and Miller, 1993). According to a United Nations report, approximately 1 billion people migrated in the last half of the 1980s (United Nations, 1994). Millions of workers from less developed countries have come to the industrialized nations in search of work opportunities and better living conditions, as either temporary labor migrants or legal or illegal immigrants, affecting the demographic and cultural make-up and economic development in both sending and receiving areas. At the same time, tens of millions of refugees have fled political persecution, civil wars, ethnic strife, and ecological disasters across national boundaries; some coming to industrialized countries, but most to other poor developing countries, generally nearby. The magnitude of refugee flows is outstripping the capacity of the receiving countries to provide for them.

Besides population movements across national boundaries, migration within nation-states, at varying geographic scales, continues unabated. Large-scale population movements from rural to urban areas, especially in Asia and Latin America, have spurred massive urbanization at rates that dwarf the urbanization experience of the USA and Europe in the late nineteenth and early twentieth centuries (Parnwell, 1993; Fan, 1996). Retirement migration to the "sunbelt" in the southern United States has accelerated growth and changed the demographic make-up of sunbelt cities (McHugh and Mings, 1996). Lack of employment opportunities in economically declining regions in the advanced industrialized countries of the West has engendered large-scale inter-regional migration to economically growing regions (Champion, 1995). Finally, the migration of families from central cities to the suburban fringe, in search of larger houses and greenfield environments, continues, particularly in the United States. Since most of the families moving to suburbs are white and better off, this has resulted in increasing social polarization within US metropolitan areas (O'Loughlin and Friedrichs, 1996).

Empirical research on migration, whether internal or international, has demonstrated that migration does not simply imply a unidirectional, permanent flow between origin and destination, but rather often consists of complex temporal and spatial patterns of moves. Migration may occur in *stages*, involving several intermediate destinations over an extended period of time, as migrants move, for example, from a village to a medium-sized city to a capital city. It may be *cyclical* (or circular), with migrants moving in relatively regular intervals in both directions between areas of destination and origin. This is frequently the case with rural to urban migration in less developed countries, where movements may be synchronized with the agricultural cycle (Parnwell, 1993). Migration may also change from being temporary into permanent migration, as has happened in the case of the migration of many foreign guestworkers from the Mediterranean basin to Western Europe in the second part of the twentieth century. Beginning in the late 1950s, as a temporary migration of single people recruited by the receiving countries to fill job vacancies on their labor markets, by the 1970s the guestworker program had turned into a quasi-permanent migration of families; employers wanted the foreign workers to stay, and workers brought their families to join them (Leitner, 1995). Finally, migration may involve *return* migration to the origin after a short or long sojourn. The extent of return international migration is often overlooked. In the United States, for example, roughly 200,000 former immigrants were leaving each year in the 1980s (reducing the net inflow by about one-fifth), the largest proportion being recent entrants.

These examples illustrate how migration escapes easy categorization and classification into such categories as permanent or temporary or short or long-distance migration often found in the literature. Rather, the spatial and temporal dimensions of migration are complex and the boundaries are fuzzy.

In the recent past the distinction between international and internal migration also has become blurred as a result of changes in the territorial structure of governance. For example, the integration of European nation-states in the European Union and the associated freedom of movement among member states have changed the migration of EU citizens from an international to an internal migration. Generally, however, the distinction between international and internal migration continues to be significant, since the two types of migration are governed by different principles. In most countries of the world there exists a legal right to freedom of movement within the state territory. In contrast, freedom of entry into another state territory is universally denied as a legal right (Hull, 1987; Leitner, 1995). The entry of foreign migrants into the territory of a nation-state is governed by immigration laws and policies, which regulate how many and who gets admitted into the national territory, thus influencing the volume and nature of migration flows across national borders (Leitner, 1995).

Nevertheless, there are commonalities in both the origins and selectivity of international and national population movements, although that is generally not born out in the literature. Instead, scholarly research on international and internal migration has been evolving separately as two different bodies of literature, with little to no dialogue occurring between the two, with the notable exceptions of Parnwell (1993) and Skeldon (1997). This has been most obvious in the area of migration theory, characterized by the absence of mutual references, despite the fact that both bodies of literature draw on similar economic and social theories as explanatory frameworks.

While I feel that transgressing the boundaries between the two bodies of literature is desirable and could stimulate progress in the field, this chapter is also guilty of perpetuating the division by engaging primarily in a discussion of the international labor migration literature. This reflects my own expertise in international labor migration to the advanced industrialized countries of Europe. Within the literature on international migration, this chapter does attempt to overcome another common division, however, between studies focusing either on the developing and/or the developed world. This is particularly important in the light of an increasingly global economy, which has heightened not only the interchange of goods and capital, but also the transfer of population across the globe, in the process reshaping societies, politics, and livelihoods in different countries and regions of the world.

The purpose of this chapter is to critically examine competing explanations for both the origins and the impacts of international migration, and their relevance in helping us understand the new realities of international migration, specifically labor migration. The chapter begins with an account of some new realities characterizing the past 40 years of international migration. This is followed by a discussion of competing theoretical perspectives on international migration and their relative explanatory power. I argue that some of these theories are too simplistic, reducing the origins of international migration to either individual choice or the structure of the capitalist world economy, and ignoring the complex combination of individual actions and social structures that engender international migration; that some are too economistic, ignoring for example the role of states in inducing and facilitating international migration; and that others are not contextual, ignoring the specific spatial and historical context within which international migration takes place.

The final two sections review empirical studies on the impacts of international migration on receiving and sending areas, critically examining competing claims about the relationship between international migration, development, and inequality, and the implications of these for policy measures. These sections make clear that in an increasingly interdependent world, impacts on receiving and sending regions cannot and should not be viewed in isolation, and that we need to recognize the complex ways in which economic, political, and cultural developments in these areas depend on one another.

New Realities in International Migration

Examining the nature of international migration during the past 40 years, scholars have identified the following new realities: acceleration, globalization, and feminization (Castles and Miller, 1993). I would add to this another tendency – an increasing institutionalization of international population movements.

Acceleration refers to a significant growth in the volume of international migration. The United Nations Population Fund (UNPF) has estimated that at least 100 million international migrants live outside the countries where they were born (without counting immigrants who have already become citizens in their country of destination). Of these, approximately two-thirds are economic migrants, 20 million are fleeing violence or environmental deterioration, and about 17 million are political refugees (*Asian Migrant*, 1993, cited in Ball, 1997, p. 1611).

Globalization refers to how more and more countries of the world are sending and/or receiving international migrants (Castles and Miller, 1993). Associated with this trend has been a remarkable increase in the national diversity of immigrants to the major destination areas. For example, Western European countries have seen increasing immigration from Africa and the Caribbean, and more migrants to North America come from different parts of Asia. This means that the advanced industrialized countries of Europe and North America, in particular, receive immigrants with increasingly diverse national origins and cultural backgrounds.

Notwithstanding such trends, there continues to be a clear *regional dimension*, with particular sending countries tending to dominate flows to the major destination areas – the global centers of economic wealth. These destinations include the highly advanced countries of Western Europe and North America, Australia, and New Zealand, the oil-rich states of the Middle East, and most recently Japan, Singapore, Hong Kong, South Korea, and Taiwan (Salt, 1989). The majority of immigrants to France, for example, still come from North Africa (Morocco, Algeria, and Tunisia), those entering the USA come from Mexico and Asia, and the oil-rich states of the Middle East largely receive foreign labor from South and Southeast Asia. Within Asia, patterns of flows are from the relatively less developed, labor-exporting countries (the Philippines, India, Bangladesh, Pakistan, and Indonesia) towards those with higher per capita incomes, low rates of natural increase, and declining labor force growth rates (Japan, Singapore, Hong Kong, South Korea, and Taiwan) (Tyner, 1998).

The *feminization* of international labor migration refers to the increasing role of women in international migration, particularly the growing number of female-dominated international labor flows. For example, of the 500,000 Filipino migrant workers deployed abroad annually, about 40 percent are women, employed primarily in service sector jobs, from domestic workers to nurses and waitresses (Tyner, 1996). This development has necessitated a revision of the generalizations that males are more likely to migrate internationally than females, and that males usually dominate the early stages of international migration, to be followed later by their dependents – women and children. Research has shown that patterns of female migration do not simply mirror those of male migration, with men as the initiators of migration and women as mere followers (Simon and Brettel, 1986). Gendered patterns of international labor migration reveal not only the gendered nature of labor markets, but also the worldwide existence of sexist stereotypes among both importers and exporters of foreign labor (Tyner, 1996, p. 408). Tyner (1996) discloses how, within the highly institutionalized Filipino labor-export system, individuals and institutions employ specific sterotypical representations of men and women to selectively market and recruit workers. For example, promotional materials designed to attract potential employees depict men in professional and construction occupations, while women are portrayed as nurses and domestic workers, thus defining the type of work for which women and men are suited. Fincher et al. (1994) reveal that the selection of immigrant settlers in Australia has been based on different expectations about men and women, and about masculinity and femininity, which in turn had differential effects on the admission and settlement of men and women.

The *institutionalization* of international migration can be considered a fourth important tendency in international labor migration. Both public and private sector institutions have been playing a heightened role in influencing who and how many will migrate across national borders, and from which origins to which destinations.

Nation-states have increased their role through more elaborate immigration control policies and through state-sponsored import and export policies (Cornelius et al., 1994; Goss and Lindquist, 1995). During the past ten years, major receiving states in Europe and North America have overhauled and extended their institutional frameworks for immigration control, immigration laws, and policies. The outcome of this legal and institutional reform has been a tightening of national controls on entry and over immigrant rights. Most recently we have seen attempts to coordinate immigration control policies at the supra-national scale, as evidenced in the newly emerging supra-national framework for immigration control for the European Union (Leitner, 1997).

At the same time major sending states, again in cooperation with private capital, are today engaged in the global marketing of workers, creating what some have called an international labor migration industry – managing, regulating and organizing the global flow of migrants (Tyner, 1998). In addition, the illegal transfer of people across national borders has become a highly organized and flourishing business activity (Salt and Stein, 1997). European and North American states in particular have become increasingly concerned about people-smuggling by highly organized criminal rings. According to an article in *The Economist* (1999), the smuggling of people into the European Union, run by Mafia rings based in such places as Istanbul, Tirana, and other cities in eastern Europe, has grown into a lucrative business in recent years, possibly worth $3–4 billion a year in Europe alone.

Theoretical Perspectives on International Migration

Much conventional theorizing on international migration has concentrated on the economic factors that determine "voluntary" international migration, on the economic impacts on both sending and receiving areas, and on immigrant incorporation into the receiving society. The last topic will not be addressed here, but see Mitchell (this volume).

The most pervasive framework employed by scholars, either explicitly or implicitly, views migration as the outcome of rational economic decisions by individual actors. According to neoclassical economic theory, labor markets are the primary mechanisms by which international flows of labor are induced. Individual rational actors decide to migrate when a cost–benefit calculation leads them to expect a positive net return in terms of higher income. At the macro-scale, international migration only occurs if there are differences in wage rates between countries. Neoclassical theory also predicts that the ensuing migration, from low-wage to high-wage countries, will progressively eliminate wage differentials, thereby eventually ending labor movements (Massey et al., 1993, 1994).

Although neoclassical theories have strongly influenced intellectual and political discourse on the origins and impacts of international migration, the theoretical reasoning as well as the empirical accuracy of these theories have become increasingly challenged. I argue that these theories have shaped the intellectual discourse on international migration primarily because of their theoretical simplicity, not because of their empirical accuracy in explaining the volume of international migration flows or migration decisions. (For an attempt to evaluate these theories in the light of empirical evidence see Massey et al., 1994.)

The "new economics of migration" framework, while still assuming rational actors, views migration as a rational economic decision taken by households rather than individuals. Through international migration, households attempt to reduce risks to family income and well-being and to ameliorate their sense of relative deprivation, rather than seeking to maximize income (Stark and Taylor, 1991). In this view, wage differentials between countries are not a prerequisite for international migration, because other conditions in the sending areas, such as the absence of a social safety net or a household's position in the local income distribution, are important in the household migration decision. This approach also goes beyond the neoclassical model in its conceptualizations of the impact of migration on the sending areas, by considering the effects of remittances, money sent back home by migrants, on the accumulation of income-producing assets, such as livestock and equipment, in sending areas. It is anticipated that remittances will allow poor families in the sending areas to gain access to scarce capital, which will translate into productive investment (Massey et al., 1994; Taylor, 1992).

Although this approach is less simplistic than neoclassical economic theories in its assumptions about the functioning of markets and migration decisions, and incorporates the effects of income transfers, it still conceptualizes the migration decision as a rational choice, following a considered evaluation of options available in terms of economic benefits. This approach has also been criticized for its assumption that households have homogenous interests, thus ignoring gendered power relations within households and how household members often pursue their own individual interests rather than that of the collectivity (Goss and Lindquist, 1995).

Since the late 1970s, new approaches have challenged the theoretical reasoning of both these theories. In contrast to their largely micro focus, the new approaches are largely macro in focus. They emphasize the role of economic structures – particularly the workings of the capitalist economy – in inducing population movements. These theorists acknowledge the importance of disparities in income, life changes, and employment opportunities, but feel it is necessary to proceed further in the chain of causation to theorize and investigate how these conditions have been generated.

Dual labor market theory argues that international migration is largely caused by a permanent demand for immigrant labor that is inherent to the economic structure of advanced industrial economies, which have segmented labor markets (Piore, 1979). Segmented labor markets are characterized by a capital-intensive primary sector, providing skilled, secure, high-paying jobs with good benefits and working conditions; and by a labor-intensive secondary sector, where workers hold unskilled, insecure, and low-paying jobs with few benefits and poor working conditions. Since conditions in the secondary sector make it unattractive to native workers, employers turn to foreign labor to fill shortfalls in demand. There exists significant empirical evidence that labor markets in some major receiving areas, notably the USA, are indeed segmented as suggested by Piore, and that many labor migration movements arise as a result of labor recruitment by the receiving countries. Yet, as Massey et al. (1994) maintain, some urban labor markets in receiving areas may be further segmented into an immigrant enclave, and recruitment represents only one of several inducements to migrate. This leads them to conclude that "it is not clear that labor market segmentation explains all or even most of the demand for immigrants" (Massey et al., 1994, p. 721). Another problem with this theory is that it assigns primary explanatory power to the labor demand conditions in receiving areas, at the expense of a fuller consideration of the processes linking sending and receiving countries.

Building on Immanuel Wallerstein's "world systems theory" (Wallerstein, 1974), a variety of theorists link the origin of international migration to the structure of the global economy rather than the bifurcation of the labor market within advanced industrial economies. They see international migration as a consequence of economic globalization and capitalist market penetration into ever larger proportions of the globe (Portes and Walton, 1981; Sassen, 1988; Zolberg, 1989). In this view, it is the penetration of capitalist economic relations into peripheral societies that creates a mobile population prone to migrate abroad. For example, "traditional" economies, such as peasant farming in the periphery, are disrupted as land, production, and labor come under the influence and control of, for example, Western agrobusinesses. The outgrowth is the displacement of large numbers of people from such traditional livelihoods as peasant farming.

Disruptions resulting from capitalist market penetration do not generate emigration by themselves, however. According to Sassen (1988, p. 20), who examines the impact of foreign direct investment in peripheral regions as a leading indicator of capitalist market penetration, "foreign direct investment is not a cause but a structure that creates certain conditions for emigration to emerge as an option." In order to explain how the pool of potential migrants created by capitalist market penetration will result in large-scale emigration, she contends that we need to theorize and investigate the variety of different economic, political, and cultural processes that link sending and receiving countries in complex ways.

Linkages between sending and receiving areas at different scales (i.e. migrant networks linking individuals or groups; political and economic relations between countries; and transnational linkages between institutions) have received much attention in empirical studies of migration flows. Numerous case studies demonstrate that the existence of migrant networks – sets of interpersonal relations that connect migrants, former migrants, and non-migrants in the origin and destination areas through ties of kinship, friendship, and other communal ties – is crucial in influencing the probability of migration and in creating a self-sustaining stream of migrants between areas of origin and destination (Findley, 1987; Massey et al., 1990; Gurak and Caces, 1992). Goss and Lindquist (1995), however, contend that the role of migrant networks in promoting international migration is inadequately theorized. "It is still not clear how these networks operate as social entities beyond the sum of the individual relationships of which they are constituted" (Goss and Lindquist, 1995, p. 331). Employing structuration theory, which stresses the interdependencies between individual behavior and societal structures, they suggest that migrant networks be "conceived as migrant institutions that articulate the individual migrant and the global economy, 'stretching' social relations across time and space to bring together the potential migrant and the overseas employer" (1995, p. 335).

Besides the informal institution of migrant networks, contemporary large-scale international migration also has been increasingly structured by formal national and international, public and private institutions, which are at least as important in influencing the nature of international migration according to some observers. Most attempts to theorize the role of institutional structures and relations in influencing international migration flows recognize that these are themselves embedded in global economic and political structures and transformations (Richmond, 1994; Goss and Lindquist, 1995). For example, Findlay et al. (1996) demonstrate how skilled international migration into Hong Kong is related to the organization and culture of large corporations, within a global context of changing forms of production and accumulation.

Similarly, state and private institutions in parts of the less developed world, particularly in South and Southeast Asia, have shown to be actively involved in the recruiting and global marketing of their workers to developed or rapidly developing economies. A particularly well-studied example is the Philippines, which has been so successful in the global deployment of its labor that it is often considered to be a model for states wishing to develop a labor-export industry (Goss and Lindquist, 1995; Tyner, 1996; Ball, 1997, p. 1616). In the Philippines, the recruitment and marketing process is highly regulated by the state, which works together with private recruitment agencies acting as labor brokers bringing together workers wanting to work overseas with overseas employers seeking temporary workers. The state-run Philippine Overseas Employment Agency (POEA) not only controls individual access to overseas employment opportunities and selects migrants, but also requires Filipino migrants to funnel a portion of their overseas earnings through government-approved banks (Goss and Lindquist, 1995, pp. 339–40). Through this labor export system, the Filipino state and private recruitment agencies have a major influence on the nature and direction of international migration flows and the channeling of remittances. The increased engagement of developing countries in exporting labor is related to the heightened opportunities for international contract labor. This in turn is a "result of sustained socioeconomic development in the [Newly Industrializing Economies] and their demand for service sector labor, the construction boom and rapid economic development in the Middle Eastern oil-producing economies, together with continued underdevelopment of the non/oil-producing and non-industrializing Third World" (Goss and Lindquist, 1995, p. 336).

Thus, in contrast to the neoclassical economic and new economics of migration approaches, structural and structurationist approaches stress, to varying degrees, the role of the public and private institutions that link sending and receiving areas in complex ways, in generating and directing transnational population flows. More importantly, the constitution of these linkages is seen as located within global economic structures and processes and in the articulation of these with individual migrants. Finally, the structural and structurationist approaches do not assume that people across space and time behave the same way, but rather acknowledge the significance of the spatial and historical context for understanding origins and impacts of international migration.

All of the theoretical frameworks discussed up to this point have recently been criticized for being insufficiently attentive to the experience of migration, to differences in migration experiences among individuals and social groups, and to the constructed nature of such social categories as migration and migrants (Silvey and Lawson, 1999). Drawing on poststructural social and cultural theory (see Gibson-Graham, this volume), migrants are conceptualized as conscious, diverse human subjects, whose actions and identities are rooted in the experience of everyday life in a specific space and time context, rather than simply as rational economic decisionmakers, or as victims of capitalist development, recruitment agencies, and the state. Focusing on the migrant experience also implies accentuating the histories and interpretations of migrants themselves, taking more seriously the voices of migrants and the diversity of their experiences (Halfacree and Boyle, 1993; Vandsemb, 1995). It is suggested that analysis of migrants' narratives of their migration history will expand our understanding of the complex economic, cultural, and political processes and transformations, which both induce and are set in motion by international migration. Concentration on migrants' narratives may also enable us to better capture the spatial and temporal complexities of the migration process mentioned above.

This re-conceptualization of the category "migrant" not only draws attention to differences among migrants along age, class, gender, and nationality lines – indeed the selectivity of migration along these axes of difference has long been recognized – but also focuses on how power relations, themselves constructed around differences of gender, age, race, and nation, shape migration decisions, processes, and experiences (Fincher et al., 1994; Kofman and England, 1997; England and Stiell, 1997; Fincher, 1997). For example, as noted above, Fincher et al.'s (1994) work on gender equity in Australian immigration policy illustrates how the selection of immigrant settlers in Australia has been based on different gendered expectations, with differential effects on admission and settlement by gender. Similarly, Tyner (1996) recounts how differential representations of men and women's work in the Filipino labor export system construct women and men in ways that influence who migrates where.

These new theoretical perspectives on international migration not only complicate the questions posed by traditional migration theories about who migrates where, but more importantly have challenged previously hegemonic epistemologies and have generated new questions and ideas. Particularly noteworthy, in my opinion, is the question of the role of discourses in constructing international migration and migrants, and the political, economic, and social implications that this involves (Walker, 2000). In the contemporary immigration debate in the USA, for example, the dominant political discourse constructs immigrants as a burden on the US economy and welfare state, and has been used to justify tighter immigration control and restricting the rights of immigrants. As will be discussed below, discourses about immigration as a burden have also drawn selectively on empirical research supporting such arguments, against the majority opinion in the academic community. This example illustrates that what we claim to understand and explain is infused with power relations, is actively contested, is changing over time, and is interpreted differently depending on the positionality and interests of those involved (White and Jackson, 1995). The same logic can be applied to understanding the changing influence of the various theoretical perspectives on international migration reviewed here.

Impacts of International Migration on Receiving and Sending Areas

Debating impacts on receiving areas

Much of the international and national political and public discourse on international migration has focused on the impact of immigration on the receiving economies and societies. This topic has assumed such political and economic currency that it was even on the agenda of the 1991 annual "G-7" meeting, of the heads of state of the seven most powerful industrial countries. These discussions have been primarily motivated by concerns about the negative impacts of large-scale immigration of people from the less developed countries on the employment and wage prospects of native workers and on economic development. "The presumption that immigrants have an adverse impact on the labor market continues to be used as a key justification for policies designed to restrict the size and composition of immigrant flows into the United States" (Borjas, 1998, p. 217).

For the United States, an enormous amount of research has been completed since the 1980s on the impact of immigration on the labor market. By contrast, little attention has been paid to the contributions of immigrants to the US economy. Most of the econometric studies (both cross-sectional and longitudinal) analyzing the relationship between the overall supply of immigrants and availability of jobs or wages have produced no strong evidence that immigration overall has a major adverse impact on the earnings and job opportunities of natives in the US labor market (Fix and Passel, 1994; Borjas, 1998). Studies examining the impact of immigration on specific sub-populations, such as African Americans and low-skilled workers in urban areas, have identified differential impacts on wages and employment for these sub-populations. Enchautegui (1993) found that African Americans in high-immigration areas fared better in terms of wage and employment growth than those in low-immigration areas. Altonji and Card's (1991) analysis of US metropolitan areas indicates that while immigration increases the share of the less-skilled African American labor force that is employed, it reduces the weekly earnings of less-skilled African American men and women. Generally, regardless of the native population sub-groups considered, most empirical evidence suggests that the earnings of native population sub-groups are barely affected by the entry of immigrants into the local labor market (Borjas, 1998). More important for explaining changes in earnings of the native population are such factors as economic restructuring at the national and local scale.

Several scholars have advanced the argument that the impact of immigrants on earnings and employment opportunities of natives is dampened by the fact that native-born workers, particularly the unskilled, respond to the economic pressures introduced by immigrants by leaving areas of high immigrant concentration (Filer, 1992; Frey, 1995). Wright, Ellis, and Reibel's (1997) study of the effects of immigrants on internal migration in the largest metropolitan areas in the USA challenges this argument, however, demonstrating that the net outmigration of native (including unskilled) workers from large metropolitan areas is more likely to be due to economic restructuring than to immigration. In their analysis, immigration flows are positively related to the net migration of the native born, prompting them to state that: "Conclusions such as immigration causes 'demographic balkanization' miss the mark and do nothing to muffle the rhetoric in this already overheated immigration debate" (Wright, Ellis, and Reibel, 1997, p. 252).

The other issue at the center of the immigration debate in the USA has been the public costs of immigrants – in particular, costs imposed on the welfare system. Adding to the fervor of this debate has been a set of studies in the early 1990s attributing enormous public costs to immigrants (Huddle, 1993). According to Fix and Passel (1994), however, these studies grossly overstate the public costs of immigrants, because their calculations of service costs and of the tax contributions of immigrants are erroneous. While acknowledging the difficulties involved in estimating the net cost of immigrants to the public sector, they contend that, when all levels of government are considered together, immigrants generate significantly more in the taxes they pay than they cost in services received. Indeed most national-level studies have found that immigrants are not a fiscal burden.

At the state and local level, however, the picture is different. In some states with high levels of immigration, immigrants can constitute a net fiscal burden to state and local authorities (Rothman and Espenshade, 1992). This uneven distribution of costs across levels of government has been further enhanced by new federal legislation passed in 1996. Provisions in the 1996 welfare reform bill deny many legal immigrants access to most federal public assistance programs (e.g. food stamps and SSI – a program providing assistance for the elderly and people with disabilities) for the first five years of residence (for further details see Espenshade, 1998). These provisions potentially shift the costs of aiding immigrants in need of assistance to state and local governments. In response to this, those states and cities receiving large numbers of needy immigrants have been suing the federal government for reimbursement of costs arising from the new federal policies.

Notwithstanding the majority of studies indicating that the overall economic effects of immigration are generally positive, citizens continue to harbor negative views and beliefs about the impact of recent immigration on the US welfare state and economy. These beliefs, combined with anxieties about the integration of increasingly culturally different immigrants into the host society, have been fueled by a nativist and anti-foreigner political discourse. The anti-foreigner/anti-immigration public and political discourse in the USA, and in the highly advanced industrialized countries of Europe, has marginalized discourses in sections of academia highlighting the positive contributions of immigrants to the wealth of the receiving societies, and the negative and positive consequences of international labor migration for the migrants themselves (Ley, 1999). With respect to the latter, studies of labor migrants from the Philippines working oversees have documented, for example, the well-known phenomenon of deskilling, with Filipino doctors and teachers working as maids, janitors, and in other menial jobs in destination countries (Tyner, 1996).

Examining impacts on sending areas

While major receiving countries have focused on the burden of immigrants on the national economy and state in recent years, sending countries have emphasized the contributions of emigrants to national economic development. In particular, remittances – the transfer of cash earnings of migrants from the receiving to the sending country – are seen as important to economic development in the sending areas. These have been found to be considerable for some less developed Asian countries, such as Pakistan where estimates indicate that remittances contributed almost 9 percent of the GDP in the mid-1980s. Many less developed countries have also come to depend on remittances as a source of foreign exchange. In Bangladesh, for example, remittances contribute 25 percent of foreign exchange earnings (for further details see United Nations, 1995).

Scholars generally agree that remittances help to redistribute income among countries and that they have increased the availability of money in sending countries. They disagree, however, as to whether remittances improve the national and local economies of sending areas (Jones, 1998a). Some scholars have raised a concern about the increasing reliance and dependence of small less developed economies on the money supplied through remittances, at the expense of indigenous economic development initiatives (Bertram and Watters, 1985). Others have maintained that such concerns underestimate the productive nature of remittance investment in local businesses, creating job opportunities, wealth, and thus development in areas of origin (Addleton, 1992; Conway and Cohen, 1998). Clearly these contrasting findings are a result of the complex web of conditions shaping migration and remittances. These include the characteristics of migrants and of job opportunities and earnings in destination areas, the purposes for which remittances are spent in sending areas, and the structure and condition of the political economy in the sending areas (Jones, 1998b).

There is also disagreement regarding the impact of remittances on economic and spatial inequalities in the sending countries. Some studies have found that remittances tend to be funneled towards families and towns in sending areas that are already better off, thus increasing social and regional inequality (Atalik and Beeley, 1993). In contrast, others have suggested that remittances reduce rural–urban income inequalities because they are spent in the low-income rural areas of out-migration (Griffin, 1976). According to Jones (1998b) these contradictory findings result in part from the differences in the geographic scale at which analyses measuring inequalities have been carried out (inter-regional, inter-urban, rural-urban, and inter-familial scale), and in part from failures to consider the influence of a place's stage of emigration (defined as the duration and quantity of emigration from a particular place). Using evidence from a case study in central Zacatecas, Mexico, he argues that in the first phase of emigration inter-familial inequities in communities decrease up to a point, but rise again during the advanced stage of emigration to the USA. "At the scale of the family, better-off families improve their status at the expense of poorer families, with advanced stages of U.S. migration. In contrast, at the rural-urban scale, advanced stages of migration result in rural places improving their income positions vis a vis urban places" (Jones, 1998b, pp. 22–3).

Remittances are only one way in which migration affects sending areas. The selectivity of migration by age and skill level – the majority of international migrants are in the economically productive age and the most dynamic members of their communities – has generally been seen as a drain on sending areas, further undermining the potential for development in these areas. According to some studies, however, these losses do not appear to have a major negative impact on the labor market in sending areas (Skeldon, 1997, p. 159).

Transnational Perspectives: Linking Sending and Receiving Areas

Debates about the impact of migration on sending or receiving regions can easily shift our attention away from the complex ways in which developments in these regions depend on one another. This has not been a central focus of research on the impacts of international migration. Such research is beginning to emerge, however, documenting the importance of these interdependencies and thinking about their implications for migration processes and policies. This is being examined both at the micro and macro scale.

Migrants as transnational economic actors

Recent studies indicate that immigrants today maintain stronger relations with the society from which they come. Migrants are increasingly transnational economic actors, involved in both the home and host societies. For example, affluent immigrants from China and Hong Kong now settled in the USA and Canada are returning as capitalist investors to their places of birth, where they use ethnic ties and *guanxi*[1] networks as channels for developing subcontracting arrangements between overseas Chinese businesses and enterprises in mainland China. Besides ethnic ties with the homeland, they are attracted by the potential profit from utilizing China's cheap, skilled labor force, and its growing consumer market. According to Ong (1998) it is impossible to disentangle the irresistible pull of potential profits from nostalgic sentiments toward the homeland.

Lessinger (1992) observes a similar behavior pattern for Indian migrants now settled primarily in the United States, Europe, Southeast Asia, and the Middle East. These Indian migrants, dubbed Non-Resident Indians (NRIs), are part of a new transnational business class which is investing capital accumulated overseas in industrial ventures and banks in India. This behavior has been facilitated by a change in India's development policy since the early 1980s from an inward orientation, emphasizing self-reliance, to a policy promoting inward foreign investment in which the Indian Government provides financial incentives to foreign investors – and treats NRIs as a favored subcategory of foreign investors. The Indian Government sees NRIs not only as investors, but also as supplying technical and managerial expertise acquired abroad, and providing access to scientific, business, and financial communities in the West. In order to accomplish this, the Indian Government has not only provided generous subsidies to NRIs, but also used the rhetoric and sentiments of Indian nationalism and cultural identity to entice them. According to Lessinger (1992, p. 78): "The old, pure capitalist imperatives of profit and self-interest are no longer enough. Ideologies of nationalism, of common history and cultural integration, as well as the emotions of love, guilt and ambivalence, are invoked to woo NRIs into participating in an economy they once rejected." At the same time, the emergence of NRI enterprises and their privileged economic position have come under enhanced scrutiny within India. There is increasing criticism of the government's NRI strategy and NRI enterprises. Although receiving large state subsidies, NRI enterprises are seen as union-busters, and some NRI enterprises, such as medical institutes and hospitals, are geared primarily the rich. These conflicts highlight existing internal class conflicts within India (Lessinger, 1992). Thus immigrant investment in India does not necessarily generate the kinds of development that improve the standard of living of the population in India at large, or reduce the income gap between the rich and the poor.

Transnational economic and political measures to reduce migration flows

Taking a global perspective, a number of authors similarly have argued that capital transfers to the less developed world, whether as traditional remittances or as immigrant entrepreneurs' investments in their home country, are unlikely to foment economic transformation and development in countries of the less developed world, thereby reducing the income gap between rich and poor countries (Jones, 1998a). If this is the case, the number of people from poor countries seeking to migrate to richer countries will continue to increase, at a time when the rich countries of the world, obsessed with real and potential immigration pressures, are less and less prepared to admit more immigrants into their territories. This has precipitated a discussion, in the advanced industrial world, about measures that would reduce pressures for emigration. The goal is to create conditions in the sending areas that would enable people "to achieve at home what they seek to achieve abroad, whether economic advancement or freedom from persecution and insecurity" (Böhning and Schloeter-Paredes, 1994, p. 4). A number of measures have been proposed as a means to promote development in poor emigration countries, such as trade liberalization, increased foreign direct investment (FDI), and international aid.

While these measures might help induce economic development in low-income sending countries in the long run, scholars generally agree that current trends are in the opposite direction. Regarding FDI, "poor countries have been receiving a shrinking share of global capital investment – down from 31 percent in 1968 to 17 percent twenty years later – suggesting that its effects on emigration pressure can only be limited" (Böhning and Schloeter-Paredes, 1994, p. 5). The picture is similarly gloomy with respect to trade liberalization, because rich countries often introduce protectionist trade policies discriminating against the labor-intensive goods that make up much of the exports of low-income sending countries. Such protectionist policies are reducing employment prospects in these countries (Böhning and Schloeter-Paredes, 1994). Foreign aid has also been declining as a percentage of the gross domestic product (GDP) of developed countries, from 0.44 percent in 1960 to 0.35 percent in 1992 (Skeldon, 1997). Foreign aid often increases dependence on donor nations because of the side conditions under which loans are made (Porter and Sheppard, 1998, Chapter 23). It also is not targeted to countries, regions, and population groups with high immigration rates, since motives for foreign aid are often geopolitical. As a consequence, foreign aid has become less effective as a way of reducing emigration pressures (Böhning and Schloeter-Paredes, 1994).

As noted earlier, several scholars have questioned seriously whether any of these measures, even if they did enhance economic growth in sending countries, actually reduce emigration pressures (Massey, 1988; Sassen, 1988). Massey (1988) argues that attempts by the USA to stimulate development in Mexico will not reduce immigration to the USA in the short run; indeed it may increase it. Sassen (1988), examining the relationship between foreign direct investment and emigration, also maintains that foreign direct investment contributes either directly or indirectly to emigration. While working within different explanatory frameworks, both authors insist that greater attention be paid to the precise nature of the varied economic, political, and social processes linking sending and receiving areas, and to their importance in generating and directing transnational population flows. According to Sassen (1988, p. 6):

Thinking and policies stemming from this recognition may carry a rather different focus from current US policies aimed at controlling the border or reducing population growth and promoting economic growth in Third World countries. Recognition of intervening processes may move the focus away from conditions in emigration countries and invite an examination of processes that link the United States to those countries and may contribute to the initiation of new migration flows to the US. And it would invite an examination of labor demand conditions in the US that may contribute to the continuation of such flows. Policies stemming from such a recognition may have to address issues not usually considered relevant to immigration.

Conclusion

As the variety of new theoretical perspectives and studies on transnational labor migration reviewed here makes clear, the migration decision cannot be simply conceived as individuals making a rational choice solely for economic reasons. The volume of population flows across national boundaries also cannot be understood simply as the result of disparities in incomes and potential earnings between sending and receiving areas, or labor market demand in the receiving areas. Rather, understanding the generation and impact of migration requires an examination of the complex, geographically and historically specific, economic, social, political, and cultural linkages between sending and receiving areas at different scales. This implies not only greater attention, for example, to migrant networks linking individuals or groups, and to economic and political relations between countries and institutions, but also a greater focus on migrants' experiences in sending and receiving areas. A focus on migrants' experiences also may enable us to better capture the spatial and temporal complexities of the migration process itself.

It is worth noting that the shifting popularity of different theoretical perspectives in academic debate reflects not only their empirical explanatory power, but also the degree to which they conform with other theoretical debates about agency, structure, and social transformation. When different theoretical perspectives are brought to bear on debates about migration policy, it is similarly clear that they are used selectively to promote particular, and often pre-conceived, discourses about the desirability of emigration and immigration. For example, evidence of the positive impact of migration on receiving regions is often ignored in debates about immigration control, and the complexity of how emigration impacts sending regions is of little concern to receiving countries. Immigrants themselves are strategic and knowledgeable, and increasingly transnational, actors, who demonstrate again and again their ability to get around the barriers that continually are raised to their desires to seek a better life elsewhere. Perhaps we need to think again about what the problem is, and what a solution might be. Instead of seeing migration as the problem, and stopping large-scale migration as the solution, perhaps we need to remember the positive impacts that can be associated with migration, for sending and receiving regions and migrants alike. Perhaps we need to move towards a world where migration is not a response to economic marginalization and political oppression, but a means for anyone to seek out elsewhere the good life that they are not able to achieve at home.

Endnote

1. *Guanxi* literally means "relations" in Chinese, and has been defined by Hwang (1987) as a set of interpersonal connections that facilitate exchange of favors between people on a dynamic basis.

Bibliography

Addleton, J. S. 1992. *Undermining the Centre: The Gulf Migration and Pakistan*. Karachi: Oxford University Press.

Altonji, J. G. and Card, D. 1991. The effects of immigration on the labor market outcomes of less-skilled natives. In John M. Abowd and R. B. Freeman (eds). *Immigration, Trade, and the Labor Market*. Chicago: University of Chicago Press.

Atalik, G. and Beeley, B. 1993. What mass migration has meant to Turkey. In R. King (ed). *Mass Migrations in Europe: The Legacy and the Future*. London: Belhaven Press, 156–73.

Ball, R. 1997. The role of the state in the globalization of labour markets: the case of the Philippines. *Environment and Planning A*, 29, 1603–28.

Bertram, I. G. and Watters, R. F. 1985. The MIRAB economy in South Pacific microstates. *Pacific Viewpoint*, 27, 47–59.

Böhning, W. R. and Schloeter-Paredes, M. (eds). 1994. *Aid in Place of Migration?* Geneva: ILO.

Borjas, G. J. 1998. The impact of immigrants on employment opportunities of natives. In D. Jacobson (ed). *The Immigration Reader – America in a Multidisciplinary Perspective*. Oxford: Blackwell Publishers, 217–30.

Castles, S. and Miller, M. 1993. *The Age of Migration*. New York: Guilford Press.

Champion, T. 1995. Internal migration, counterurbanization and changing population distribution. In R. Hall and P. White (eds). *Europe's Population – Towards the Next Century*. London: UCL Press, 99–129.

Conway, D. and Cohen, J. H. 1998. Consequences of migration and remittances for transnational communities. *Economic Geography*, 74, 1, 26–44.

Cornelius, W. A., Martin, P. L., and Hollifield, J. F. (eds). 1994. *Controlling Immigration – A Global Perspective*. Stanford, CA: Stanford University Press.

Enchautegui, M. E. 1993. *Immigration and County Employment Growth*. Policy Discussion Paper PRIP-UI-23. Program for Research on Immigration Policy. Washington D.C.: The Urban Institute.

England, K. and Stiell, B. 1997. "They think you're as stupid as your English is": constructing foreign domestic Workers in Toronto. *Environment and Planning A*, 29, 195–215.

Espenshade, T. J. 1998. U.S. immigration and the new welfare state. In D. Jacobson (ed). *The Immigration Reader – America in a Multidisciplinary Perspective*. Oxford: Blackwell Publishers, 231–50.

Fan, C. C. 1996. Economic opportunities and internal migration: a case study of Guangdong Province, China. *The Professional Geographer*, 48, 28–45.

Filer, R. 1992. The effect of immigrant arrivals on migratory patterns of native workers. In G. J. Borjas and R. B. Freeman (eds). *Immigration and the Work Force: Economic Consequences for the United States and Source Areas*. Chicago: University of Chicago Press, 245–70.

Fincher, R. 1997. Gender, age, and ethnicity in immigration for an Australian nation. *Environment and Planning A*, 29, 217–36.

Fincher, R., Foster, L., and Wilmot, R. 1994. *Gender Equity and Australian Immigration Policy*. Canberra: Australian Government Publishing Service.

Findlay, A. M., Li, F. L. N., Jowett, A. J., and Skeldon, R. 1996. Skilled international migration and the global city: a study of expatriates in Hong Kong. *Transactions of the Institute of British Geographers*, NS 21, 49–61.

Findley, S. E. 1987. An Interactive Contextual model of migration in Ilocos Norte, the Philippines. *Demography*, 23, 163–90.

Fix, M. and Passel, J. S. 1994. *Immigration and Immigrants – Setting the Record Straight*. Washington D.C.: The Urban Institute.

Frey, W. H. 1995. Immigration and internal migration "flight" from US metropolitan areas: toward a new demographic balkanization. *Urban Studies*, 32, 733–57.

Goss, J. and Lindquist, B. 1995. Conceptualizing international labor migration: a structuration perspective. *International Migration Review*, 29, 317–51.

Griffin, K. 1976. On the emigration of the peasantry. *World Development*, 4, 353–61.

Gurak, D. T. and Caces, F. 1992. Migration networks and the shaping of migration systems. In M. M. Kritz, L. L. Lim, and H. Zlotnick (eds). *International Migration Systems: A Global Approach*. Oxford: Clarendon Press, 150–76.

Halfacree, K. and Boyle, P. J. 1993. The challenge facing migration research: The case for a biographical approach. *Progress in Human Geography*, 17, 333–48.

Hwang, K. 1987. Face and favor: The Chinese power game. *American Journal of Sociology*, 92, 944–74.

Huddle, D. 1993. The cost of immigration. *Carrying Capacity Network*. Revised July 1993. Washington, D.C.

Hull, H. 1987. Population and the present world structure. In W. Alonso (ed). *Population in an Interacting World*. Cambridge, MA: Harvard University Press, 74–94.

Jones, R. C., 1998a. Introduction: The renewed role of remittances in the new world order. *Economic Geography*, 74, 1–7.

Jones, R. C. 1998b. Remittances and inequality: a question of migration stage and geographic scale. *Economic Geography*, 74, 8–25.

Kofman, E. and England, K. 1997. Citizenship and international migration: taking account of gender, sexuality, and race. *Environment and Planning A*, 29, 191–94.

Leitner, H. 1995. International migration and the politics of admission and exclusion in postwar Europe. *Political Geography*, 14, 259–78.

Leitner, H. 1997. Reconfiguring the spatiality of power – the construction of a supra-national migration framework for the European Union. *Political Geography*, 16, 123–43.

Lessinger, J. 1992. Investing or going home? A transnational strategy among Indian immigrants in the United States. *Annals of the New York Academy of Sciences*, 645, 53–80.

Ley, D. 1999. Myths and meanings of immigration and the metropolis. *The Canadian Geographer*, 43, 2–19.

Massey, D. S. 1988. Economic development and international migration in comparative perspective. *Population and Development Review*, 14, 383–413.

Massey, D. S., Alarcon, R., Durand, J. and Gonzales, H. 1990. *Return to Aztlan: The Social Process of International Migration from Western Mexico*. Berkeley and Los Angeles, CA: University of California Press.

Massey, D. S., Arango, J., Hugo, G., Kouaouci, A., Pellegrino, A., Taylor, J. E. 1993. Theories of international migration: a review and appraisal. *Population and Development Review*, 19, 431–66.

Massey, D. S., Arango, J., Hugo, G., Kouaouci, A., Pellegrino, A., and Taylor, J. E. 1994. An evaluation of international migration theory: The North American case. *Population and Development Review*, 20, 699–751.

McHugh, R. and Mings, R. C. 1996. Attachment to place in aging. *Annals of the Association of American Geographers*, 86, 530–50.

O'Loughlin, J. and Friedrichs, J. (eds). 1996. *Social Polarization in Post-Industrial Metropolises*. Berlin, New York: Walter de Gruyter.

Ong, A. 1998. Flexible citizenship among Chinese cosmopolitans. In P. Cheah and B. Robbins (eds). *Cosmopolitics – Thinking and Feeling Beyond the Nation*. Minneapolis: University of Minnesota Press, 134–62.

Parnwell, M. 1993. *Population Movements and the Third World*. London and New York: Routledge.

Piore, M. J. 1979. *Birds of Passage: Migrant Labor and Industrial Societies.* Cambridge: Cambridge University Press.

Porter, P. W. and Sheppard, E. 1998. *A World of Difference – Society, Nature, Development.* New York and London: Guilford Press.

Portes, A. and Walton, J. 1981. *Labor, Class and the International System.* New York: Academic Press.

Richmond, A. H. 1994. *Global Apartheid – Refugees, Racism, and the New World Order.* Oxford: Oxford University Press.

Rothman, E. S. and Espenshade, T. J. 1992. Fiscal impacts of immigration to the United States. *Population Index*, 58 (3), 381–415.

Salt, J. 1989. A comparative overview of international trends and types, 1950–1980. *International Migration Review*, 23, 431–56.

Salt, J. and Stein, J. 1997. Migration as a business: the case of trafficking. *New Community – The Journal of the European Research Centre on Migration and Ethnic Relations*, 23, 467–91.

Sassen, S. 1988. *The Mobility of Labor and Capital: A Study in International Investment and Labor Flow.* Cambridge: Cambridge University Press.

Silvey, R. and Lawson, V. 1999. Placing the migrant. *Annals of the Association of American Geographers*, 89, 121–32.

Simon, R. J. and Brettel, C. B. (eds). 1986. *International Migration: The Female Experience.* Totowa, NJ: Rowman and Allanheld.

Skeldon, R. 1997. *Migration and Development: A Global Perspective.* London: Longman.

Stark, O. and Taylor, J. E. 1991. Relative deprivation and migration: theory, evidence, and policy implications. In S. Diaz-Briquets and S. Weintraub (eds). *Determinants of Emigration from Mexico, Central America, and the Caribbean.* Boulder: Westview Press, 121–44.

Taylor, J. E. 1992. Remittances and inequality reconsidered: direct, indirect, and intertemporal effects. *Journal of Policy Modeling*, 14, 187–208.

The Economist 1999. Europe's smuggled Masses. February 20, 45–6.

Tyner, J. A. 1996. The gendering of Philippine international labor migration. *The Professional Geographer*, 48, 405–16.

Tyner, J. A. 1998. Asian labor recruitment and the world wide web. *The Professional Geographer*, 50, 331–44.

United Nations 1994. Population distribution and migration. Proceedings of the United Nations Expert Meeting on Population Distribution and Migration, Santa Cruz, Bolivia, January 18–22, 1993. New York.

United Nations 1995. *International Migration Policies 1995.* New York: United Nations, Department of Economic and Social Information and Policy Analysis.

Vandsemb, B. H. 1995. The place of narrative in the study of third world migration: the case of spontaneous rural migration in Sri Lanka. *The Professional Geographer*, 47, 411–25.

Walker, J. 2000. *The State, Labor Import/Export, and Economic Restructuring in Taiwan.* Athens: University of Georgia. Ph.D. Dissertation.

Wallerstein, I. 1974. *The Modern World System: Capitalist Agriculture and the Origins of the European World-Economy in the Sixteenth Century.* New York: Academic Press.

White, P. and Jackson, P. 1995. (Re)theorising population geography. *International Journal of Population Geography*, 1, 111–23.

Wright, R. A., Ellis, M. and Reibel, M. 1997. The linkage between immigration and internal migration in large metropolitan areas in the United States. *Economic Geography*, 73, 234–54.

Zolberg, A. 1989. The next waves: migration theory for a changing world. *International Migration Review*, 23, 403–30.

Chapter 28

Transportation: Hooked on Speed, Eyeing Sustainability

Susan Hanson

Travel figures centrally in Charles Frazier's *Cold Mountain* (1997), a novel set during the American Civil War. With the warrior hero, Inman, slowly making his way back to Ada, who is busy creating the home that is his destination, the story, like the *Odyssey*'s, is built around the contrast between movement (Inman's travel adventures) and domestic stability (the home place that Ada is making). In the exchange recounted here, Ada, now subsistence farming with another woman, Ruby, in the mountains of North Carolina, has just put together a garden scarecrow from remnants of clothing left from her earlier, cosmopolitan life in Charleston, S.C. Ruby, who has never been outside her mountain community of birth, begins with a comment on the scarecrow's hat:

"That hat in particular's a fine touch," she said.
"It came from France," Ada said.
"France?" Ruby said. "We've got hats here. A man up East Fork weaves straw hats and will swap them for butter and eggs. Hatter in town makes beaver and wool but generally wants money."

This business of carrying hats halfway around the world to sell made no sense to her. It marked a lack of seriousness in a person that they could think about such matters. There was not one thing in a place like France or New York or Charleston that Ruby wanted. And little she even needed that she couldn't make or grow or find on Cold Mountain. She held a deep distrust of travel, whether to Europe or anywhere else. Her view was that a world properly put together would yield inhabitants so suited to their lives in their assigned place that they would have neither need nor wish to travel. No stagecoach or railway or steamship would be required; all such vehicles would sit idle. Folks would, out of utter contentment, choose to stay home since the failure to do so was patently the root of many ills, current or historic. In such a stable world as she envisioned, some might live many happy years hearing the bay of a distant neighbor's dog and yet never venture out far enough from their own fields to see whether the yawp was from hound or setter, plain or pied (p. 192).

These musings of an Appalachian woman more than a century ago capture several enduring themes of transportation geography. First, there is the double-edged power of transportation (or movement) to create economic benefit for some people and places (Parisian hatmakers, in this case), to the possible detriment of others (local hatters). A second and related point is that travel – the ability to traverse distance – often breeds specialization, as in the production of stylish headgear in Paris. Third, at the core of Ruby's reflections is the idea that travel, though clearly held in disdain by herself, has value to some; to put this idea in economic terms, travel has utility for some people at some times, but disutility for Ruby (at all times). A related point, illustrated precisely by Ruby's travel preferences, is that differential levels of mobility lead to distinct cultures of consumption. Fourth is the association of travel with vehicles, which is related to a fifth theme, that of geographic scale: Ruby's "deep distrust of travel" is aimed at movement beyond the bounds of her local community; she travels extensively on foot around her farm, the surrounding forests, and into the village. Finally, Ruby is so accustomed to her everyday space-traversing activities that she doesn't even think of them as travel.

A life truly without travel – lived within a circumference of, say, five feet – is unthinkable and is, of course, unsustainable, especially if everyone else in the world is also living a transportation-free existence. Even Ruby depended on movement, on mobility, on transportation – however local – for access to water, food, wood for fuel, herbs for health. But mobility is more than simply the means to survival; it can also be, as it was for Inman, the means to encounter different landscapes and cultures, in short, more dimensions of diversity than those that exist within one's immediate surroundings. When walking was the only available means of transportation, people like Ruby could, and often did, live their whole lives interacting only with a small circle of well-known others, people who shared many of the same life experiences as oneself. Mobility broadens the scope of experience and opportunity within one's geographic reach, often with disastrous consequences for individuals and society, as Ruby implies.

How much mobility we need – how far we need to travel to find food, water, friends, books, diversity – depends, of course, on where these items are located relative to where we happen to be. Like Ruby, one can have access to people, goods, and services without having a great deal of mobility. Today, we so take for granted the ability to move, to traverse distance, to have access to places other than where we happen to be that we have become numb to the notion that mobility – indeed, transportation – is at the heart of contemporary economic, social, political, and cultural life. How many Americans realize that transportation accounts for about one-fifth of their household expenditures, more than any other single item except housing and as much as is spent for food and health care combined (http://stats. bls.gov)? People concerned about the unsustainability of contemporary settlement systems argue that less mobility might actually be desirable.

Yet transportation geography as such has become a quiet, some might say moribund, corner of our discipline. How has such an important area of inquiry become so marginalized? It has not always been this way: in the 1950s and 1960s transportation questions were central not just to economic geography but also to human geography. My goals in this chapter are twofold. First, I sketch out the nature of what I refer to as traditional transportation geography as it has developed since the 1950s and 1960s, making clear why transportation analysis once was so central to the discipline of geography. Second, I outline some of the emerging themes and approaches that an enlarged and more critical transportation geography might more fully embrace. I argue that transportation geography has fallen outside the mainstream of human geography because much transportation work remains within the paradigm of the 1960s. Attention to new questions and openness to new ways of creating knowledge could and should enable transportation geography to contribute vitally to contemporary economic and human geography.

Traditional Transportation Geography

Transportation analysis was in many ways at the core of the dominant paradigm in the human geography of the English-speaking world from the 1950s to the mid-1970s. As the contemporary transportation geography agenda still strongly bears the imprint of that time, it is worth examining the central concerns of this subfield then and considering why they have been so enduring. Because transportation studies tend to be divided along lines of geographic scale, the first part of this section examines such studies at the regional and national scales, with emphasis on transportation's starring role in the economic and human geographies of the 1950s and 1960s; the second part focuses on studies of transportation at the metropolitan scale.

Transportation in economic geography

The reigning paradigm in geography in the 1960s was spatial analysis, whose goal was to identify or discover general spatial patterns and, more importantly, to understand the processes creating these patterns. Within human geography, the search was for generalizable processes governing the spatial organization of human activity, and the friction of distance – along with the ability to overcome it – were seen as central organizing principles. With distance representing a spatial and temporal barrier to be overcome, spatial separation acquired a monetized cost, and space became *commodified* – equated with the cost of moving people or goods over distance.

Models like those of Christaller and Weber, which dominated economic geography at that time, had the friction of distance, and hence transportation costs, at their core. Christaller sought to explain the spatial pattern of different-sized settlements in an agricultural landscape: why do settlements of similar size tend to be equidistant from each other; why are smaller settlements more numerous than larger ones; and what governs the nature of goods and services available in settlements of different sizes? He saw the answer in the distance people are willing to travel to obtain a particular bundle of goods and services. Because, under the model's assumptions, transportation costs are proportionate to distance, another way of stating Christaller's answer is that it lies in the varying transportation costs people are willing to bear in different circumstances. Weber's goal was to explain the spatial pattern of industry: why are certain types of industrial activity located where they are? At the center of Weber's explanation, like Christaller's, are transportation costs; in this case the costs of moving various kinds of raw materials to the site of manufacturing, relative to the cost of moving finished products to market.

Transportation costs were therefore at the heart of the space-economy that human geographers sought to understand. These costs dominated the calculus of the rational economic actors whose decisions fashioned the spatial organization of retail activity (the geography of consumption), and of industrial activity (the geography of production).

In an industrialized capitalist system, time is money. This is as true for shoppers and workers, who must traverse space to reach stores and jobs, as it is for manufacturers, who need to bring raw materials together at the point of production and then ship a product to market. The commodification of space means that a unit of travel time has a money value. For an individual, travel time is valued at the person's marginal wage rate; it is the cost of foregoing a certain number of extra minutes of wage earning in order to travel. For the manufacturer, the volume of production and hence profits depend, inter alia, on how quickly the transportation system can move raw materials and finished product. In each case, economic prosperity is associated with speed. An economic geographic analysis grounded in the commodification of space – setting a money value on the time it takes to traverse distance – has, therefore, fed the construction of transportation systems dedicated above all else to promoting increasingly rapid mobility. In short, capitalism thrives on speed.

How, pragmatically, can speed be increased? The brief answer is, via networks and technology. Movement is faster and cheaper along corridors designed to promote speed; even walking across an open meadow is easier if there is a footpath. These corridors are linked together into networks, which serve to bring places on those networks closer together in space-time than are places that are not part of those networks. The other important element to consider in speeding up movement is the mode of travel, especially the technology associated with a mode: is it shoe leather, roller skates, a human-powered bicycle, a Model T Ford, a bullet train, a propeller airplane, or a supersonic jet? Because of the value it places on "saving time," the geographic-economic analysis on which transportation decisions are based favors the development of ever-faster travel modes. I believe that the obsession with speed of movement is the main reason that transportation has tended to be thought of in terms of technology and infrastructure – roads, bridges, buses, trains – rather than in more general terms as an enabling and constraining facet of life, a source of pleasure and exasperation, power and control.

Because movement between places reflects and creates their interdependence, economic and transportation geographers have been fascinated by the relationship between transportation infrastructure and the prosperity of places, between network characteristics and what goes on at network nodes. For example, the construction or improvement of transportation networks has long been at the core of regional development strategies, whether in Appalachia or Amazonia. The logic is that transportation networks shape the accessibility of places (how quickly and easily they can be reached), and accessibility shapes economic activity: more-accessible locations have more prosperous economies because lower transportation costs reduce both production and consumption costs there. Accessibility also enables economic specialization, another facet of wealth creation. Changes in transportation technology, such as the move from a footpath to a paved road, from bicycles to automobiles, or from rails to trucks, differentially affect the accessibility of places and thereby alter the space-economy. For example, the advent of paved roads and motorcars, which meant that dispersed rural populations could cover more distance in less time, spelled the demise of the smallest places, which had provided only convenience goods.

Because accessibility is seen as the *sine qua non* of economic prosperity, regional development strategies have sought to improve the accessibility of economically marginal areas like Appalachia by integrating them into broader-scale transportation networks. The problem is that prosperity has not always followed the construction of the transportation infrastructure designed to spark it. Some of the reasons for the failure of this relationship between transportation and economic activity to materialize will be considered in the next section.

Because of the close associations of economic prosperity with accessibility, accessibility with mobility, and mobility with large-scale infrastructure, governments have assumed a central role in the planning, financing, and building of transportation improvements. Clearly, private economic interests benefit differentially from these huge public investments, and some scholars (e.g. Whitt and Yago, 1985) have pointed to corporations as the initiators – and the prime beneficiaries – of transportation policy; arguing that the state has been ineffective in advancing and supporting the public interest. Transportation projects are ultimately about the public distribution of amenities (e.g. access, jobs) and disamenities (e.g. noise and air pollution, marginalization), and therefore such projects centrally involve issues of differential political power (Wachs, 1995). Because huge amounts of public money are at stake, transportation politics are lively. The politics of transportation funding were abundantly clear in the USA during the process preceding the 1998 reauthorization of the federal transportation legislation known as TEA-21 (Transportation Equity Act for the 21st Century, a six-year $215 billion bill).

Within economic geography, scholars have tended to think about transportation in utilitarian, instrumentalist ways; the pragmatic goal has been to move people, goods, or information from origin to destination as efficiently (fast) as possible. From the outset, the field has been heavily influenced by economics, with its premium on parsimonious explanation and its focus on the monetized value of time. Transportation geographers have tended to focus on the analysis of networks and the relationship between location in a network (and associated accessibility) and economic activity. Since its inception in the 1950s, the broader field of transportation analysis has been dominated by civil engineering and to a lesser extent by economists: they have seen their mission as informing policymakers of the best way to solve pressing and very real transportation problems, primarily congestion and the slower speeds that congestion brings. Nowhere has the search for a solution to congestion been more intense than in urbanized areas.

The study of movement in cities

According to those who focus on the economic advantages of agglomeration, the very existence of cities is due to the accessibility edge they provide over non-urban places. The ease of access offered by close spatial proximity yields positive technological and monetary externalities known as economies of agglomeration; that is, with greater density and diversity of economic activity, average production and consumption costs decline and the likelihood of technological innovations grows (Anas et al., 1998; Harrison et al., 1996). Agglomeration economies require high levels of accessibility and hence speed-enhancing networks. Congestion on those networks is the dark side, or the "diseconomy," of agglomeration. As in the geographic-economic analysis of the broader economy, the urban transportation problem to solve thus becomes how to increase speed, especially during the morning and evening rush hours when transportation networks are clogged with people going to and from work. Transportation analysts have seen this as a technical problem requiring a technical solution, to increase network capacity, and often themselves have had a vested interest in promoting solutions that require the building of additional transportation infrastructure (Wachs, 1995).

The enshrinement of increased speed as the overriding goal of urban transportation planning is evident in one of the earliest US transportation studies, the 1959 Chicago Area Transportation Study: "The dominant objective of a transportation facilities plan... is to reduce travel frictions by the construction of new facilities so that people and vehicles... can move about within the study area as rapidly as possible." The legacy of this message remains strong, with analysts advocating policies that increase mobility rather than restricting it (e.g. Wachs, 1996; Dunn, 1998). This policy aim of increasing speed has guided the type of analysis conceived, the type of data collected, and the range of options considered.

In a refreshing critique of this view, John Whitelegg (1993) identifies the monetized value of travel time as the main culprit in urban transportation analysis and models of urban travel. These models are used to assess the costs and benefits of proposed transportation projects, including the costs and benefits associated with the "no-build" option. By placing a high monetary value on very small increments (e.g. five minutes) of time "saved," the models overemphasize the benefits accruing from the "time savings" resulting from additional lanes of freeway and have, as a result, promoted massive urban highway building.

The focus on congestion as the problem, and more lanes of highway as the solution, has indeed changed the physical and social infrastructure of metropolitan areas. This conceptualization of "the urban transportation problem" has not only fed decentralization but has led above all else to higher levels of automobile dependence, with attendant increased energy consumption and air pollution. Additional corollaries of abundant highway construction have been reduced use of public transportation, destruction of open space and wildlife habitats, more-dispersed journey-to-work patterns, increased racial and socioeconomic residential segregation, and greater inequities among social groups in access to employment (Cervero, 1995).

These changed patterns of urban form and urban travel also help explain the seeming paradox that despite increasing trips per capita, and growing trip distances, travel times have not changed. Suburbanization, supported with high-speed roads at the urban periphery, together with the shift of trips from transit (a slower mode) to auto (a faster mode) have helped to keep average travel times on the journey to work virtually unchanged from 1983 to 1990 (Gordon and Richardson, 1995). Downs (1992, p. 30) points to the ways that individual auto drivers, faced with recurring peak-hour road congestion, will change their behavior so as to keep their travel times from lengthening: they will shift the timing of their trips from peak to non-peak times, shift the routes they take, or shift from auto to transit during the peak.

Despite Gordon and Richardson's evidence of stable travel times, other researchers find substantial increases in urban highway congestion, and congestion remains a top concern in surveys of urban residents (Transportation Research Board, 1994).

The growth in number of households has outstripped population growth, contributing to increase in personal travel, and the growth of women's waged employment has fueled the demand for mobility. These factors have contributed to the dramatic growth in vehicle miles traveled (VMT) in the USA, up 62 percent from 1975 to 1990 (49 percent for passenger cars only) (Downs 1992, p. 10). Increased VMT together with the dearth of new highways being built are the main ingredients for highway congestion (Downs, 1992; Hodge, 1992).

Road building is no longer accepted as the panacea for congestion, as research has shown that added highway capacity becomes congested by the increased travel demand it stimulates (Sheppard, 1995). A good deal of attention is now directed, therefore, toward the question of how to reduce congestion without building more roads. Economists like Anthony Downs (1992) favor making drivers pay for road use, especially for the marginal cost of driving during times of peak congestion. He suggests using the pricing of roads, fuel, and parking to change people's behavior, especially to increase auto occupancy rates (i.e. more people per auto). Geographer David Hodge (1992) advocates reducing auto dependency through changing land use configurations, so that transit becomes a more attractive alternative. Other ways of reducing congestion that Hodge proposes are Transportation System Management (TSM), which aims to increase travel speeds through managing traffic light timing and ramp signals, and Transportation Demand Management (TDM), which involves reducing the number of autos on the road especially during peak hours via flextime, car pools, telecommuting, road pricing, and more costly parking fees.

Hodge's proposal to reduce congestion by making land-use patterns more supportive of public transportation reflects the fact that transit requires relatively high land-use densities. It also reflects a long-standing bias in favor of transit among urban analysts. During the 1960s and 1970s large public investments were made in public transportation systems, especially light rail systems, because transit was heralded as the solution to a variety of urban problems. It would create higher densities, reduce energy consumption, cut pollution, curb sprawl, eliminate congestion, enhance urban redevelopment, keep downtowns alive, and provide mobility for those without autos (Fielding, 1995). In short, planners wanted to use transit "to reshape the American landscape" (Snow, 1986, p. 45). In part because of this belief in transit as the city's salvation, transportation analysts have devoted substantial effort to studying the conditions under which people choose or might choose public transit over the auto, in order to identify policy changes that might induce a larger share of the traveling public to use public transportation.

Except in extremely dense urban environments, however, the accessibility advantage of the auto has meant that transit's share of riders continues to fall in most US cities. Although transit's share is higher for work trips than for other trip purposes, by 1990 only about 5 percent of US work trips were made on public transportation (US Dept. of Transportation, 1994). In the San Francisco area, for example, despite the construction of BART (Bay Area Rapid Transit), auto travel times to major employment centers have remained about 35 percent shorter than transit times (Giuliano, 1995). The light rail systems built in the 1970s have not succeeded in persuading large numbers of auto drivers to abandon their cars in favor of transit; in fact, most of those attracted to new rail transit facilities had previously ridden the bus (Fielding, 1995). Transit use in New York City, however, has recently increased, owing in part to immigration and in part to a policy permitting free transfers between subway and bus.

Analysts have accepted almost without question that transportation technology shapes land use and urban form (Muller, 1995) as well as the broader settlement system (Borchert, 1967). Certainly urban and economic theories that base their explanations on accessibility predict that any change in the transportation system will have an impact on the spatial arrangement of economic activity. Yet studies at varying scales, from the sub-metropolitan to the regional, have had difficulty demonstrating that transportation investments such as the building of a light rail transit system, a highway, or an additional lane on an existing road, actually do transform land use patterns.

Genevieve Giuliano (1995) has carefully reviewed these studies and identified the many reasons why it is so difficult to show that a specific transportation improvement is responsible for a particular land-use change. First, initial investments (like building the first road through an area) produce bigger impacts than do subsequent investments (like adding a highway lane or another link in an already-developed network). The latter often make only marginal changes in accessibility and therefore can be expected to have only a marginal influence on land-use change (see also US Congressional Budget Office, 1998). Second, the availability (or lack) of developable land affects the magnitude of the impact that any transportation change can have: land use controls such as zoning regulations can constrain the land use changes that are possible. Third, the health of the regional economy is an important factor in determining whether a transportation investment will lead to higher-value-added economic activity. The transportation investments in Appalachia over the past several decades, for example, have not led to an economic boom there (Glasmeier and Fuellhart, 1999). Finally, it is often difficult to know whether any resulting changes in land use represent a net gain in wealth-producing activity within a region or simply a relocation of already-existing activity. This difficulty is part of a related problem, namely that of identifying the geographic scale at which any land use changes are likely to occur. Theoretically, a change in any part of a transportation network differentially alters accessibility throughout the entire network, facilitating land use changes in places quite distant from the site of the transportation investment.

Urban transportation studies have always had robust ties to transportation planning, land use, and engineering, which may account for their strongly pragmatic and positivist nature. A large portion of transportation studies has been aimed, quite literally, at telling decisionmakers where to lay the concrete. I do not wish to essentialize men or women, but I think it is not inconsequential that transportation analysis has been almost entirely a male domain. Perhaps this reflects the age-old association of travel with male adventures like those of Odysseus in the *Odyssey* or Inman in *Cold Mountain*; perhaps it reflects the association of transportation with infrastructure and heavy equipment, or maybe it's the mathematical bent of traditional transportation analysis. Nevertheless, as Kingwell (1998, p. 48) has noted, a fascination with speed speaks of desire for control: "Speed, we might admit, is our preeminent trope of control and domination." Traditional economic-geographic analysis has been hooked on speed, a legacy of its origins in the spatial analysis of the 1960s. Some transportation geographers have begun to question the valuing of speed above all else, to ask who is served by such a goal, to examine the societal costs associated with a focus on increasing speed, and to propose alternative bases for thinking about transportation. In the next section I outline some of these emerging themes and argue for expanding the approaches to knowledge creation in transportation geography.

Emerging Themes and Approaches in Transportation Geography

How analysts define, or frame, a problem affects how they will design their analysis and, in the case of transportation issues, how policy options will be imagined. Although the transportation problem has been defined primarily in terms of the need for increasingly rapid mobility (or the need to conquer congestion), one might also envision a range of other ways of defining the problem. A few possibilities are: unsustainability, fossil fuel dependence, restricted choice, inequity, inflexibility, and poor health. Clearly, values are important: traditional transportation analysis has valued motorized over non-motorized modes, longer travel distances over shorter ones, independence and privacy over more-communal forms of living and traveling, mobility over access, efficiency over inequity. Some of the emerging themes discussed below are grounded in values that differ from that of speed and reflect, in part, the changed societal context within which transportation studies now take place. Before considering some of the emerging and potential research areas and alternative research approaches in transportation geography, I highlight a few of the ways in which the analytical context has changed.

Changed context

The societal context for transportation studies has shifted dramatically in a number of ways since the 1960s. Most important are changes in socio-demographics, attitudes toward the environment, the role of activists, the impact of information technology (IT), and concern about sustainability. Transportation-relevant socio-demographic changes in industrialized countries include aging populations, declining household size, altered household structures, changed gender relations, increased employment instability, and growing income polarization. Each of these shifts has implications for mobility needs and should affect how analysts think about transportation.

Environmental concerns, especially those of air quality and biodiversity, have become far more important in defining the context for transportation and land use modeling than they were in the 1960s. These concerns are reflected in an altered regulatory context, with the national government in the USA, for example, requiring metropolitan areas to meet federal air quality standards. Garrett and Wachs (1996, p. 25) maintain that air quality is now the prime focus: "Congress and the Environmental Protection Agency have concluded that air quality concerns should replace mobility considerations as the overriding factor in highway planning."

A third dimension of the changed analytical context is in the level and nature of activism around transportation issues. Grassroots citizen organizing and lobbying are far more sophisticated now than they were 30 or 40 years ago. The shift in US transportation priorities evident in the 1990s federal transportation bills (e.g. the

1998 TEA-21) is due in no small part to citizen activism. Although the bulk of federal moneys still go for highways, very little of that (less than 5 percent in TEA-21) goes for new roads, and the proportions for bicycle, pedestrian, and transit facilities are rising.

New information technologies describe a fourth dimension of the changed context. The growth of personal computers and the advent of the Worldwide Web have the potential to alter mobility needs and the demand for travel. The geographic and social unevenness of access to these technologies raises serious questions for transportation geography. In addition, GIS technologies, particularly their graphic capabilities, may change the nature of transportation research and the relationship between planners and citizens (Elwood and Leitner, 1998).

A final element of the changed context within which transportation analysis now takes place is that the sustainability of current transportation practices is now a subject of debate. In particular, the many dimensions of extreme auto dependence and the increased scale of movements are increasingly questioned. Taken together, these aspects of the changed context open up new possibilities for transportation geography. In particular, they point to the importance of cultural discourses in shaping transportation-related behaviors, decisions, and policies.

Emerging themes

The changed transportation context makes clear how little we know about some very pressing transportation-related questions and how great is the need for a reinvigorated, critical, transportation geography. In this and the next section I outline a few of the themes and the approaches to knowledge creation that might form the basis of such investigations. The themes can be broadly grouped into issues of sustainability and IT impacts. Crosscutting these is the need for research on transportation issues in the so-called developing world and on the crucial role of mobility in constructions of class and gender.

What are the full costs and benefits of a society predicated on the auto, and who bears these costs? In traditional planning calculations of the expected costs and benefits of a proposed transportation investment, social costs or externalities (such as air, noise, and water pollution; death or injury from accidents; the loss of open space) do not appear on the cost side. In recent years scholars have tried to assess the social costs of the automobile by putting a monetary value on these externalities (see Murphy and Delucchi, 1998, for a review and evaluation of these studies). Delucchi (1997) has estimated that the social cost of the auto in the USA in 1991 ranged between $8,791 and $17,352 per registered vehicle. These figures include costs not borne by drivers, such as those associated with congestion, air pollution, accidents, noise, police, and climate change. Although many drivers believe that their gasoline taxes fully pay for highway construction and maintenance, in the USA gas (fuel) taxes and other user fees cover only about 60 percent of federal, state, and local spending on highways and roads; the remainder comes from general funds, property taxes, and the like. In other words, non-drivers (who are more likely to have very low incomes) subsidize drivers. When social costs are included in the calculations, drivers bear only about 25 percent of the total cost of their transportation; the rest is borne by employers (especially for parking at the workplace), tax payers, and unborn generations. These studies and figures pose a host of research questions not only about the equity of the current system but also about its long-term sustainability (see Transportation Research Board, 1997; Vasconcellos, 1997a, 1997b).

What are the prospects for sustainable transportation? Which elements of the current system are more or less sustainable, and why? Despite their importance, these questions have received relatively little research attention. Whitelegg (1993) sees sustainability as essentially an issue of scale, requiring slower, smaller-scale movements and the consumption of locally produced products. He would consider Ruby's community-contained world sustainable! Arguing that the dominant transportation paradigm amounts to the "conquest of distance by the destruction of time," Whitelegg (1993, p. 77) also decries the "time pollution" of a speed-craving, and in some ways speed-dependent, society. That is, the more we try to save time via faster movement, the less time we have and the more harassed we feel. In addition, the more we emphasize time savings, the more the transportation system becomes skewed toward meeting the needs of the wealthy (car owners) at the expense of those who rely on more-sustainable, non-motorized means of travel. Vasconcellos (1997a) stresses that equity, especially for non-motorized modes, is a necessary component of any truly sustainable transportation system. Activists, too, have resisted the dominant philosophy of speed and promoted the benefits of slow mobility through "traffic calming" (infrastructure changes like road bumps and narrowed streets that enforce slower vehicular speeds) and through bicycle- and pedestrian-oriented settlements (see the newsletter of the Surface Transportation Policy Project, a US lobbying organization, at www.transact.org).

Activists have been instrumental in placing another sustainability concern – air quality and the air pollution stemming from transportation sources – on the research agenda. Worldwide, the transportation sector contributes about one-quarter to one-third of all carbon dioxide emissions (Transportation Research Board, 1997). Current US laws, requiring metropolitan areas to meet air quality standards, put pressure on transportation agencies to increase auto occupancy rates, raise public transport ridership, boost bicycle and pedestrian travel, and reduce automobile vehicle miles traveled. In a recent San Francisco legal case, brought by the Sierra Club and other citizen groups, the court ruled that local planners must predict the air quality impacts associated with transportation policies (Garrett and Wachs, 1996, provide a full discussion of this fascinating and important case). Current transportation models are not designed to make accurate and reliable air quality predictions, however, and incorporating such predictions into the modeling process will be extremely difficult.

Policy questions also swirl around the air quality issue. Policy analysts see two ways of reducing emissions: change transportation technology (via cleaner fuels, electric cars, lighter vehicles) or change travel behavior (reduce vehicular miles traveled, shift people from single-occupancy to high-occupancy vehicles, increase transit ridership and walk or bike trips). Debate thrives over how best to bring about behavioral change, whether through the market (mainly via the pricing of road use, parking, and fuel) or via regulation (e.g. gas rationing, restricting auto use at certain times or places. See Downs, 1992). Critics like Whitelegg, however, argue that neither regulation nor market-based solutions will yield more sustainable transportation because they will not create exclusive spaces for cyclists and pedestrians. I would argue, moreover, that the role of discourse in shaping policy options needs more explicit recognition. Radical activist groups like Critical Mass, which aims precisely at improving conditions for cyclists, fully understand the policy relevance of altering the public discourse around transportation, and strategize accordingly (Blickstein and Hanson, 2000).

The energy (fossil fuels) consumed in transportation is another source of both concern about the sustainability of the current system and of important unanswered questions. Because aggregate data show a relationship between land use density and distances traveled, and between density and energy consumed, the creation of higher-density settlements via land use planning is commonly viewed as a way to reduce energy consumption. Breheny (1995) asks how much urban decentralization in the UK increased energy consumption between 1960 and 1990. In one simulation, he estimates that if all 1960–1990 growth had been as evenly distributed over the settlement system as it was in 1960 (i.e. if no urban decentralization had taken place), the transportation energy saved would amount to only 2.5 percent. He concludes that the energy savings associated with urban compaction, a policy widely favored in the UK but difficult to implement, are relatively small. Surprisingly little is known about the relationship between land use configurations and energy consumption, making this an important emerging area of research, especially for those interested in sustainability.

Another vital area of emerging research concerns the intersection of information technology (IT) and grounded social relations (those requiring face-to-face contact): how will these two forms of interaction shape each other (Hanson, 1998)? Questions abound: What is the impact of IT on the need for mobility? Does IT enable access without mobility? Is distance dead? As more interactions take place in virtual (cyber) space, how can we conceptualize and measure accessibility in an information age? It is clear that such measures must take into account access to information technologies (both hardware and software) as well as access to cyberplaces within those technologies (Hodge and Janelle, 2000). An important set of questions surrounds the equity impacts of IT (see also Warf, this volume). How, for example, is IT related to geographic and social marginalization: does it alleviate or reinforce inequities? I have argued that answering this question requires recognizing the embeddedness of physical and cyber interactions in (usually grounded) social relations, and exploring how new technologies intersect with, as well as help to shape, these social relations (Hanson, 1998).

As the Worldwide Web expands exponentially and people increasingly shop, visit, work, and entertain themselves on the Internet, what kind of geographies will these kinds of interactions create? In a world where the friction of distance held sway, speed shaped commerce and commerce, in large part, shaped places. Kingwell (1998, p. 41) suggests that the kind of speed that IT underwrites stands in a different relation to place-creation: "Speed's annihilation of time and place means, finally, speedy annihilation of places – and times. The inner logic of technology is not technical but commercial, and within the logic of commerce, location is an increasingly meaningless concept. We confront, now, a new topology, a world of instant and direct contact between every point on the globe." Who, if not geographers – and especially transportation geographers – will explore these momentous processes?

Sheppard (2001) and the articles in Hodge and Janelle (2000) are excellent points of departure for this exploration.

The final two emerging themes cut across and through the others. The first of these is that of examining access and mobility questions in the so-called developing world (Leinbach, 2000; Vasconcellos, 1997a, 1997b). Such studies are needed for a more truly international understanding of transportation and mobility. The world's population is predicted to grow from the current six billion to nine billion people by 2050, and almost all of this increase is expected to be concentrated in cities outside Europe and North America (United Nations, 1993). All of the sustainability issues raised above are thrown into relief by these stark – and extremely likely – forecasts. As Vasconcellos (1997b) points out, South American cities have suffered from the wholesale adoption of North American approaches to transportation. Leinbach (2000) argues for rethinking the relationship between mobility and development, and suggests several high-priority questions for research. Economic and transportation geographers could contribute enormously to the sustainability transition by helping to increase understanding of alternatives to the US transportation-settlement system. What might US planners and decisionmakers learn from the experiences of the so-called developing countries? Enriching the research base outside of Europe, North America, and Australia could be part of a more self-consciously geographic approach, focused on the importance of context and place-to-place differences.

The other cross-cutting theme is the need to comprehend how mobility and transportation issues shape understandings of class and gender. Vasconcellos (1997b), for example, argues that transportation technology is embedded in contemporary patterns of social reproduction, and that consequently, especially in developing countries, the car is essential to the reproduction of the middle class as a class: if middle-class people were to use their legs or transit to conduct their daily activities, they would no longer be middle class. In their study of local labor markets, Hanson and Pratt (1995) show how mobility is implicated in the construction of gender. For example, the many women who had accepted primary responsibility for child rearing in their families chose waged jobs located close to home and schools so that they could be easily accessible to their children in an emergency. Employers reinforced such gendered mobility constructs by preferentially hiring women with shorter commutes and men with longer ones. Are women's lives truly more localized, as the *Odyssey* and *Cold Mountain* imply? How are changes in mobility related to shifting conceptions of gender and class?

Approaches to Creating Knowledge

If transportation geography is to become central again to economic and human geography, it must broaden its methodological and epistemological repertoires. The economic and engineering approaches that have long dominated the analysis of transportation issues need to be supplemented with other ways of understanding, such as those grounded in historical and cultural analyses. For example, the story about transportation and land use changes is often one starring technology (see, for example, Muller, 1995), but where do these technologies come from? Why are certain ones adopted while others are neglected? What determines how a technology will be used? In tracing out the origins of Los Angeles' low-density, freeway-reliant settlement pattern, Martin Wachs (1996) highlights the role of an ideology that associated modernity with the antithesis of the dense industrial city and equated the modern with a low-rise urban profile – the city in a garden. Also important were historically specific conceptions of what is environmentally sound: in 1920s LA, the auto and the bus were presented as more environmentally benign than railways. Wachs's analysis raises questions about the origins, development, and impact of such ideologies and discourses. That is, being open to new modes of analysis means opening up new areas of inquiry.

As another example, cultural, feminist, and historical analyses have begun to expand understanding of the role of movement and mobility in identity formation (e.g. McDowell, 1999, chapter 8). In the quote from *Cold Mountain* at the outset of this chapter, we saw in Ruby an example of an individual who constructed her identity in terms of [im]mobility. As scholars have shifted their view of the subject from one based in a unitary and stable identity to one assuming multiple and fluid identities, movement has become key. Excavating the values embedded in current transportation practices – and proposing practices based in different values – is another important form of cultural and historical analysis that deviates significantly from the traditional positivistic approaches of transportation geography (Whitelegg, 1993; Vasconcellos, 1997a). The transportation research agenda should include opening up and critically evaluating core concepts like accessibility, mobility, density, and sustainability, as well as understanding the role of cultural and political discourses in shaping transportation alternatives.

These and many of the emerging themes discussed in the previous section are best tackled from a variety of epistemological and methodological perspectives, including qualitative approaches as well as the quantitative ones that have long dominated the field. Opening transportation geography to cultural, historical, and feminist modes of analysis will deepen understanding, and holds out the hope of pointing the way toward more-sustainable transportation practices.

Conclusion

Issues of access and mobility always have been and will continue to be central to understanding space and place. When geography sought universal explanations, built around the friction of distance, transportation geography was at the core of economic and human geography. As the discipline has moved from a concern with universal theories to mid-level theories, and from embracing logical positivism to post-positivist and interpretive epistemologies, transportation geography has lost its disciplinary centrality, largely because it has remained within the analytical framework of the 1960s. Although this may in some small part reflect the lower relative importance of transportation costs now compared to 40 years ago, I think it is mostly due to a reluctance to adopting a critical stance vis à vis the reigning transportation paradigm of ever-increasing speed and embracing a broader spectrum of epistemologies and methodologies.

A changed research context – most notably the need to understand the access- and mobility-related dimensions of sustainability and IT and transportation processes outside of OECD countries – presses new questions upon transportation geography. Recognizing the complexity of transportation-related issues and the importance of context calls for deploying diverse approaches to creating knowledge. Scholars and activists have begun speaking for values other than those of speed and wealth creation, and for strategies other than road building. By focusing on far more than the friction of distance, geographers have much to contribute to the many important questions that hinge on movement, access, and mobility – in short, on transportation.

Acknowledgment

My thanks to Gen Giuliano, David Hodge, Don Janelle, Eric Sheppard, and Phil Steinberg for their comments on an earlier draft of this chapter.

Bibliography

Anas, A., Arnott, R., and Small, K. 1998. Urban spatial structure. *Journal of Economic Literature*, 36, 1426–64.

Blickstein, S. and Hanson, S. 2000. Critical mass: Forging a politics of sustainable mobility in the information age. *Transportation* (forthcoming).

Borchert, J. 1967. American metropolitan evolution. *Geographical Review*, 57, 301–32.

Breheny, M. 1995. The compact city and transport energy consumption. *Transactions of the Institute of British Geographers*, 20, 81–101.

Cervero, R. 1995. Changing live-work spatial relationships: Implications for metropolitan structure and mobility. In J. Brotchie, M. Batty, E. Blakely, P. Hall, and P. Newton (eds). *Cities in Competition: Productive and Sustainable Cities for the 21st Century*. Longman: Australia, 330–47.

Delucchi, M. 1997. The social cost of motor vehicle use. *Annals, AAPSS*, 553, 130–42.

Downs, A. 1992. *Stuck in Traffic: Coping with Peak-Hour Traffic Congestion*. Washington, D.C.: The Brookings Institution.

Dunn, J. 1998. *Driving Forces: The Automobile, Its Enemies, and the Politics of Mobility*. Washington, D.C.: The Brookings Institution.

Elwood, S. and Leitner, H. 1998. GIS and community-based planning: Exploring the diversity of neighborhood perspectives and needs. *Cartography and Geographic Information Systems*, 25, 77–88.

Fielding, G. 1995. Transit in American cities. In S. Hanson (ed). *The Geography of Urban Transportation*. New York: Guilford Press, 287–304.

Frazier, C. 1997. *Cold Mountain*. New York: Atlantic Monthly Press.

Garrett, M. and Wachs, M. 1996. *Transportation Planning on Trial: The Clean Air Act and Travel Forecasting*. Thousand Oaks: Sage Publications.

Giuliano, G. 1995. Land use impacts of transportation investments: Highway and transit. In S. Hanson (ed). *The Geography of Urban Transportation*. New York: Guilford Press, 305–41.

Glasmeier, A. K. and Fuellhart, K. G. 1999. Building on past experience: Creating a new future for distressed counties. Institute for Policy Research, Pennsylvania State University.

Gordon, P. and Richardson, H. W. 1995. Sustainable congestion. In J. Brotchie, M. Batty, E. Blakely, P. Hall, and P. Newton (eds). *Cities in Competition: Productive and Sustainable Cities for the 21st Century*. Longman: Australia, 345–58.

Hanson, S. 1998. Off the road? Reflections on transportation geography in the information age. *Journal of Transport Geography*, 6, 241–9.

Hanson, S. and Pratt, G. 1995. *Gender, Work, and Space*. NY: Routledge.

Harrison, B., Kelley, M. R., and Gant, J. 1996. Innovative firm behavior and local milieu: Exploring the intersection of agglomeration, firm effects, and technological change. *Economic Geography*, 72, 233–58.

Hodge, D. 1992. Urban congestion: Reshaping urban life. *Urban Geography*, 13, 577–88.

Hodge, D. and Janelle, D. 2000. *Measuring and Representing Accessibility in an Information Age*. Berlin: Springer-Verlag.

Kingwell, M. 1998. Fast forward: Our highspeed chase to nowhere. *Harper's Magazine*, May, 37–48.

Leinbach, T. 2000. Mobility in development context: Changing perspectives, new interpretations, and the real issues. *Journal of Transport Geography* (forthcoming).

McDowell, L. 1999. *Gender, Identity, and Place: Understanding Feminist Geographies*. Cambridge: Polity Press.

Muller, P. 1995. Transportation and urban form: Stages in the spatial evolution of the American metropolis. In S. Hanson (ed). *The Geography of Urban Transportation*. New York: Guilford Press, 26–52.

Murphy, J. J. and Delucchi, M. A. 1998. A review of the literature on the social cost of motor vehicle use in the United States. *Journal of Transportation and Statistics*, 1, 15–42.

Sheppard, E. 1995. Modelling and predicting aggregate flows. In S. Hanson (ed). *The Geography of Urban Transportation*. New York: Guilford Press, 100–28.

Sheppard, E. 2001. Information society, geography of. *International Encyclopedia of the Social and Behavioral Sciences*. London: Elsevier, (forthcoming).

Snow, T. 1986. The great train robbery. *Policy Review*, 10, 44–9.

Transportation Research Board 1994. *Curbing Gridlock: Peak-Period Fees to Relieve Traffic Congestion*. Washington, D.C.: National Academy Press.

Transportation Research Board 1997. *Toward a Sustainable Future: Addressing the Long-Term Effects of Motor Vehicle Transportation on Climate and Ecology*. Washington, D.C.: National Academy Press.

United Nations 1993. *World Population Prospects*. New York: United Nations Press.

United States Congressional Budget Office 1998. *The Economic Effects of Federal Spending on Infrastructure and Other Investments*. Washington, D.C.: U.S. Government Printing Office.

United States Department of Transportation 1994. *Transportation Statistics Annual Report 1994*. Washington, D.C.: U.S. Government Printing Office.

Vasconcellos, E. 1997a. The urban crisis in developing countries: Alternative policies for an equitable society. *World Transport Policy & Practice*, 3, 3, 4–10.

Vasconcellos, E. 1997b. The demand for cars in developing countries. *Transportation Research A*, 31, 245–58.

Wachs, M. 1995. The political context of transportation policy. In S. Hanson (ed). *The Geography of Urban Transportation*. New York: Guilford Press, 269–86.

Wachs, M. 1996. The evolution of transportation policy in Los Angeles. In A. J. Scott and E. W. Soja (eds). *The City: Los Angeles and Urban Theory as the End of the Twentieth Century*. Berkeley: University of California Press, 106–59.

Whitelegg, J. 1993. *Transport for a Sustainable Future: The Case for Europe*. London: Belhaven Press.

Whitt, J. A. S. and Yago, G. 1985. Corporate strategies and the decline of transit in U.S. cities. *Urban Affairs Quarterly*, 21, 37–65.

Chapter 29

Telecommunications and Economic Space

Barney Warf

The vast majority of employment in industrialized nations, including the United States – particularly well-paying, white-collar employment – consists of information collection, processing, and transmission in one form or another. These functions have increased in importance as computers have dramatically declined in cost and risen in power, as the production and marketing of goods and services have steadily become more information-intensive, as technological changes have accelerated, as product cycles have shortened, and as a deregulated, worldwide market has increased uncertainty and increased the competition among places for investment and jobs. Economic activities have become stretched over ever-larger distances, including the worldwide spaces of the global economy, and the means to transmit information have grown accordingly (Akwule, 1992). As a result, contemporary economic landscapes are closely tied to the deployment and use of telecommunications systems.

Telecommunications are important in another capacity as well. If geography is the study of how human beings are distributed over the Earth's surface, a vital part of that process is how we know and feel about space and time. Although space and time appear as "natural" and outside of society, they are in fact social constructions: every society develops different ways of dealing with and perceiving them (Harvey, 1990). In this reading, time and space are socially created, plastic, mutable institutions that profoundly shape individual perceptions and social relations. Telecommunications have been critical to this process for more than 150 years, accelerating the flow of information across distance and bringing places closer to one another in relative space through time–space compression (Brunn and Leinbach, 1991). As we shall see, this is quite different from the annihilation of space that some writers predict.

Telecommunications are not a new phenomenon. Beginning with the invention of the telegraph in 1844, the transmission of information over long distances was made possible; communications became detached from transportation. For decades after the invention of the telephone in 1876, telecommunication was synonymous with the simple telephone service (Brooks, 1975; de Sola Pool, 1977; Marvin, 1988). Just as the telegraph was instrumental to the colonization of the American West, in the late nineteenth century the telephone became critical to the growth of the American city-system (Abler, 1977), allowing firms to centralize their headquarters functions while they spun-off branch plants to smaller towns. Even today, despite the proliferation of several new technologies, the telephone remains by far the most commonly used form of telecommunications for businesses and households.

During the early and mid-twentieth century, the American Telegraph and Telephone Company (AT&T) enjoyed a monopoly over the US telephone industry and had few incentives to change; the primary focus was upon guaranteeing universal access, resulting in a 95 percent penetration rate among US households. The widespread deregulation of industry extended to telecommunications, and in 1984 AT&T was broken up into one long-distance and several local service providers ("Baby Bells"), and new firms such as MCI and Sprint entered the field (Warf, 1998). Faced with mounting competition, telephone companies have steadily upgraded their copper cable systems to include fiber optics lines, which allow large quantities of data to be transmitted rapidly, securely, and virtually error-free (Kahin, 1993).

In the 1980s and 1990s, as the cost of computing capacity dropped and the power increased rapidly with the microelectronics revolution, new technologies, particularly fiber optics and satellites, drastically increased the capacity of telecommunications. With the digitization of information in the late twentieth century, telecommunications steadily merged with computers to form integrated networks, most spectacularly through the Internet. New technologies such as fiber optics have complemented and at times substituted for telephone lines. Fax services and 800 number free toll calls are now standard for virtually all companies, and even newer technologies such as Electronic Data Interchange and wireless services are becoming increasingly popular. A substantial literature has demonstrated the importance of telecommunications in the globalization of producer services, particularly finance (Langdale, 1989; Warf, 1995). Within cities, digital networks have contributed to an ongoing reconstruction of urban space (Rheingold, 1993; Mitchell, 1995).

This chapter offers an overview of the role of telecommunications in economic geography. It begins by dispelling some popular myths about information systems, particularly simple post-industrial, technological determinist views that argue telecommunications entail a uniform decentralization of economic activity and the end of the importance of proximity. Second, it dwells on the role of such systems in financial markets, focusing on their impacts in global banking and securities. Third, it traces their importance to the global cities that are intimately bound to the global economy, such as New York, London, and Tokyo. Fourth, it examines the decentralization of clerical jobs at the local, regional, and global scales. Fifth, it focuses on urban infrastructural investments in this area, including telecommuting. Finally, it emphasizes the growing importance of the Internet, and social discrepancies in access to it.

Misconceptions about Telecommunications and Spatial Economic Change

There exists considerable confusion about the real and potential impacts of telecommunications on urban structure, in part due to the long history of exaggerated claims made in the past, particularly by those subscribing to "post-industrial" theory (e.g. Toffler, 1980). Often such views, which are widespread among many academics and planners, hinge upon a simplistic, utopian technological determinism that ignores the complex, often contradictory, relations between telecommunications and local economic, social, and political circumstances.

For example, repeated proclamations that telecommunications would allow everyone to work at home via telecommuting, dispersing all functions and spelling the obsolescence of cities (O'Brien, 1992; Cairncross, 1997), have fallen flat in the face of the persistence of growth in dense urbanized places. In fact, telecommunications are generally a poor substitute for face-to-face meetings, the medium through which most sensitive corporate interaction occurs, particularly when the information involved is irregular, proprietary, and unstandardized in nature. Most managers spend the bulk of their working time engaged in face-to-face contact, and no electronic technology can yet allow for the subtlety and nuances critical to such encounters. For this reason, a century of technological change, from the telephone to fiber optics, has left most high-wage, white-collar, administrative command and control functions clustered in downtown areas. In contrast, telecommunications are ideally suited for the transmission of routinized, standardized forms of data, facilitating the dispersal of functions involved with their processing (i.e. back offices) to low-wage regions. In short, there is no *a priori* reason to believe that telecommunications inevitably lead to the dispersal or deconcentration of functions: by allowing the decentralization of routinized ones, information technology may actually enhance the comparative advantage of inner cities for non-routinized functions (albeit with jobs generally filled by suburban commuters). Telecommunications facilitate the simultaneous concentration and deconcentration of economic activities (Moss, 1987).

Popular notions that "telecommunications will render geography meaningless" (see O'Brien, 1992; Cairncross, 1997) are simply naïve. While the costs of communications have decreased, other factors have risen in importance, including local regulations, the cost and skills of the local labor force, and infrastructural investments. Economic space, in short, will not evaporate because of the telecommunications revolution. It is true that networks such as the Internet allow some professionals to move into rural areas, where they can conduct most of their business on-line (Kirn et al., 1990), gradually permitting them to escape from their long-time reliance upon large cities where they needed face-to-face contact. Yet the full extent to which these systems facilitate decentralization is often countered by other forces that promote the centralization of activity. Exactly how telecommunications are deployed is a contingent matter of local circumstances, public policy, and local niche within the national and world economy. In short, telecommunications may facilitate, but do not determine, economic development; an advanced communications infrastructure is a necessary but not sufficient precondition to stimulate growth.

Telecommunications and Finance

Telecommunications have probably had their most important economic impacts in financial markets. Banks, insurance companies, and securities firms, which are very information-intensive activities, have been at the forefront of the construction of an extensive worldwide network of leased and private communication networks. Electronic funds transfer systems form the nerve center of the international financial economy, allowing banks to move capital around at a moment's notice, arbitraging interest rate differentials, taking advantage of favorable exchange rates, and avoiding political unrest (Langdale, 1989; Warf, 1989). With the breakdown of the Bretton Woods Agreement in 1971 and the collapse of fixed currency exchange rates, electronic trade in national currencies sky-rocketed (Solomon, 1997). Citicorp, for example, erected its Global Telecommunications Network to allow it to trade $200 billion daily in foreign exchange markets around the world. Such networks give banks an ability to move money – by some estimates, more than $1.5 trillion daily – around the globe at stupendous rates. Reuters, with 200,000 interconnected terminals worldwide, linked through systems such as Instinet and Globex, alone accounts for 40 percent of global financial trades each day (Kurtzman, 1993, p. 47). In securities markets, global telecommunications systems have facilitated the linking of stock and bond dealers through computerized trading programs (Hepworth, 1991). The volume and volatility of these markets rose accordingly. Trade on the New York Stock Exchange, the world's largest, rose from 10 million shares a day in the 1960s to more than 1 billion a day in the 1990s, and brokers buy and sell to foreign clients with as much ease as they do with those next door. Subject to digitization, information and capital have become two sides of the same coin. This process was essential to the rapid internationalization of finance that has occurred since the 1970s.

The ascendancy of electronic money has shifted the function of finance from investing to transacting, enhancing the attractiveness of speculation (e.g. in national currencies) rather than direct investments in productive capacity, and institutionalizing volatility in the process. Traveling at the speed of light, as nothing but assemblages of zeros and ones, global money performs an electronic dance around the world's neural networks in astonishing volumes. The world's currency markets, for example, trade roughly $800 billion every day, dwarfing the $25 billion that changes hands daily to cover global trade in goods and services. Every two weeks the total dollar volume of funds that pass through New York's fiber optic lines surpasses the annual product of the entire planet (Kurtzman, 1993). The boundaries of nation-states have little significance in this context: it is much easier, say, to move $1 billion from London to New York than a truckload of oranges from Florida to Georgia.

As large quantities of funds cross borders with mounting ease, national financial policies have become increasingly questionable in their effectiveness, making monetary controls over exchange, interest, and inflation rates ever harder to sustain. In the USA, for example, the Federal Reserve, worried about possible inflation, increased the reserve ratio of banks (the proportion legally required to be kept on hand rather than loaned out) seven times in 1994 and 1995, but found that its ability to restrict the national money supply had diminished severely. Now, it would be a mistake to exaggerate this phenomenon: clearly it is not the case that nation-states possess no leverage whatsoever over flows of capital. Yet telecommunications have obviously eased the manner in which financial capital transcends national borders.

Telecommunications have also affected the financial industries through the growth of offshore banking. Usually in response to highly favorable tax laws implemented to attract foreign firms, offshore banking has become important to many micro-states in the Caribbean (e.g. the Cayman Islands, Bahamas), Europe (e.g. Luxembourg, Jersey, Gibraltar, San Marino, Liechtenstein, and the Isle of Man), the Middle East (e.g. Cyprus, Bahrain), and the south Pacific (e.g. Vanuatu) (Roberts, 1994). As the technological barriers to moving money around internationally have fallen, legal and regulatory ones have increased in importance, and financial firms have found the topography of regulation to be of the utmost significance in choosing locations.

Among US cities, telecommunications have accelerated the spatial reorganization of financial services. By relying upon economies of scale, large firms can combine services in a few centralized database management systems. American Express, for example, shifted its credit card processing to three facilities, and Aetna Insurance consolidated 55 claims adjustment centers into 22 metropolitan regions. Allstate Insurance consolidated 28 policy processing centers into three (Charlotte, Dallas, and Columbus), CNA centralized theirs in Reading, Pennsylvania, and Travelers Insurance established two in Knoxville, Tennessee, and Albany, New York. Other insurers are developing online marketing via the Internet. Among telecommunications carriers, US WEST is consolidating its customer service workers from 563 sites into 26, AT&T has six mega-centers, and Sprint opted for lower-cost places such as Jacksonville, Florida; Dallas, Texas; Kansas City, Missouri; Phoenix, Arizona; and Winona, Minnesota. Meanwhile, local sales offices in small towns have experienced a steady decline. This phenomenon exemplifies the manner in which telecommunications can simultaneously centralize as well as decentralize different economic activities.

Telecommunications and Global Cities

One of the most significant repercussions of the internationalization of financial markets has been the growth of "global cities," particularly London, New York, and Tokyo (Moss, 1987; Sassen, 1991). Global cities act as the "command and control" centers of the world system, serving as the home to massive complexes of financial firms, producer services, and corporate headquarters of multinational corporations. In this capacity, they operate as arenas of interaction, allowing face-to-face contact, political connections, artistic and cultural activities, and opportunities for elites to rub shoulders easily. While other cities (e.g. Paris, Toronto, Los Angeles, Osaka, Hong Kong, and Singapore) certainly can lay claim to being national cities in a global economy, the trio of New York, London, and Tokyo has played a disproportionate role in the production and transformation of international economic relations in the late twentieth century.

At the top of the international urban hierarchy, global cities are simultaneously: (a) centers of creative innovation, news, fashion, and culture industries; (b) metropoles for raising and managing investment capital; (c) centers of specialized expertise in advertising and marketing, legal services, accounting, computer services, etc., and (d) the management, planning and control centers for corporations and nongovernmental organizations (NGOs) that operate with increasing ease over the entire planet (Knox, 1995). At their core, global cities allow the generation of specialized expertise upon which so much of the current global economy depends. Each city is tied through vast tentacles of investment, trade, migration, and telecommunications to clients and markets, suppliers and competitors, scattered around the world. All three metropolises are endowed with enormous telecommunications infrastructures that allow corporate headquarters to stay in touch with global networks of branch plants, back offices, customers, subcontractors, subsidiaries, and competitors. This phenomenon again illustrates how geographic centralization can be facilitated by telecommunications (Moss, 1987).

However, telecommunications simultaneously threaten the agglomeration advantages of urban areas, particularly those obtained through face-to-face communications. This trend is particularly evident in finance, the bread-and-butter of global cities. For example, the National Associated Automated Dealers Quotation System (NASDAQ) has emerged as the world's fourth largest stock market; but unlike many other exchanges, NASDAQ lacks a trading floor, connecting half a million traders worldwide through telephone cable and fiber optic lines. Similarly, Paris, Brussels, Madrid, Vancouver, and Toronto recently abolished their trading floors in favor of screen-based trading. Whether this trend will lead to the dispersal of financial trading altogether in the future remains to be seen.

Telecommunications and Back Office Relocations

Another economic activity that has been heavily affected by telecommunications is routinized back office functions (Howland, 1993), which currently employ about 250,000 people in the USA. Back offices essentially perform data entry of office records, telephone books, library catalogues, stock transfers, payroll and billing, bank checks, insurance claims, and magazine subscriptions. These tasks involve unskilled or semi-skilled labor, primarily women, and frequently operate on a 24-hour-per-day basis. Back offices have few of the interfirm linkages associated with headquarters activities: they require extensive data processing facilities, reliable sources of electricity, and sophisticated telecommunications networks.

Historically, back offices have located next to headquarters activities in downtown areas to insure close management supervision and rapid turnaround of information. However, as central city rents rose in the 1980s and 1990s and companies faced shortages of qualified (i.e. computer-literate) labor, many firms began to uncouple their headquarters and back office functions, moving the latter out of the downtown areas to cheaper locations on the urban periphery (Moss and Dunau, 1986; Nelson, 1986).

Recently, given the increasing locational flexibility afforded by satellites and interurban fiber optics lines, back offices have also begun to relocate on a much broader, continental scale, making them increasingly footloose. Many financial and insurance firms and airlines moved their back offices from New York, San Francisco, and Los Angeles to low-wage communities in the Midwest and South. Phoenix, Atlanta, and Kansas City have been particularly significant beneficiaries of this trend. Omaha, Nebraska, claims to have created 100,000 tele-generated jobs in the last decade, in part because of its location at the crossroads of the national fiber optic infrastructure (Richardson, 1994). Similarly, with abundant cheap labor, San Antonio and Wilmington, Delaware, have become well-known centers of telemarketing (*Business Facilities*, 1995). Boulder, Colorado, and Columbus, Ohio, have moved in much the same direction, in part because of their centralized geographic location.

Internationally, this trend has taken the form of the offshore office. Offshore back office operations remained insignificant until transoceanic fiber optics lines enabled their relocation on an international scale (Warf, 1993). The capital investments in such operations are minimal and they possess great locational flexibility, maximizing their ability to choose among places based on slight variations in labor costs or profitability.

The primary motivation for offshore relocation is low labor costs, although other considerations include worker productivity, skills, turnover, and benefits. Offshore back offices are established not to serve foreign markets, but to generate cost savings for US firms by tapping Third World labor pools where wages are as low as one-fifth of those in the USA. Notably, many firms with offshore back offices are in industries facing powerful competitive pressures to enhance productivity, including insurance, publishing, and airlines. Several New York-based life insurance companies, for example, relocated back office facilities to Ireland (Lohr, 1988; McGahey et al., 1990). Often situated near Shannon Airport, they ship in documents by Federal Express and export the digitized records back via satellite or one of the numerous fiber optic lines that connect New York and London. Likewise, the Caribbean, particularly Anglophone countries such as Jamaica and Barbados, has become a particularly important locus for American back offices. American Airlines paved the way in the Caribbean when it moved its data processing center from Tulsa to Barbados in 1981. Through its subsidiary Caribbean Data Services (CDS), it expanded operations to Montego Bay, Jamaica, and Santo Domingo, Dominican Republic, in 1987. Manila, in the Philippines, has emerged as a back office center for British firms, with wages 20 percent of those in the UK (Money, 1992). Such trends indicate that telecommunications may accelerate the off-shoring of many low-wage, low-value-added jobs from the USA, with dire consequences for unskilled workers.

Telecommunications and Urban Space

Telecommunications have had important effects on the urban organization of space. Urban infrastructure investments in communications technologies have remained surprisingly unimportant to policymakers (Graham and Marvin, 1996). For example, only 5 percent of US municipalities have explicit plans for telecommunications. One reason is that there is no statistical correlation between local investment in telecommunications and economic growth (Schwartz, 1990; Cronin et al., 1993; Gibbs, 1993; Gibbs and Tanner, 1997); the widespread notion that "if you build it, they will come" is not necessarily true. However, while the telecommunications industry *per se* generates relatively few jobs, and while telecommunications do not guarantee economic development, such systems have become necessities for many firms.

Although large cities typically have much better developed telecommunications infrastructures than do small ones, the technology has rapidly diffused through the urban hierarchy (Alles et al., 1994). In the future, therefore, the competitive advantages based on telecommunications will diminish, forcing competition among localities to occur on the cost and quality of labor, taxes, and local regulations. Regions with an advantage in telecommunications generally succeed because they have attracted successful firms for other reasons.

Telecommunications affect the urban infrastructure in other, less obvious ways. One increasingly important effect of new information systems is "telework" or "telecommuting," in which workers substitute some or all of their working day at a remote location (almost always home) for time usually spend at the office (Grantham and Nichols, 1994–95). The self-employed do not count as teleworkers because they do not substitute it for commuting. Telework is most appropriate for jobs involving mobile activities or routine information handling such as data entry or directory assistance (Moss and Carey, 1995). Proponents of telework claim that it enhances productivity and morale, reduces employee turnover and office space, and leads to reductions in traffic congestion (especially at peak hours), air pollution, energy use, and accidents (Handy and Mokhtarian, 1995; Van Sell and Jacobs, 1994). For example, Ernst and Young reduced its office space in Chicago's Sears Tower by 10 percent through this method, and several federal government offices are experimenting with flexible workplace arrangements around Washington, D.C. (Office of Technology Assessment, 1995). The US Department of Transportation (1993) estimated that two million people in the USA (1.6 percent of the national labor force) telecommuted one to two days per week in 1992, while the Department of Energy (1994) estimated that 4.2 million (3.3 percent) did so, a volume expected to rise to 7.5 to 15.0 million (5 to 10 percent) of the US labor force by 2002. Further growth of telecommuting will encourage more decentralization of economic activity in suburban areas.

However, as Graham and Marvin (1996) point out, there are countervailing reasons why telecommunications may *increase* the demand for transportation rather than decrease it. First, while telecommuters spend fewer days at their workplace, it is not at all clear that they have shorter *weekly* commutes overall; indeed, by allowing them to live farther from their home, the total distances traveled may actually rise. Second, time freed from commuting may be spent traveling for other purposes, such as shopping or recreation. Telecommuting may alter the reasons for travel, but not necessarily the frequency or volume. Third, by reducing congestion, telecommuting may lead to significant induced effects, whereby others formerly inhibited from driving may be induced to do so. In short, whether or not an actual trade-off between telecommunications and commuting occurs (their substitutability rather than complementarity) is not clear.

Another potential, and growing, impact of information systems on urban form concerns transportation informatics, including a variety of improvements in surface transportation such as smart metering, electronic road pricing, synchronized traffic lights, automated toll payments and turnpikes, automated road maps, information for trip planning and navigation, travel advisory systems, electronic tourist guides, remote traffic monitoring and displays, and computerized traffic management and control systems, all of which are designed to minimize congestion and optimize traffic flow (particularly at peak hours), enhancing the efficiency, reliability, and attractiveness of travel (Office of Technology Assessment, 1995). Wireless technologies such as cellular phones allow more productive use of time otherwise lost to congestion. Such systems do not so much comprise new technologies as the enhancement of existing ones. In-vehicle navigation systems with inboard computers, such as the Intelligent Vehicle Highway System currently underway in southern California, represent the next generation of this technology. Finally, many buildings are implementing information systems to replace manual reading of meters for utility consumption, monitor wind shear, or computerize the control of heating, ventilation, lighting, and security.

The widespread use of telecommunications has eased the locational mobility of firms and reinforced changes in the nature of urban planning. Desperate for jobs, many localities compete with one another with ever-greater concessions to attract firms, forming an auction that resembles a zero-sum game. The effects of such a competition are hardly beneficial to those with the least purchasing power and political clout. Left to sell themselves to the highest corporate bidder, localities frequently find themselves in a "race to the bottom" in which entrepreneurial governments promote growth – but do not regulate its aftermath – via tax breaks, subsidies, training programs, looser regulations, low-interest loans, infrastructure grants, and zoning exemptions (Hall and Hubbard, 1996). As a result, local planners have become increasingly less concerned with issues of social redistribution, compensation for negative externalities, provision of public services, and so forth, and more enthralled with questions of economic competitiveness, attracting investment capital, and the production of a favorable "business climate."

The Geopolitics of the Internet

Among the various networks that comprise the nation's and world's telecommunications infrastructure, the largest and most famous is the Internet, an unregulated electronic network connecting an estimated 100 million people in more than 100 countries (Warf, 1995). From its military origins in the USA in the 1960s, the Internet emerged on a global scale through the integration of existing telephone, fiber optic, and satellite systems, which was made possible by the technological innovation of packet switching and Integrated Services Digital Network (ISDN), in which individual messages may be decomposed, the constituent parts transmitted by various channels, and then reassembled, virtually instantaneously, at the destination. Spurred by declining prices of services and equipment, the Internet has grown worldwide at a rapid rate, the number of users doubling roughly every year. Popular access systems in the USA, such as Compuserve, Prodigy, and America On-Line allow any individual with a microcomputer and modem to "plug in" to cyberspace.

Despite exaggerations regarding the potential impacts of the Internet and the World Wide Web, such systems will clearly have substantial, if unanticipated, effects upon the social fabric over time (Negroponte, 1995). As Graham and Aurigi (1997, p. 26) note, "Large cities, based, in the past, largely on face-to-face exchange in public spaces, are dissolving and fragmenting into webs of indirect, specialized relationships." Information systems such as the Internet may reinforce existing disparities in wealth, connecting elites in different nations who may be increasingly disconnected from the local environments of their own cities and countries. Indeed, in a socio-psychological sense, cyberspace may allow for the reconstruction of "communities without propinquity," groups of users who share common interests but not physical proximity (Anderson and Melchior, 1995). In the age of the "City of Bits" (Mitchell, 1995), in which social life is increasingly mediated through computer networks, the reconstruction of interpersonal relations around the digitized spaces of cyberspace is of the utmost significance.

Significant discrepancies exist in terms of access to the Internet, largely along the lines of wealth, gender, and race: while one-third of US households have personal computers, only 12 percent have modems at home. Access to computers linked to the Internet, either at home or at work, is highly correlated with income; the median household income of Internet users is $79,000, almost twice the national average (*Internet World*, 1996). The Internet is also segregated by gender: roughly 80 percent of all US Internet users and 82 percent of World Wide Web users are male (Doctor, 1994; Doheny-Farina, 1996; Miller, 1996). A survey by the *Los Angeles Times* of 1,200 networked computer users in southern California found that they are overwhelmingly wealthy, young, and white (Harmon, 1996). A survey by the National Telecommunications and Information Administration (1995) revealed that white households used networked computers three times more frequently than did black or Latino ones (NTIA 1995). The elderly likewise often find access to the Internet to be intimidating and unaffordable, although they comprise the fastest-growing demographic group of users. American Internet users thus tend to be overwhelmingly white and middle class, well educated, younger than average, and employed in professional occupations demanding college degrees.

Social and spatial differentials in access to the skills, equipment, and software necessary to get onto the electronic highway threaten to create a large, predominantly minority underclass deprived of the benefits of cyberspace (Wresch, 1996). This phenomenon must be viewed in light of the growing inequalities throughout industrialized nations generated by labor market polarization (i.e. deindustrialization and growth of low-income, contingent service jobs). Modern economies are increasingly divided between those who are comfortable and proficient with digital technology and those who neither understand it nor trust it, disenfranchising the latter group from the possibility of citizenship in cyberspace. Despite the falling prices for hardware and software, basic entry-level machines for Internet access cost roughly $1,000 – an exorbitant sum for low-income households. Internet access at work is also difficult for many: for employees in poorly paying service jobs that do not offer access to the Internet at their place of employment (the most rapidly growing category of employment), the obstacles to access are formidable.

Nor does the public educational system offer an easy remedy. Even in the USA the wide discrepancies in funding and the quality of education among school districts, particularly between suburban and central city schools, may reproduce this inequality rather than reduce it (Kozol, 1991). While some public libraries offer free access to the Internet, there are currently only about 80 networked ones in operation in 29 states (Guthrie and Dutton, 1992; Schuler, 1996). Mounting financial constraints in many municipalities, moreover, have curtailed the growth of these systems (Norris and Kraemer, 1996). Even within the most digitized of cities there remain large pockets of "off-line" poverty (Thrift, 1995; Resnick and Rusk, 1996; Sawicki and Craig, 1996). Those who need the Internet the least, already living in information-rich environments with access through many non-Internet channels (e.g. newspapers and cable TV), may have the most access to it, while those who may benefit the most (e.g. through electronic job banks) may have the smallest chance to log on.

To some degree, public policy can ameliorate these social and spatial discrepancies (Bowe, 1993; Kahin, 1993). The Clinton administration launched the National Information Infrastructure (NII), promising to connect every classroom, library,

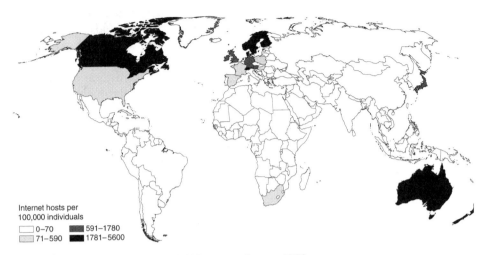

Figure 29.1   Internet hosts per 100,000 persons, January 1998
Source: Compiled by author from data at http://www.nw.com/zone/WWW/dist-byname.html

hospital, and clinic in the nation by the year 2000 (NTIA, 1995). In 1996, the Telecommunications Act sought to reduce spatial inequalities by guaranteeing access to basic telecommunications services, an important step for low-income rural areas where the marginal costs of installation and maintenance tend to be high.

These discrepancies in access are replicated at the international level, where, given the American and European dominance, the "World Wide Web" hardly lives up to its name. Inequalities in access to the Internet internationally, measured in terms of hosts per 100,000 people (figure 29.1), reflect the long-standing bifurcation between the First and Third Worlds. The best-connected nations are in Scandinavia, Canada, and Australia; the USA, surprisingly, is ranked relatively low in this regard, a reflection of its sizeable, poorly served population. Yet due to the enormous size of the American economy, 90 percent of all international Internet traffic is either to or from the USA. Even Japan remains relatively poorly connected in this regard. Outside of North America and Europe, the vast bulk of the world's people, particularly in the Third World, have little to no Internet access.

Access to the Internet is deeply conditioned by the density, reliability, and affordability of national telephone systems. Most Internet communications occurs along lines leased from telephone companies, many of which are state-regulated (in contrast to the largely unregulated state of the Internet itself). Prices for access vary by length of the phone call, distance, and the degree of monopoly: in nations with telecommunications monopolies, prices are 44 percent higher than in those with deregulated systems (*Economist*, 1996). The global move toward deregulation in telecommunications will likely lead to more use-based pricing (the so-called "pay-per" revolution), in which users must bear the full costs of their calls, and fewer cross-subsidies among different groups of users (e.g. between commercial and residential ones), a trend that will likely make access to cyberspace even less affordable to low-income users.

Conclusions

Economic activity today is overwhelmingly dominated by the production of intangibles, in which knowledge, information, and communications are critically important. The geography of the world economy rests heavily upon invisible flows of data and capital, binding places unevenly to the world-system. Telecommunications have also transformed the ways in which economic landscapes are constructed, suturing localities to global "spaces of flows" that move at the speed of light and giving firms unprecedented mobility and flexibility (Castells, 1996). At the end of the millennium, telecommunications had ushered in a vast round of time–space compression, linking places together to an historically unprecedented degree. Markets, labor processes, transportation routes, planning policies, and spatial structures have changed accordingly.

Cities and regions search for a competitive advantage in this world in several ways. In large metropolitan areas with dense complexes of firms bound together by face-to-face interactions, telecommunications have left skilled, high-wage functions largely intact, but not unchallenged – as the electronic trading of stocks suggests. By allowing firms to stay in contact with operations around the world, telecommunications have contributed to the centralization of key activities in global cities such as New York, London, and Tokyo, which rely upon their extensive connections to the global telecommunications infrastructure to serve as nerve centers of the world economy. For such tasks, telecommunications have been largely unable to substitute for face-to-face interactions. Other services that process routinized information, however, and rely upon unskilled, low-wage labor, are highly vulnerable to substitution by new telecommunications systems and have decentralized (e.g. back offices). A key theme in understanding the new economic landscapes of the information economy is that telecommunications tend to reinforce the agglomeration of high-wage, high-value-added, white-collar functions while decentralizing low-wage, low-value-added, blue (or pink)-collar ones.

In short, telecommunications have a variety of impacts upon cities and regions, both positive and negative, local and global, centralizing and decentralizing. To appreciate the complexity of these effects requires a step back from the simple utopianism and technological determinism that tends to pervade much public opinion and policymaking. Whether they be the telephone, Geographic Information Systems (Pickles, 1995), or the Internet, telecommunications systems are a social product, interwoven with relations of class, race, gender, and power (Jones, 1995; Shields, 1996). A growing body of critical literature has detailed how electronic systems are used to monitor everyday life, including credit cards, visas and passports, tax records, medical data, police reports, telephone calls, utility records, automobile registration, crime statistics, and sales receipts (Lyon, 1994). In this light, telecommunications can be used against people as well as for them. The unfortunate tendency in the popular media to engage in technocratic utopianism largely obscures these power relations. *Contra* this post-industrial utopian perspective, social categories of wealth and power and geographical categories of core and periphery continue to be reinscribed in cyberspace.

Given the rapid rate of technological change in the late twentieth century, predictions about the future of telecommunications are hazardous at best. Capitalism has had a very long history of technological and economic changes that periodically refashion local and global landscapes: the information revolution of the late twentieth century is the latest chapter in this story.

Bibliography

Abler, R. 1977. The telephone and the evolution of the American metropolitan system. In I. de Sola Pool (ed). *The Social Impact of the Telephone*. Cambridge: MIT Press, 318–41.

Akwule, R. 1992. *Global Telecommunications: The Technology, Administration, and Policies*. Boston: Focal Press.

Alles, P., Esparza, A., and Lucas, S. 1994. Telecommunications and the large city-small city divide: Evidence from Indiana cities. *Professional Geographer*, 46, 307–15.

Anderson, T. and Melchior, A. 1995. Assessing telecommunications as a tool for urban community building. *Journal of Urban Technology*, 3, 29–44.

Bowe, F. 1993. Access to the information age: Fundamental decisions in telecommunications policy. *Policy Studies Journal*, 21, 2–19.

Brooks, J. 1975. *Telephone: The First Hundred Years*. New York: Harper and Row.

Brunn, S. and Leinbach, T. (eds). 1991. *Collapsing Space and Time: Geographic Aspects of Communications and Information*. London: HarperCollins.

Business Facilities. 1995 Business Services. February, 18–22.

Cairncross, F. 1997. *The Death of Distance: How the Communications Revolution will Change Our Lives*. Boston: Harvard Business School Press.

Castells, M. 1996. *The Rise of the Network Society*. Oxford: Blackwell.

Cronin, F., Parker, E., Colleran, E., and Gold, M. 1993. Telecommunications infrastructure investment and economic development. *Telecommunications Policy*, August, 415–30.

De Sola Pool, I. (ed). 1977. *The Social Impact of the Telephone*. Cambridge, MA: MIT Press.

Doctor, R. 1994. Seeking equity in the national information infrastructure. *Internet Research*, 4, 9–22.

Doheny-Farina, S. 1996. *The Wired Neighborhood*. New Haven: Yale University Press.

Economist. 1996. The economics of the internet. October 19, 21–4.

Gibbs, D. 1993. Telematics and urban development policies. *Telecommunications Policy*, May/June, 250–56.

Gibbs, D. and Tanner, K. 1997. Information and communication technologies and local economic development policies: the British case. *Regional Studies*, 31, 768–74.

Graham, S. 1992a. Electronic infrastructures and the city: some emerging municipal policy roles in the U.K. *Urban Studies*, 29, 755–81.

Graham, S. 1992b. The role of cities in telecommunications development. *Telecommunications Policy*, April, 187–93.

Graham, S. and Aurigi, A. 1997. Virtual cities, social polarization, and the crisis in urban public space. *Journal of Urban Technology*, 4, 19–52.

Graham, S. and Marvin, S. 1996. *Telecommunications and the City: Electronic Spaces, Urban Places*. London: Routledge.

Grantham, C. and Nichols, L. 1994–95. Distributed work: learning to manage at a distance. *The Public Manager*, winter, 31–4.

Guthrie, K. and Dutton, W. 1992. The politics of citizen access technology: the development of public information utilities in four cities. *Journal of Policy Studies*, 20, 574–96.

Hall, T. and Hubbard, P. 1996. The entrepreneurial city: new urban politics, new urban geographies? *Progress in Human Geography*, 20, 153–74.

Handy, S. and Mokhtarian, P. 1995. Planning for telecommuting – measurement and policy issues. *Journal of the American Planning Association*, 61, 99–111.

Harmon, A. 1996. Computing in the '90s: the great divide. *Los Angeles Times*, October 7, 1.

Harvey, D. 1990. Between space and time: reflections on the geographical imagination. *Annals of the Association of American Geographers*, 80, 418–34.

Hepworth, M. 1990. *Geography of the Information Economy*. London: Guilford Press.

Hepworth, M. 1991. Information technology and the global restructuring of capital markets. In S. Brunn and T. Leinbach (eds). *Collapsing Space and Time: Geographic Aspects of Communications and Information*. London: HarperCollins, 132–48

Howland, M. 1993. Technological change and the spatial restructuring of data entry and processing services. *Technological Forecasting and Social Change*, 43, 185–96.

Internet World. 1996. "The State of the Net." 7, 1.

Jones, S. (ed). 1995. *CyberSociety: Computer-Mediated Communication and Community*. Beverly Hills, CA: Sage.

Kahin, B. 1993. *Building Information Infrastructure: Issues in the Development of the National Research and Education Network*. Boston: McGraw-Hill.

Kirn, T., Conway, R., and Beyers, W. 1990. Producer services development and the role of telecommunications: a case study of rural Washington. *Growth and Change*, 21, 33–50.

Kozol, J. 1991. *Savage Inequalities: Children in America's Schools*. New York: HarperPerennial.

Knox, P. 1995. World cities and the organization of global space. In R. Johnston, P. Taylor and M. Watts (eds). *Geographies of Global Change*. Oxford: Blackwell, 232–47.

Kurtzman, J. 1993. *The Death of Money*. Boston: Little, Brown and Co.

Langdale, J. 1989. The geography of international business telecommunications: the role of leased networks. *Annals of the Association of American Geographers*, 79, 501–22.

Lohr, S. 1988. The growth of the global office. *New York Times*, Oct. 18, D1.

Lyon, D. 1994. *The Electronic Eye: The Rise of Surveillance Society*. Minneapolis: University of Minnesota Press.

Marvin, C. 1988. *When Old Technologies Were New: Thinking about Electric Communication in the Late Nineteenth Century*. Oxford: Oxford University Press.

McGahey, R., Malloy, M., Kazanas, K., and Jacobs, M. 1990. *Financial Services, Financial Centers: Public Policy and the Competition for Markets, Firms and Jobs*. Boulder: Westview Press.

Miller, S. 1996. *Civilizing Cyberspace: Policy, Power and the Information Superhighway*. New York: ACM Press.

Mitchell, W. 1995. *City of Bits: Space, Place, and the Infobahn*. Cambridge, MA: MIT Press.

Money, J. 1992. White collar jobs flow from Britain as data processors are lured by cheap labour. *The Guardian*, August 25, 1.

Moss, M. 1987. Telecommunications, world cities, and urban policy. *Urban Studies*, 24, 534–46.

Moss, M. and Carey, J. 1995. Information technologies, telecommuting, and cities. In J. Brotchie, M. Batty, E. Blakely, P. Hall, and P. Newton (eds). *Cities in Competition: Productive and Sustainable Cities for the 21st Century*. Sydney: Longman, 181–99.

Moss, M. and Dunau, A. 1986. Offices, information technology, and locational trends. In J. Black, K. Roark, and L. Schwartz (eds). *The Changing Office Workplace*. Washington, D.C.: Urban Land Institute, 171–82.

National Telecommunications and Information Administration, 1995. *Falling through the Net: A Survey of the Have Nots in Rural and Urban America*. U.S. Department of Commerce, Washington, D.C.: U.S. Government Printing Office.

Negroponte, N. 1995. *Being Digital*. New York: Knopf.

Nelson, K. 1986. Labor demand, labor supply and the suburbanization of low-wage office work. In A. Scott and M. Storper (eds). *Production, Work, Territory*. Boston: Allen & Unwin, 149–71.

Norris, D. and Kraemer, K. 1996. Mainframe and PC computing in American cities: myths and realities. *Public Administration Review*, 56, 568–76.

O'Brien, R. 1992. *Global Financial Integration: The End of Geography.* Washington: Council on Foreign Relations.

O'Connell, V. 1995. Brokerage firms are moving into cyberspace. *Wall Street Journal*, July 6, C1.

Office of Technology Assessment. 1995. *The Technological Reshaping of Metropolitan America.* Washington, D.C.: U.S. Government Printing Office.

Pickles, J. (ed). 1995. *Ground Truth: The Social Implications of Geographic Information Systems.* New York: Guilford Press.

Resnick, M. and Rusk, N. 1996. Access is not enough: computer clubhouses in the inner city. *The American Prospect*, July–August, 60–8.

Rheingold, H. 1993. *The Virtual Community: Homesteading on the Electronic Frontier.* New York: HarperPerennial.

Richardson, R. 1994. Back officing front office functions – organisational and locational implications of new telemediated services. In R. Mansell (ed). *Management of Information and Communication Technologies.* London: Aslib, 309–55.

Roberts, S. 1994. Fictitious capital, fictitious spaces: the geography of offshore financial flows. In S. Corbridge, R. Martin, and N. Thrift (eds). *Money, Power and Space.* Oxford: Blackwell.

Sassen, S. 1991. *The Global City: New York, London, Tokyo.* Princeton, NJ: Princeton University Press.

Sawicki, D. and Craig, W. 1996. The democratization of data: bridging the gap for community groups. *Journal of the American Planning Association*, 68, 219–31.

Schuler, D. 1996. *New Community Networks: Wired for Change.* New York: Addison-Wesley.

Schwartz, G. 1990. Telecommunications and economic development policy. *Economic Development Quarterly*, 4, 88–91.

Shields, R. (ed). 1996. *Cultures of Internet: Virtual Spaces, Real Histories, Living Bodies.* London: Sage Publications.

Solomon, E. 1997. *Virtual Money: Understanding the Power and Risks of Money's High-Speed Journey into Electronic Space.* Oxford: Oxford University Press.

Thrift, N. 1995. A hyperactive world. In R. Johnston, P. Taylor, and M. Watts (eds). *Geographies of Global Change.* Oxford: Blackwell, 18–35.

Toffler, A. 1980. *The Third Wave.* New York: William Morrow.

U.S. Department of Energy. 1994. *Energy, Emissions and the Social Consequences of Telecommuting: Energy Efficiency in the U.S. Economy.* Technical Report One, Doe/Po-0026: Washington, D.C.: U.S. Government Printing Office.

U.S. Department of Transportation. 1993. *Transportation Implications of Telecommuting.* Washington, D.C.: U.S. Department of Transportation.

Van Sell, M. and Jacobs, S. 1994. Telecommuting and quality of life: a review of the literature and a model for research. *Telematics and Informatics*, 11 (2), 81–93.

Warf, B. 1989. Telecommunications and the globalization of financial services. *Professional Geographer*, 41, 257–71.

Warf, B. 1993. Back office dispersal: implications for urban development. *Economic Development Commentary*, 16, 11–16.

Warf, B. 1995. Telecommunications and the changing geographies of knowledge transmission in the late 20th century. *Urban Studies*, 32, 361–78.

Warf, B. 1998. Reach out and touch someone: AT&T's global operations in the 1990s. *Professional Geographer*, 50, 255–67.

Wilson, R. and Teske, P. 1990. Telecommunications and economic development: the state and local role. *Economic Development Quarterly*, 4, 158–74.

Wresch, W. 1996. *Disconnected: Haves and Have-nots in the Information Age.* New Brunswick, NJ: Rutgers University Press.

Chapter 30

International Political Economy

Michael Webber

The large-scale questions of economic geography center on the evolution of the world economy. By *economy* I mean the assemblage of rules and practices by which the nature of the productive effort is determined, by which labor is allocated to the different tasks of that effort, and by which the rewards of that labor are allocated to individuals. Understanding the geography of the world economy generates a vast array of questions about the relations between the rules and practices within different places; about the relationship between the history of economy within a particular place and general theory; and about the relations between economic characteristics, politics, and other arenas of social life. The intellectual terrains within which such forms of research are conducted include economic history and international political economy.

International political economy (IPE) is an approach to (or a branch of) international relations.[1] Within political science, it is distinguished by two claims. The first, defining, characteristic is its insistence that understanding international political relations means also understanding the economic relations between places, and that to understand the economic relations between places it is necessary to understand their political relations. IPE is therefore about the interrelationships between politics and economics at a world scale. The second distinguishing interest of IPE is the relationship between what happens inside states and what happens outside them – the relations between internal and international events. So the central problematic of IPE is the intersection of a pair of themes (figure 30.1).

In the past, the rules and practices of economy have been clearly nationally specific, as well as regionally variable. Now, economic geography, like other forms of human geography, is dominated by claims that the nationally specific economy is breaking down as a globalized form of economy takes shape. In this new globalized form of economy, the capacities of states to articulate and to shape national social goals are being renegotiated as social and economic policies seem to become increasingly homogeneous. A central task of modern economic geography is to understand how the geography of the world economy is being reworked. The principles that IPE emphasizes can assist geographers in this task. On the one hand, debates over

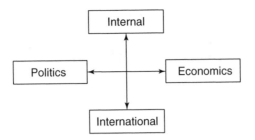

Figure 30.1 The problematic of international political economy
Source: Author

globalization must be conducted at scale-intersecting levels, for to understand global means to understand local events; and to understand what is happening here means to understand what is global (and going on there). This framework disarms the distinction between scales. On the other hand, the emergence of a globalized economy exemplifies the intersection of economics and politics. This prescription is not widespread among all varieties of social scientists, many of whom regard economics as a matter of pure rationality: into decisions about the economy, matters of mere politics should not intrude. It is the insistence of IPE on these two principles that makes it such a useful framework within which to think about and study the emergence of a new globalized economic geography.

This chapter seeks to demonstrate the validity of this assertion through five major sections. First, since IPE is part of another discipline, its language is different from that of economic geography; so the chapter opens by summarizing the key theoretical positions within international relations. On that basis, the second section identifies the implications of a variety of post-positivist theories of social science for recent international relations. Then the chapter goes on to portray the central ideas of IPE and the main lines of debate within it over the past decade or so. The fourth section surveys the recent intellectual development of some aspects of globalization within IPE. To a large extent, as the diagram above makes clear, globalization is a theme that is necessarily embedded within IPE. The conclusion of the chapter uses this information to state a research agenda for geographers about the evolution of the global economy. I hope to demonstrate that this agenda together with the central themes of IPE provides an important and useful framework within which to think about bigger questions than those that drive a lot of economic geography nowadays.

The Language of International Relations

The three classic theoretical traditions of economics are the subjective preference theory of value (presently, neoclassical theory), the cost of production theory of value (derived from Ricardo and Keynes), and the abstract labor theory of value (drawing on Marx).[2] Within international relations, these theoretical traditions are loosely reflected in what political scientists call realist theories, liberal theories, and Marxist theories respectively.[3] This section introduces the classic theoretical distinctions as a prelude to more recent thinking.

Realists believe that states are the only actors that really count in international politics.[4] The state in this view is understood as a unitary, strategic actor (following the "national interest"), rather than fragmented and buffeted by a variety of competing domestic interests. The state is a territorially bounded political community united by a central agency capable of expressing and enforcing common interests and goals.

States are, according to realists, crucial foci in the distribution of power across the world. Internally, states are regarded as having a monopoly over the legitimate use of physical force within a given territory, a legitimacy that is granted by the citizens of the state and by other states. Internationally, all states coexist in an anarchic system (i.e. lacking any common power over states). Since no global government has power over all states, cooperation between states can only be based on promises; therefore states must in the end rely on themselves to achieve their key objective – survival.[5] So states seek to maximize their relative power or capability. In other words, the anarchical structure of international relations (rather than the historical and social circumstances of individual states) determines the form, functions, and behavior of states. Differences in state goals at the international level reflect differences in their power.

Realism is *the* theory of power politics. Since its adherents have close ties to governments, it has been the principal theoretical position within international relations, particularly in the USA. Challenges to it come first from those who argue that world politics ought to be conducted on a more moral basis. There are also empirical challenges, however, principally from those who claim that other actors share international power with states – transnational corporations, international human rights organizations, and such economic organizations as the International Monetary Fund and the World Trade Organization, for instance.[6] The theoretical critics of realism argue that the central concept of power is well-nigh undefined, and that an anarchic international order does not imply that states can only rely on themselves.

The principal theoretical alternative to realism in international relations has been *liberalism*, at least until recently. In many respects the key principle of liberalism is its rejection of realism.[7] In a crucial analogy to the invisible hand mechanism of Adam Smith, the natural order envisaged by such nineteenth-century liberals as Jeremy Bentham was one of peace, individual liberty, free trade, and prosperity. So the arbitrary use of power by states – in opposition to the wishes of people – underpinned much international conflict. Such liberals argued that, just as the behavior of individuals was regulated within a state, so the international behavior of states should be regulated by a system of legal rights and duties. One outcome of this belief was President Wilson's advocacy of the League of Nations after World War I.

Despite the history of the League of Nations, liberal ideas re-emerged after World War II. They underpinned attempts to create transnational organizations to resolve common problems (see the list below). As they have championed in practical politics the emergence of such organizations, so liberals have increasingly theorized the international system as one in which states no longer have exclusive power. Power has to be shared between states, transnational corporations, and other organizations (the *pluralist* thesis); and international *regimes* offer frameworks within which states can learn to cooperate.[8]

The conditions under which regimes can emerge and be effective comprises one of the central liberal research agendas of recent times. Regime theory is one of the battlegrounds between realists and idealists. Regime theory also provides the principal theoretical lens through which international negotiation over environmental governance is understood. Another important item in the liberal research agenda concerns the democratic peace thesis: that liberal democracies do not go to war against other liberal democracies. Since there seems to be empirical evidence in support of this thesis, the agenda asks why this is so (people's abhorrence of war? relative wealth?) and what policy implications should be drawn (how should liberal democratic values be spread?).

Within international relations, Marxist ideas have been reflected mainly through *world systems theory* (Wallerstein, 1974, 1980, 1989; Frank, 1979).[9] Crudely put, world systems theory claims:

- that the events of world politics occur within a structure that shapes, defines, and determines them;
- that structure is the world economy, organized according to the logic of global capitalism.

This capitalist world economy emerged in Europe in the sixteenth century, integrating states in a system of core, peripheral, and semiperipheral relations. Core regions (Western Europe and North America) exploit peripheral regions (Africa, Latin America, and South Asia); the semiperiphery (including Australia) is simultaneously exploiter and exploited.

In world systems theory, the form of the capitalist world economy and the structure of the inter-state system are intertwined. Internally, states perform important functions for capitalism: for example, guaranteeing property rights and seeking to ameliorate such contradictions as that between wages as a source of demand and wages as a cost of production. The structure of the world economy means that states are competing with each other, so they do not have the power to dominate "their" capitalists: competition between states for political and military power depends partly on their economic power, which in turn rests on the strength of their domestic corporations, and that implies that states have broadly to support rather than restrict those corporations. This idea is a key insight into contemporary international political economy: what is the relationship between deregulation (especially of capital flows) and the power of states over corporations?

The Recent Debates: Importing Post-positivism

Broader movements within social science have been integrated with these three traditions to produce the recent debates about international relations. Rather later than in many other social sciences, *constitutive*[10] and *anti-foundationalist*[11] theories have appeared within international relations during the last decade. In some ways, the recent history of international relations as a discipline can be represented as the unification of liberal and realist traditions against attacks from a variety of post-positivist approaches to social science. Thus, what is now called the "rationalist synthesis" represents a common, neoliberal, and neorealist approach to common questions. While neoliberals and neorealists offer slightly different answers to the questions, they now share fundamental assumptions about the inter-state system: that the state is the most important actor; and that actors are utility maximizers (Baldwin, 1993). Although rationalism reflects the thinking of the foreign policy establishment, especially of the USA, it has not comprehended the nature of the challenge laid down by post-positivist theories. So rationalism is not interesting as a framework *for* study.[12] More vital are the alternatives.

Women have finally made an appearance. Feminist research follows a number of approaches to international relations:

- What are the roles played by women in international relations, and why is the importance of those roles hidden? (Enloe, 1990, 1993).
- How do the world capitalist system and patriarchy combine to produce systematic disadvantage for women as compared to men? (Bunch and Reilly, 1994; Institute for Women, Law and Development, 1993).
- In what respects is existing knowledge about international relations male knowledge, and what is female knowledge of international relations? (Weber, 1994; Marchand, 1996; Shapiro and Bilmayer, 1999).

Postmodernists have also surfaced in international relations theory. International relations would seem an especially fruitful field within which to develop examples of Foucault's claim that the knowledge that we produce depends on and reinforces existing power relations. To think seriously of big concepts like "the liberal international economic order" or "the anarchic inter-state system" or even of lower-level concepts such as "free trade" and "deregulation" is to think about the manner in which ideology is normalized and power hidden by the way in which words are used. Derrida's method of double-reading can be applied to deconstruct such concepts as anarchy (Ashley, 1988). In like manner, Weber (1995) and Biersteker and Weber (1996) have examined the manner in which the concept of state sovereignty is constructed socially.

Within international relations, historical sociologists have studied the evolution of the relationship between the form of domestic societies and the international system. They question the manner in which structures are produced by social processes. Tilly (1990), for example, has identified how the nation-state emerged in Europe as the predominant state form: after about 1500, the nation-state was the only form of state that could afford to fight the types of war that occurred. Structuration theory provides a basis for this work (Reus-Smit, 1996):

- social structures (the rules and resources of social reproduction) define what constitutes a legitimate actor and the actor's realm of justifiable action; but
- social structures do not exist independently of individuals' practices – they are created by the persistence of routine practices.

So, the structural properties of social systems are both the medium and the outcome of the practices they organize. Such ideas generate questions about the relations between domestic and international policy practices, about the emergence of agents of international economic governance, and about the relative power of different agencies of governments.

Finally, other varieties of Marxism than world systems theory have become important in international relations, especially through the work of Cox (1987, 1994).[13] Examples of work inspired by Marxism will be discussed below.

Recent Debates within IPE

In my introductory remarks IPE was distinguished by its insistence on the need to understand interrelationships between politics and economics at a world scale, and by its concern with the relations between what happens inside states and what happens outside them. We need to flesh out these definitional remarks with some substantive claims, before examining some recent debates in IPE.

Commenting that the study of international relations pays far too much attention to the interests and actions of governments and far too little to the interests of people, Strange (1994a, pp. 17–18) begins her study of IPE by observing that human beings seek four critical values through societies: individual wealth; personal security; freedom (or the right to choose); and justice (or equality of opportunity). (Many add that environmental sustainability has become another critical value: see Diesendorf and Hamilton, 1997.) She goes on to define IPE as concerned with "the social, political and economic arrangements affecting the global systems of production, exchange and distribution, and the mix of values reflected therein" (Strange, 1994a, p. 18). (The values are the critical four: wealth, security, freedom, and justice.) International political economists must therefore study the way things are managed, how they came to be managed in that way, and the choices that remain for the future. Strange's book is an extended essay on the manner in which state–market relations embody the weights given in different societies to the four critical values, and therefore the emphases placed on those values by states and institutions as they go about constructing the international political economy anew.[14] The role of markets in social life depends on the role they are allowed by those with authority.

Power is the capacity of one institution or state to get other institutions or states to do what it wants – by virtue of coercive force, wealth, or moral authority. Power is used

- directly, by making an explicit or tacit threat; or
- indirectly, by constructing the rules of the game in such a way that others freely choose to do what is desired.[15]

Any critical study of IPE must therefore identify who holds power and why. Marxists identify the structure of production as *the* source of social power, so the state is the political embodiment of class power.[16] Cox (1987) provides an historical and wide-ranging study of the evolution of the international political economy along Marxist lines that traces changes in modes of production, in the social organization of states, in the relations between national economies, and in the structure of the inter-state system. Strange (1994a, pp. 26–32) also recognizes the structure of production as one source of power, but adds control over security, control over the finance system, and control over knowledge as others. The nature and consequences of US military power, the financial events in Southeast and East Asia since mid-1997, and Foucault-like studies of power and knowledge all certainly indicate that security, finance, and knowledge are sources of power. Marxists would ask, however, about the extent to which those sources of power are independent of the power that derives from the production system. Issues of internal domestic power and international power, of authority, and of markets are vital to the current preoccupation of IPE: globalization.

Globalization, as an issue in IPE, generates a huge array of questions:

- What are the characteristics of the emerging global economy? To what degree can we perceive globalization rather than simply internationalization? What is the degree of novelty involved? What are the new balances of wealth, security, freedom, and justice – for whom, and where?
- Whose power has been deployed to underpin the emergence of this global economy? What are the crucial struggles of interest inherent in globalization?
- Is the emergence of this economy altering earlier balances of power between states and markets? What does a new political economy imply for future distributions of wealth, security, freedom, and justice?

The global preoccupation underlies research on most of the immediate questions of IPE. In the following section I will explore some aspects of this preoccupation a little more carefully. During this exploration, I shall present a perspective that draws on recent trends in Marxist IPE, and contrast that to the views that draw upon rationalism. What I want to demonstrate is that differences in *a priori* theoretical perspective dominate empirical and interpretative research. Even the act of defining globalization separates theoretical persuasions. If definitions are different, so are the events one looks for. If the events are different, then so are the interpretations. Rationalists tend to regard changes in the organization of the global economy as having causes that are technological (communications technologies, especially) and political (states learn about appropriate policy settings); in a constructivist-Marxist mode, the organization of the global economy is created by state actions (in the context of previous actions) that themselves are encouraged by dominant social groups within countries (especially business groupings).

Globalization and State Capacity

Defining globalization

Globalization is often defined empirically: authors identify the forms of organization of the international political economy that constitute globalization, and contrast them with mere internationalization.

The *international* economy can be defined as the economic relations between and among nation-states (Hirst and Thompson, 1996, pp. 8–9). The basic units of the international economy are national economies; the significant activities of the international economy are trade, the consequent international division of labor, the international payments system, and international investment. International events filter into the national economy through national policies. Multinational corporations, for example, generally have a clear home base, whence they are regulated.

Internationalization, then, is the intensification of interactions between nation-states.

However, superimposed on the international economy is an integrated economy, which comprises the activities of transnational actors. These transnational actors include transnational corporations, transnational labor organizations (such as the ILO), institutions of global economic and environmental governance (such as the IMF), global social movements (such as Oxfam) and global business organizations (such as the International Standards Organisation and the World Economic Forum). The global economy might be defined as the international economy together with the integrated economy, and *globalization* as the intensification of borderless (or across border) social activities (see also Cox, 1994).[17] This definition would be acceptable to many rationalists.

In such empirical definitions, a global economy is identified by the organizations that it contains. Two other categories of phenomena are sometimes added to those identified by Hirst, Thompson and Cox:

- Environmental catastrophes now affect broader and broader areas. Humans have invaded physical and ecological relations to such an extent that nowhere on earth escapes our interference: the whole world has been Disney-fied in our attempt to manipulate nature (McKibben, 1990). Furthermore, the impacts of processes in individual places are increasingly felt abroad: when rainforests burn in Indonesia, the social and ecological ramifications extend around the world. This move would be acceptable to rationalists.
- The concept of globalization can also refer to the tendency for the market to assimilate spheres of social life that have in the past been organized by non-market relations (Hinkson, 1996): government services are corporatized; health, education, and welfare marketized; services privatized; and sport and other cultural activities professionalized and commercialized. This use would not be acceptable to rationalists, who do not distinguish theoretically between state and privately provided services.[18]

However, empirical definitions, like this definition of globalization, fail to identify whether the various observed trends are related to each other (cf. Webber, 2000). Is globalization merely the conjunction of observed trends, or is it a process? If globalization is to be taken seriously as a category of social change, then it must be defined theoretically, not just as a collection of purported empirical trends. Radical social scientists claim that social, cultural, and economic changes are to be identified, understood, and assessed by the way they alter social relations. Marxists argue that the social relations of a commodity producing system are summarized in the concept of value. To Marxists, then, globalization has to be understood in terms of the difference it makes to the ways in which commodity values are determined and calculated.

One such definition is: globalization is the enlargement of the sphere of a common value system (cf. Swyngedouw, this volume). In sectoral terms, globalization occurs when market forms of valuation are applied to ever wider areas of daily life as the commodity form is generalized. People whose work previously lay outside the sphere of capitalist labor relations would be increasingly subjected to the demands for surplus value. In spatial terms, globalization occurs when values in different places tend to become similar. If places compete more, if corporations demand equal productivity and production quality in different places, if states are increasingly subject to the requirements of the institutions of global governance – then the demands for surplus value would be made more equally in different societies. In such a definition, derived from the theory of value, globalization is an economic process, though not one that applies only to those areas of social life that we call the economy.

Whether one prefers an empirical definition or a theoretical one, there arises the question: what is the extent of globalization? Hirst and Thompson (1996) are critical of those who think that globalization is well advanced; they claim that there is a lot of internationalization but little globalization. Cox (1994), on the other hand, regards globalization as an important characteristic of social life, for he, like others, is much impressed by the new power of finance capital to graze the globe for profit (Leyshon, this volume). The problem for this debate is that we know how to study international economic relations (international trade and capital flows are measured routinely), but not how to study and measure global economic relations (how would we measure the degree of global integration of transnational businesses?). Actually, a solution to this empirical problem is provided by the theoretical definition: are measured values becoming more similar in different sectors and places?[19]

Even those who think that globalization is well advanced (such as Bryan and Rafferty, 1999) recognize some limits on its progress. Globalization is not progressing at the same rate everywhere (much of Africa, South Asia, and Central Asia have been left largely behind); nor does globalization mean increasing homogeneity in social life. Furthermore, the fact that the reproduction of labor power remains largely outside the scope of capitalist social relations impedes processes by which values are equalized across societies.

With a sense of the variable nature and extent of globalization, however, the concept can be used to understand the new balances of wealth, security, freedom, and justice (for whom, where).

Whose is the power to go global?

One facet of the global economy concerns the growth of international economic relations. World trade has grown from about 7 percent of world production in 1950 to 22 percent in the mid-1990s (WTO, 1995).[20] Though difficult to measure, it is supposed that foreign investment and transnational business have grown similarly: Howell (1998), for example, has estimated that private financial flows into developing countries increased six-fold between 1990 and 1997. This expansion of international and global activity has gone hand in hand with the emergence of a system to regulate and promote it.

One existing agency of global economic governance predated World War II:

1. The Bank for International Settlements, established in 1930 and headquartered in Basel, Switzerland, monitors monetary policy and money flows and is seeking international regulation of banking. Its members are national central banks.

However, the principal elements of the system of regulation were agreed by the USA and its allies at Bretton Woods in 1944 (Leyshon, this volume):

2. The IMF, established in 1945 with headquarters in Washington, D.C., supervises short-term money flows and foreign exchange. Since 1979 it has formulated Structural Adjustment Programs for many poorer countries in exchange for assistance in refinancing debt. Members are countries; their votes are proportional to their financial contributions, so the IMF is dominated by the USA.

3. The World Bank, established in 1945, also with headquarters in Washington, D.C., provides long-term development finance and has cooperated with the IMF in Structural Assistance Programs. Members are countries; their votes are proportional to their financial contributions (also dominated by the USA).

4. The General Agreements on Tariffs and Trade (GATT), established in 1947, has offices in Geneva.[21] Members are states, with one state–one vote.

5. The Bretton Woods agreement fixed the price of gold in US dollars and enshrined the principle of fixed exchange rates against the US dollar. This system disintegrated in the early 1970s.

Since the late 1940s, other organizations have emerged:

6. The Organization of Economic Cooperation and Development (OECD), founded in 1962 and headquartered in Paris, is a principal source of research and propaganda on economic questions. Its members are advanced industrial states (currently 29).

7. The United Nations Conference on Trade and Development was established in 1964, with headquarters in Geneva. Its members are states; it examines the implications of trade for development.

8. The G8, first established as the G5 in 1975, provides a forum for macroeconomic coordination by the eight leading industrial states.

9. The International Organization of Securities Commissions was established in 1984 with headquarters in Montreal. Its members are official securities regulators. It seeks to formalize the transborder supervision of securities firms.

10. The World Trade Organization, WTO, was established in 1995 with headquarters in Geneva. It incorporates GATT, but has a wider agenda (including agricultural goods, services, and intellectual property as well as manufactured commodities) and greater powers of supervision.

The emergence and existence of these organizations have prompted wide ranging debates within IPE.

Have these organizations and their policies affected rates and forms of world economic growth? Some argue that the Bretton Woods system underpinned the rapid expansion of the world economy between 1950 and 1975 and that when the exchange rate agreements were undone, growth faltered. Others, like Strange (1994a) and Webber and Rigby (1996), are more sceptical of the influence of the Bretton Woods agreement. Similarly, rationalist commentators effectively claim that free trade causes faster economic growth. The World Bank (1993), for example, claims that rapid economic growth in the newly industrialized and industrializing economies of East Asia was associated with government policies that freed up markets or mimicked markets. Wade (1990) and Hamilton (1986), however, document continued government intervention in the economies of Taiwan and Korea, and argue that the strategic qualities of this intervention allowed these two places to escape the position in the international division of labor to which they seemed to be assigned in early 1950s.

What have the organizations and their policies done for individual countries? The sources of global policies, and the implications of structural adjustment programs since the Third World debt crisis of the early 1980s, have been debated by Lever and Huhne (1985), Lombard (1985), Branford and Kucinski (1988), Sachs (1989), Ghai (1992), George (1992), Volcker and Gyohten (1992), Lehman (1993), and Ould-Mey (1994). Those debates have been reignited by the currency crisis in East Asia since mid-1997. The causes of the East Asian currency crisis have been widely argued. Was the crisis fundamentally a matter of:

- policy mistakes, corruption and non-transparent government–business relations in the Third World (as argued by the IMF) *or*
- the increasing availability of European and American short-term finance and reduced controls over its use, which have reoriented policy towards the short term (rather than the long) and have privileged speculation in assets rather than investment in production (as argued by Wade, 1998) *or*
- a shift in the model of development from one in which exports provided the engine of growth, to one in which exports and imports grow at approximately the same rate while capital inflows fuel rates of growth of productive capacity that exceed the rate of growth of demand (as I would argue)?

Or should we redirect our attention away from East Asia as the site of the cause toward the organization of the financial system and the manner in which it encourages speculation in and manipulation of the currencies of other peoples' daily lives? These questions are not merely academic, for the IMF has deployed its answer to justify its demand that Indonesia, Korea, and Thailand all liberalize their economies in return for assistance. Critics who follow Wade would focus on the need to control short-term capital flows more carefully. Or perhaps the system of currency speculation needs to be regulated? Peoples' lives depend on such policy choices.[22]

Which states have sought to develop these organizations and have been most influential in directing their evolution and policies? Do these policies represent an emerging consensus within the organizations or of the power of the few? Whence comes this power and what are the bargains involved in implementing it?

Why have (some) states encouraged free flows of goods, services and capital? Is such support a matter of rational politics, following "the national interest"? This is the position of the IMF (Fischer, 1998) as well as of the WTO (1995), which argued that protectionism in the USA in the 1930s converted a recession into a depression. Is it a matter of sectional interests (and whose)? (See also the next section.) And if so, why have those sectional interests become more powerful lately? What ideologies have underpinned the development of free trade ideas? What is the language of the free trade agenda? Quiggin (1996), Dunkley (1997) and Argy (1998) argue these issues.

States and markets

One crucial debate within IPE over globalization concerns its implications for state power. Some regard globalization as inevitable, a force imposed on countries from outside (Cox, 1994; Catley, 1996). The state is being imposed upon and so losing its ability to set and to follow national agendas (Ohmae, 1990 is a forceful exponent of this view). Typically, rationalists are not perturbed by this purported trend. Reus-Smit (1996) argues, however, that globalization is not a process imposed on the state from outside. Rather, the global economy is a structure that is being created by the process of globalization. The structure invites compliance from states, corporations, and other organizations in the global economy; in turn, the actions that comprise that compliance enhance the process of globalization and so strengthen the structure of the global economy. The global economy, then, is a structure that is being produced by the actions of states, corporations, and other organizations; actions that are in response to existing conditions. In this interpretation, state actions and the global economy are causing one another.

For example, states' financial policies and the integration of a global financial system have proved mutually constitutive. The histories of the rise and demise of Bretton Woods provided by Daly and Logan (1989), Ingham (1994) and Strange (1994b) all point to the crucial role of the growth of Eurodollar markets in the 1960s and 1970s in forcing the US Government to abandon the gold–dollar standard and to deregulate the value of the $US. Certainly, the financial market here exerted power over US Government policies. However, past political decisions strengthened the hand of markets now: the Eurodollar market was not an amorphous, global entity, outside the control of any state. Rather, it was constituted by a series of transactions, orchestrated primarily through British banks, which (being in dollars) were not regulated by the British Government. In other words, the Eurodollar market was made possible by the particular structure of financial regulations put in place by states after 1945. Furthermore, the growth of the Eurodollar market owed much to economic and military policies of the United States, including the huge military expenditures abroad in the 1950s and the deficit financing of the Vietnam War.

This argument implies that the policy choices of states depend not only on the global economy (Weiss, 1998 makes an extended argument that the powerless state is a myth). Internal conditions within states influence policy towards the global economy, because state, corporate, and organizational responses to the global economy depend on internal conditions, strategic decisions, and future expectations.

In Australia, for example, the internal conditions affecting the response of the state to the global economy include the structure of class and fractional interests inherited from past economic conditions (Webber, 2000 has a fuller account). After World War II, Australia's growth strategy involved *import-substitution*[23] by industrialization (initially behind tariff barriers) and a program of immigration that provided both labor and markets for its new manufacturing industries. The strategy also involved centralized, state-regulated wage bargaining. These policies "taxed"

Australia's export-oriented rural sector through high exchange rates, which lowered the prices received abroad for rural products while raising the prices paid by the export sector for domestic manufactured products. At the same time the wage bargaining system provided an institutional means whereby the "taxes" were passed on, in part, to workers. In effect this became a national growth strategy, agreed by all the major political players.

This strategy led to the development of a marked dual economy, which has dominated debate over Australia's growth strategy for the last 20 years. One economy is outward oriented, originally comprising the traditional rural industries on which Australia's prosperity initially rested – then mining, and more recently tourism. The finance and communications sectors have also joined this group. Retailing, though oriented to domestic sales, has increasingly found that its interests lie with the export sector, because it depends more and more upon imports. The protected, domestic-oriented economy is dominated by manufacturing. This sector has few trade or financial ties to the export sector, for manufacturing did not develop from local processing of raw materials but from corporations that were enticed to serve a domestic market. The labor movement has been largely aligned with the domestic sector.

By the 1960s, two factors served to splinter the agreement about a growth strategy. In Australia, as in the North Atlantic, world growth faltered (Webber and Rigby, 1996). Secondly, the prospects of the export sector improved relative to those of the domestically oriented sector. A new minerals boom coincided with the emergence of profitability and cost-competitiveness problems among manufacturers. The minerals boom strengthened the forces within the economy (the export-oriented sector) which favored lower levels of protection for manufacturing.

The decades after the mid-1960s saw a wide-ranging debate about the wisdom of continuing the post-war strategy of import-substituting industrialization. The intellectual arguments in that debate were underlain by a change in the industrial structure of the country, and in the relative fortunes of the two strands of the dual economy. Much of Australia's changing policy direction since the early 1980s can be understood not so much in terms of the demands of the global economy as in the light of these shifting power structures within the society. The need to change direction was understood intellectually in the mid-1970s, long before responses to globalization became issues for governments. The demands for change, from newly strengthening power blocs, emerged from old-fashioned demands for competitiveness in export markets, not from new questions posed by globalization.[24]

Certainly, the global financial system constrains state action. Yet the internationalization of Australia's manufacturing sector since the mid-1980s has reflected a state-led strategy that evolved gradually since the mid-1960s in response to the failure of the old strategy, new configurations of social power, and a new popularity of hyper-liberal views. There remains evidence of vacillation about the precise form of the new strategy. Nevertheless, it has been the state that has orchestrated the internationalization of manufacturing in Australia: globalization produced by the state.

Such changes in policy are less a retreat of the state in the face of global pressure, and more the creation of conditions under which capitalist organizations can use global forms to develop and prosper. Globalization is a process in which business and state activities operate across national borders, take advantage of differences created by the existence of borders, and thereby equalize values more completely. It thus reflects the intersection of state regulation and corporate response (Piccioto, 1989). The Australian state, like other states, has participated in creating the conditions under which the borders of Australian economic identity have become fluid and permeable. As this has happened, new powerful actors – especially global financial institutions – have emerged to challenge the state's capacity to set goals and tactics. Other actors have clearly lost power: over the last 30 years, outwardly-oriented forces have gained power at the expense of the domestically oriented sector and, more recently, the union movement has been a notable casualty. But this does not imply that the state has diminished capacity.

Globalization and values

The two major camps of IPE offer quite different assessments of the implications of the evolution of a global economy. The argument between these extremes is clouded, however, by the fact they give different priorities to the different values that Strange has identified.

The primary rationalist justification given for the changes promoted by the institutions of economic governance is increased *wealth*. For example, the OECD (1997) argues that a new global age is taking shape, in which all countries can participate. Provided governments liberalize trade and investment, maintain macroeconomic discipline, reform product and labor markets, and strengthen financial systems and implement effective environmental policies, then we are promised a large increase in prosperity and welfare across the world. The OECD has also promised greater political *security* after liberalization and integration. For those who live on the edge of survival, the World Bank (1996) is also pointing the way to greater personal security.

The standard structural adjustment program, currently being implemented in Indonesia and propagated in other countries, also offers more personal political *freedom*, at least of a certain kind. Ouattara[25] (1988) indicates the IMF's current interpretation of the needs: transparency and accountability in government; stronger banking systems that are less driven by personal favoritism; liberal capital flows; transparent regulation of business; reductions in unproductive government expenditure; more expenditure on health care and education; social protection; and dialogue with opposition groups. This is classic rationalism. Such political programs finally promise improved economic justice for people.

Yet, there remain doubts. The distribution of the benefits of globalization has been highly uneven, both between countries and within them. The benefits of economic integration are unequally distributed between developing and industrial societies (as groups) and between individual developing countries, and the global distribution of income has deteriorated over the past few decades (Griffin and Khan, 1992). The same is true within individual countries (Argy, 1998), as class relations are changing (Wilkin, 1996). The relative power of men and women is also changing, often not for the better (see reviews in Connelly, 1996; Ward and Pyle, 1995; Barber, 1996). Globalization is doing little for the environment in the Third World (Mukta, 1995; Salleh, 1995) or the First (Dunkley, 1997), and the development program of the institutions of governance seems little but the reimposition of Western imperial power (Mehmet, 1995). Marxists thus would argue that, to a large degree, globalization represents the emergence of global markets, increasingly dominated by capitalist modes of production. Their valuation of this tendency would reflect their preoccupation with justice, personal security, and freedom – as mediated by the social relations of production – rather than with wealth.

Conclusion

The geography of the world economy is being remade. Understanding this new economic geography must become one of the central tasks of new generations of economic geographers. There are, of course, important questions about the internal economic geography of countries; but these internal questions are increasingly framed in the context of presumed global changes. Geographers thus need to give their answers to questions such as:

- Are states losing their capacity to act? Many conservatives (seeking reduced state capacity) and radicals (who deplore it) argue that globalization necessarily means reduced state capacity. The evidence is not good, however, and geographical variation implies a theoretical basis for explaining states that lies outside the bounds of traditional thinking in international political economy. It is not at all clear that this question is even well framed.
- How are we to understand the emergence of institutions of international environmental and economic governance? These have both radical (e.g. globalized green organization) and conservative (e.g. global economic institution) elements. But do such institutions constrain states or popular movements and under what circumstances?
- What are we to make of the variable progress of new arrangements for regional economic and political integration? The political, economic, and scale-altering implications of these arrangements are in some cases clear, but others have a less obvious political economic geography.
- What have been the history and implications of global trade, production, (and labor)? Our understandings of these elements have long been dominated by orthodox thinking, based on ideas that assume equilibrium outcomes and homogenous industries (see Plummer, this volume). What if we understand change as occurring away from equilibrium, and industries as heterogenous and locally variable?
- What are the theoretical and practical implications of the increasing dominance of finance over production? Perhaps the key event of the 1990s – in terms of its impact on our understanding that the world economy is being remade – was the East Asian financial crisis that emerged in mid-1997. It demonstrated that finance and financial flows dominate trade and production capital. How has this new economy emerged and what does it mean for our lives?
- What are the political economy of poverty, hunger, and development, and the international political economy of race, sexuality, and gender? These are not new questions, but they have received precious few answers from economic geographers. They need to become central both within the discipline and in answers to the questions posed above.

The questions are huge and they are asked both outside as well as within academia. If geographers do not give geographical answers to these questions, then the difference that geography makes will be ignored in social understandings of the new world economy. The questions are not simply about the economy, however: they concern politics as well – the deployment of power. Nor are they simply global issues: they reflect concerns about local, daily lives. These characteristics mean – as I hope you can see from the argument of this chapter – that international political economy is a useful theoretical framework from which to begin to seek answers to such questions.

Endnotes

1. There are many good introductions to international relations and IPE. An excellent text, which is fair to all major approaches, is Baylis and Smith (1997). A more challenging introduction to IPE is Strange (1994a). For a political geographical contribution to IPE, which is, however, gender-blind, see Agnew and Corbridge (1995). See, too, Agnew (1998).
2. An excellent introduction to economic theories is to be found in Cole et al. (1983).
3. However, realist and liberal thinking has recently converged and they are now simply varieties of international relations' equivalents of subjective preference theory.
4. For good surveys of realist thought, see Smith (1986), Hollis and Smith (1990: Chapter 5).
5. Some realists regard human nature as they key principle of international politics: individuals are first and foremost self-interested (Morgenthau, 1978).
6. On the development and significance of transnational organizations, see Dicken (1998); on human rights organizations, see Willetts (1996). A good introduction to the ideology and activities of these organizations is in their WWW pages.
   See: http://www.wto.org/, http://www.worldbank.org/, http://www.imf.org/, http://www.oecd.org/publications/observer/index_en.html.
   For a directory of humanitarian organizations,
   see http://www.reliefweb.int/library/contacts/dirhomepage.
   You do, though, need to read between the lines of what are public relations documents.
7. On the debate between the two, see Kegley (1995).
8. A regime is a set "of implicit or explicit principles, norms, rules, and decisionmaking procedures around which actors' expectations converge in a given area of international relations" (Krasner, 1983, p. 2). One of the most significant of these has been the liberal international economic regime, under US leadership, underpinned by free trade, stable monetary systems, and stable domestic economies. This regime has been implemented through the General Agreement on Tariffs and Trade (GATT), the IMF (IMF), and the International Bank for Reconstruction and Development (World Bank). On the liberal theory of regimes, see Zacher (1996).
9. For a geographical critique, see Harvey (1987).
10. Constitutive views of social science believe that theoretical and empirical research is itself part of the social world and helps to construct that world: theory is not separate from social reality; therefore appeals to social reality are not independent tests of a theory's validity. By contrast, explanatory views regard the world as external to the theories that describe them.
11. Foundationalism is the thesis that all truth claims can be judged true or false, either logically (derived from prior theoretical statements, themselves believed to be true) or empirically (by reference to the external world).

12. Rationalism is, however, an important topic *of* study – see the views of the international institutions discussed in the section on globalization and state capacity.

13. Walker (this volume) introduces such ideas. See also Cole et al. (1983).

14. Like all definitions in the social sciences, this definition of power is theoretically loaded. Other definitions can be found in Baylis and Smith (1997), Cox (1987), Stubbs and Underhill (1994).

15. On the meaning and sources of state power, see Weiss (1998).

16. Theories of the capitalist state differ widely; for a survey, see Painter (this volume). For a discussion of class, see Sadler (this volume).

17. There are many definitions. Robertson (1992) identifies globalization as the development of a global society or of the consciousness of the world as a single place. Giddens (1996) identifies globalization as the intensification of action at a distance, so that local happenings are shaped by events occurring far away.

18. That is: Marxists argue that the social relations involved in state production are quite different from those involved in private, capitalist production. Rationalists do not regard Marxists' identification of the social relations as either accurate or important and instead make pragmatic distinctions between the profit-oriented motives of business people and the bureaucratic motives of state officials.

19. This is one test of an adequate theoretical definition: does it lead to useful empirical work?

20. Hirst and Thompson (1996, p. 20), cynical about the extent of globalization, observe that world trade exceeded 30 percent of global output just before the outbreak of World War I. Understanding the institutional structure that underpinned the internationalization of economic activities in the late nineteenth century remains an important research task.

21. Actually, the Bretton Woods agreement called for a somewhat different organization, the International Trade Organization. GATT was a weaker organization than originally foreseen. On the emergence of GATT, see Wilcox (1949), Spero (1985) and Dryden (1995).

22. Both the earlier crises and the East Asian experience since 1997 lead to additional questions. Concerning the internal politics of structural adjustment: why did some states accept responsibility for large private debts and then agree to IMF/World Bank terms whereas others (such as Malaysia) sought to remain independent of the IMF? What have been the implications of structural adjustment for different production sectors and classes, political and social cultures, environments?

23. Import-substitution is a model of industrialization in which the state seeks to encourage corporations to invest by offering a protected domestic market. The market may be protected by tariffs or quotas or (less commonly) by agreements that government departments will purchase preferentially from domestic corporations. Sometimes, import-substitution is contrasted to *export-orientation*, in which the state encourages corporations by providing incentives for them to export their products.

24. The period after 1980 saw changes in the form of the state, as bureaucrats trained in economics came to dominate the bureaucracy, as business-derived management practices replaced the careerist public service ethos of the 1950s bureaucracy, and as the central departments of finance, treasury, and the prime minister came to power over the "line" departments of trade, social services, and industries. See also Cox (1987, pp. 211–67).

25. Alassane D. Ouattara was Deputy Managing Director of the IMF.

Bibliography

Agnew, J. 1998. *Geopolitics*. London: Routledge.

Agnew, J. and Corbridge, S. 1995. *Mastering Space: Hegemony, Territory, and International Political Economy*. London: Routledge.

Argy, F. 1998. *Australia at the Crossroads*. Sydney: Allen & Unwin.

Ashley, R. 1988. Untying the sovereign state: a double reading of the anarchy problematique. *Millennium*, 17.

Baldwin, D. (ed). 1993. *Neoliberalism and Neorealism: The Contemporary Debate*. New York: Columbia University Press.

Barber, P. G. 1996. Modes of resistance: gendered responses to global impositions in coastal Philippines. *Asia Pacific Viewpoint*, 37, 181–94.

Baylis, J. and Smith, S. (eds). 1997. *The Globalization of World Politics*. Oxford: Oxford University Press.

Biersteker, T. J. and Weber, C. (eds). 1996. *State Sovereignty as Social Construct*. Cambridge: Cambridge University Press.

Branford, S. and Kucinski, B. 1988. *The Debt Squads: The US, the Banks and Latin America*. London: Zed Books.

Bryan, D. and Rafferty, M. 1999. *The Global Economy in Australia*. Sydney: Allen & Unwin.

Bunch. C. and Reilly, N. 1994. *Demanding Accountability*. New Jersey: Rutgers University Center for Women's Global Leadership.

Catley R. 1996. *Globalising Australian Capitalism*. Cambridge: Cambridge University Press.

Cole, K., Cameron J., and Edwards, C. 1983. *Why Economists Disagree*. London: Longman.

Connelly, M. P. 1996. Gender matters: global restructuring and development. *Social Politics*, 3, 1, 12–31.

Cox, R. W. 1987. *Production, Power, and World Order*. New York: Columbia University Press.

Cox, R. W. 1994. Global restructuring: making sense of the changing international political economy. In R. Stubbs and G. R. D. Underhill (eds). 1994. *Political Economy and the Changing Global Order*. New York: St Martin's Press, 45–59.

Daly, M. and Logan, M. I. 1989. *The Brittle Rim*. Ringwood, Victoria: Penguin.

Dicken, P. 1998. *Global Shift: Transforming the World Economy*. London: Chapman.

Diesendorf, M. and Hamilton, C. (eds). 1997. *Human Ecology, Human Economy*. Sydney: Allen & Unwin.

Dryden, S. 1995. *Trade Warriors*. New York: Oxford University Press.

Dunkley, G. 1997. *The Free Trade Adventure*. Melbourne: Melbourne University Press.

Enloe, C. 1990. *Bananas, Beaches and Bases*. Berkeley: University of California Press.

Enloe, C. 1993. *The Morning After: Sexual Politics at the End of the Cold War*. Berkeley, CA: University of California Press.

Fischer, S. 1998. The IMF and the Asian Crisis. Los Angeles: Forum Funds Lecture, UCLA, March 20 1998 (http://www.imf.org/external/np/speeches/1998/032098.HTM).

Frank, A. G. 1979. *Dependent Accumulation and Underdevelopment*. New York: Monthly Review Press.

George, S. 1992. *The Debt Boomerang*. Boulder, CO: Westview Press.

Ghai, D. 1992. Structural adjustment, global integration and social democracy. Discussion Paper of the United Nations Research Institute for Social Development 37. New York: United Nations.

Giddens, A. 1996. Affluence, poverty and the idea of a post-scarcity society. *Development and Change*, 27, 365–77.

Griffin, K. and Khan, A. R. 1992. *Globalization and the Developing World: An Essay on the International Dimensions of Development in the Post-Cold War Era*. New York: United Nations Research Institute for Social Development Report 92/3.

Hamilton, C. 1986. *Capitalist Industrialization in Korea*. Boulder, CO: Westview Press.

Harvey, D. W. 1987. The world systems theory trap. *Studies in Comparative International Development*, 22, 12–39.

Hinkson, J. 1996. The state of postmodernity: beyond cultural nostalgia or pessimism. In P. James (ed). *The State in Question*. Sydney: Allen & Unwin, 196–223.

Hirst, P. and Thompson, G. 1996. *Globalization in Question: The International Economy and the Possibilities of Governance*. Cambridge: Polity.

Hollis, M. and Smith, S. 1990. *Explaining and Understanding International Relations*. Oxford: Clarendon Press.

Howell, M. 1998. Asia's "Victorian" financial crisis. Paper presented at the East Asia Crisis workshop, Institute of Development Studies (http://www.ids.ac.uk/ids/research/howell.pdf).

Ingham, G. 1994. States and markets in the production of world money: sterling and the dollar. In S. Corbridge, R. Martin, and N. Thrift (eds). 1996. *Money, Power and Space*. Oxford: Blackwell, 29–48.

Institute for Women, Law and Development, 1993. *Claiming our Place*. Washington: Institute for Women, Law and Development.

Kegley, C. (ed). 1995. *Controversies in International Relations: Realism and the Neoliberal Challenge*. New York: St Martin's Press.

Krasner, S. D. (ed). 1983. *International Regimes*. Ithaca, NY: Cornell University Press.

Lehman, H. P. 1993. *Indebted Development: Strategic Bargaining and Economic Adjustment in the Third World*. New York: St Martin's Press.

Lever, H. and Huhne, C. 1985. *Debt and Danger*. Harmondsworth: Penguin.

Lombard, R. W. 1985. *Debt Trap: Rethinking the Logic of Development*. New York: Praeger.

Marchand, M. H. 1996. Reconceptualising gender and development in an era of globalisation. *Millennium – Journal of International Studies*, 25, 577–604.

McKibben, B. 1990. *The End of Nature*. London: Penguin.

Mehmet, O. 1995. *Westernizing the Third World: The Eurocentricity of Economic Development Theories*. London: Routledge.

Morganthau, H. J. 1978. *Politics among Nations: The Struggle for Power and Peace*. New York: Knopf.

Mukta, P. 1995. Wresting riches, marginalizing the poor, criminalizing dissent: the building of the Narmada dam in western India. *South Asia Bulletin*, 15, 99–108.

OECD, 1997. *Towards a New Global Age: Challenges and Opportunities*. Policy Report 38. Paris: Organisation for Economic Cooperation and Development.

Ohmae, K. 1990. *The Borderless World: Power and Strategy in the Interlinked Economy*. London: Fontana.

Ouattara, A. D. 1998. The Asian crisis: origins and lessons. Address to the Royal Academy of Morocco Seminar, Fez, on *Why Have the Asian Dragons Caught Fire?* (http://www.imf.org/external/np/speeches/1998/050498A.HTM).

Ould-Mey, M. 1994. Global adjustment: implications for peripheral states. *Third World Quarterly*, 15, 319–36.

Piccioto, S. 1989. Slicing a shadow: taxation in an international framework. In L. Hancher and M. Moran (eds). *Capitalism, Culture, and Economic Regulation*. Oxford: Clarendon Press, 11–47.

Quiggin, J. 1996. *Great Expectations*. Sydney: Allen & Unwin.

Reus-Smit, C. 1996. Beyond foreign policy: state theory and the changing global order. In P. James (ed). *The State in Question*. Sydney: Allen & Unwin, 161–95.

Robertson, R. 1992. *Globalization: Social Theory and Global Culture*. London: Sage.

Sachs, J. D. 1989. *Developing Country Debt and the World Economy*. Chicago: University of Chicago Press.

Salleh, A. 1995. Nature, woman, labor, capital: living the deepest contradiction. *Capitalism, Nature, Socialism*, 6, 21–39.

Shapiro, I. and Bilmayer, L. 1999. *Global Justice*. New York: New York University Press.

Smith, M. J. 1986. *Realist Thought from Weber to Kissinger*. Baton Rouge: Louisiana State University Press.

Spero, J. 1985. *The Politics of International Relations*. Sydney: Allen & Unwin.

Strange, S. 1994a. *States and Markets*, second edition. London: Pinter.

Strange, S. 1994b. From Bretton Woods to the casino economy. In S. Corbridge, R. Martin, and N. Thrift (eds). *Money, Power and Space*. Oxford: Blackwell, 49–62.

Stubbs, R. and Underhill, G. R. D. (eds). 1994. *Political Economy and the Changing Global Order*. New York: St Martin's Press.

Tilly, C. 1990. *Coercion, Capital, and European States, AD 990–1990*. Oxford: Blackwell.

Volcker, P. A. and Gyohten, T. 1992. *Changing Fortunes: The World's Money and the Threat to American Leadership*. New York: Times Books.

Wade, R. 1990. *Governing the Market*. Princeton: Princeton University Press.

Wade, R. 1998. The Asian crisis: debt deflation, vulnerabilities, moral hazard or panic? Paper presented at the East Asia Crisis workshop at the Institute of Development Studies (http://www.ids.ac.uk/ids/research/ wade.pdf).

Wallerstein, I. 1974, 1980, 1989. *The Modern World-System* (volumes 1, 2, 3). San Diego: Academy Press.

Ward, K. B. and Pyle, J. L. 1995. Gender, industrialization and development. *Development*, 1, 67–71.

Webber, M. 2000. Globalisation: local agency, the global economy and Australia's industrial policy *Environment and Planning A*, forthcoming.

Webber, M. J. and Rigby, D. L. 1996. *The Golden Age Illusion: Rethinking Postwar Capitalism*. New York: Guilford.

Weber, C. 1994. Good girls, little girls and bad girls – male paranoia in R. Keohane's critique of feminist international relations. *Millennium – Journal of International Studies*, 23, 337–49.

Weber, C. 1995. *Simulating Sovereignty*. Cambridge: Cambridge University Press.

Weiss, L. 1998. *The Myth of the Powerless State*. Cambridge: Polity.

Wilcox, C. 1949. *A Charter for World Trade*. New York: Macmillan.

Wilkin, P. 1996. New myths for the South: globalisation and the conflict between private power and freedom. *Third World Quarterly*, 17, 227–38.

Willetts, P. (ed). 1996. *The Conscience of the World*. London: Hurst and Co.

World Bank. 1993. *The East Asian Miracle*. New York: Oxford University Press.

World Bank. 1996. *Global Economic Prospects and the Developing Countries*. Washington, D.C.: World Bank.

World Trade Organisation. 1995. *International Trade: Trends and Statistics*. Geneva: WTO.

Zacher, M. W. 1996. *Governing Global Networks: International Regimes for Transportation and Communications*. Cambridge: Cambridge University Press.

Index